the incredible illustrated ELECTRICITY BOOK

Edited by: Don Richard Riso

A Pathmark Book
Pathfinder Publications
Boston

OTHER PATHMARK BOOKS

The Incredible Illustrated Tool Book
Net Results
How To Get What You Want From The U.S. Government
Roget's Thesarus
Webster's Dictionary
The Holy Bible
The Constitution and the Declaration of Independence
Chess in Thirty Minutes

Soon To Be Released:

On Sail, The Complete Sailing Handbook
The Great Struggle, A History of the Civil War
The Signers of the Declaration
Enjoy Weekend Adventuring, The Complete Camping Handbook

International Standard Book Number: 0-913390-09-7
Library of Congress Catalogue Number 74-31555

Printed in the United States of America

Published by Pathmark Books, Inc.
108 Massachusetts Avenue
Boston, Massachusetts 02123

CONTENTS

INTRODUCTION

For decades electricity has played an important part in our lives, and it is being used more today than ever before. Whether it be in computers, pinball machines, satellites or stereo components, electricity is the basis of their conception and design.

THE INCREDIBLE ILLUSTRATED ELECTRICITY BOOK will serve as an excellent guide to understanding the electrical phenomena that surround us. As a text for the home repair amateur, the hobbyist, or the technical apprentice, this handbook lives up to its name. It distinguishes itself as one of the most clearly written, well-paced, extensively illustrated expositions ever published. With all the unnecessary complexities omitted, it is especially well-suited for use as a self-teaching manual.

THE INCREDIBLE ILLUSTRATED ELECTRICITY BOOK is designed to quickly instruct the reader in a wide-ranging understanding of the subject. He is thoroughly introduced to voltage, currents, resistance, direct current circuits and their analysis. He then progresses to a knowledge of the fundamental principles (primarily Ohm's Law) which show how each idea and its applications harmonize with other concepts into a logical whole. The middle portion of the book takes up inductance, capacitance and Kirchhoff's Laws, which are crucial for an understanding of alternating current circuits. Chapter 13 includes an extensive group of exercises which test and sharpen the reader's understanding of alternating current circuits.

This guide also supplies a great deal of practical information. Chapter 7 introduces electrical conductors and wiring techniques; Chapters 14 and 15 give more details on circuit protection methods, control devices and wiring techniques suitable for home construction or alteration. The necessary electrical data as well as convenient summaries of formulas and numerical tables are appended.

Learning electrical theory was once thought to be a difficult undertaking, but this is no longer true. In the past, tedious calculations by pencil and slide-rule tended to obscure theory and dull the inquisitive. However, today's pocket calculators take the drudgery out of the calculations, illuminating the mathematical principles relevant to electricity, thereby making it easier to understand. Moreover, the cost of electrical components has actually declined. Even the most exotic kind can be bought literally for pennies. For the imaginative beginner, hundreds of components are available in old television sets and radios at no cost.

Finally, as in all books of this kind, no matter how comprehensive the text, the reader will not realize the full value of the information unless he reinforces his study with a reasonable amount of practical work. Given the number of inexpensive kits and electrical components (particularly used or salvaged items) available for performing experiments, it is easier than ever to learn what was once a difficult subject.

As a general interest text, as a reference work for home use, or even as a resource for more advanced electronics and electrical work, THE INCREDIBLE ILLUSTRATED ELECTRICITY BOOK is hard to surpass.

THOMAS EDSON
Cambridge, 1975

CHAPTER 1

SAFETY

Anyone dealing with electricity and electrical equipment is exposed to many potentially dangerous conditions and situations. No home manual, no set of rules, no list of Federal regulations or catalogue of hazards can make every possible condition perfectly safe. However, it is possible to complete hundreds of jobs around the home or office without any injury whatsoever. The attainment of this goal only requires that anyone working with electricity be aware of the main sources of danger, and that he remain constantly alert to those dangers. Not knowing the possible sources of danger, or worse, being inattentative to those sources are the main reasons why there are injuries among those working with electricity. If you are going to work with electricity, you must take the proper precautions and practice the basic rules of safety— one must be safety conscious at all times, and this safety consciousness must become second nature to you.

The purpose of this chapter is to indicate some of the major hazards encountered under normal working conditions and to indicate some of the basic precautions that should be observed. Although many of these hazards and precautions are general, some of them are especially applicable to electrical and electronic maintenance and working conditions which may be encountered in the home or shop.

Most accidents can be prevented if care is exercised to eliminate unsafe acts and conditions. Safety precautions in this chapter are not intended to replace the information given in instruction manuals or first aid guides. If there is ever any doubt about the safety of a procedure, consult a physician.

A number of general safety principles might be kept in mind; these principles apply to a wide range of situations which anyone working with electricity or electronic equipment might find himself in.

1. Take note of, repair or discard any piece of equipment which is unsafe. Do not take chances with doubtful equipment. If you are not completely sure that a piece of equipment is safe, don't use it.

2. If you are working with others, warn them of anything which might be unsafe. Working with electricity is no time to keep secrets. Let others know the safety status of every piece of equipment which will be used.

3. Wear and use protective clothing or equipment of the type approved for the specific job you are working on. If you don't have the correct piece of safety equipment or protective clothing, do not attempt to "make do" with something which won't protect you. Do it right the first time or there might not be another time to learn from your mistake.

4. Take immediate action on any injury which might have occurred. This might seem so obvious as to not require mentioning and for serious accidents there is no need to remind people to get to a doctor or first aid station. But minor injuries can also be serious, especially when working with electricity. Even a minor cut can reduce your conductive resistance, so be careful and attend to injuries immediately.

5. Lastly, be cautious at all times. If you can foresee all your problems, you won't have any.

THE EFFECTS OF ELECTRIC SHOCK

The amount of current that may pass through the body without danger depends on the individual and the quantity, type, path, and the length of contact time.

Body resistance varies from 1,000 to 500,000 ohms for unbroken, dry skin. Resistance is lowered by moisture and high voltage and conversely, is highest with dry skin and low voltage. Cuts or burns lower body resistance. A current of 1 milliampere can be felt and will cause a person to avoid it. (The term "milliampere" is discussed later in this book; however, it is sufficient for now to define a milliampere as a very small amount of current, or

NOTES:

1--1000 of an ampere. Current as low as 5 milliamperes can be dangerous. If the palm of the hand makes contact with the conductor, a current of about 12 milliamperes will tend to cause the hand muscles to contract, freezing the body to the conductor. Such a shock may or may not cause serious damage, depending on the contact time and your physical condition, particularly the condition of your heart. However, a current of only 25 milliamperes has been known to be fatal; 100 milliamperes can possibly be fatal and is certainly very dangerous.

Due to the physiological and chemical nature of the human body, five times more direct current than alternating current is needed to freeze the same body to a conductor. Also, a 60-hertz (cycles per second) alternating current is the most dangerous frequency — even though it is the frequency most often used in residential, commercial and industrial situations.

The damage from shock is also proportional to the number of vital organs transversed, especially the percentage of current that reaches the heart. Currents between 100 and 200 milliamperes are lethal. Ventricular fibrillation of the heart occurs when the current through the body approaches 100 milliamperes. Ventricular fibrillation is the uncoordinated actions of the walls of the heart's ventricles. This causes the loss of the pumping action of the heart until some force restores the coordination of the heart's actions.

Severe burns and unconsciousness are also produced by currents of 200 milliamperes or higher. These currents usually do not cause death if the victim is given immediate attention, most often in the form of artificial respiration. Death is not immediate because the 200 milliampere current is so strong that it paralyzes the heart muscles, preventing them from going into ventricular fibrillation. If a person is rendered unconscious by a large current passing through the body, and if breathing has stopped, artificial respiration must be applied immediately.

GENERAL SAFETY PRECAUTIONS

When any electrical equipment is to be overhauled or repaired, the main supply switches or cutout switches in each circuit from which power could possible be fed should be secured in the open position. The covers of fuse boxes and junction boxes should be kept securely closed except when work is being done. Safety devices such as interlocks, overload relays, and fuses should never be altered or disconnected except for replacement.

The interlock switch is ordinarily wired in series with the powerline leads to the electronic power supply unit, and is installed on the lid or door of the enclosure to break the circuit when the lid or door is opened. A true interlock switch is entirely automatic in action; it does not have to be manipulated by an operator. Multiple interlock switches, connected in series, may be used for increased safety. One switch may be installed on the access door of a device, and another on the cover or the power-supply section.

Because electrical and electronic equipment may have to be serviced without de-energizing the circuits (even in home or business situations) interlock switches are constructed so that they can be disabled only by someone who knows how to do so. However, to minimize the danger of disabling them accidentally, they are generally located in such a way that a certain amount of manipulating is necessary in order to disable them.

Fuses should be removed and replaced only after the circuit has been de-energized. When a fuse blows, it should be replaced only with a fuse of the same current and voltage rating. When possible, the circuit should be carefully checked before making the replacement, since the burnt-out fuse is often the result of a circuit fault.

HIGH VOLTAGE SAFETY PRECAUTIONS

It is human nature to become careless with routine procedures. This tendency is very dangerous when one is working around electrical equipment which employs voltages which are dangerous and even fatal if contacted. Practical safety precautions have been incorporated into home and industrial electrical systems; when the basic rules of safety are ignored, however, the built-in protections become useless.

The following rules are basic and should be followed by anyone working with or near high voltage circuits:

1. CONSIDER THE CONSEQUENCES OF EACH ACT — there is absolutely no reason for an individual to take chances that will endanger his life or the lives of others.
2. KEEP AWAY FROM LIVE CIRCUITS — do not change parts or make adjustments inside the equipment with high voltages on.
3. DO NOT SERVICE DANGEROUS EQUIPMENT ALONE — always service equipment in the presence of another person capable of rendering assistance of first aid in an emergency.
4. DO NOT TAMPER WITH INTERLOCKS — do not depend on interlocks for protection; always

NOTES:

shut down equipment. Never remove, short circuit, or tamper with interlocks except to repair the switch.

5. DO NOT GROUND YOURSELF — make sure you are not grounded when adjusting equipment or using measuring equipment. Use only one hand when servicing energized equipment. Keep the other hand behind you.

6. WATCH FOR WATER — do not energize equipment if there is any evidence of water leakage; repair the leak and wipe up the water before energizing.

These rules, teamed with the ideas that voltage shows no favoritism and that personal caution is your greatest safeguard, may prevent serious injury or even death.

WORKING ON ENERGIZED CIRCUITS

Insofar as it is practicable, repair work on energized circuits should not be undertaken. When repairs on operating equipment must be made because of emergency conditions every known safety precaution should be carefully observed. Ample light for good illumination should be provided; you should be insulated from ground with some suitable nonconducting material such as several layers of dry canvas, dry wood, or a rubber mat of approved construction. You should, if possible, use only one hand in accomplishing the necessary repairs. A friend should be stationed near the main switch or the circuit breaker so that the equipment can be de-energized immediately in case of an emergency. Someone qualified in first aid for electric shock should also stand by during the entire period of the repair, especially in highly dangerous situations.

GROUNDING EQUIPMENT

A poor safety ground, or one that is wired incorrectly, is more dangerous than no ground at all. The poor ground is dangerous because it does not offer full protection, while the user is lulled into a false sense of security. The incorrectly wired ground is a hazard because one of the line wires and the safety ground are transposed, making the shell of the tool "hot" the instant the plug is connected. Thus the unwary user is trapped, unless by pure chance the safety ground is connected to the grounded side of the line on a single-phase grounded system or no grounds are present on a ungrounded system. In this instance, the user again goes blithely along using the tool until he encounters a receptacle which has its wires transposed or a ground appears on the system.

Because there is no absolutely foolproof method of insuring that all tools are safely grounded (and because of the tendency of the average home electrician to ignore the use of the grounding wire), the old method of using a separate external grounding wire has been discontinued. Instead, a 3-wire, standard, color-coded cord with a polarized plug and a ground in is required. In this manner, the safety ground is made of a part of the connecting cord and plug. Since the polarized plug can be connected only to a mating receptacle, the user has no choice but to use the safety ground.

All new tools, properly connected, use the green wire as the safety ground. This wire is attached to the metal case of the tool at one end and to the polarized grounding pin in the connector at the other end. It normally carries no current, but is used only when the tool insulation fails, in which case it short circuits the electricity around the user to ground and protects him from shock. The green lead must never be mixed with the black or white leads which are the true current-carrying conductors.

Check the resistance of the grounding system with a low reading ohmmeter to be certain that the grounding is adequate (less than 1 ohm is acceptable). Ohmmeter and its use is discussed in chapter 15. If the resistance indicates greater than 1 ohm, use a separate ground strap.

Some old installations are not equipped with receptacles that will accept the grounding plug. In this event, use one of the following methods:

1. Use an adapter fitting.
2. Use the old type plug and bring the green ground wire out separately.
3. Connect an independent safety ground line.

When using the adapter, be sure to connect the ground lead extension to a good ground. (Do not use the center screw which holds the cover plate on the receptable.) Where the separate safety ground leads are externally connected to a ground, be certain to first connect the ground and then plug in the tool. Likewise, when disconnecting the tool, first remove the line plug and then disconnect the safety ground. The safety ground is always connected first and removed last.

BATTERY SAFETY PRECAUTIONS

The principal hazard in connection with batteries is the danger of acid burns when refilling or when handling them. These burns can be prevented by the proper use of eyeshields, rubber gloves, rubber aprons, and rubber boots with nonslip soles.

NOTES:

Rubber boots and apron need be worn only when batteries are being refilled. It is a good practice, however, to wear the eyeshield whenever working around batteries to prevent the possibility of acid burns of the eyes. Wood slat floorboards, if kept in good condition, are helpful in preventing slips and falls as well as electric shock from the high-voltage side of charging equipment.

Another hazard is the danger of explosion due to the ignition of hydrogen gas given off during battery charging operation. This is especially true where the accelerated charging method is used. Open flames or smoking is prohibited in the battery charging room, and the charging rate should be held at a point that will prevent the rapid liberation of hydrogen gas. Manufacturers' recommendations as to the charging rates for various size batteries should be closely followed and a shop exhaust system should be used.

Particular care should be taken by technicians to prevent short circuits while batteries are being charged, tested, or handled. Hydrogen gas, which is accumulated while charging, is highly explosive and a spark from a shorted circuit could easily ignite the gas, causing serious damage to personnel and equipment.

Extreme caution should be exercised when installing and removing batteries. The nature of battery construction is such that the batteries are heavy for their size and are somewhat awkward to handle. These charactgeristics dictate the importance of using proper safety precautions. There is the possibility of acid causing damage to equipment or injury to personnel and the danger of an explosion that may be caused from the gas that is produced as the battery is charged. Follow the prescribed safety precautions in working with batteries.

PRECAUTIONS WITH CHEMICALS

Volatile fluids are liquids which evaporate rapidly at relatively low temperatures. In this discussion, the term will be used to include pressurized gases escaping from their containers. In safety considerations, two main types of fluids must be considered: First, those which result in explosive vapors; and second, those which are toxic.

Explosive Vapors

Fuels, alcohol, painting materials and supplies, insulating varnish, certain cleaning supplies, and many industrial gases produce potentially explosive vapors when allowed to accumulate in closed spaces. The hazards relating to these materials are associated with the FLASHPOINT of the liquid. This is the lowest temperature at which the liquid gives off vapor which accumulates near the surface in sufficient quantity to form a combustible mixture with air. Although liquid oxygen does not have a flashpoint, the explosive effect is the same. Almost any material that will oxidize becomes highly explosive when in the presence of liquid oxygen.

The flashpoint varies with specific materials. Anyone working with volatile materials should become familiar with the characteristics of the particular materials with which they are associated. You should know the flashpoint and also the concentration which constitutes a combustible mixture.

A general safety rule regarding explosive vapors is to always provide adequate ventilation to prevent accumulation of vapors, or to dispel accumulated vapors prior to operating electrical equipment in that space. The presence or absence of an odor of gasoline or other flammable or explosive vapors is not a reliable indicator of flammability since the ability to detect odors varies between observers.

TOXIC VAPORS

Some liquids, upon evaporating, produce vapors which are harmful. Carbon tetrachloride is an effective solvent, an excellent cleaning agent, and an excellent fire extinguisher — but it produces vapors so toxic that it is highly unsafe to use. This is true of many other commonly found substances which you might have in your work area. In most instances, precautions are listed on the container; they should be rigidly adhered to and safer materials used whenever possible.

Safety considerations required that everyone working in a home electrical situation make himself familiar with the hazards involved in the use of all materials. Toxic materials and vapors may be easily detectable, or they may be almost completely undetectable; they may act slowly, or they may act almost instantly; they may cause discomfort, temporary damage, permanent injury, or even death; they may or may not be explosive in addition to being toxic. Toxic vapors may produce headache, dizziness, nausea, and a general feeling of illness. They may cause a gradual loss of interest, energy, or awareness. They may result in unconsciousness. They may cause temporary or permanent damage to the respiratory system, eyes

NOTES:

or skin. They may cause paralysis or death. Avoid prolonged exposure to vapors not known to be safe.

CONFINED SPACES

When you are working in confined spaces, adequate ventilation is essential. This includes oxygen for normal breathing, cooling to prevent heat exhaustion, air movement and exchange to prevent hazardous accumulations of vapors, and an additional or alternate source of ventilation for use in the event of emergency.

When you are working in confined spaces another important consideration must be attended to: provision should be made for space necessary for your rescue if that should be necessary. Provisions should include the use of a safety line for retrieving anyone in danger. Some means of communication must be established so that the conditions existing inside and outside the space may be made known to everyone concerned. In dangerously confined spaces, a safety man should be nearby to keep a constant check on the condition of the space and the worker, and he must be prepared to sound the alarm for additional help or to render assistance to the worker in the confined space, if that should be required.

Fumes tend to collect in confined spaces, so the condition of the space should be checked prior to entry, and monitored during your stay. Constant communication with someone outside the confined space should be maintained to inform him of any abnormal conditions which might arise during the time of an emergency.

The equipment you use in a confined space is of considerable importance. Enough light should be provided so that you can see clearly what you are doing. The light should be insulated so that it does not present a shock hazard (confined spaces are usually quite warm, and a safety light produces additional heat, so perspiration may become a serious problem). When possible, explosion-proof equipment should be used, and protective clothing should be used if toxic fumes are known or suspected to exist within the space.

PREVENTING ELECTRICAL FIRES

General cleanliness of the work area and of electrical apparatus is essential for the prevention of electrical fires. Oil, grease and carbon dust can be ignited by electrical arcing. Therefore, electrical and electronic equipment should be kept absolutely clean and free of all such deposits.

Wiping rags and other flammable waste material must always be placed in tightly closed metal containers, which must be emptied at the end of the work period.

Containers holding paints, varnishes, cleaners, or any volatile solvents should be kept tightly closed when not in actual use. They must be stored in a separate building or in a fire-resisting room which is well-ventilated and where they will not be exposed to excessive heat or to the direct rays of the sun.

FIREFIGHTING

In case of electrical fires, the following steps should be taken:

1. De-energize the circuit.
2. Call the Fire Department.
3. Control or extinguish the fire, using the correct type of fire extinguisher.

For combating electrical fires, use a CO_2 (carbon dioxide) fire extinguishers and direct it toward the base of the flame. Carbon tetyrachloride should never be used for firefighting since it changes to phosgene (a poisonous gas) upon contact with hot metal, and even in open air this creates a hazardous condition. The application of water to electrical fires is dangerous; the foam-type fire extinguishers should not be used since the foam is electrically conductive.

In cases of cable fires in which the inner layers of insulation or insulation covered by armor are burning, the only positive method of preventing the fire from running the length of the cable is to cut the cable and separate the two ends. All power to the cable should be secured and the cable should be cut with a wooden handled ax or insulated cable cutter. Keep clear of the ends after they have been cut.

When selenium rectifiers (discussed later) burn out, fumes of selenium dioxide are liberated which cause an overpowering stench. The fumes are poisonous and should not be breathed. If a rectifier burns out, de-energize the equipment immediately and ventilate the area. Allow the damaged rectifier to cool before attempting any repairs. If possible, move the equipment containing it out of doors. Do not touch or handle the defective rectifier while it is hot since a skin burn might result, through which some of the selenium compound could be absorbed.

Fires involving wood, paper, cloth, or explosives should be fought with water. Water works well on them. Therefore, advantage is taken of its inexpensiveness, availability, and safety in han-

A steady stream of water does not work in extinguishing fires involving substances like oil, gasoline, kerosene, or paint, because these sub-

NOTES:

stances will float on top of the water and keep right on burning. Also, a stream of water will scatter the burning liquid and spread the fire. For this reason, foam or fog must be used in fighting such fires.

SAFETY PRECAUTIONS WHEN USING ELECTRICAL TOOLS

As a general precaution, be sure that all tools used conform to industry and home-use standards as to quality and type, and use them only for the purposes for which they were intended. All tools in active use should be maintained in good repair, and all damaged or nonworking tools should be repaired or replaced.

When using a portable power drill, grasp it firmly during the operation to prevent it from bucking or breaking loose, thereby causing injury to yourself or damage to the tool.

Use only straight, undamaged, and properly sharpened drills. Tighten the drill securely in the chuck, using the key provided; never with wrenches or pliers. It is important that the drill be set straight and true in the chuck. The work should be firmly clamped and, if of metal a center punch should be used to score the material before the drilling operation is started.

In selecting a screwdriver for electrical work, be sure that it has a nonconducting handle. The screwdriver should not be used as a substitute for a punch or a chisel, and care should be taken that one is selected of the proper size to fit the screw.

When using a fuse puller, make certain that it is the proper type and size for the particular fuse being pulled.

The soldering iron is a fire hazard and a potential source of burns. Always assume that a soldering iron is hot; never rest the iron anywhere but on a metal surface or rack provided for that purpose. Keep the iron holder in the open to minimize the danger of fire from accumulated heat. Do not shake the iron to dispose of excess solder — a drop of hot solder may strike someone, or strike the equipment and cause a short circuit. Hold small soldering jobs with pliers or clamps.

When cleaning the iron, place the cleaning rag on a suitable surface and wipe the iron across it — do not hold the rag in the hand. Disconnect the iron when leaving the work, even for a short time — the delay may be longer than planned.

PORTABLE POWER TOOLS

All portable power tools should be carefully inspected before being used to see that they are clean, well-oiled, and in a proper state of repair. The switches should operate normally, and the cords should be clean and free of defects. The case of all electrically driven tools should be grounded. Sparking portable electric tools should not be used in any place where flammable vapors, gases, liquids, or exposed explosives are present.

Be sure that power cords do not come in contact with sharp objects. The cords should not be allowed to kink, nor be left where they might be run over. They should not be allowed to come in contact with oil, grease, hot surfaces, or chemicals; and when damaged, should be replaced instead of being patched with tape. When unplugging power tools from receptacles, grasp the plug, not the cord.

SHOP MACHINERY

Much electrical work requires the use of certain pieces of shop machinery, such as a power grinder or drill press. In addition to the general precautions on the use of tools, there are a few other precautions which should be observed when working with machinery. The most important ones are:

1. Never operate a machine with a guard or cover removed.
2. Never operate mechanical or powered equipment unless thoroughly familiar with the controls. When in doubt, consult the appropriate instruction manual or ask someone who knows.
3. Always make sure that everyone is clear before starting or operating mechanical equipment.
4. Never try to clear jammed machinery without first cutting off the source of power.
5. Never plug in electric machinery without insuring that the source voltage is the same as that called for on the nameplate of the machine.

ELECTRICAL HAZARDS

Every person who works with electrical equipment should be constantly alert to the hazards of the equipment to which he may be exposed, and also be capable of rendering first aid to injured friends and workers. Carelessness can result in serious injury or death due to electric shock, falls, burns and flying objects. After an accident has happened, investigation almost invariably shows that it could have been prevented by the exercising of simple safety precautions and procedures.

Everyone concerned with electrical equipment should make it his personal responsibility to read and become thoroughly familiar with the safety

NOTES:

practices and procedures contained in applicable safety directives, manuals, and other publications, and in equipment technical manuals prior to performing work on electrical equipment. It is the individual's responsibility to identify and eliminate unsafe conditions and unsafe acts which cause accidents.

SHOCK

Electric shock is a jarring, shaking sensation resulting from contact with electric circuits or from the effects of lightning. The victim usually feels that he has received a sudden blow; if the voltage and resulting current is sufficiently high, the victim may become unconscious. Severe burns may appear on the skin at the place of contact; muscular spasm may occur, causing the victim to clasp the apparatus or wire which caused the shock and be unable to turn it loose.

The following procedure is recommended for rescue and care of shock victims:

1. Remove the victim from electrical contact at once, but do not endanger yourself. This can be accomplished by: (1) Throwing the switch if it is nearby; (2) cutting the cable or wires to the apparatus, using an ax with a wooden handle while taking care to protect your eyes from the flash when the wires are severed; (3) using a dry stick, rope, belt, coat, blanket, or any other nonconductor of electricity, to drag or push the victim to safety.
2. Determine whether the victim is breathing. Keep him lying down in a comfortable position and loosen the clothing about his neck, chest, and abdomen so that he can breath freely. Protect him from exposure to cold, and watch him carefully.
3. Keep him from moving about. In this condition, the heart is very weak and any sudden muscular effort or activity on the part of the patient may result in heart failure.
4. Do not give stimulants or opiates. Send for a physician at once and do not leave the patient until he has adequate medical care.
5. If the victim is not breathing, it will be necessary to apply artificial respiration without delay, even though he may appear to be lifeless. Do not stop artificial respiration until a physician pronounces the victim beyond help.

BURNS AND WOUNDS

In administering first aid for burns, the objectives are to relieve the pain, to make the patient as comfortable as possible to prevent infection and to guard against shock which often accompanies burns of a serious nature.

Any break in the skin is dangerous because it allows germs to enter the wound. Although infection may occur in any wound, it is of particular danger in wounds which do not bleed freely, wounds in which torn tissue or skin falls back into place and so prevents the entrance of air, and wounds which involve crushing of tissues.

Minor wounds should be washed immediately with soap and clean water, dried and painted with a mild, nonirritating antiseptic. Benzalkonium chloride tincture is a good antiseptic; it does not irritate the skin but it effectively destroys many of the microorganisms which cause infection. (Benzalkonium chloride is commonly sold under the name of Zephiran or Zepherin chloride.) Apply a dressing if necessary.

Larger wounds should be treated only by a doctor. Merely cover the wound with a dry sterile compress and fasten the compress in place with a bandage.

Since the saving of a person's life often depends on prompt and correct first aid treatment, everyone working with electrical equipment should become thoroughly familiar with first aid.

SAFETY PRECAUTIONS

Take time to be safe when working on electrical circuits and equipment. Carefully study the schematics and wiring diagrams of the entire system, noting what circuits must be de-energized in addition to the main power supply. Remember that electrical equipments frequently have more than one source of power. Be certain that all power sources are de-energized before servicing the equipment. Do not service any equipment with the power on unless it is necessary.

It must be borne in mind that de-energizing main supply circuits by opening supply switches will not necessarily "kill" all circuits in a given piece of equipment. A source of danger that has often been neglected or ignored — sometimes with tragic results — is the inputs to electrical equipment from other sources, such as synchros, remote control circuits, and so forth. For example, turning off the antenna safety switches will disable the antenna, but it may not turn off the antenna synchro voltages from other sources. Moreover, the rescue of a victim shocked by the power input from a remote source is often hampered because of the time required to determine the source of power and turn it off. Therefore, turn off all power inputs before working on equipment.

Do not work on high voltage circuits alone;

NOTES:

have another person (safety observer), who is qualified in first aid for electrical shock, present at all times. A friend or helper stationed nearby should also know the circuits and switches controlling the equipment, and should be given instructions to pull the switch immediately if anything unforeseen happens.

Always be aware of the nearness of high voltage lines or circuits. Use rubber gloves where applicable, and stand on approved rubber matting. Remember, not all so-called rubber mats are good insulators.

Equipment containing metal parts, such as brushes and brooms, should not be used in an area within 4 feet of high voltage circuits or any electric wiring having exposed surfaces.

Keep clothing, hands, and feet dry if at all possible. When it is necessary to work in wet or damp locations, use a dry platform or wooden stool to sit or stand on, and place a rubber mat or other nonconductive material on top of the wood. Use insulated tools and insulated flashlights of the molded type when required to work on exposed parts.

Do not wear loose or flapping clothing. The use of thin-soled shoes with metal plates or hobnails is prohibited. Safety shoes with nonconducting soles should be worn if available. Flammable articles, such as celluloid cap visors, should not be worn.

When working on an electrical apparatus, you should first remove all rings, wristwatches, bracelets, and similar metal items. Care should be taken that the clothing does not contain exposed zippers, metal buttons, or any type of metal fastener.

Do NOT work on energized circuits unless absolutely necessary. Be sure to take time to lock out (or block out) the switch and tag it. Locks for this purpose should be readily available; if a lock cannot be obtained, remove the fuse and tag it.

Use one hand when turning switches on or off. Keep the doors to switch and fuse boxes closed except when working inside or replacing fuses. Use a fuse puller to remove cartridge fuses after first making certain that the circuit is dead.

Never short out, tamper with, or block open an interlock switch.

Keep clear of exposed equipment; when it is necessary to work on it, use one hand only as much as possible.

Warning signs and suitable guards should be provided to prevent personnel from coming into accidental contact with high voltages.

Avoid reaching into enclosures except when absolutely necessary; when reaching into an enclosure, use rubber blankets to prevent accidental contact with the enclosure.

Do not use bare hands to remove hot parts from their holders. Use asbestos gloves if necessary.

Use a shorting stick, similar to the one shown in figure 1-1 to discharge all high voltage charges. Before a worker touches a capacitor or any part of a circuit which is known or likely to be connected to a capacitor (whether the circuit is de-energized or disconnected entirely), he should short circuit the terminals to make sure that the capacitor is completely discharged. Grounded shorting sticks should be permanently attached to workbenches where electrical equipment using high voltages are regularly serviced.

Make certain that the equipment is properly grounded. Ground all test equipment to the equipment under test.

Turn off the power before connecting alligator clips to any circuit.

When measuring circuits over 300 volts, do not hold the test prods.

+++

Safety is everyone's responsibility. With a reasonable regard for safety rules and a sensible amount of precaution, work in one's home on electrical equipment should not be dangerous. But it will take constant vigilance.

NOTES:

CHAPTER 2

FUNDAMENTAL CONCEPTS OF ELECTRICITY

The word "electric" is actually a Greek-derived word meaning AMBER. Amber is a translucent (semitransparent) yellowish mineral, which, in the natural form, is composed of fossilized resin. The ancient Greeks used the words "electric force" in referring to the mysterious forces of attraction and repulsion exhibited by amber when it was rubbed with a cloth. They did not understand the fundamental nature of this force. They could not answer the seemingly simple question, "What is electricity?". This question is still unanswered. Though electricity might be defined as "that force which moves electrons," this would be the same as defining an engine as "that force which moves an automobile." The effect has been described, not the force.

Presently little more is known than the ancient Greeks knew about the fundamental nature of electricity, but tremendous strides have been made in harnessing and using it. Elaborate theories concerning the nature and behavior of electricity have been advanced, and have gained wide acceptance because of their apparent truth and demonstrated workability.

From time to time various scientists have found that electricity seems to behave in a constant and predictable manner in given situations, or when subjected to given conditions. These scientists, such as Faraday, Ohm, Lenz, and Kirchhoff, to name only a few, observed and described the predictable characteristics of electricity and electric current in the form of certain rules. These rules are often referred to as "laws." Thus, though electricity itself has never been clearly defined, its predictable nature and easily used form of energy has made it one of the most widely used power sources in modern time. By learning the rules, or laws, applying to the behavior of electricity, and by understanding the methods of producing, controlling, and using it, electricity may be "learned" without ever having determined its fundamental identity.

THE MOLECULE

One of the oldest, and probably the most generally accepted, theories concerning electric current flow is that it is comprised of moving electrons. This is the ELECTRON THEORY. Electrons are extremely tiny parts, or particles, of matter. To study the electron, therefore, the structural nature of matter itself must first be studied. (Anything having mass and inertia, and which occupies any amount of space, is composed of matter.) To study the fundamental structure or composition of any type of matter, it must be reduced to its fundamental fractions. Assume the drop of water in figure 2-1 (A) was halved again and again. By continuing the process long enough, the smallest particle of water possible—the molecule—would be obtained. All molecules are composed of atoms.

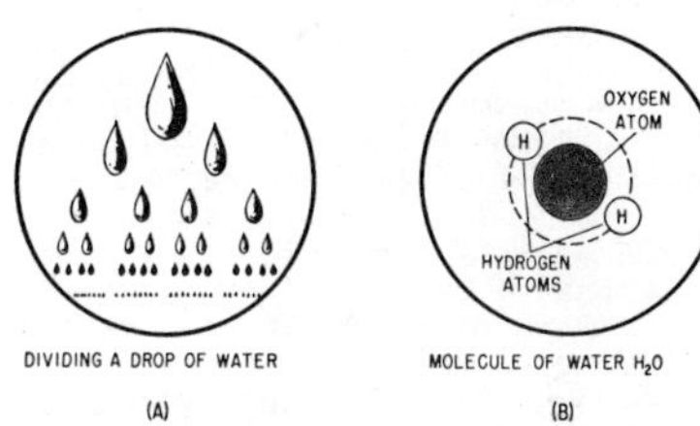

Figure 2-1.—Matter is made up of molecules.

A molecule of water (H_2O) is composed of one atom of oxygen and two atoms of hydrogen, as represented in figure 2-1 (B). If the molecule of water were further subdivided, there would

NOTES:

remain only unrelated atoms of oxygen and hydrogen, and the water would no longer exist as such. This example illustrates the following fact—the molecule is the smallest particle to which a substance can be reduced and still be called by the same name. This applies to all substances—liquids, solids, and gases.

When whole molecules are combined or separated from one another, the change is generally referred to as a PHYSICAL change. In a CHEMICAL change the molecules of the substance are altered such that new molecules result. Most chemical changes involve positive and negative ions and thus are electrical in nature. All matter is said to be essentially electrical in nature.

THE ATOM

In the study of chemistry it soon becomes apparent that the molecule is far from being the ultimate particle into which matter may be subdivided. The salt molecule may be decomposed into radically different substances—sodium and chlorine. These particles that make up molecules can be isolated and studied separately. They are called ATOMS.

The atom is the smallest particle that makes up that type of material called an ELEMENT. The element retains its characteristics when subdivided into atoms. More than 100 elements have been identified. They can be arranged into a table of increasing weight, and can be grouped into families of material having similar properties. This arrangement is called the PERIODIC TABLE OF THE ELEMENTS.

The idea that all matter is composed of atoms dates back more than 2,000 years to the Greeks. Many centuries passed before the study of matter proved that the basic idea of atomic structure was correct. Physicists have explored the interior of the atom and discovered many subdivisions in it. The core of the atom is called the NUCLEUS. Most of the mass of the atom is concentrated in the nucleus. It is comparable to the sun in the solar system, around which the planets revolve. The nucleus contains PROTONS (positively charged particles) and NEUTRONS which are electrically neutral.

Most of the weight of the atom is in the protons and neutrons of the nucleus. Whirling around the nucleus are one or more smaller particles of negative electric charge. THESE ARE THE ELECTRONS. Normally there is one proton for each electron in the entire atom so that the net positive charge of the nucleus is balanced by the net negative charge of the electrons whirling around the nucleus. THUS THE ATOM IS ELECTRICALLY NEUTRAL.

The electons do not fall into the nucleus even though they are attracted strongly to it. Their motion prevents it, as the planets are prevented from falling into the sun because of their centrifugal force of revolution.

The number of protons, which is usually the same as the number of electrons, determines the kind of element in question. Figure 2-2 shows a simplified picture of several atoms of different materials based on the conception of planetary electrons describing orbits about the nucleus. For example, hydrogen has a nucleus consisting of 1 proton, around which rotates 1 electron. The helium atom has a nucleus containing 2 protons and 2 neutrons with 2 electrons encircling the nucleus. Near the other extreme of the list of elements is curium (not shown in the figure), an element discovered in the 1940's, which has 96 protons and 96 electrons in each atom.

The Periodic Table of the Elements is an orderly arrangement of the elements in ascending atomic number (number of planetary electrons) and also in atomic weight (number of protons and neutrons in the nucleus). The various kinds of atoms have distinct masses or weights with respect to each other. The element most closely approaching unity (meaning 1) is hydrogen whose atomic weight is 1.008 as compared with oxygen whose atomic weight is 16. Helium has an atomic weight of approximately 4, lithium 7, fluorine 19, and neon 20, as shown in figure 2-2.

Figure 2-3 is a pictorial summation of the discussion that has just been presented. Visible matter, at the left of the figure, is broken down first to one of its basic molecules, then to one of the molecule's atoms. The atom is then further reduced to its subatomic particles—the protons, neutrons, and electrons. Subatomic particles are electric in nature. That is, they are the particles of matter most affected by an electric force. Whereas the whole molecule or a whole atom is electrically neutral, most subatomic particles are not neutral (with the exception of the neutron). Protons are inherently positive, and electrons are inherently negative. It is these inherent characteristics which make subatomic particles sensitive to electric force.

When an electric force is applied to a conducting medium, such as copper wire, electrons in the outer orbits of the copper atoms are forced

NOTES:

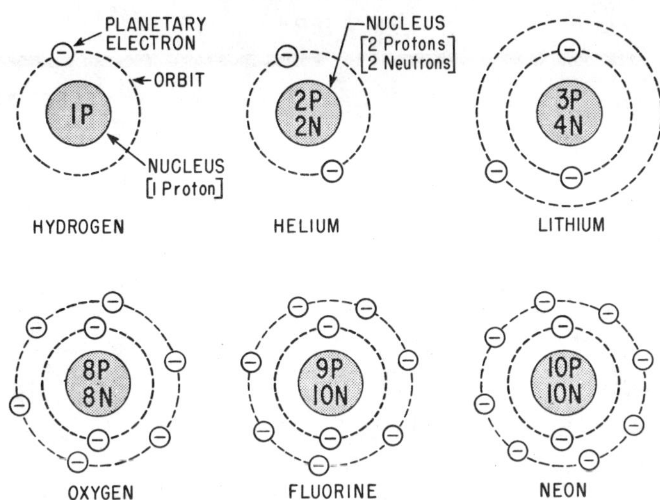

Figure 2-2.—Atomic structure of elements.

out of orbit and impelled along the wire. The direction of electron movement is determined by the direction of the impelling force. The protons do not move, mainly because they are extremely heavy. The proton of the lightest element, hydrogen, is approximately 1,850 times heavier than its electron. Thus, it is the relatively light electron that is moved by electricity.

When an orbital electron is removed from an atom it is called a FREE ELECTRON. Some of the electrons of certain metallic atoms are so loosely bound to the nucleus that they are comparatively free to move from atom to atom. Thus, a very small force or amount of energy will cause such electrons to be removed from the atom and become free electrons. It is these free electrons that constitute the flow of an electric current in electrical conductors.

If the internal energy of an atom is raised above its normal state, the atom is said to be EXCITED. Excitation may be produced by causing the atoms to collide with particles that are impelled by an electric force. In this way, energy is transferred from the electric source to the atom. The excess energy absorbed by an atom may become sufficient to cause loosely bound outer electrons to leave the atom against the force that acts to hold them within. An atom that has thus lost or gained one or more electrons is said to be IONIZED. If the atom loses electrons it becomes positively charged and is referred to as a POSITIVE ION. Conversely, if the atom gains electrons, it becomes negatively charged and is referred to as a NEGATIVE ION. An ion may then be defined as a small particle of matter having a positive or negative charge.

CONDUCTORS, SEMICONDUCTORS, AND INSULATORS

Substances that permit the free motion of a large number of electrons are called CONDUCTORS. Copper wire is considered a good conductor because it has many free electrons. Electrical energy is transferred through conductors by means of the movement of free electrons that migrate from atom to atom inside the conductor. Each electron moves a very short distance to the neighboring atom where it replaces one or more electrons by forcing them out of their orbits. The replaced electrons repeat the process in other nearby atoms until the

NOTES:

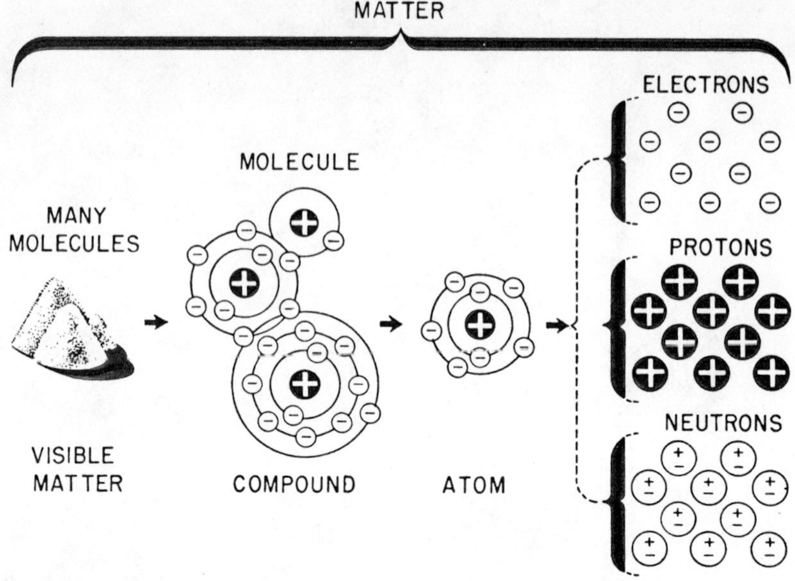

Figure 2-3.—Breakdown of visible matter to electric particles.

movement is transmitted throughout the entire length of the conductor. The greater the number of electrons that can be made to move in a material under the application of a given force the better are the conductive qualities of that material. A good conductor is said to have a low opposition or low resistance to the current (electron) flow. Among the most commonly known metals used as conductors are silver, copper, and aluminum. The best conductor is silver, copper, and aluminum, in that order. However, copper is used more extensively because it is less expensive. A material's ability to conduct electricity also depends on its dimensions.

In contrast to good conductors, some substances such as rubber, glass, and dry wood have very few free electrons. In these materials large amounts of energy must be expended in order to break the electrons loose from the influence of the nucleus. Substances containing very few free electrons are called POOR CONDUCTORS, NONCONDUCTORS, or INSULATORS. Actually, there is no sharp dividing line between conductors and insulators, since electron motion is known to exist to some extent in all matter. Electricians simply use the best conductors as wires to carry current and the poorest conductors as insulators to prevent the current from being diverted from the wires.

Listed below are some of the best conductors and best insulators arranged in accordance with their respective abilities to conduct or to resist the flow of electrons.

Conductors	Insulators
Silver	Dry air
Copper	Glass
Aluminum	Mica
Zinc	Rubber
Brass	Asbestos
Iron	Bakelite

NOTES:

A semiconductor is a material that is neither a good conductor nor a good insulator. Germanium and silicon are substances that fall into this category. These materials, due to their peculiar crystalline structure, may under certain conditions act as conductors; under other conditions, as insulators. As the temperature is raised, however, a limited number of electrons become available for conduction.

STATIC ELECTRICITY

In a natural, or neutral state, each atom in a body of matter will have the proper number of electrons in orbit around it. Consequently, the whole body of matter comprised of the neutral atoms will also be electrically neutral. In this state, it is said to have a "zero charge," and will neither attract nor repel other matter in its vicinity. Electrons will neither leave nor enter the neutrally charged body should it come in contact with other neutral bodies. If, however, any number of electrons are removed from the atoms of a body of matter, there will remain more protons than electrons, and the whole body of matter will become electrically positive. Should the positively charged body come in contact with another body having a normal charge, or having a negative (too many electrons) charge, an electric current will flow between them. Electrons will leave the more negative body and enter the positive body. This electron flow will continue until both bodies have equal charges.

When two bodies of matter have unequal charges, and are near one another, an electric force is exerted between them because of their unequal charges. However, since they are not in contact, their charges cannot equalize. The existence of such an electric force, where current cannot flow, is referred to as static electricity. "Static" means "not moving." This is also referred to as an ELECTROSTATIC FORCE.

One of the easiest ways to create a static charge is by the friction method. With the friction method, two pieces of matter are rubbed together and electrons are "wiped off" one onto the other. If materials that are good conductors are used, it is quite difficult to obtain a detectable charge on either. The reason for this is that equalizing currents will flow easily in and between the conducting materials. These currents equalize the charges almost as fast as they are created. A static charge is easier to obtain by rubbing a hard nonconducting material against a soft, or fluffy, nonconductor. Electrons are rubbed off one material and onto the other material. This is illustrated in figure 2-4.

When the hard rubber rod is rubbed in the fur, the rod accumulates electrons. Since both fur and rubber are poor conductors, little equalizing current can flow, and an electrostatic charge is built up. When the charge is great enough, equalizing currents will flow regardless of the material's poor conductivity. These currents will cause visible sparks, if viewed in darkness, and produce a cracking sound.

CHARGED BODIES

One of the fundamental laws of electricity is that LIKE CHARGES REPEL EACH OTHER and UNLIKE CHARGES ATTRACT EACH OTHER. A positive charge and negative charge, being unlike, tend to move toward each other. In the atom the negative electrons are drawn toward the positive protons in the nucleus. This attractive force is balanced by the electron's centrifugal force caused by its rotation about the nucleus. As a result, the electrons remain in orbit and are not drawn into the nucleus. Electrons repel each other because of their like negative charges, and protons repel each other because of their like positive charges.

The law of charged bodies may be demonstrated by a simple experiment. Two pith (paper pulp) balls are suspended near one another by threads, as shown in figure 2-5.

If the hard rubber rod is rubbed to give it a negative charge, and then held against the right-hand ball in part (A), the rod will impart a negative charge to the ball. The right-hand ball will be charged negative with respect to the left-hand ball. When released, the two balls will be drawn together, as shown in figure 2-5 (A). They will touch and remain in contact until the left-hand ball acquires a portion of the negative charge of the right-hand ball, at which time they will swing apart as shown in figure 2-5 (C). If, positive charges are placed on both balls (fig. 2-5 (B)), the balls will also be repelled from each other.

COULOMB'S LAW OF CHARGES

The amount of attracting or repelling force which acts between two electrically charged bodies in free space depends on two things—(1) their charges, and (2) the distance between them. The relationship of charge and distance to

NOTES:

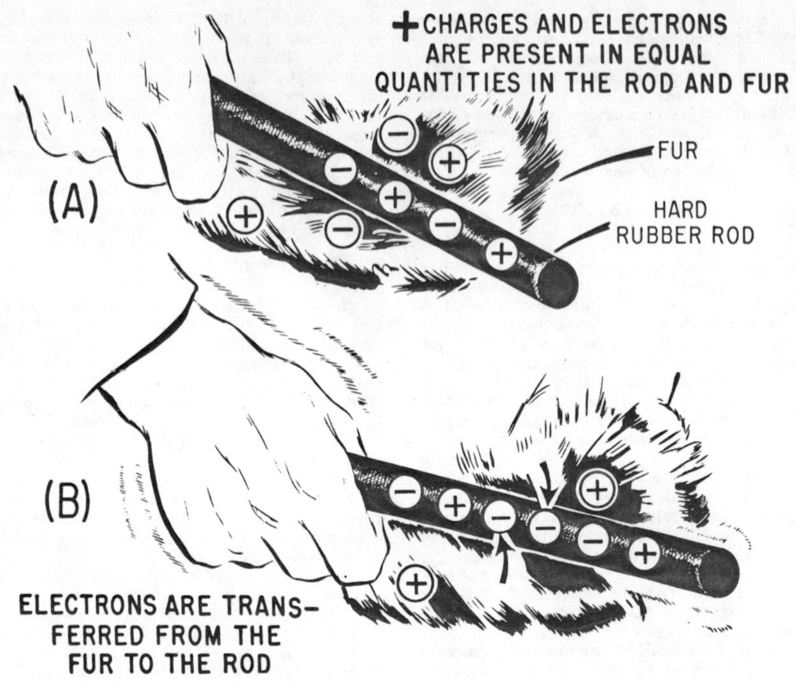

Figure 2-4.—Producing static electricity by friction.

electrostatic force was first discovered and written by a French scientist named Charles A. Coulomb. Coulomb's Law states that CHARGED BODIES ATTRACT OR REPEL EACH OTHER WITH A FORCE THAT IS DIRECTLY PROPORTIONAL TO THE PRODUCT OF THEIR CHARGES, AND IS INVERSELY PROPORTIONAL TO THE SQUARE OF THE DISTANCE BETWEEN THEM.

ELECTRIC FIELDS

The space between and around charged bodies in which their influence is felt is called an ELECTRIC FIELD OF FORCE. The electric field is always terminated on material objects and extends between positive and negative charges. It can exist in air, glass, paper, or a vacuum. ELECTROSTATIC FIELDS and DIELECTRIC FIELDS are other names used to refer to this region of force.

Fields of force spread out in the space surrounding their point of origin and, in general, DIMINISH IN PROPORTION TO THE SQUARE OF THE DISTANCE FROM THEIR SOURCE.

The field about a charged body is generally represented by lines which are referred to as ELECTROSTATIC LINES OF FORCE. These lines are imaginary and are used merely to represent the direction and strength of the field. To avoid confusion, the lines of force exerted by a positive charge are always shown leaving the

NOTES:

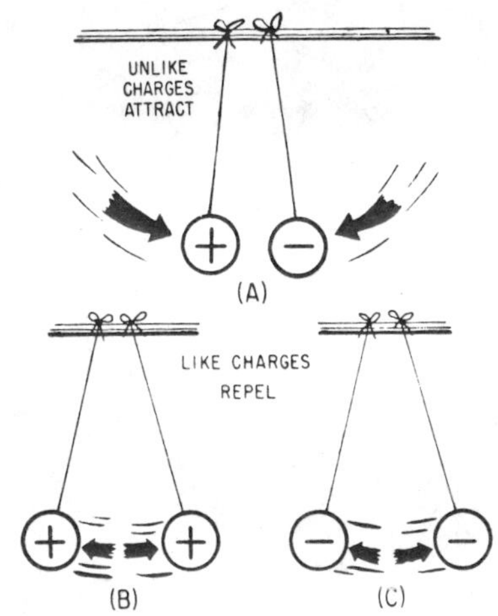

Figure 2-5.—Reaction between charged bodies.

charge, and for a negative charge they are shown as entering. Figure 2-6 illustrates the use of lines to represent the field about charged bodies.

Figure 2-6 (A) represents the repulsion of like-charged bodies and their associated fields. Part (B) represents the attraction between unlike-charged bodies and their associated fields.

MAGNETISM

A substance is said to be a magnet if it has the property of magnetism—that is, if it has the power to attract such substances as iron, steel, nickel, or cobalt, which are known as MAGNETIC MATERIALS. A steel knitting needle, magnetized by a method to be described later, exhibits two points of maximum attraction (one at each end) and no attraction at its center. The points of maximum attraction are called MAGNETIC POLES. All magnets have at least two poles. If the needle is suspended by its middle so that it rotates freely in a horizontal plane about its center, the needle comes to rest in a approximately north-south line of direction. The same pole will always point to the north, and the other will always point toward the south. The magnetic pole that points northward is called the NORTH POLE, and the other the SOUTH POLE.

A MAGNETIC FIELD exists around a simple bar magnet. The field consists of imaginary lines along which a MAGNETIC FORCE acts. These lines emanate from the north pole of the magnet, and enter the south pole, returning to the north pole through the magnet itself, thus forming closed loops.

A MAGNETIC CIRCUIT is a complete path through which magnetic lines of force may be established under the influence of a magnetizing force. Most magnetic circuits are composed largely of magnetic materials in order to contain the magnetic flux. These circuits are similar to the ELECTRIC CIRCUIT, which is a complete path through which current is caused to flow under the influence of an electromotive force.

Magnets may be conveniently divided into three groups.

1. NATURAL MAGNETS, found in the natural state in the form of a mineral called magnetite.

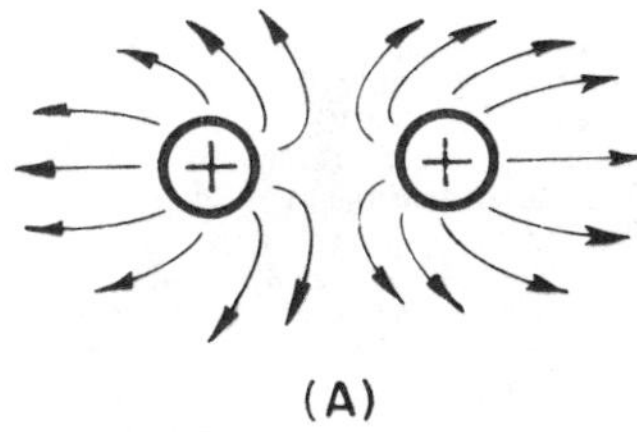

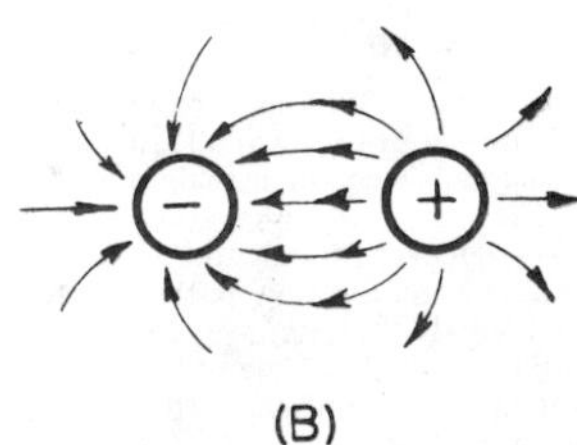

Figure 2-6.—Electrostatic lines of force.

NOTES:

2. PERMANENT MAGNETS, bars of hardened steel (or some form of alloy such as alnico) that have been permanently magnetized.

3. ELECTROMAGNETS, composed of soft-iron cores around which are wound coils of insulated wire. When an electric current flows through the coil, the core becomes magnetized. When the current ceases to flow, the core loses most of its magnetism.

Permanent magnets and electromagnets are sometimes called ARTIFICIAL MAGNETS to further distinguish them from natural magnets, and are discussed in this manual under the heading of "Artificial Magnets."

NATURAL MAGNETS

For many centuries it has been known that certain stones (magnetite, Fe_3O_4) have the ability to attract small pieces of iron. Because many of the best of these stones (natural magnets) were found near Magnesia in Asia Minor, the Greeks called the substance MAGNETITE, or MAGNETIC.

Before this, ancient Chinese observed that when similar stones were suspended freely, or floated on a light substance in a container of water, they tended to assume a nearly north-and-south position. Probably Chinese navigators used bits of magnetite floating on wood in a liquid-filled vessel as crude compasses. At that time it was not known that the earth itself acts like a magnet, and these stones were regarded with considerable superstitious awe. Because bits of this substance were used as compasses they were called LOADSTONES (or lodestones), which means "leading stones."

Natural magnets are also found in the United States, Norway, and Sweden. A natural magnet, demonstrating the attractive force at the poles, is shown in figure 2-7 (A).

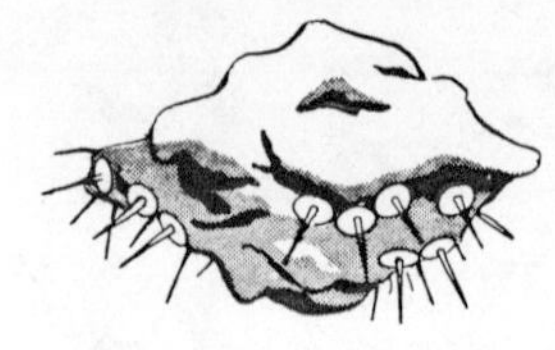

(A)
NATURAL

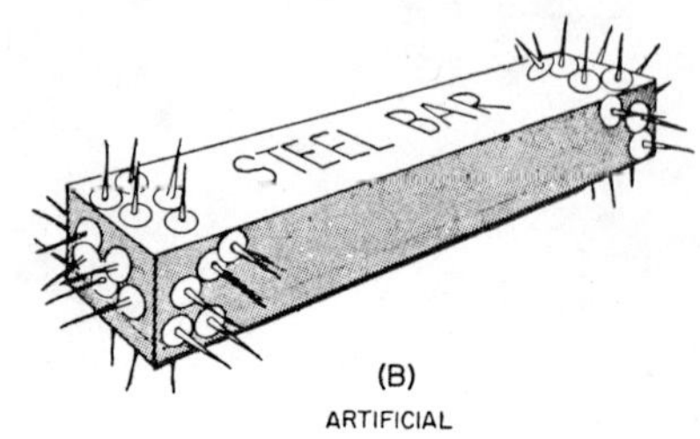

(B)
ARTIFICIAL

Figure 2-7.—(A) Natural magnet; (B) artificial magnet.

ARTIFICIAL MAGNETS

Natural magnets no longer have any practical value because more powerful and more conveniently shaped permanent magnets can be produced artificially. Commercial magnets are made from special steels and alloys—for example, alnico, made principally of aluminum, nickel, and cobalt. The name is derived from the first two letters of the three principal elements of which it is composed. An artificial magnet is shown in figure 2-7 (B).

An iron, steel, or alloy bar can be magnetized by inserting the bar into a coil of insulated wire and passing a heavy direct current through the coil, as shown in figure 2-8 (A). This aspect of magnetism is treated later in the chapter. The same bar may also be magnetized if it is stroked with a bar magnet, as shown in figure 2-8 (B). It will then have the same magnetic property that the magnet used to induce the magnetism has—namely, there will be two poles of attraction, one at either end. This process produces a permanent magnet by INDUCTION—that is, the magnetism is induced in the bar by the influence of the stroking magnet.

Artificial magnets may be classified as "permanent" or "temporary" depending on their ability to retain their magnetic strength after the magnetizing force has been removed. Hardened steel and certain alloys are relatively difficult to magnetize and are said to have a LOW PERMEABILITY because the magnetic lines of force do not easily permeate, or distribute themselves readily through the steel. (Permeability is a measure of the relative ability of a substance to conduct magnetic lines of force as compared with

NOTES:

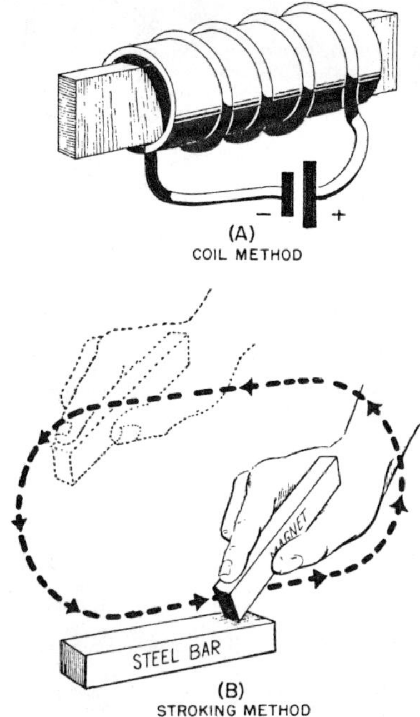

Figure 2-8.—Methods of producing artificial magnets.

air. It is discussed in greater detail later in this manual.) Once magnetized, however, these materials retain a large part of their magnetic strength and are called PERMANENT MAGNETS. Permanent magnets are used extensively in electric instruments, meters, telephone receivers, permanent-magnet loudspeakers, and magnetos. Conversely, substances that are relatively easy to magnetize—such as soft iron and annealed silicon steel—are said to have a HIGH PERMEABILITY. Such substances retain only a small part of their magnetism after the magnetizing force is removed and are called TEMPORARY MAGNETS. Silicon steel and similar materials are used in transformers where the magnetism is constantly changing and in generators and motors where the strength of the fields can be readily changed.

The magnetism that remains in a temporary magnet after the magnetizing force is removed is called RESIDUAL MAGNETISM. The fact that temporary magnets retain even a small amount of magnetism is an important factor in the build-up of voltage in self-excited d-c generators.

NATURE OF MAGNETISM

A popular theory of magnetism considers the molecular alinement of the material. This is known as Weber's Theory. This theory assumes all magnetic substances to be composed of tiny molecular magnets. All unmagnetized materials have the magnetic forces of its molecular magnets neutralized by adjacent molecular magnets thereby eliminating any magnetic effect. A magnetized material will have most of its molecular magnets lined up so that the north pole of each molecule points in one direction, and the south pole faces the opposite direction. A material with its molecules thus alined will then have one effective north pole, and one effective south pole. An illustration of Weber's Theory is shown in figure 2-9 (A) where a steel bar is magnetized by stroking. When a steel bar is stroked several times in the same direction by a magnet, the magnetic force from the north pole of the magnet causes the molecules to aline themselves. The polarity of the magnet formed is dependent upon the direction of the magnetizing force as it is brought over the random magnetic molecules.

Some justification of Weber's Theory occurs when a magnet is split in half. It is found that each half possess both a north and a south magnetic pole as shown in figure 2-9 (B). The polarities of the poles are in the same respective directions as the poles of the original magnet. If a magnet is further divided into small parts, it will be found that each part, down to its last molecule, will all have similar north and south poles. Each part would exhibit its own magnetic properties.

Further support of Weber's Theory comes from the fact that when a bar magnet is held out of alinement with the earth's magnetic field and repeatedly jarred or heated, the molecular alinement is disarranged and the material becomes demagnetized. For example, measuring devices which make use of permanent magnets become inaccurate when subjected to severe jarring or exposure to opposing magnetic fields.

NOTES:

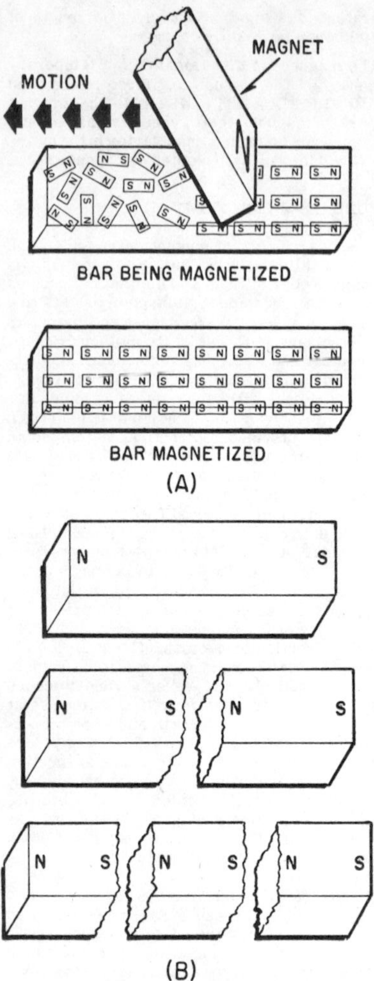

Figure 2-9.—(A) Molecular magnets; (B) broken magnets.

NOTES:

DOMAIN THEORY

A more modern theory of magnetism is based on the electron spin principle. From the study of atomic structure it is known that all matter is composed of vast quantities of atoms, each atom containing one or more orbital electrons. The electrons are considered to orbit in various shells and subshells depending upon their distance from the nucleus. The structure of the atom has previously been compared to the solar system, wherein the electrons orbiting the nucleus correspond to the planets orbiting the sun. Along with their orbital motion about the sun, these planets also revolve on their axes. It is believed that the electron also revolves on its axis as it orbits the nucleus of an atom.

It has been experimentally proven that an electron has a magnetic field about it along with an electric field. The effectiveness of the magnetic field of an atom is determined by the number of electrons spinning in each direction. If an atom has equal numbers of electrons spinning in opposite directions, the magnetic fields surrounding the electrons cancel one another, and the atom is unmagnetized. However, if more electrons spin in one direction than another, the atom is magnetized. An atom such as iron with an atomic number of 26 has 26 protons in the nucleus and 26 revolving electrons orbiting its nucleus. If 13 electrons are spinning in a clockwise direction and 13 electrons are spinning in a counterclockwise direction, the opposing magnetic fields will be neutralized. When more than 13 electrons spin in either direction, the atom is magnetized. An example of a magnetized atom of iron is shown in figure 2-10. Note that in this specific illustration the electrons magnetic fields in all except the M shell neutralize each other. As illustrated in the diagram, there exists 15 electrons spinning in one direction and only 11 electrons spinning in an opposite direction. Therefore, the unopposed magnetic fields of 4 electrons will cause this iron atom to become an infinitely small magnet.

When a number of such atoms are grouped together to form an iron bar, there is an interaction between the magnetic forces of various atoms. The small magnetic force of the field surrounding an atom affects adjacent atoms, thus producing a small group of atoms with parallel magnetic fields. This group of from 10^{14} to 10^{15} magnetic atoms, having their magnetic poles orientated in the same direction, is known as a DOMAIN. Throughout a domain

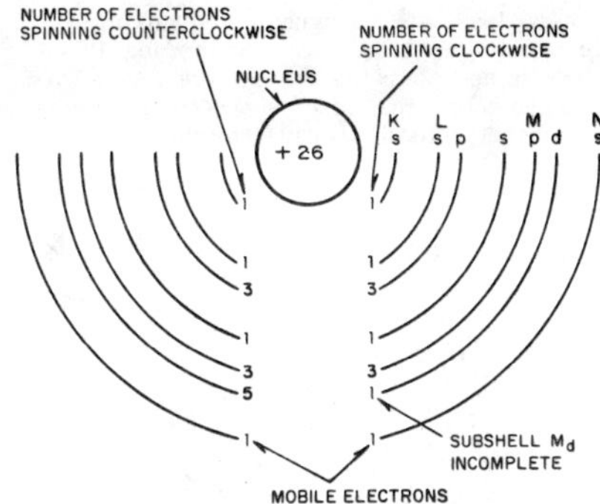

Figure 2-10.—Iron atom.

there is an intense magnetic field without the influence of any external magnetic field. Since about 10 million tiny domains can be contained in 1 cubic millimeter, it is apparent that every magnetic material is made up of a large number of domains. The domains in any substance are always magnetized to saturation but are randomly orientated throughout a material. Thus, the strong magnetic field of each domain is neutralized by opposing magnetic forces of other domains. When an external field is applied to a magnetic substance the domains will line up with the external field. Since the domains themselves are naturally magnetized to saturation, the magnetic strength of a magnetized material is determined by the number of domains alined by the magnetizing force.

MAGNETIC FIELDS AND LINES OF FORCE

If a bar magnet is dipped into iron filings, many of the filings are attracted to the ends of the magnet, but none are attracted to the center of the magnet. As mentioned previously, the ends of the magnet where the attractive force is the greatest are called the POLES of the magnet. By using a compass, the line of direction of the magnetic force at various points near the magnet may be observed. The compass needle itself is a magnet. The north end of the compass needle always points toward the south pole, S, as shown in figure 2-11 (A), and thus the sense of direction (with respect to the polarity of the bar magnet) is also indicated. At the center, the compass needle points in a direction that is parallel to the bar magnet.

When the compass is placed successively at several points in the vicinity of the bar magnet the compass needle alines itself with the field at each position. The direction of the field is indicated by the arrows and represents the direction in which the north pole of the compass needle will point when the compass is placed in this field. Such a line along which a compass needle alines itself is called a MAGNETIC LINE OF FORCE. (This magnetic line of force does not actually exist but is an imaginary line used to illustrate and describe the pattern of the magnetic field. As mentioned previously, the magnetic lines of force are assumed to emanate from the north pole of a magnet, pass through the surrounding space, and enter the south pole. The lines of force then pass from the south pole to the north pole inside the magnet to form a closed loop. Each line of force forms an independent closed loop and does not merge with or cross other lines of force. The lines of force between the poles of a horseshoe magnet are shown in figure 2-11 (B).

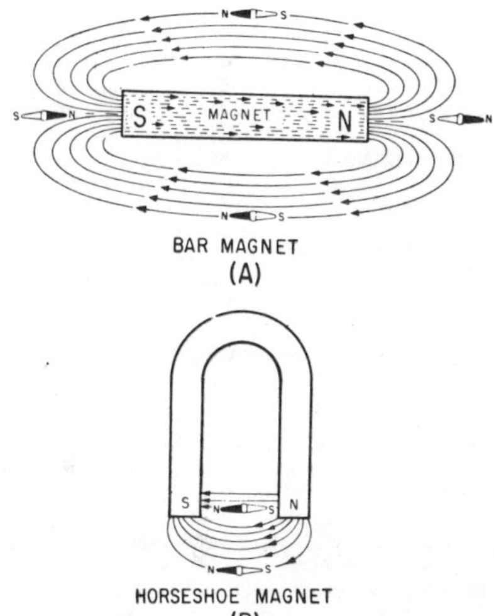

Figure 2-11.—Magnetic lines of force.

NOTES:

Although magnetic lines of force are imaginary, a simplified version of many magnetic phenomena can be explained by assuming the magnetic lines to have certain real properties. The lines of force can be compared to rubberbands which stretch outward when a force is exerted upon them and contract when the force is removed. The characteristics of magnetic lines of force can be described as follows:

1. Magnetic lines of force are continuous and will always form closed loops.
2. Magnetic lines of force will never cross one another.
3. Parallel magnetic lines of force traveling in the same direction repel one another. Parallel magnetic lines of force traveling in opposite directions tend to unite with each other and form into single lines traveling in a direction determined by the magnetic poles creating the lines of force.
4. Magnetic lines of force tend to shorten themselves. Therefore, the magnetic lines of force existing between two unlike poles cause the poles to be pulled together.
5. Magnetic lines of force pass through all materials, both magnetic and nonmagnetic.

The space surrounding a magnet, in which the magnetic force acts, is called a MAGNETIC FIELD. Michael Faraday was the first scientist to visualize the magnet field as being in a state of stress and consisting of uniformly distributed lines of force. The entire quantity of magnetic lines surrounding a magnet is called MAGNETIC FLUX. Flux in a magnetic circuit corresponds to current in an electric circuit.

The number of lines of force per unit area is called FLUX DENSITY and is measured in lines per square inch or lines per square centimeter. Flux density is expressed by the equation

$$B = \frac{\Phi}{A},$$

where B is the flux density, Φ (Greek phi) is the total number of lines of flux, and A is the cross-sectional area of the magnetic circuit. If A is in square centimeters, B is in lines per square centimeter, or GAUSS. The terms FLUX and FLOW of magnetism are frequently used in textbooks. However, magnetism itself is not thought to be a stream of particles in motion, but is simply a field of force exerted in space.

A visual representation of the magnetic field around a magnet can be obtained by placing a plate of glass over a magnet and sprinkling iron filings onto the glass. The filings arrange themselves in definite paths between the poles. This arrangement of the filings shows the pattern of the magnetic field around the magnet, as in figure 2-12.

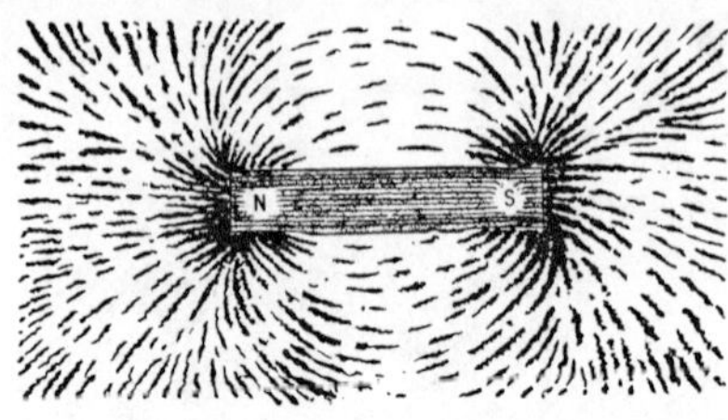

Figure 2-12.—Magnetic field pattern around a magnet.

The magnetic field surrounding a symmetrically shaped magnet has the following properties:

1. The field is symmetrical unless disturbed by another magnetic substance.
2. The lines of force have direction and are represented as emanating from the north pole and entering the south pole.

LAWS OF ATTRACTION AND REPULSION

If a magnetized needle is suspended near a bar magnet, as in figure 2-13, it will be seen that a north pole repels a north pole and a south pole repels a south pole. Opposite poles, however, will attract each other. Thus, the first two laws of magnetic attraction and repulsion are:

1. LIKE magnetic poles REPEL each other.
2. UNLIKE magnetic poles ATTRACT each other.

The flux patterns between adjacent UNLIKE poles of bar magnets, as indicated by lines, are shown in figure 2-14 (A). Similar patterns for adjacent LIKE poles are shown in figure 2-14 (B). The lines do not cross at any point and they act as if they repel each other.

Figure 2-15 shows the flux pattern (indicated by lines) around two bar magnets placed close together and parallel with each other. Figure 2-15 (A) shows the flux pattern when opposite

NOTES:

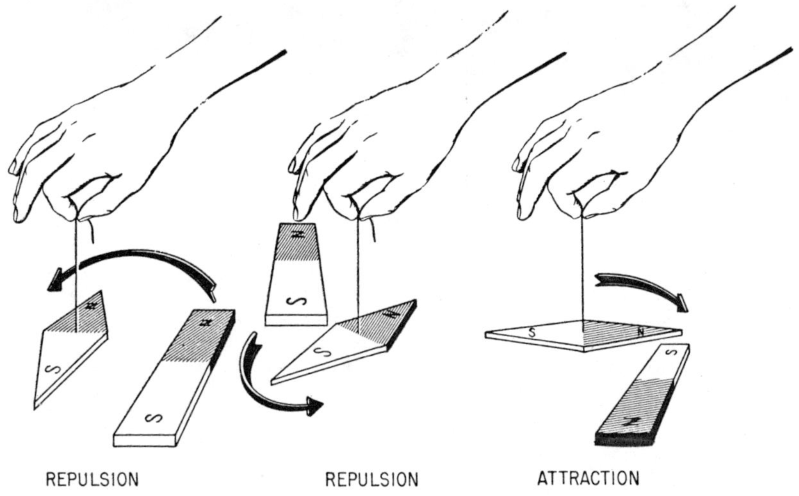

Figure 2-13.—Laws of attraction and repulsion.

poles are adjacent; and figure 2-15 (B) shows the flux pattern when like poles are adjacent.

The THIRD LAW of magnetic attraction and repulsion states in effect that the force of attraction or repulsion existing between two magnetic poles decreases rapidly as the poles are separated from each other. Actually, the force of attraction or repulsion varies directly as the product of the separate pole strengths and inversely as the square of the distance separating the magnetic poles, provided the poles are small enough to be considered as points. For example, if the distance between two north poles is increased from 2 feet to 4 feet, the force of repulsion between them is decreased to one-fourth of its original value. If either pole strength is doubled, the distance remaining the same, the force between the poles will be doubled.

THE EARTH'S MAGNETISM

As has been stated, the earth is a huge magnet; and surrounding the earth is the magnetic field produced by the earth's magnetism. The magnetic polarities of the earth are as indicated in figure 2-16. The geographic poles are also shown at each end of the axis of rotation of the earth. The magnetic axis does not coincide with the geographic axis, and therefore the magnetic and geographic poles are not at the same place on the surface of the earth.

The early users of the compass regarded the end of the compass needle that points in a northerly direction as being a north pole. The other end was regarded as a south pole. On some maps the magnetic pole of the earth towards which the north pole of the compass pointed was designated a north magnetic pole. This magnetic pole was obviously called a north pole because of its proximity to the north geographic pole.

When it was learned that the earth is a magnet and that opposite poles attract, it was necessary to call the magnetic pole located in the northern hemisphere a SOUTH MAGNETIC POLE and the magnetic pole located in the southern hemisphere a NORTH MAGNETIC POLE. The matter of naming the poles was arbitrary. Therefore, the polarity of the compass needle that points toward the north must be opposite to the polarity of the earth's magnetic pole located there.

NOTES:

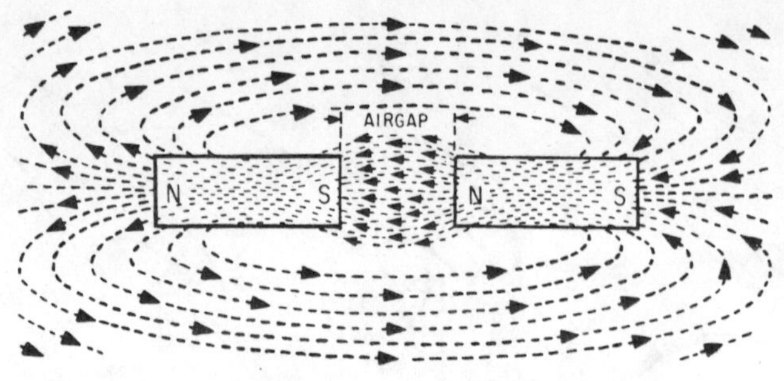

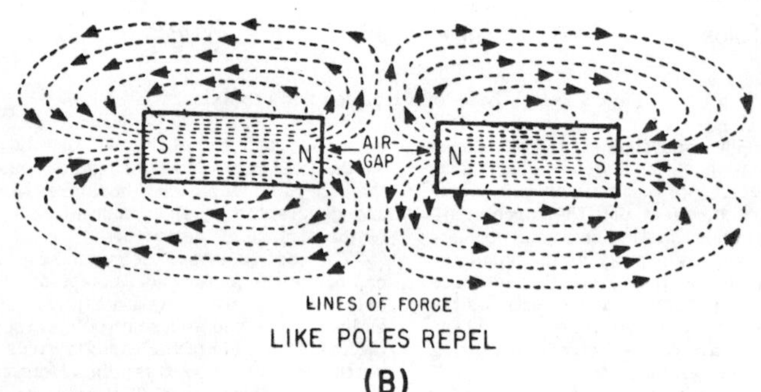

Figure 2-14.—Lines of force between unlike and like poles.

As has been stated, magnetic lines of force are assumed to emanate from the north pole of a magnet and to enter the south pole as closed loops. Because the earth is a magnet, lines of force emanate from its north magnetic pole and enter the south magnetic pole as closed loops. The compass needle alines itself in such a way that the earth's lines of force enter at its south pole and leave at its north pole. Because the north pole of the needle is defined as the end that points in a northerly direction it follows that the magnetic pole in the vicinity of the north geographic pole is in reality a south magnetic pole, and vice versa.

Because the magnetic poles and the geographic poles do not coincide, a compass will not (except at certain positions on the earth) point in a true (geographic) north-south direction—that is, it will not point in a line of direction that passes through the north and south

NOTES:

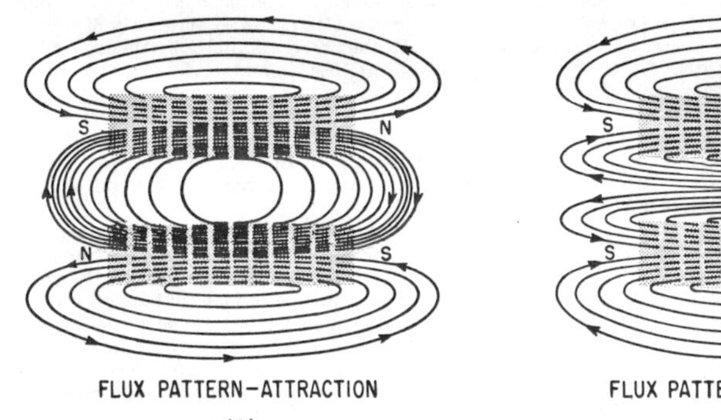

Figure 2-15.—Flux patterns of adjacent parallel bar magnets.

geographic poles, but in a line of direction that makes an angle with it. This angle is called the angle of VARIATION OR DECLINATION.

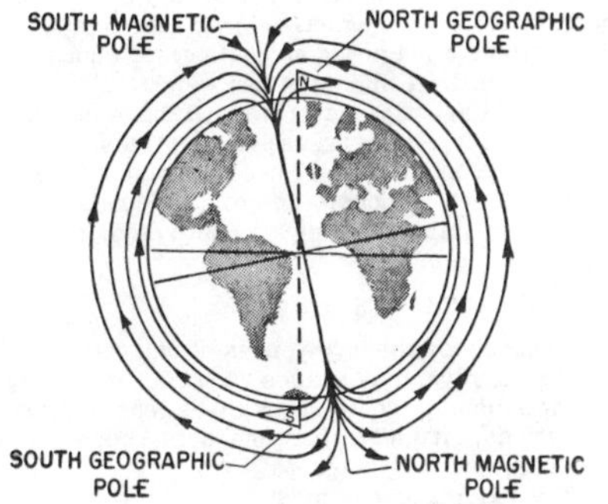

Figure 2-16.—Earth's magnetic poles.

MAGNETIC SHIELDING

There is not a known INSULATOR for magnetic flux. If a nonmagnetic material is placed in a magnetic field, there is no appreciable change in flux—that is, the flux penetrates the nonmagnetic material. For example, a glass plate placed between the poles of a horseshoe magnet will have no appreciable effect on the field although glass itself is a good insulator in an electric circuit. If a magnetic material (for example, soft iron) is placed in a magnetic field, the flux may be redirected to take advantage of the greater permeability of the magnetic material as shown in figure 2-17. Permeability, as discussed earlier, is the quality of a substance which determines the ease with which it can be magnetized.

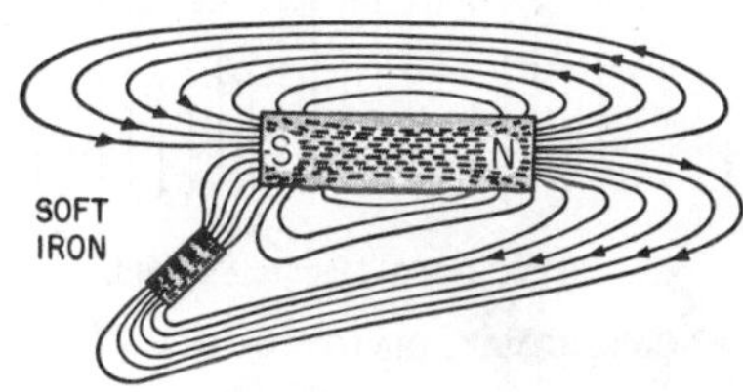

Figure 2-17.—Effects of a magnetic substance in a magnetic field.

The sensitive mechanism of electric instruments and meters can be influenced by stray magnetic fields which will cause errors in their readings. Because instrument mechanisms

NOTES:

cannot be insulated against magnetic flux, it is necessary to employ some means of directing the flux around the instrument. This is accomplished by placing a soft-iron case, called a MAGNETIC SCREEN OR SHIELD, about the instrument. Because the flux is established more readily through the iron (even though the path is longer) than through the air inside the case, the instrument is effectively shielded, as shown by the watch and soft-iron shield in figure 2-18.

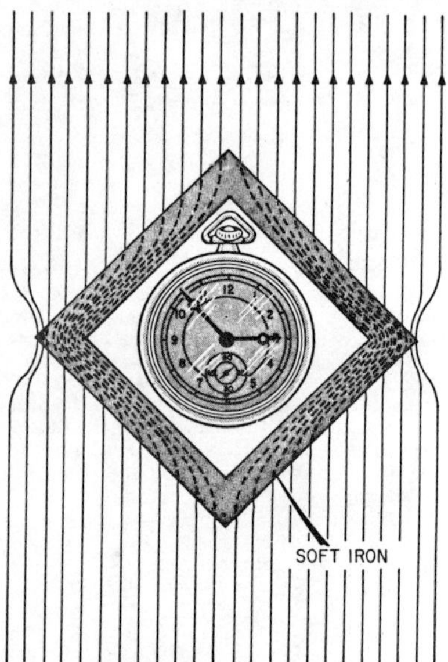

Figure 2-18.—Magnetic shield.

MAGNETIC MATERIALS

Early magnetic studies classified materials merely as being magnetic and nonmagnetic. Present studies classify materials into one of three groups; namely, paramagnetic, diamagnetic, and ferromagnetic.

PARAMAGNETIC materials are those that become only slightly magnetized even though under the influence of a strong magnetic field. This slight magnetization is in the same direction as the magnetizing field. Materials of this type are aluminum, chromium, platinum, and air.

DIAMAGNETIC materials can also be only slightly magnetized when under the influence of a very strong field. These materials, when slightly magnetized, are magnetized in a direction opposite to the external field. Some diamagnetic materials are copper, silver, gold, and mercury.

Paramagnetic and diamagnetic materials have a very low permeability. Paramagnetic materials have a permeability slightly greater than one; diamagnetic materials have a permeability less than one. Because of the difficulty in obtaining some magnetization of paramagnetic and diamagnetic materials, these materials are considered for all practical purposes as nonmagnetic materials.

The most important group of materials for applications of electricity and electronics are the FERROMAGNETIC MATERIALS. Ferromagnetic materials are those which are relatively easy to magnetize such as iron, steel, cobalt, Alnico, and Permalloy, the latter two being alloys. Alnico consists primarily of aluminum, nickel, and cobalt. These new alloys can be very strongly magnetized with Alnico capable of obtaining a magnetic strength great enough to lift five hundred times its own weight.

Ferromagnetic materials all have a high permeability. However, as previously discussed, a material, such as steel used to make a permanent magnet, is considered to have a relatively low permeability in comparison to other ferromagnetic materials.

MAGNETIC SHAPES

Because of the many uses of magnets, they are found in various shapes and sizes. However, magnets usually come under three general classifications; namely, bar magnets, horseshoe magnets, and ring magnets.

The bar magnet is most often used in schools and laboratories for studying the properties and effects of magnetism. In the preceding test material, the bar magnet proved very helpful in demonstrating magnetic effects.

Another type of magnet is the ring magnet used for computer memory cores. A common application for a temporary ring magnet would be the shielding of electrical instruments as previously discussed.

The shape of the magnet most frequently used in electrical or electronic equipment is

NOTES:

called the horseshoe magnet. A horseshoe magnet is similar to a bar magnet but is bent in the shape of a horseshoe. The horseshoe magnet provides much more magnetic strength than a bar magnet of the same size and material because of the closeness of the magnetic poles. The magnetic strength from one pole to the other is greatly increased due to the concentration of the magnetic field in a smaller area. Electrical measuring devices quite frequently use horseshoe type magnets.

CARE OF MAGNETS

A piece of steel that has been magnetized can lose much of its magnetism by improper handling. If it is jarred or heated, there will be a disalinement of its domains resulting in the loss of some of its effective magnetism. Had this steel formed the horseshoe magnet of a meter, the meter would no longer be operable or would give inaccurate readings. Therefore, care must be exercised when handling instruments containing magnets. Severe jarring or subjecting the instrument to high temperature will damage the device.

A magnet may also become weakened from loss of flux. Thus, when storing magnets one should always try to avoid excess leakage of magnetic flux. A horseshoe magnet should always be stored with a keeper, a soft iron bar used to join the magnetic poles. By use of the keeper while the magnet is being stored, the magnetic flux will continuously circulate through the magnet and not leak off into space.

When storing bar magnets, the same principle must be remembered. Therefore, bar magnets should always be stored in pairs with a north pole and a south pole placed together. This provides a complete path for the magnetic flux without any flux leakage.

The study of electricity and magnetism and how they interact with each other is given more thorough coverage in later chapters in this manual. The discussion of magnetism up to this point has been mainly intended to clarify terms and meanings, such as "polarity," "fields," "lines of force," etc. Only one fundamental relationship between magnetism and electricity is discussed in this chapter. This relationship pertains to magnetism as used to generate a voltage and it is discussed in the material that follows.

DIFFERENCE IN POTENTIAL

The force that causes free electrons to move in a conductor as an electric current may be referred to as follows:

1. Electromotive force (emf).
2. Voltage.
3. Difference in potential.

When a difference in potential exists between two charged bodies that are connected by a conductor, electrons will flow along the conductor. This flow is from the negatively charged body to the positively charged body until the two charges are equalized and the potential difference no longer exists.

An analogy of this action is shown in the two water tanks connected by a pipe and valve in figure 2-19. At first the valve is closed and all the water is in tank A. Thus, the water pressure across the valve is at maximum. When the valve is opened, the water flows through the pipe from A to B until the water level becomes the same in both tanks. The water then stops flowing in the pipe, because there is no longer a difference in water pressure between the two tanks.

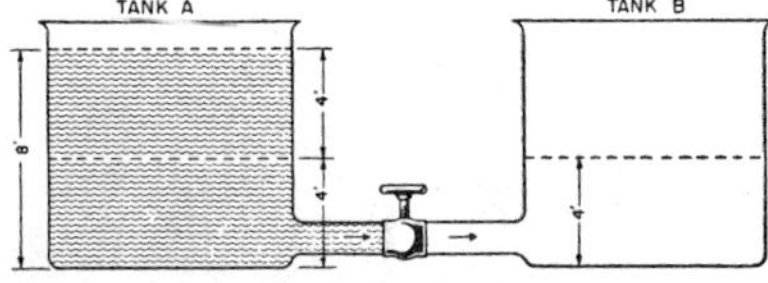

Figure 2-19.—Water analogy of electric difference in potential.

Current flow through an electric circuit is directly proportional to the difference in potential across the circuit, just as the flow of water through the pipe in figure 2-19 is directly proportional to the difference in water level in the two tanks.

A fundamental law of current electricity is that the CURRENT IS DIRECTLY PROPORTIONAL TO THE APPLIED VOLTAGE; that is, if the voltage is increased, the current is increased. If the voltage is decreased, the current is decreased.

PRIMARY METHODS OF PRODUCING A VOLTAGE

Presently, there are six commonly used methods of producing electromotive force (emf). Some of these methods are much more widely used than others. The following is a list of the six most common methods of producing electromotive force.

NOTES:

1. FRICTION.—Voltage produced by rubbing two materials together.

2. PRESSURE (Piezoelectricity).—Voltage produced by squeezing crystals of certain substances.

3. HEAT (Thermoelectricity).—Voltage produced by heating the joint (junction) where two unlike metals are joined.

4. LIGHT (Photoelectricity).—Voltage produced by light striking photosensitive (light sensitive) substances.

5. CHEMICAL ACTION.—Voltage produced by chemical reaction in a battery cell.

6. MAGNETISM.--Voltage produced in a conductor when the conductor moves through a magnetic field, or a magnetic field moves through the conductor in such a manner as to cut the magnetic lines of force of the field.

VOLTAGE PRODUCED BY FRICTION

This is the least used of the six methods of producing voltages. Its main application is in Van de Graf generators, used by some laboratories to produce high voltages. As a rule, friction electricity (often referred to as static electricity) is a nuisance. For instance, a flying aircraft accumulates electric charges from the friction between its skin and the passing air. These charges often interfere with radio communication, and under some circumstances can even cause physical damage to the aircraft. Most individuals are familiar with static electricity and have probably received unpleasant shocks from friction electricity upon sliding across dry seat covers or walking across dry carpets, and then coming in contact with some other object.

VOLTAGE PRODUCED BY PRESSURE

This action is referred to as piezoelectricity. It is produced by compressing or decompressing crystals of certain substances. To study this form of electricity, the meaning of the word "crystal" must first be understood. In a crystal, the molecules are arranged in an orderly and uniform manner. A substance in its crystallized state and its noncrystallized state is shown in figure 2-20.

For the sake of simplicity, assume that the molecules of this particular substance are spherical (ball-shaped). In the noncrystallized state, in (A), note that the molecules are arranged irregularly. In the crystallized state, (B), the molecules are arranged in a regular and uniform manner. This illustrates the major physical difference between crystal and noncrystal forms of matter. Natural crystalline matter is rare; an example of matter that is crystalline in its natural form is diamond, which is crystalline carbon. Most crystals are manufactured.

Crystals of certain substances, such as Rochelle salt or quartz, exhibit peculiar electrical characteristics. These characteristics, or effects, are referred to as "piezoelectric." For instance, when a crystal of quartz is

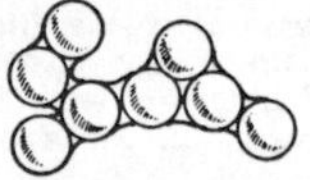
MOLECULES OF NON-CRYSTALLIZED MATTER
(A)

MOLECULES OF CRYSTALLIZED MATTER
(B)

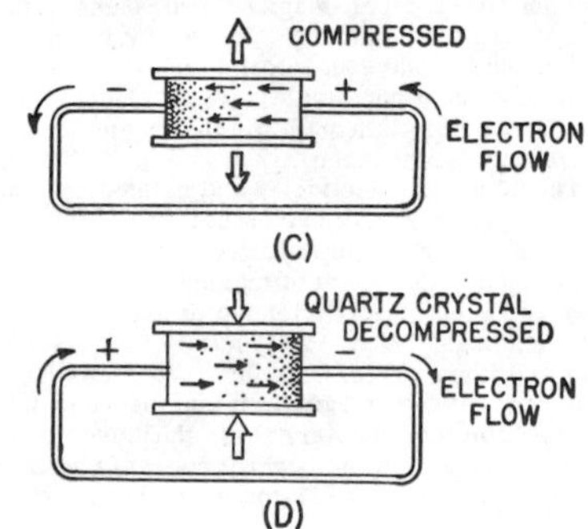

Figure 2-20.—(A) Noncrystallized structure; (B) crystallized structure; (C) compression of a crystal; (D) decompression of a crystal.

NOTES:

compressed, as in figure 2-20 (C), electrons tend to move through the crystal as shown. This tendency creates an electric difference of potential between the two opposite faces of the crystal. (The fundamental reasons for this action are not known. However, the action is predictable, and therefore useful.) If an external wire is connected while the pressure and emf are present, electrons will flow. If the pressure is held constant, the electron flow will continue until the charges are equalized. When the force is removed, the crystal is decompressed, and immediately causes an electric force in the opposite direction (D). Thus, the crystal is able to convert mechanical force, either pressure or tension, to electrical force.

The power capacity of a crystal is extremely small. However, they are useful because of their extreme sensitivity to changes of mechanical force or changes in temperature. Due to other characteristics not mentioned here, crystals are most widely used in communication equipment.

VOLTAGE PRODUCED BY HEAT

When a length of metal, such as copper, is heated at one end, electrons tend to move away from the hot end toward the cooler end. This is true of most metals. However, in some metals, such as iron, the opposite takes place and electrons tend to move TOWARD the hot end. These characteristics are illustrated in figure 2-21. The negative charges (electrons) are moving through the copper away from the heat and through the iron toward the heat. They cross from the iron to the copper at the hot junction, and from the copper through the current meter to the iron at the cold junction. This device is generally referred to as a thermocouple.

Thermocouples have somewhat greater power capacities than crystals, but their capacity is still very small if compared to some other sources. The thermoelectric voltage in a thermocouple depends mainly on the difference in temperature between the hot and cold junctions. Consequently, they are widely used to measure temperature, and as heat-sensing devices in automatic temperature control equipment. Thermocouples generally can be subjected to much greater temperatures than ordinary thermometers, such as the mercury or alcohol types.

VOLTAGE PRODUCED BY LIGHT

When light strikes the surface of a substance, it may dislodge electrons from their orbits around the surface atoms of the substance. This occurs because light has energy, the same as any moving force.

Some substances, mostly metallic ones, are far more sensitive to light than others. That is, more electrons will be dislodged and emitted from the surface of a highly sensitive metal, with a given amount of light, than will be emitted from a less sensitive substance. Upon losing electrons, the photosensitive (light sensitive) metal becomes positively charged, and an electric force is created. Voltage produced in this manner is referred to as "a photoelectric voltage."

The photosensitive materials most commonly used to produce a photoelectric voltage are various compounds of silver oxide or copper oxide. A complete device which operates on the photoelectric principle is referred to as a "photoelectric cell." There are many sizes and types of photoelectric cells in use, each of which serves the special purpose for which it was designed. Nearly all, however, have some of the basic features of the photoelectric cells shown in figure 2-22.

The cell (fig. 2-22 (A)) has a curved light-sensitive surface focused on the central anode. When light from the direction shown strikes the sensitive surface, it emits electrons toward the anode. The more intense the light, the greater is the number of electrons emitted. When a wire is connected between the filament and the back, or dark side, the accumulated electrons will flow to the dark side. These electrons will eventually pass through the metal of the reflector and replace the electrons leaving the light-sensitive surface. Thus, light energy is converted to a flow of electrons, and a usable current is developed.

The cell (fig. 2-22 (B)) is constructed in layers. A base plate of pure copper is coated with light-sensitive copper oxide. An additional semitransparent layer of metal is placed over the copper oxide. This additional layer serves two purposes:

1. It is EXTREMELY thin to permit the penetration of light to the copper oxide.
2. It also accumulates the electrons emitted by the copper oxide.

An externally connected wire completes the electron path, the same as in the reflector type cell. The photocell's voltage is utilized as needed by connecting the external wires to some other device, which amplifies (enlarges) it to a usable level.

NOTES:

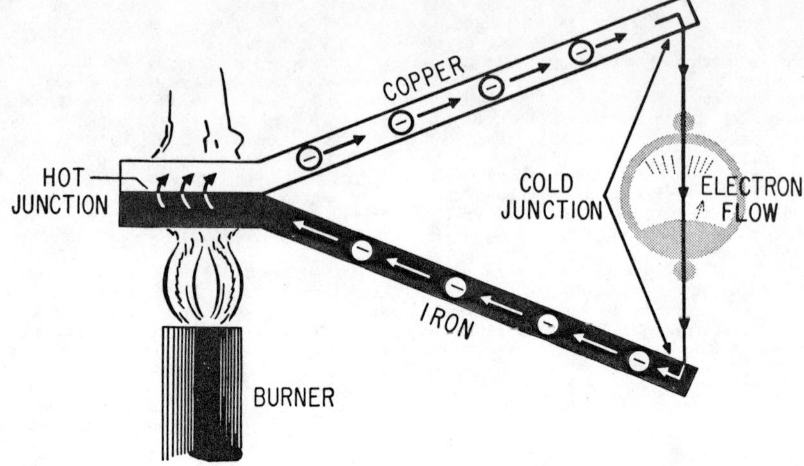

Figure 2-21.—Voltage produced by heat.

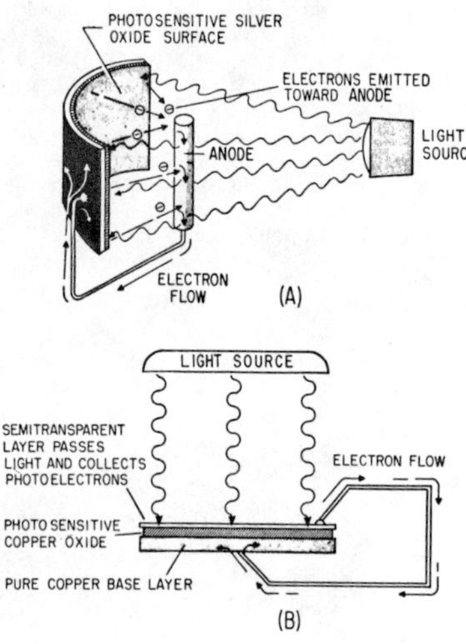

Figure 2-22.—Voltage produced by light.

A photocell's power capacity is very small. However, it reacts to light-intensity variations in an extremely short time. This characteristic makes the photocell very useful in detecting or accurately controlling a great number of processes or operations. For instance, the photoelectric cell, or some form of the photoelectric principle, is used in television cameras, automatic manufacturing process controls, door openers, burglar alarms, and so forth.

VOLTAGE PRODUCED BY CHEMICAL ACTION

Up to this point, it has been shown that electrons may be removed from their parent atoms and set in motion by energy derived from a source of friction, pressure, heat, or light. In general, these forms of energy do not alter the molecules of the substances being acted upon. That is, molecules are not usually added, taken away, or split-up when subjected to these four forms of energy. Only electrons are involved.

When the molecules of a substance are altered, the action is referred to as CHEMICAL. For instance, if the molecules of a substance combines with atoms of another substance, or

NOTES:

gives up atoms of its own, the action is chemical in nature. Such action always changes the chemical name and characteristics of the substance affected. For instance, when atoms of oxygen from the air come in contact with bare iron, they merge with the molecules of iron. This iron is "oxidized." It has changed chemically from iron to iron oxide, or "rust." Its molecules have been altered by chemical action.

In some cases, when atoms are added to or taken away from the molecules of a substance, the chemical change will cause the substance to take on an electric charge. The process of producing a voltage by chemical action is used in batteries and is explained in chapter 3.

VOLTAGE PRODUCED BY MAGNETISM

Magnets or magnetic devices are used for thousands of different jobs. One of the most useful and widely employed applications of magnets is in the production of vast quantities of electric power from mechanical sources. The mechanical power may be provided by a number of different sources, such as gasoline or diesel engines, and water or steam turbines. However, the final conversion of these source energies to electricity is done by generators employing the principle of electromagnetic induction. These generators, of many types and sizes, are discussed in later chapters of this manual. The important subject to be discussed here is the fundamental operating principle of ALL such electromagnetic-induction generators.

To begin with, there are three fundamental conditions which must exist before a voltage can be produced by magnetism. They are as follows:

1. There must be a CONDUCTOR, in which the voltage will be produced.
2. There must be a MAGNETIC FIELD in the conductor's vicinity.
3. There must be relative motion between the field and the conductor. The conductor must be moved so as to cut across the magnetic lines of force, or the field must be moved so that the lines of force are cut by the conductor.

In accordance with these conditions, when a conductor or conductors MOVE ACROSS a magnetic field so as to cut the lines of force, electrons WITHIN THE CONDUCTOR are impelled in one direction or another. Thus, an electric force, or voltage, is created.

In figure 2-23, note the presence of the three conditions needed for creating an induced voltage:

1. A magnetic field exists between the poles of the C-shaped magnet.
2. There is a conductor (copper wire).
3. There is relative motion. The wire is moved back and forth ACROSS the magnetic field.

In figure 2-23 (A), the conductor is moving TOWARD the front of the page. This occurs because of the magnetically induced emf acting on the electrons in the copper. The right-hand end becomes negative, and the left-hand end positive. The conductor is stopped (B), motion is eliminated (one of the three required conditions), and there is no longer an induced emf. Consequently, there is no longer any difference in potential between the two ends of the wire. The conductor at (C) is moving away from the front of the page. An induced emf is again created. However, note carefully that the REVERSAL OF MOTION has caused a REVERSAL OF DIRECTION in the induced emf.

If a path for electron flow is provided between the ends of the conductor, electrons will leave the negative end and flow to the positive end. This condition is shown in part (D). Electron flow will continue as long as the emf exists. In studying figure 2-23, it should be noted that the induced emf could also have been created by holding the conductor stationary and moving the magnetic field back and forth.

The more complex aspects of power generation by use of mechanical motion and magnetism are discussed in later chapters of this manual under the heading "Generators."

ELECTRIC CURRENT

The drift or flow of electrons through a conductor is called electric current or electron flow. During the early study of electricity, electric current was erroneously assumed to be a movement of electrons from a positive potential to a negative potential. This assumption, termed conventional current flow, is a concept that became entrenched in the minds of many scientists, technicians, and writers. Consequently, conventional current flow is indicated in many textbooks, and this concept of electron movement should be realized. However, since this early concept, it has been positively determined that the direction of electron movement is from a region of negative potential to a region of less negative potential or more positive potential. Various terms may be used in this manual and other textbooks to describe current flow. The terms current, current flow, electron flow,

NOTES:

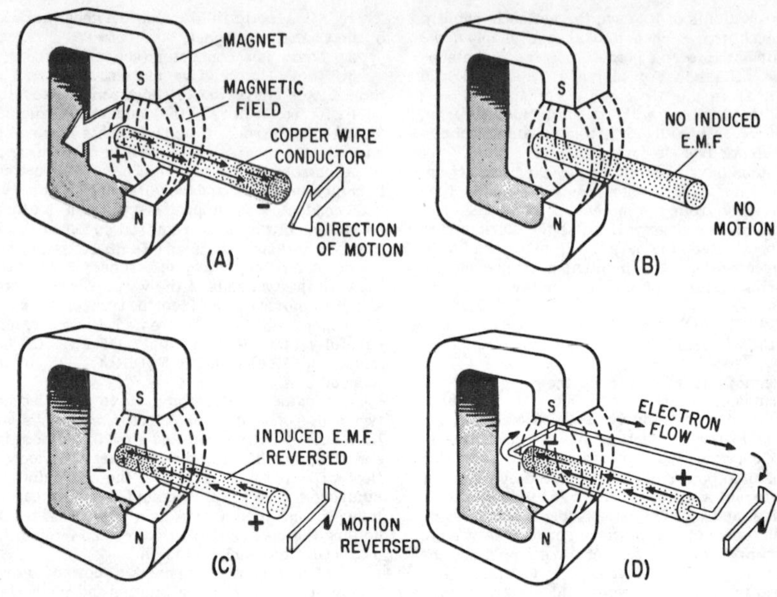

Figure 2-23.—Voltage produced by magnetism.

electron current, etc., may be used to describe the same phenomenon; however, the reader should realize that regardless of the term used, the movement of electrons will be from a negative potential to a positive potential.

Electric current is generally classified into two general types—direct current and alternating current. Direct current flows in the same direction whereas an alternating current periodically reverses direction. These two types of current are discussed in greater detail later in this manual.

In order to determine the amount (number) of electrons flowing in a given conductor, it is necessary to adopt a unit of measurement of current flow. The term AMPERE is used to define the unit of measurement of the rate at which current flows (electron flow). The symbol for current flow is I. Current flow is measured in amperes. The abbreviation for ampere is amp. One ampere may be defined as the flow of 6.28×10^{18} electrons per second past a fixed point in a conductor.

A unit quantity of electricity is moved through an electric circuit when 1 ampere of current flows for 1 second of time. This unit is equivalent to 6.28×10^{18} electrons, and is called the COULOMB. The coulomb is to electricity as the gallon is to water. The symbol for the coulomb is Q. The rate of flow of current in amperes and the quantity of electricity moved through a circuit are related by the common factor of time. Thus, the quantity of electric charge, in coulombs, moved through a circuit is equal to the product of the current in amperes, I, and the duration of flow in seconds, t. Expressed as an equation, $Q = It$.

For example, if a current of 2 amperes flows through a circuit for 10 seconds the quantity of electricity moved through the circuit is 2 x 10,

NOTES:

or 20 coulombs. Conversely, current flow may be expressed in terms of coulombs and time in seconds. Thus, if 20 coulombs are moved through a circuit in 10 seconds, the average current flow is $\frac{20}{10}$, or 2 amperes. Note that the current flow in amperes implies the rate of flow of coulombs per second without indicating either coulombs or seconds. Thus a current flow of 2 amperes is equivalent to a rate of flow of 2 coulombs per second. Frequently, the ampere is much too large a unit for practical utilization. Therefore, the milliampere (ma), one-thousandth of an ampere (or the microampere, one-millionth of an ampere), is used. The device used to measure current is called an ammeter and is discussed later in this manual.

RESISTANCE

Every material offers some resistance, or opposition, to the flow of electric current through it. Good conductors, such as copper, silver, and aluminum, offer very little resistance. Poor conductors, or insulators, such as glass, wood, and paper, offer a high resistance to current flow.

The size and type of material of the wires in an electric circuit are chosen so as to keep the electrical resistance as low as possible. In this way, current can flow easily through the conductors, just as water flows through the pipe between the tanks in figure 2-19. If the water pressure remains constant the flow of water in the pipe will depend on how far the valve is opened. The smaller the opening, the greater the opposition to the flow, and the smaller will be the rate of flow in gallons per second.

In the electric circuit, the larger the diameter of the wires, the lower will be their electrical resistance (opposition) to the flow of current through them. In the water analogy, pipe friction opposes the flow of water between the tanks. This friction is similar to electrical resistance. The resistance of the pipe to the flow of water through it depends upon (1) the length of the pipe, (2) the diameter of the pipe, and (3) the nature of the inside walls (rough or smooth). Similarly, the electrical resistance of the conductors depends upon (1) the length of the wires, (2) the diameter of the wires, and (3) the material of the wires (copper, aluminum, etc.).

Temperature also affects the resistance of electrical conductors to some extent. In most conductors (copper, aluminum, iron, etc.) the resistance increases with temperature. Carbon is an exception. In carbon the resistance decreases as temperature increases. Certain alloys of metals (manganin and constantan) have resistance that does not change appreciably with temperature.

The relative resistance of several conductors of the same length and cross section is given in the following list with silver as a standard of 1 and the remaining metals arranged in an order of ascending resistance:

Silver	1.0
Copper	1.08
Gold	1.4
Aluminum	1.8
Platinum	7.0
Lead	13.5

The resistance in an electrical circuit is expressed by the symbol R. Manufactured circuit parts containing definite amounts of resistance are called RESISTORS. Resistance (R) is measured in OHMS. One ohm is the resistance of a circuit element, or circuit, that permits a steady current of 1 ampere (1 coulomb per second) to flow when a steady emf of 1 volt is applied to the circuit.

CONDUCTANCE

Electricity is a study that is frequently explained in terms of opposites. The term that is exactly the opposite of resistance is conductance. Conductance (G) is the ability of a material to pass electrons. The unit of conductance is the Mho, which is ohm spelled backwards. Whereas the symbol used to represent resistance is the Greek letter omega (Ω), the symbol used to represent conductance is the Greek letter omega upside down (℧). The relationship that exists between resistance and conductance is the reciprocal. A reciprocal of a number is obtained by dividing the number into one. In terms of resistance and conductance:

$$R = \frac{1}{G}$$

$$G = \frac{1}{R}$$

If the resistance of a material is known, dividing its value into one will give its conductance. Similarly, if the conductance is known, dividing its value into one will give its resistance.

NOTES:

CHAPTER 3

BATTERIES

Batteries are widely used as sources of direct-current electrical energy in automobiles, boats, aircraft, ships, portable electric/electronic equipment, and lighting equipment. In some instances, they are used as the only source of power; while in others, they are used as a secondary or standby power source.

A battery consists of a number of cells assembled in a common container and connected together to function as a source of electrical power.

CELL

A cell is a device that transforms chemical energy into electrical energy. The simplest cell, known as either a galvanic or voltaic cell, is shown in figure 3-1. It consists of a piece of carbon (C) and a piece of zinc (Zn) suspended in a jar that contains a solution of water (H_2O) and sulfuric acid (H_2SO_4).

The cell is the fundamental unit of the battery. A simple cell consists of two strips, or electrodes, placed in a container that holds the electrolyte.

ELECTRODES

The electrodes are the conductors by which the current leaves or returns to the electrolyte. In the simple cell, they are carbon and zinc strips that are placed in the electrolyte; while in the dry cell (fig. 3-2), they are the carbon rod in the center and the zinc container in which the cell is assembled.

ELECTROLYTE

The electrolyte is the solution that acts upon the electrodes which are placed in it. The electrolyte may be a salt, an acid, or an alkaline solution. In the simple galvanic cell and in the automobile storage battery, the electrolyte is in a liquid form; while in the dry cell, the electrolyte is a paste.

PRIMARY CELL

A primary cell is one in which the chemical action eats away one of the electrodes, usually the negative. When this happens, the electrode must be replaced or the cell must be discarded. In the galvanic type cell, the zinc electrode and the liquid solution are usually replaced when this happens. In the case of the dry cell, it is usually cheaper to buy a new cell. Some primary cells have been developed to the state where they can be recharged.

SECONDARY CELL

A secondary cell is one in which the electrodes and the electrolyte are altered by the chemical action that takes place when the cell delivers current. These cells may be restored to their original condition by forcing an electric current through them in the opposite direction to that of discharge. The automobile storage battery is a common example of the secondary cell.

BATTERY

As was previously mentioned, a battery consists of two or more cells placed in a common container. The cells are connected in series, in parallel, or in some combination of series and parallel, depending upon the amount of voltage and current required of the battery. The connection of cells in a battery is discussed in more detail later in this chapter.

BATTERY CHEMISTRY

If a conductor is connected externally to the electrodes of a cell, electrons will flow under

NOTES:

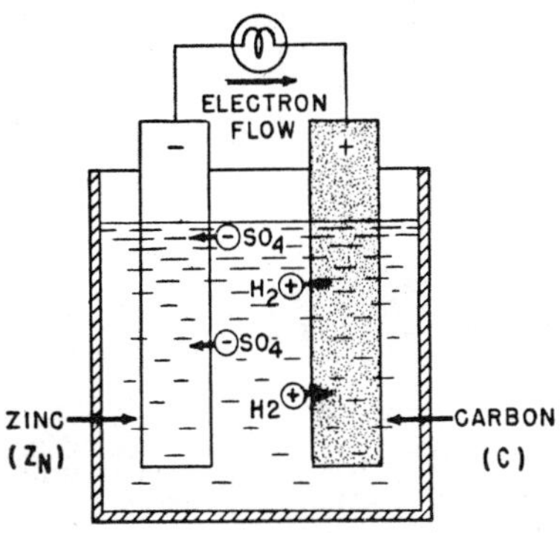

Figure 3-1.—Simple voltaic cell.

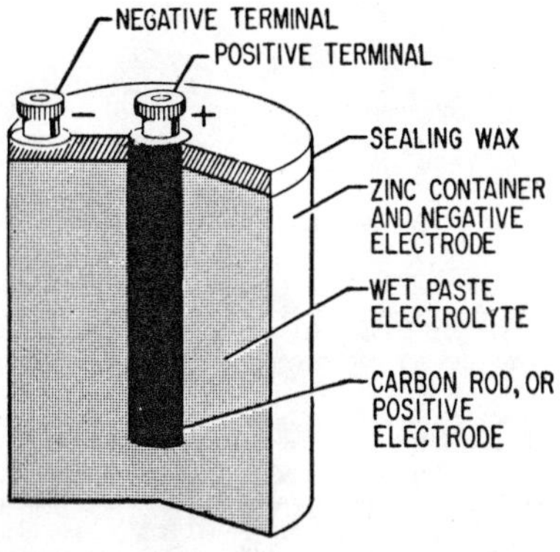

Figure 3-2.—Dry cell, cross-sectional view.

the influence of a difference in potential across the electrodes from the zinc (negative) through the external conductor to the carbon (positive), returning within the solution to the zinc. After a short period of time, the zinc will begin to waste away because of the "burning" action of the acid. If zinc is surrounded by oxygen, it will burn (become oxidized) as a fuel. In this respect, the cell is like a chemical furnace in which energy released by the zinc is transformed into electrical energy rather than heat energy.

The voltage across the electrodes depends upon the materials from which the electrodes are made and the composition of the solution. The difference of potential between carbon and zinc electrodes in a dilute solution of sulfuric acid and water is about 1.5 volts.

The current that a primary cell may deliver depends upon the resistance of the entire circuit, including that of the cell itself. The internal resistance of the primary cell depends upon the size of the electrodes, the distance between them in the solution, and the resistance of the solution. The larger the electrodes and the closer together they are in solution (without touching), the lower the internal resistance of the primary cell and the more current it is capable of supplying to a load.

When current flows through a cell, the zinc gradually dissolved in the solution and the acid is neutralized. A chemical equation is sometimes used to show the chemical action that takes place. The symbols in the equation represent the different materials that are used. The symbol for carbon is C and zinc Zn. The equation is quantitative and equates the number of parts of the materials used before and after the zinc is oxidized. As stated previously in chapter 2, all matter is composed of atoms and molecules, with the atom being the smallest part of an element and the molecule the smallest part of a compound.

A compound is a chemical combination of two or more elements in which the physical properties of the compound are different from those of the elements comprising it. For instance, a molecule of water (H_2O) is composed of two atoms of hydrogen (H_2) and one atom of oxygen (O). Ordinarily, hydrogen and oxygen are gases; but when combined, as stated above, they form water, which normally is a liquid. However, sulfuric acid (H_2SO_4) and water (H_2O) form a mixture (not a compound), because the identity of both liquids is preserved when they are in solution together.

When a current flows through a primary cell having carbon and zinc electrodes and a dilute solution of sulfuric acid and water, the

NOTES:

chemical reaction that occurs can be expressed as

$$Zn + H_2SO_4 + H_2O \xrightarrow[\text{discharge}]{} ZnSO_4 + H_2O + H_2\uparrow$$

The expression indicates that as current flows, a molecule of zinc combines with a molecule of sulfuric acid to form a molecule of zinc sulfate ($ZnSO_4$) and a molecule of hydrogen (H_2). The zinc sulfate dissolves in the solution and the hydrogen appears as gas bubbles around the carbon electrode. (A gas is designated by the arrow pointing upward in the equation.) As current continues to flow, the zinc is gradually consumed and the solution changes to zinc sulfate and water. The carbon electrode does not enter into the chemical changes taking place but simply provides a return path for the current.

In the process of oxidizing the zinc, the solution breaks up into positive and negative ions that move in opposite directions through the solution (fig. 3-1). The positive ions are hydrogen ions that appear around the carbon electrode (positive terminal). They are attracted to it by the free electrons from the zinc that are returning to the cell by way of the external load and the positive carbon terminal. The negative ions are SO_4 ions that appear around the zinc electrode. Positive zinc ions enter the solution around the zinc electrode and combine with the negative SO_4 ions to form zinc sulfate ($ZnSO_4$), a grayish-white substance that dissolves in water. At the same time that the positive and negative ions are moving in opposite directions in the solution, electrons are moving through the external circuit from the negative zinc terminal, through the load, and back to the positive carbon terminal. When the zinc is used up, the voltage of the cell is reduced to zero. There is no appreciable difference in potential between zinc sulfate and carbon in a solution of zinc sulfate and water.

Polarization

The chemical action that occurs in the cell (fig. 3-1) while the current is flowing causes hydrogen bubbles to form on the surface of the positive carbon electrode in great numbers until the entire surface is surrounded. This action is called polarization. Some of these bubbles rise to the surface of the solution and escape into the air. However, many of the bubbles remain until there is no room for any more to be formed.

The hydrogen tends to set up an electromotive force in the opposite direction to that of the cell, thus increasing the effective internal resistance, reducing the output current, and lowering the terminal voltage.

A cell that is heavily polarized has no useful output. There are several ways to prevent polarization from occurring or to overcome it after it has occurred. The very simplest method might be to remove the carbon electrode and wipe off the hydrogen bubbles. When the electrode is replaced in the electrolyte, the emf and current are again normal. This method is not practicable because polarization occurs rapidly and continuously in the simple voltaic cell. A commercial form of voltaic cell, known as the dry cell, employs a substance rich in oxygen as a part of the positive carbon electrode, which will combine chemically with the hydrogen to form water, H_2O. One of the best depolarizing agents used is manganese dioxide (MnO_2), which supplies enough free oxygen to combine with all of the hydrogen so that the cell is practically free from polarization.

The chemical action that occurs may be expressed as

$$2MnO_2 + H_2 \longrightarrow Mn_2O_3 + H_2O$$

The manganese dioxide combines with the hydrogen to form water and a lower oxide of manganese. Thus the counter emf of polarization does not exist in the cell, and the terminal voltage and output current are maintained normal.

Local Action

When the external circuit is opened, the current ceases to flow, and theoretically all chemical action within the cell stops. However, commercial zinc contains many impurities, such as iron, carbon, lead, and arsenic. These impurities form many small cells within the zinc electrode in which current flows between the zinc and its impurities. Thus the zinc is oxidized even though the cell itself is an open circuit. This wasting away of the zinc on open circuit is called local action. For example, a small local cell exists on a zinc plate containing impurities of iron, as shown in figure 3-3. Electrons flow between the zinc and iron and the solution around the impurity becomes ionized. The negative SO_4 ions combine with the positive Zn ions to form $ZnSO_4$. Thus the acid is depleted in solution and the zinc consumed.

NOTES:

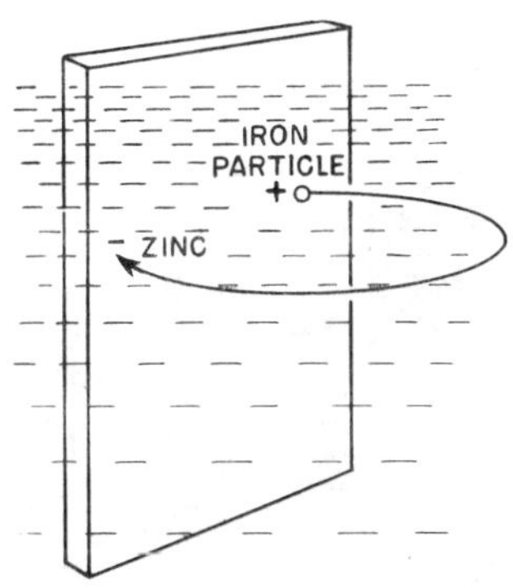

Figure 3-3.—Local action on zinc electrode.

Local action may be prevented by using pure zinc (which is not practical), by coating the zinc with mercury, or by adding a small percentage of mercury to the zinc during the manufacturing process. The treatment of the zinc with mercury is called amalgamating (mixing) the zinc. Since mercury is 13.6 times as heavy as an equal volume of water, small particles of impurities having a lower relative weight than that of mercury will rise (float) to the surface of the mercury. The removal of these impurities from the zinc prevents local action. The mercury is not readily acted upon by the acid; and even when the cell is delivering current to a load, the mercury continues to act on the impurities in the zinc, causing them to leave the surface of the zinc electrode and float to the surface of the mercury. This process greatly increases the life of the primary cell.

TYPES OF BATTERIES

The development of new and different types of batteries in the past decade has been so rapid that it is virtually impossible to have a complete knowledge of all of the various types currently being developed or now in use. A few recent developments are the silver-zinc, nickel-zinc, nickel-cadmium, silver-cadmium, magnesium-magnesium perchlorate, mercury, thermal, and water-activated batteries.

The lead-acid battery has been in service for a relatively long period of time; however, there are still various improvements being incorporated into the battery to improve its efficiency and life span. The material presented in this chapter, though not all inclusive, provides the reader with a knowledge of various types of batteries.

DRY (PRIMARY) CELL

The dry cell is so called because its electrolyte is not in a liquid state. Actually, the electrolyte is a moist paste. If it should become dry, it would no longer be able to transform chemical energy to electrical energy. The name dry cell, therefore, is not strictly correct in a technical sense.

Construction of the Dry Cell

The construction of a common type of dry cell is shown in figure 3-4. The internal parts of the cell are located in a cylindrical zinc container. This zinc container serves as the negative electrode of the cell. The container is lined with a nonconducting material, such as blotting paper, to insulate the zinc from the paste. A carbon electrode is located in the center, and it serves as the positive terminal of the cell. The paste is a mixture of several substances. Its composition may vary, depending on its manufracturer. Generally, however, the paste will contain some combination of the following substances: ammonium chloride (sal ammoniac), powdered coke, ground carbon, maganese dioxide, zinc chloride, graphite, and water.

This paste, which is packed in the space between the carbon and the blotting paper, also serves to hold the carbon electrode rigid in the center of the cell. When packing the paste in the cell, a small expansion space is left at the top. The cell is then sealed with asphalt-saturated cardboard.

Binding posts are attached to the elctrodes so that wires may be conveniently connected to the cell.

Since the zinc container is one of the electrodes, it must be protected with some insulating material. Therefore, it is common practice for the manufacturer to encluse the cells in cardboard containers.

NOTES:

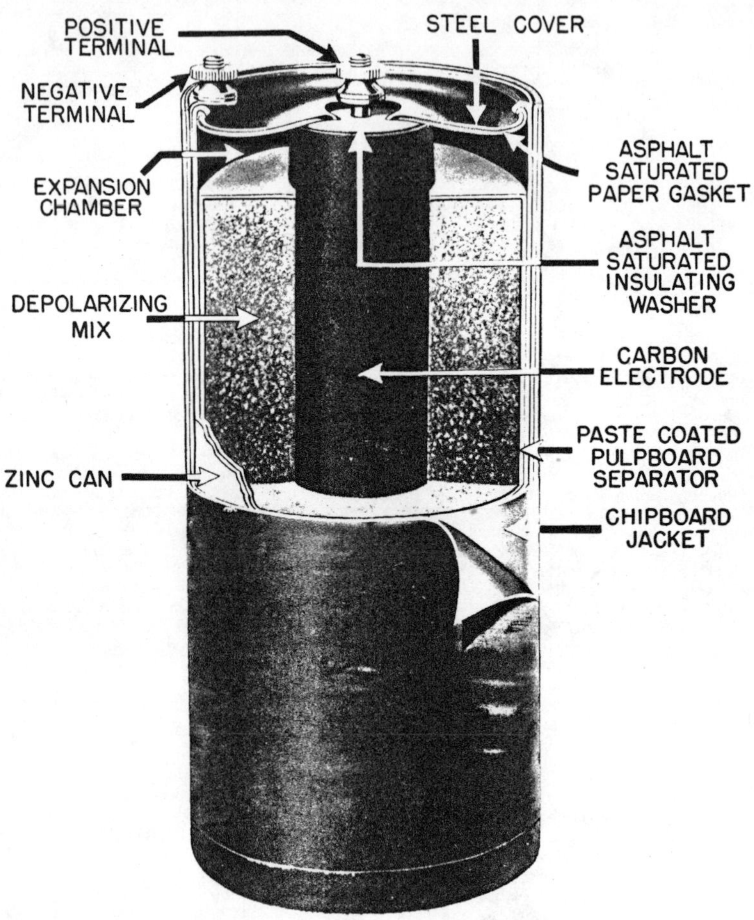

Figure 3-4.—Cutaway view of the general-purpose dry cell.

NOTES:

Chemical Action of the Dry Cell

The dry cell (fig. 3-4) is fundamentally the same as the simple voltaic cell (wet cell) described earlier, as far as its internal chemical action is concerned. The action of the water and the ammonium chloride in the paste, together with the zinc and carbon electrodes, produces the voltage of the cell. The manganese dioxide is added to reduce the polarization when line current flows and zinc chloride reduces local action when the cell is idle. The blotting paper (paste coated pulpboard separator) serves two purposes, one being to keep the paste from making actual contact with the zinc container and the other being to permit the electrolyte to filter through to the zinc slowly. The cell is sealed at the top to keep air from entering and drying the electrolyte. Care should be taken to prevent breaking this seal.

Rating of the Standard Size Cell

One of the popular sizes in general use is the standard, or No. 6, dry cell. It is approximately 2 1/2 inches in diameter and 6 inches in length. The voltage is about 1 1/2 volts when new but decreases as the cell ages. When the open-circuit voltage falls below 0.75 to 1.2 volts (depending upon the circuit requirements), the cell is usually discarded. The amount of current that the cell can deliver, and still give satisfactory service, depends upon the length of time that the current flows. For instance, if a No. 6 cell is to be used in a portable radio, it is likely to supply current constantly for several hours. Under these conditions, the current should not exceed 1/8 ampere, the rated constant-current capacity of a No. 6 cell. If the same cell is required to supply current only occasionally, for only short periods of time, it could supply currents of several amperes without undue injury to the cell. As the time duration of each discharge decreases, the interval of time between discharges increases, the allowable amount of current available for each discharge becomes higher, up to the amount that the cell will deliver on short circuit.

The short-circuit current test is another means of evaluating the condition of a dry cell. A new cell, when short circuited through an ammeter, should supply not less than 25 amperes. A cell that has been in service should supply at least 10 amperes if it is to remain in service.

Rating of the Unit Size Cell

A popular size of dry cell, the size D, is 1 3/8 inches in diameter and 2 3/4 inches in length. It is also known as the unit cell. The size D cell voltage is 1.5 volts when new. A discharged cell may expand, allowing the electrolyte to leak and cause corrosion. Some manufacturers place a steel jacket around the zinc container to prevent this action.

Shelf Life

A cell that is not being used (sitting on the shelf) will gradually deteriorate because of slow internal chemical actions (local action) and changes in moisture content. However, this deterioration is usually very slow if cells are properly stored. Highgrade cells of the larger sizes should have a shelf life of a year or more. Smaller size cells have a proportionately shorter shelf life, ranging down to a few months for the very small sizes. If unused cells are stored in a cool place, their shelf life will be greatly increased; therefore, to minimize deterioration, they should be stored in refrigerated spaces (10° F to 35° F) that are not dehumidified (dry).

MERCURY CELLS

With the advent of the space program and the development of small transceivers and miniaturized equipment, a power source of miniaturized size was needed. Such equipment requires a small battery which is capable of delivering maximum electrical capacity per unit volume while operating in varying temperatures and at a constant discharge voltage. The mercury battery, which is one of the smallest batteries, meets these requirements.

Present mercury batteries are manufactured in three basic structures. The wound anode type (fig. 3-5) (A)) has its anode composed of a corrugated zinc strip with a paper absorbent wound in an offset manner so that it protrudes at one end. The zinc is amalgamated (mixed) with mercury (10 percent), and the paper is impregnated with the electrolyte which causes it to swell and produce a positive contact pressure.

NOTES:

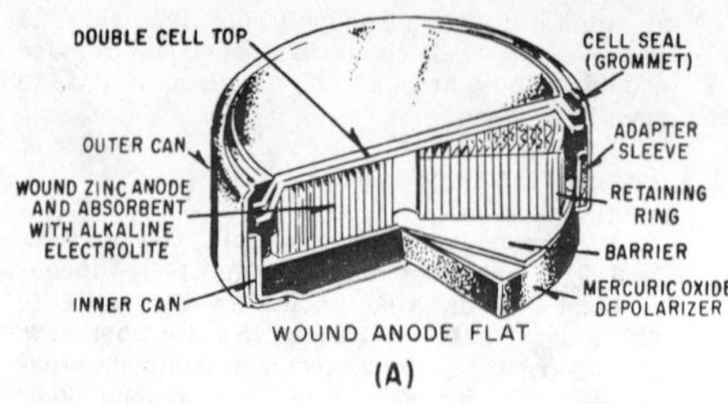

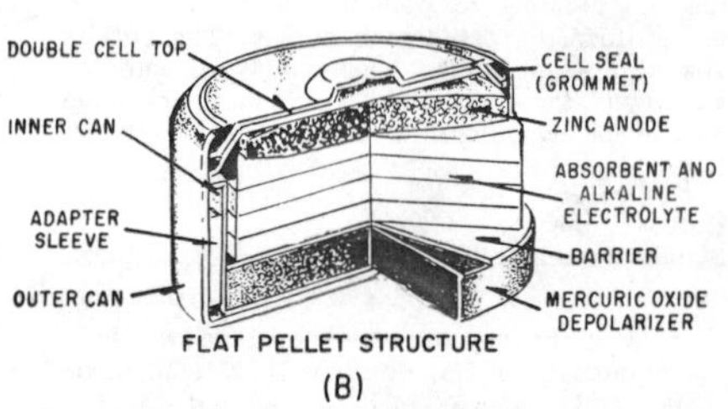

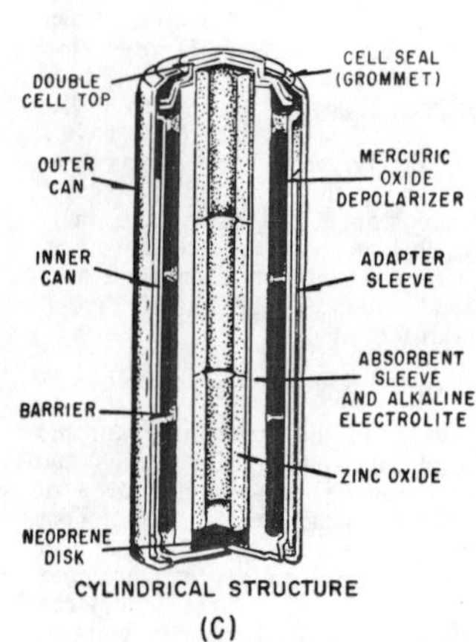

Figure 3-5.—Mercury cells.

In the pressed powder cells (fig. 3-5 (B) and (C)), the zinc powder is preamalgamated prior to pressing into shape; its porosity allows electrolyte impregnation with oxidation in depth when current is discharged. A double can structure is used in the larger sized cells. The space between the inner and outer containers provide passage for any gas generated by an improper chemical balance or impurities present within the cell. The construction is such that, if excessive gas pressures are experienced, the compression of the upper part of the grommet by internal pressure allows the gas to escape into the space between the two cans. A paper tube surrounds the inner can so that any liquid carried by discharging gas will be absorbed, maintaining a

NOTES:

leak resistant structure. Release of excessive gas pressure automatically reseals the cell.

NOTE: Mercury batteries have been known to explode with considerable force when shorted. Caution should be exercised to insure that the battery is not accidentally shorted.

The overall chemical action by which the mercury cell produces electricity is given by the following chemical formula:

$$Zn + H_2O + HgO \longrightarrow ZnO + H_2O + Hg$$

This action, the same as in other type cells, is a process of oxidation. The alkaline electrolyte is in contact with the zinc electrode. The zinc oxidizes (Zn changes to ZnO), thus taking atoms of oxygen from water molecules in the electrolyte. This leaves positive hydrogen ions, which move toward the mercuric oxide pellet, causing polarization. These hydrogen ions take oxygen from the mercuric oxide (thus changing HgO to Hg). Where one molecule of water is destroyed at the negative electrode, one molecule is produced at the positive electrode, maintaining the net amount of water constant. By absorbing oxygen, the zinc electrode accumulates excess electrons, making it negative. By giving up oxygen, the mercuric oxide electrode loses electrons, making it positive. In the discharged state, the negative electrode is zinc oxide, and the positive electrode is ordinary mercury.

RESERVE CELL

A reserve cell is one in which the elements are kept dry until the time of use; the electrolyte is then admitted and the cell starts producing current. In theory, this means that a reserve cell should be able to be stored for an indefinite period of time before it is activated.

One new reserve cell (fig. 3-6) is the alkaline manganese cell of the standard D size (flashlight battery). This reserve cell exhibits a high efficiency over a wide temperature range and is capable of momentary high current pulses in the range of 12 to 15 amperes.

The reserve cell is manufactured in a dry state, the electrolyte being contained in a plastic vial within the cell. When stored in this manner, the cell has a shelf life capability of over 10 years. To activate the cell, the activating mechanism is rotated 35° in either direction. This releases a spring-loaded plunger which breaks the plastic vial of electrolyte. Continued

Figure 3-6.—Reserve cell.

rotation permits the activating mechanism to be removed and discarded, resulting in a D size cell. A safety device is incorporated to prevent accidental activation during handling and transit.

Activation time is approximately 2 seconds when the cell is not under load. When under a 4-ohm load, the activation time (to reach a 1.35-volt level) is less than 5 seconds at 70° F and less than 30 seconds at 30° F.

The cell has been designed so that it is not position-sensitive during either the activation or the discharge period, and after activation can be handled and used as a standard D cell. After activation, shelf life of the reserve cell is approximately 2 years less than that of the standard alkaline manganese cell.

Reserve cells are used for emergency lighting and communications equipment, military ordnance devices, and in any other use where long storage ability is of prime importance.

COMBINING CELLS

In many cases, a battery-powered device may require more electrical energy than one cell can provide. The device may require either a higher voltage or more current, and in some cases both. Under such conditions it is necessary to combine, or interconnect, a sufficient number of cells to meet the higher requirements. Cells connected in series provide a higher voltage, while cells connected in parallel provide a higher

NOTES:

current capacity. To provide adequate power when both voltage and current requirements are greater than the capacity of one cell, a combination series-parallel network of cells must be interconnected.

Series Connected Cells

Assume that a load requires a power supply with a potential of 6 volts and a current capacity of 1/8 ampere. Since a single cell normally supplies a potential of only 1.5 volts, more than one cell is obviously needed. To obtain the higher potential, the cells are connected in series as shown in figure 3-7 (A).

In a series hookup, the negative electrode of the first cell is connected to the positive electrode of the second cell, the negative electrode of the second to the positive of the third, etc. The positive electrode of the first cell and negative electrode of the last cell then serve as the power takeoff terminals of the battery. In this way, the potential is boosted 1.5 volts by each cell in the series line. There are four cells, so the output terminal voltage is 1.5 x 4 = 6 volts. When connected to the load, 1/8 ampere flows through the load and each cell of the battery. This is within the capacity of each cell. Therefore, only four series-connected cells are needed to supply this particular load.

Parallel Connected Cells

In this case, assume an electrical load requires only 1.5 volts, but will draw 1/2 ampere of current. (Assume that a cell will supply only 1/8 ampere.) To meet this requirement, the cells are connected in parallel, as shown in figure 3-8 (A). In a parallel connection, all positive cell electrodes are connected to one line, and all negative electrodes are connected to the other. No more than one cell is connected between the lines at any one point; so the potential between the lines is the same as that of one cell, or 1.5 volts. However, each cell may contribute its maximum allowable current of 1/8 ampere to the line. There are four cells, so the total line current is 1/8 x 4 = 1/2 ampere. Hence, four cells in parallel have enough capacity to supply a load requiring 1/2 ampere at 1.5 volts.

Series-Parallel Connected Cells

Figure 3-9 depicts a battery network supplying power to a load requiring both a voltage and current greater than one cell can provide. To provide the required 4.5 volts, groups of three 1.5-volt cells are connected in series. To provide the required 1/2 ampere of current, four series groups are connected in parallel, each supplying 1/8 ampere of current.

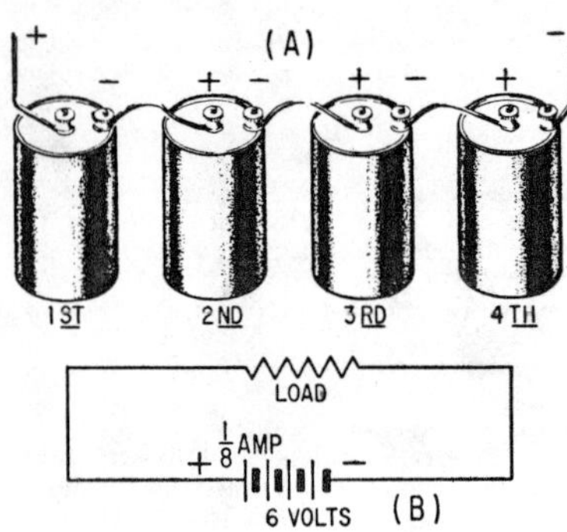

Figure 3-7.—(A) Pictorial view of series connected cells; (B) schematic of series connection.

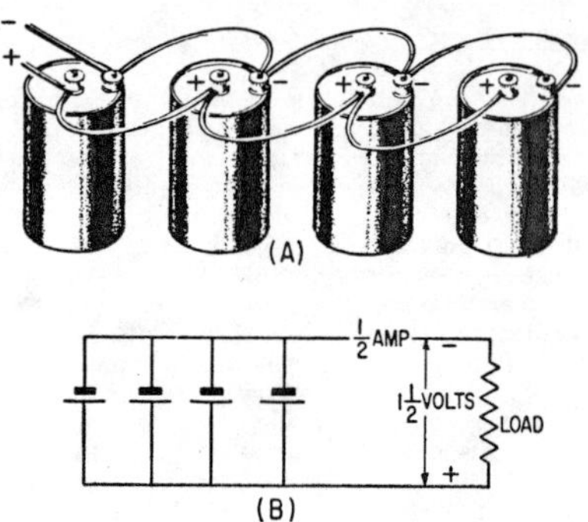

Figure 3-8.—(A) Pictorial view of parallel-connected cells; (B) schematic of parallel correction.

NOTES:

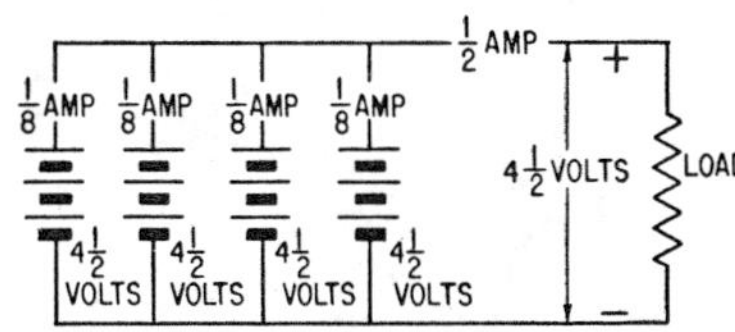

Figure 3-9.—Series-parallel connected cells.

SECONDARY (WET) CELLS

Secondary cells function on the same basic chemical principles as primary cells. They differ mainly in that they may be recharged, whereas the primary cell is not normally recharged. (As mentioned earlier, some primary cells have been developed to the state where they may be recharged.) Some of the materials of a primary cell are consumed in the process of changing chemical energy to electrical energy. In the secondary cell, the materials are merely transferred from one electrode to the other as the cell discharges. Discharged secondary cells may be restored (charged) to their original state by forcing an electric current from some other source through the cell in the opposite direction to that of discharge.

The storage battery consists of a number of secondary cells connected in series. Properly speaking, this battery does not store electrical energy, but is a source of chemical energy which produces electrical energy. There are various types of storage cells—the lead-acid type, which has an emf of 2.2 volts per cell; the nickel-iron alkali type; the nickel-cadmium alkali type, with an emf of 1.2 volts per cell; and the silver-zinc type, which has an emf of 1.5 volts per cell. Of these types, the lead-acid type is the most widely used, and is described first.

LEAD-ACID BATTERY

The lead-acid battery is an electrochemical device for storing chemical energy until it is released as electrical energy. Active materials within the battery react chemically to produce a flow of direct current whenever current consuming devices are connected to the battery terminal posts. This current is produced by chemical reaction between the active material of the plates (electrodes) and the electrolyte (sulfuric acid). The lead-acid battery is used extensively throughout the world. The parts of a lead-acid battery are illustrated in figure 3-10 and are discussed in the following paragraphs.

BATTERY CONSTRUCTION

A lead-acid battery consists of a number of cells connected together, the number needed depending upon the voltage desired. Each cell produces approximately 2 volts.

A cell consists of a hard rubber, plastic, or bituminous material compartment into which is placed the cell element, consisting of two types of lead plates, known as positive and negative plates. (See fig. 3-11.) These plates are insulated from each other by suitable separators (usually made of plastic, rubber, or glass) and submerged in a sulfuric acid solution (electrolyte).

There are a variety of plates used in the lead-acid battery—pasted plates, spun-lead (Plante) plates, Gould plates, and ironclad plates. These plates are each designed to fulfill a specific purpose. The most commonly used plates (the pasted plates) are discussed briefly.

The pasted plates are formed by applying lead-oxide pastes to a grid (fig. 3-12) made of lead-antimony alloy. The grid is designed to give the plates mechanical strength, hold the active material in place, and provide adequate conductivity for the electric current created by the chemical action. The active material (lead oxide) is applied to the grids in paste form and allowed to dry. The plates are then put through an electrochemical process that converts the active material of the positive plates into lead peroxide, and that of the negative plates into sponge lead. This action is caused by immersing the plates in electrolyte and passing a current through them in the proper direction. This type of plate is relatively light in weight compared to the other plates that are more rugged and durable in construction.

After the plates have been formed, they are built into positive and negative groups. The negative group of plates always has one more plate than the positive group so that both sides of the positive plates are acted upon chemically. This keeps the expansion and contraction that takes place in the positive plates the same on both sides and prevents buckling. These groups

NOTES:

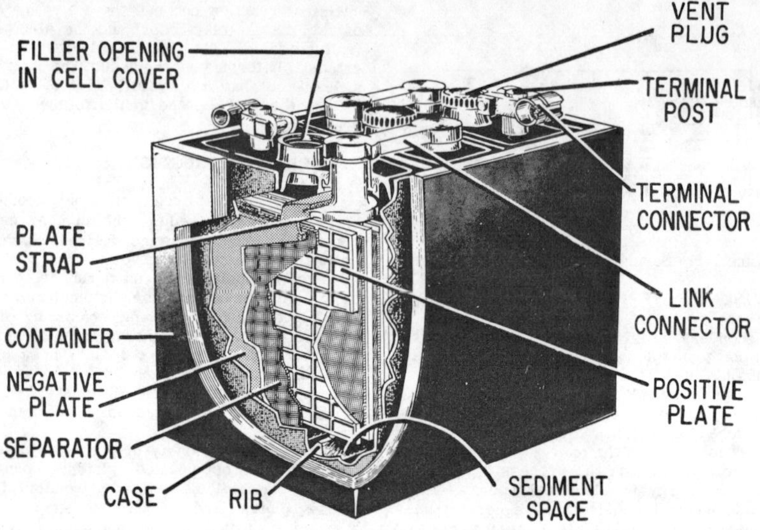

Figure 3-10.—Lead-acid battery construction.

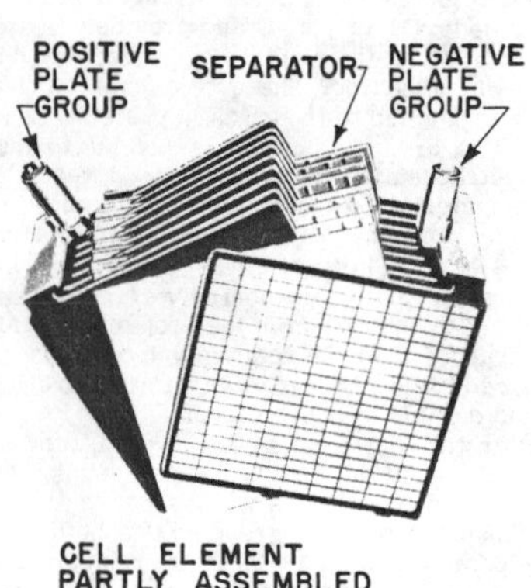

Figure 3-11.—Plate arrangement.

are then assembled together with separators to become cell elements. The separators are grooved vertically on one side and smooth on the other. The grooved side is placed next to the positive plate to permit free circulation of the electrolyte around the active material.

The positive plates which are lead peroxide and the negative plates which are spongy lead are referred to as the active material of the battery. However, these materials alone in a container will cause no chemical action unless there is a path for interaction between them. To provide this path for interaction and to carry the electric current within the battery are the functions of the electrolyte.

A battery container is the receptacle for the cells that make up the battery. Most containers are made from hard rubber, plastic, or bituminous composition that is resistant to acid and mechanical shock, and is able to withstand extreme weather conditions. Most batteries are assembled in a one-piece container with compartments for each individual cell. The bottom of the container has ribs

NOTES:

Figure 3-12.—Grid structure.

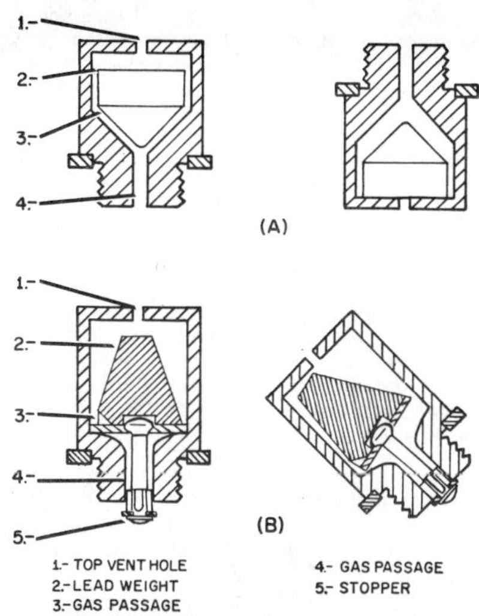

Figure 3-13.—Nonspill vent plugs.

molded into it to provide support for the elements and a sediment space for the flakes of active material that drop off the plates during the life of the battery.

The battery or cell covers and the battery container are usually made of the same material. The cell covers provide openings for the two-element terminals and a vent plug.

Cell connectors are used to connect the cell of a battery in series. The element in each cell is placed so that the negative terminal of one cell is physically located adjacent to the positive terminal of the next cell; they are connected both physically and electrically by a cell connector. Connectors must be of sufficient size to carry the current demands of the battery without over-heating.

Vent plugs are made of various designs to function in conjunction with the cover vent openings to permit the escape of gases that form within the cells while preventing leakage or loss of the electrolyte. A typical vent plug used in an automobile battery is depicted in figure 3-10.

Some batteries utilize a nonspill type of vent plug which makes it possible to place the battery in any position without loss of the electrolyte. (See fig. 3-13.) This type vent plug has found wide use in aircraft.

Sealing compound, generally made of a bituminous substance, is used to form a seal between the cell cover and the container. The compound is an acid resistant material that must conform to rigid vibration and heat standards. This insures that the sealing compound does not melt or flow at summer temperatures and does not crack at winter temperatures. Batteries with a polystyrene jar use a polystyrene cement as a sealer.

The terminals of a lead-acid battery are normally distinguishable from one another by their physical size and the marking by the manufacturer. The positive terminal marked (+) is slightly larger than the negative terminal marked (-).

BATTERY OPERATION

In its charged condition, the active materials in the lead-acid battery are lead peroxide (used as the positive plate) and sponge lead (used as the negative plate). The electrolyte is a mixture of sulfuric acid and water. The strength (acidity) of the electrolyte is measured in terms of its specific gravity. Specific gravity is the ratio of the weight of a given volume

NOTES:

of electrolyte to an equal volume of pure water. Concentrated sulfuric acid has a specific gravity of about 1.830; pure water has a specific gravity of 1.000. The acid and water are mixed in a proportion to give the specific gravity desired. For example, an electrolyte with a specific gravity of 1.210 requires roughly one part of concentrated acid to four parts of water.

In a fully charged battery, the positive plates are pure lead peroxide and the negative plates are pure lead. Also, in a fully charged battery, all acid is in the electrolyte so that the specific gravity is at its maximum value. The active materials of both the positive and negative plates are porous, and have absorptive qualities similar to a sponge.

The pores are therefore filled with the battery solution (electrolyte) in which they are immersed. As the battery discharges, the acid in contact with the plates separates from the electrolyte. It forms a chemical combination with the active material of the plate, changing it to lead sulfate. Thus, as the discharge continues, lead sulfate forms on the plates, and more acid is taken from the electrolyte. The water content of the electrolyte becomes progressively higher; that is, the ratio of water to acid increases. As a result, the specific gravity of the electrolyte will gradually decrease during discharge.

When the battery is being charged, the reverse takes place. The acid held in the sulfated plate material is driven back into the electrolyte; further charging cannot raise its specific gravity any higher. When fully charged, the material of the positive plates is again pure lead peroxide and that of the negative plates is pure lead.

Electrical energy is derived from a cell when the plates react with the electrolyte. As a molecule of sulfuric acid separates, part of it combines with the negative sponge lead plates. Thus, it makes the sponge lead plates negative, and at the same time forms lead sulfate. The remainder of the sulfuric acid molecule, lacking electrons, has thus become a positive ion. The positive ions migrate through the electrolyte to the opposite (lead peroxide) plates, and take electrons from them. This action neutralizes the positive ions, forming ordinary water. It also makes the lead peroxide plates positive, by taking electrons from them. Again, lead sulfate is formed in the process.

The action just described is represented in more detail by the following chemical equation:

$$Pb + PbO_2 + 2H_2SO_4 \xrightleftharpoons{\text{discharging}} 2PbSO_4 + 2H_2O$$

The left side of the expression represents the cell in the charged condition, and the right side represents the cell in the discharged condition.

In the charged condition, the positive plate contains lead peroxide, PbO_2; the negative plate is composed of sponge lead, Pb; and the solution contains sulfuric acid, H_2SO_4. In the discharged condition, both plates contain lead sulfate, $PbSO_4$, and the solution contains water, H_2O. As the discharge progresses, the acid content of the electrolyte becomes less and less because it is used in forming lead sulfate, and the specific gravity of the electrolyte decreases. A point is reached where so much of the active material has been converted into lead sulfate that the cell can no longer produce sufficient current to be of practical value. At this point the cell is said to be discharged (fig. 3-14 (C)). Since the amount of sulfuric acid combining with the plates at any time during discharge is in direct proportion to the ampere-hours (product of current in amperes and time in hours) of discharge, the specific gravity of the electrolyte is a guide in determining the state of discharge of the lead-acid cell.

If the discharged cell is properly connected to a direct-current charging source (the voltage of which is slightly higher than that of the cell), current will flow through the cell, in the opposite direction to that of discharge, and the cell is said to be charging (fig. 3-14 (D)). The effect of the current will be to change the lead sulfate on both the positive and negative plates back to its original active form of lead peroxide and sponge lead, respectively. At the same time, the sulfate is restored to the electrolyte with the result that the specific gravity of the electrolyte increases. When all the sulfate has been restored to the electrolyte, the specific gravity will be maximum. The cell is then fully charged and is ready to be discharged again.

It should always be remembered that the addition of sulfuric acid to a discharged lead-acid cell does not recharged the cell. Adding acid only increases the specific gravity of the electrolyte and does not convert the lead sulfate on the plates back into active material (sponge lead and lead peroxide), and consequently does

NOTES:

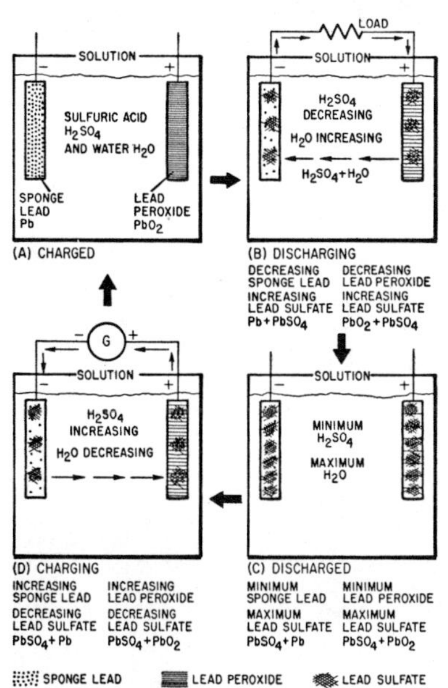

Figure 3-14.—Chemical action in lead-acid cell.

not bring the cell back to a charged condition. A charging current must be passed through the cell to do this.

As a cell charge nears completion, hydrogen gas (H_2) is liberated at the negative plate and oxygen gas (O_2) is liberated at the positive plate. This action occurs because the charging current is greater than the amount that is necessary to reduce the small remaining amount of lead sulfate on the plates. Thus, the excess current ionizes the water in the electrolyte. This action is necessary to assure full charge to the cell.

SPECIFIC GRAVITY

The ratio of the weight of a certain volume of liquid to the weight of the same volume of water is called the specific gravity of the liquid. The specific gravity of pure water is 1.000. Sulfuric acid has a specific gravity of 1.830; thus sulfuric acid is 1.830 times as heavy as water. The specific gravity of a mixture of sulfuric acid and water varies with the strength of the solution from 1.000 to 1.830.

As a storage battery discharges, the sulfuric acid is depleted and the electrolyte is gradually converted into water. This action provides a guide in determining the state of discharge of the lead-acid cell. The electrolyte that is usually placed in a lead-acid battery has a specific gravity of 1.350 or less. Generally, the specific gravity of the electrolyte in standard storage batteries is adjusted between 1.210 and 1.220. On the other hand the specific gravity of the electrolyte in submarine batteries when charged is from 1.250 to 1.265, while in aircraft batteries when fully charged it is from 1.285 to 1.300.

Hydrometer

The specific gravity of the electrolyte is measured with a hydrometer. In the syringe type hydrometer (fig. 3-15), part of the battery electrolyte is drawn up into a glass tube by means of a rubber bulb at the top.

The hydrometer float consists of a hollow glass tube weighted at one end and sealed at both ends. A scale calibrated in specific gravity is laid off axially along the body (stem) of the tube. The hydrometer float is placed inside the glass syringe and the electrolyte to be tested is drawn up into the syringe, thus immersing the hydrometer float into the solution. When the syringe is held approximately in a vertical position, the hydrometer float will sink to a certain level in the electrolyte. The extent to which the hydrometer stem protudes above the level of the liquid depends upon the specific gravity of the solution. The reading on the stem at the surface of the liquid is the specific gravity of the electrolyte in the syringe.

The Navy uses two types of hydrometer bulbs, or floats, each having a different scale. The type-A hydrometer is used with submarine batteries and has two different floats with scales from 1.960 to 1.240 and from 1.120 to 1.300. The type-B hydrometer is used with portable storage batteries and aircraft batteries. It has a scale from 1.100 to 1.300.

NOTES:

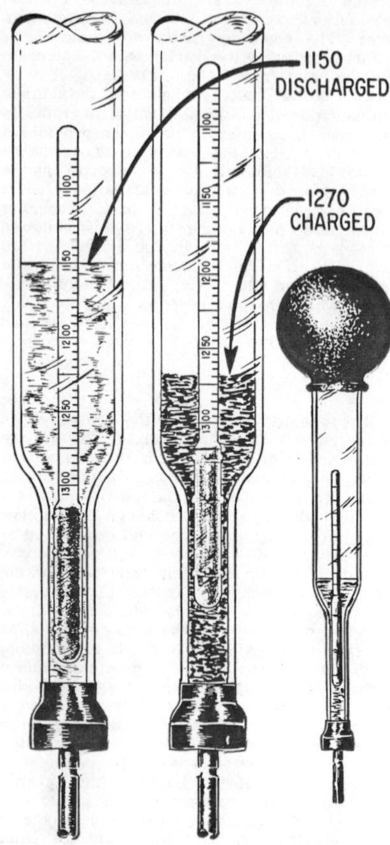

Figure 3-15.—Type-B hydrometer.

CAUTION: Hydrometers should be flushed daily with fresh water to prevent inaccurate readings. Storage battery hydrometers must not be used for any other purpose.

NOTES:

Corrections

The specific gravity of the electrolyte is affected by its temperature. The electrolyte expands and becomes less dense when heated and its specific gravity reading is lowered. On the other hand, the electrolyte contracts and becomes denser when cooled and its specific gravity reading is raised. In both cases the electrolyte may be from the same fully charged storage cell. Thus, the effect of temperature is to distort the readings.

Most standard storage batteries use 80° F as the normal temperature to which specific gravity readings are corrected. To correct the specific gravity reading of a storage battery, add 4 points to the reading for each 10° F above 80° and subtract 4 points for each 10° F below 80°. The electrolyte in a cell should be at the normal level when the reading is taken. If the level is below normal, there will not be sufficient fluid drawn into the tube to cause the float to rise. If the level is above normal there is too much water, the electrolyte is weakened, and the reading is too low. A hydrometer reading is inaccurate if taken immediately after water is added, because the water tends to remain at the top of the cell. When water is added, the battery should be charged for at least an hour to mix to electrolyte before a hydrometer reading is taken.

Adjusting Specific Gravity

Only authorized personnel should add acid to a battery. Acid with a specific gravity above 1.350 is never added to a battery.

If the specific gravity of a cell is more than it should be, it can be reduced to within limits by removing some of the electrolyte and adding distilled water. The battery is charged for 1 hour to mix the solution, and then hydrometer readings are taken. The adjustment is continued until the desired true readings are obtained.

MIXING ELECTROLYTES

The electrolyte of a fully charged battery usually contains about 38 percent sulfuric acid by weight, or about 27 percent by volume. In preparing the electrolyte, distilled water and sulfuric acid are used. New batteries may be delivered with containers of concentrated sulfuric acid of 1.830 specific gravity or electrolyte of 1.400 specific gravity, both of which must be diluted with distilled water to make electrolyte of the proper specific gravity. The container used for diluting the acid should be made of glass, earthenware, rubber, or lead.

When mixing electrolyte, ALWAYS POUR ACID INTO WATER—never pour water into acid. Pour the acid slowly and cautiously to prevent excessive heating and splashing. Stir the solution continuously with a nonmetallic rod to mix the heavier acid with the lighter water and to keep the acid from sinking to the bottom. When concentrated acid is diluted, the solution becomes very hot.

TREATMENT OF ACID BURNS

If acid or electrolyte from a lead-acid battery comes into contact with the skin, the affected area should be washed as soon as possible with large quantities of fresh water, after which a salve such as petrolatum, boric acid, or zinc ointment should be applied. If none of these salves are available, clean lubricating oil will suffice. When washing, large amounts of water should be used, since a small amount of water might do more harm than good in spreading the acid burn.

Acid spilled on clothing may be neutralized with dilute ammonia or a solution of baking soda and water.

CAPACITY

The capacity of a battery is measured in ampere-hours. As mentioned before, the ampere-hour capacity is equal to the product of the current in amperes and the time in hours during which the battery is supplying this current. The ampere-hour capacity varies inversely with the discharge current. The size of a cell is determined generally by its ampere-hour capacity. The capacity of a cell depends upon many factors, the most important of these are as follows:

1. The area of the plates in contact with the electrolyte.
2. The quantity and specific gravity of the electrolyte.
3. The type of separators.
4. The general condition of the battery (degree of sulfating, plates buckled, separators warped, sediment in bottom of cells, etc.).
5. The final limiting voltage.

RATING

Storage batteries are rated according to their rate of discharge and ampere-hour capacity. Most batteries, except aircraft and some used for radio and sound systems, are rated according to a 20-hour rate of discharge—that is, if a fully charged battery is completely discharged during a 20-hour period, it is discharged at the 20-hour rate. Thus if a battery can deliver 20 amperes continuously for 20 hours, the battery has a rating of 20 x 20, or 400 ampere-hours. Thus the 20-hour rating is equal to the average current that a battery is capable of supplying without interruption for an interval of 20 hours. (NOTE: Aircraft batteries are rated according to a 1-hour rate of discharge.) Some other ampere-hour ratings used are 6-hour and 10-hour ratings.

All standard batteries deliver 100 percent of their available capacity if discharged in 20 hours or more, but they will deliver less than their available capacity if discharged at a faster rate. The faster they discharge, the less ampere-hour capacity they have.

The low-voltage limit, as specified by the manufacturer, is the limit beyond which very little useful energy can be obtained from a battery. For example, at the conclusion of a 20-hour discharge test on a battery, the closed-circuit voltmeter reading is about 1.75 volts per cell and the specific gravity of the electrolyte is about 1.060. At the end of a charge, its closed-circuit voltmeter reading, while the battery is being charged at the finishing rate, is between 2.4 and 2.6 volts per cell. The specific gravity of the electrolyte corrected to 80°F is between 1.210 and 1.220. In climates of 40°F and below authority may be granted to increase the specific gravity to 1.280. Other batteries, of higher normal specific gravity, may also be increased.

TEST DISCHARGE

The test discharge is the best method of determining the capacity of a battery. Most battery switchboards are provided with the necessary equipment for giving test discharges. If proper equipment is not available, a tender, repair ship, or shore station may make the test. To determine the battery capacity, a battery is normally given a test discharge once every 6 months. Test discharges are also given whenever any cell of a battery after

NOTES:

charge cannot be brought within 10 points of full charge, or when one or more cells is found to have less than normal voltage after an equalizing charge.

A test discharge must always be preceded by an equalizing charge. Immediately after the equalizing charge, the battery is discharged at its 20-hour rate until either the total battery voltage drops to a value equal to 1.75 times the number of cells in series or the voltage of any individual cell drops to 1.65 volts, whichever occurs first. The rate of discharge should be kept constant throughout the test discharge. Because standard batteries are rated at the 20-hour capacity, the discharge rate for a 200 ampere-hour battery is 200/20, or 10 amperes. If the temperature of the electrolyte at the beginning of the charge is not exactly 80° F, the time duration of the discharge must be corrected for the actual temperature of the battery.

A battery of 100-percent capacity discharges at its 20-hour rate for 20 hours before reaching its low-voltage limit. If the battery or one of its cells reaches the low-voltage limit before the 20-hour period has elapsed, the discharge is discontinued immediately and the percentage of capacity is determined from the equation

$$C = \frac{H_a}{H_t} \times 100$$

where C is the percentage of ampere-hour capacity available, H_a the total hours of discharge, and H_t the total hours for 100-percent capacity. The date for each test discharge should be recorded on the storage battery record sheet.

For example, a 200-ampere-hour 6-volt battery delivers an average current of 10 amperes for 20 hours. At the end of this period the battery voltage is 5.25 volts. On a later test the same battery delivers an average current of 10 amperes for only 14 hours. The discharge was stopped at the end of this time because the voltage of the middle cell was found to be only 1.65 volts. The percentage of capacity of the battery is now $\frac{14}{20} \times 100$, or 70 percent. Thus, the ampere-hour capacity of this battery is reduced to $0.7 \times 200 = 140$ ampere hours.

STATE OF CHARGE

After a battery is discharged completely from full charge at the 20-hour rate, the specific gravity has dropped about 150 points to about 1.060. The number of points that the specific gravity drops per ampere-hour can be determined for each type of battery. For each ampere-hour taken out of a battery a definite amount of acid is removed from the electrolyte and combined with plates.

For example, if a battery is discharged from full charge to the low-voltage limit at the 20-hour rate and if 100 ampere-hours are obtained with a specific gravity drop of 150 points, there is a drop of $\frac{150}{100}$, or 1.5 points per ampere-hour delivered. If the reduction is specific gravity per ampere-hour is known, the drop in specific gravity for this battery may be predicted for any number of ampere-hours delivered to a load. For example, if 70 ampere-hours are delivered by the battery at the 20-hour rate or any other rate or collection of rates, the drop in specific gravity is 70×1.5, or 105 points.

Conversely, if the drop in specific gravity per ampere-hour and the total drop in specific gravity are known, the ampere-hours delivered by a battery may be determined. For example, if the specific gravity of the previously considered battery is 1.210 when the battery is fully charged and 1.150 when it is partly discharged, the drop in specific gravity is 1,210 - 1,150, or 60 points, and the number of ampere-hours taken out of the battery is $\frac{60}{1.5}$, or 40 ampere-hours. Thus the number of ampere-hours expended in any battery discharge can be determined from the following items.

1. The specific gravity when the battery is fully charged.
2. The specific gravity after the battery has been discharged.
3. The reduction is specific gravity per ampere-hour.

Voltage alone is not a reliable indication of the state of charge of a battery except when the voltage is near the low-voltage limit on discharge. During discharge the voltage falls. The higher the rate of discharge the lower will be the terminal voltage. Open-circuit voltage is of little value because the variation between full charge and complete discharge is so small—only about 0.1 volt per cell. However, abnormally low voltage does indicate injurious sulfation or some other serious deterioration of the plates.

NOTES:

TYPES OF CHARGES

The following types of charges may be given to a storage battery, depending upon the condition of the battery:

1. Initial charge.
2. Normal charge.
3. Equalizing charge.
4. Floating charge.
5. Fast charge.

Initial Charge

When a new battery is shipped dry, the plates are in an uncharged condition. After the electrolyte has been added it is necessary to convert the plates into the charged condition. This is accomplished by giving the battery a long low-rate initial charge. The charge is given in accordance with the manufacturer's instructions, which are shipped with each battery. If the manufacturer's instructions are not available, reference should be made to the detailed instruction in current directives.

Normal Charge

A normal charge is a routine charge that is given in accordance with the nameplate data during the ordinary cycle of operation to restore the battery to its charged condition. The following steps should be observed:

1. Determine the starting and finishing rate from the nameplate data.
2. Add water, as necessary, to each cell.
3. Connect the battery to the charging panel and make sure the connections are clean and tight.
4. Turn on the charging circuit and set the current through the battery at the value given as the starting rate.
5. Check the temperature and specific gravity of pilot cells hourly.
6. When the battery begins to gas freely, reduce the charging current to the finishing rate.

A normal charge is complete when the specific gravity of the pilot cell, corrected for temperature, is within 5 points (0.005) of the specific gravity obtained on the previous equalizing charge.

Equalizing Charge

An equalizing charge is an extended normal charge at the finishing rate. It is given periodically to insure that all the sulfate is driven from the plates and that all the cells are restored to a maximum specific gravity. The equalizing charge is continued until the specific gravity of all cells, corrected for temperature, shows no change for a 4-hour period. Readings of all cells are taken every half hour.

Floating Charge

A battery may be maintained at full charge by connecting it across a charging source that has a voltage maintained within the limits of from 2.13 to 2.17 volts per cell of the battery. In a floating charge, the charging rate is determined by the battery voltage rather than by a definite current value. The voltage is maintained between 2.13 and 2.17 volts per cell with an average as close to 2.15 volts as possible.

Fast Charge

A fast charge is used when a battery must be recharged in the shortest possible time. The charge starts at a much higher rate than is normally used for charging. It should be used only in an emergency as this type charge may be harmful to the battery.

CHARGING RATE

Normally, the charging rate of storage batteries is given on the battery nameplate. If the available charging equipment does not have the desired charging rates, the nearest available rates should be used. However, the rate should never be so high that violent gassing occurs. NEVER ALLOW THE TEMPERATURE OF THE ELECTROLYTE IN ANY CELL TO RISE ABOVE 125°F.

CHARGING TIME

The charge must be continued until the battery is fully charged. Frequent readings of specific gravity should be taken during the charge. These readings should be corrected to 80°F and compared with the reading taken before the battery was placed on charge. If the rise in specific gravity in points per ampere-hour is known, the

NOTES:

approximate time in hours required to complete the charge is as follows:

$$\frac{\text{rise in specific gravity in points to complete charge}}{\text{rise in specific gravity in points per ampere-hour} \times \text{charging rate in amperes}}$$

GASSING

When a battery is being charged, a portion of the energy is dissipated in the electrolysis of the water in the electrolyte. Thus, hydrogen is released at the negative plates and oxygen at the positive plates. These gases bubble up through the electrolyte and collect in the air space at the top of the cell. If violent gassing occurs when the battery is first placed on charge, the charging rate is too high. If the rate is not too high steady gassing which develops as the charging proceeds, indicates that the battery is nearing a fully charged condition. A mixture of hydrogen and air can be dangerously explosive. No smoking, electric sparks, or open flames should be permitted near charging batteries.

NICKEL-CADMIUM BATTERIES

The nickel-cadmium batteries are far superior to the lead-acid type. Some are physically and electrically interchangeable with the lead-acid type, while some are sealed units which use standard plug and receptacle connections which are used on other electrical components. These batteries generally require less maintenance than lead-acid batteries throughout their service life in regard to the adding of electrolyte or water.

The nickel-cadmium and lead-acid batteries have capacities that are comparable at normal discharge rates, but at high discharge rates the nickel-cadmium battery can:

1. Be charged in a short time.
2. Deliver a large amount of power.
3. Stay idle in any state of charge for an indefinite time and keep a full charge when stored for a long time.
4. Be charged and discharged any number of times without any appreciable damage.
5. The individual cells may be replaced if a cell wears out; the rest of the cells do not have to be replaced.

Due to their superior capabilities, nickel-cadmium batteries are being used extensively in many military applications that require a battery with a high discharge rate. A prime example is the aircraft storage battery.

Some lead-acid batteries are equipped with the same quick-disconnect receptacle and plug used on nickel-cadmium batteries. In distinguishing a lead-acid battery from a nickel-cadmium battery or a silver-zinc battery, the nameplate of each battery should be checked, since the physical appearance could be the same. (See fig. 3-16.)

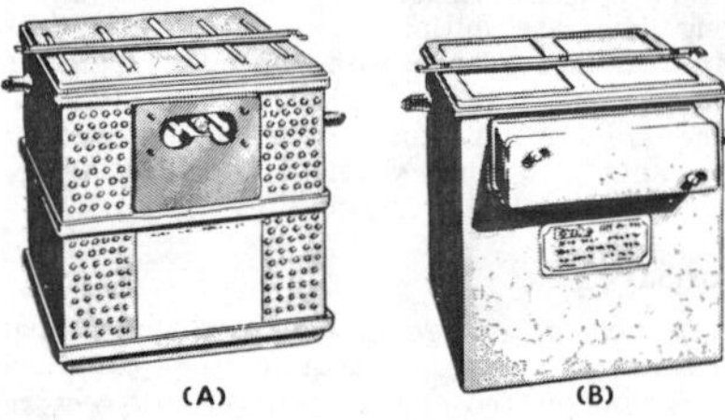

Figure 3-16.—(A) Nickel-cadmium battery; (B) lead-acid battery.

The nickel-cadmium battery plates are constructed of nickel powder sintered to a nickel wire screen. The active materials (nickel-hydroxide on the positive plate and cadmium-hydroxide on the negative plate) are electrically bonded to the basic plate structure. The separators are constructed of plastic, nylon cloth, or a special type of cellophane, and assembled as a cell core with plates. (See fig. 3-17.)

The construction of the sintered-plate cell is accomplished by a powder metallurgy process. Carbonyl nickel powder is lightly compressed in a mold and then is subjected either to a temperature of about 1,600° F in a sintering furnace or to a sudden heavy electric current. Either process causes the individual grains of nickel to weld at their points of contact, producing a porous plaque which is approximately 80 percent open holes and 20 percent solid nickel. The plaques are then impregnated with

NOTES:

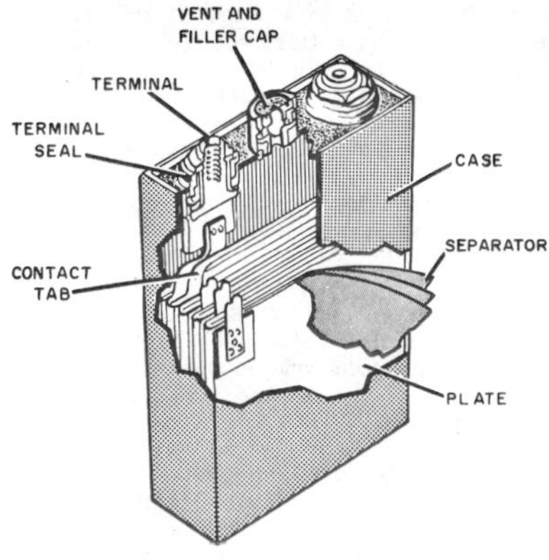

Figure 3-17.—Nickel-cadmium cell.

active materials. They are then soaked in a solution of nickel salts to make the positive plates and in a solution of cadmium salts to make negative plates. The bath is repeated until the plaques contain the amount of active material necessary to give them the capacity desired. When the plaques are impregnated, they are classified as plates.

The electrolyte used in a nickel-cadmium battery is a 30-percent-by-weight solution of potassium hydroxide in distilled water. Chemically speaking, this is just about the exact opposite to the diluted sulfuric acid used in the lead battery. As with lead-acid batteries, there are limitations on the concentration of electrolyte solution that can be used in nickel-cadmium cells. The specific gravity of the solution should not be outside the range of 1.240 to 1.300 at 70 F. The electrolyte in the nickel-cadmium battery does not chemically react with the plates as the electrolyte does in the lead battery. It acts only as a conductor of current between plates—therefore, no flaking or shredding of active material. Consequently the plates do not deteriorate, nor does the specific gravity of the electrolyte appreciably change. For this reason, it is not possible to determine the charge state of a nickel-cadmium battery by checking the electrolyte with a hydrometer; neither can the charge be determined by a voltage test because of the inherent characteristic that the voltage remains constant during 90 percent of the discharge cycle.

No external vent is required since gassing of this type battery is practically negligible. As a safety precaution, however, relief valves have been installed in the fill hole cap of each cell (fig. 3-18) in order to release any excess gas that is formed when the battery is charged improperly.

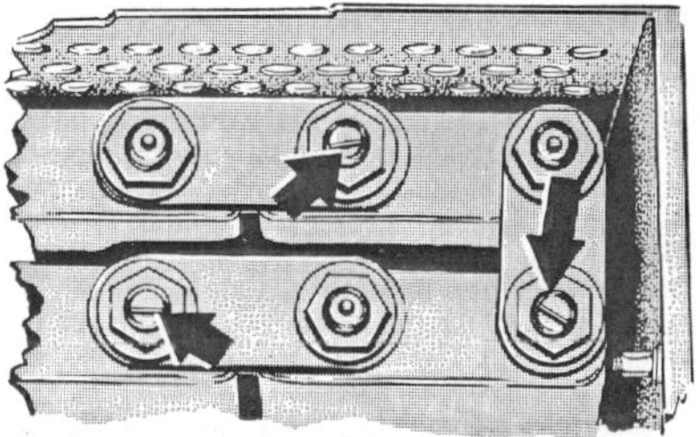

Figure 3-18.—Relief valves in negative post of cells.

CHARGE DETERMINATION

At the present time, a simple method of determining the exact state of charge has not been developed. In an aircraft battery, the only practical method is to measure the open circuit battery voltage. If the open circuit voltage exceeds 1.4 volts per cell and the battery voltage is 26.0 volts (normally 28.0 volts) or more, it can be assumed that the battery is fully charged. If the cell voltage is less than 1.0 volt under load, or the battery voltage measures 25 volts or less, it can be assumed that the battery has expended over 90 percent of its capacity and is in need of charging.

In a battery shop, either of two methods may be used for determining the state of charge of a nickel-cadmium battery; namely, the constant potential method or the discharge method. The constant potential method consists of connecting a constant potential of 28.5 ± 0.3 volts across the battery and observing the charging current. If the current falls to 3 amperes or less within 5 minutes, the battery is charged.

The discharge method consists of placing a 15-ampere load across the battery for 5 minutes. If the voltage does not drop below 22 volts during the discharge period, the battery may be returned to service after being recharged.

NOTES:

The available ampre-hour capacity cannot be accurately determined. Therefore, it is recommended that any battery whose charge is unknown or subject to doubt be discharged to or beyond the manufacturer's set end point of 1.0 or 1.1 volts, and then be recharged in accordance with the appropriate instructions. This process will prevent possible damage to the cells from overcharge.

CHARGING

Nickel-cadmium batteries should preferably be charged at an ambient temperature of 70° to 80°F. Never allow a battery on charge to exceed 100°F as this may cause overcharging and gassing. In the battery shop a thermometer should be placed between the central cells in such a manner that the bulb of the thermometer is located below the top of the cell. Whenever the temperature of battery is 100°F or higher, the battery should not be charged.

The rate of charging of a nickel-cadmium battery is dependent upon two factors, the first being the charging voltage, and the second being the temperature of the battery. In hot weather where the ground air temperature approaches 90°F or higher, the battery can be adequately charged at 27 volts. In mild ground air temperatures ranging from 35° to 85°F, the battery can be satisfactorily charged at 27.5 volts. In cold, subfreezing weather the battery requires a charging voltage of 28.5 volts.

The nickel-cadmium battery was designed and constructed to operate without gassing of the cells. The charging voltage should be maintained below the gassing voltage (approximately 29.4 volts at 80°F) so that the life of the battery is prolonged. Therefore, on constant potential charging in the battery shop, the voltage should be set at 28 volts or less. Under no circumstances should this voltage exceed 28.5 volts.

If the battery has never been placed in service, follow the manufacturer's instructions accompanying the battery for the initial charge. If possible, the battery should be charged by the constant potential method.

For constant potential charging, maintain the battery at 28 volts for 4 hours, or until the current drops below 3 amperes. Do not allow battery temperature to exceed 100°F.

For constant current charging, start the charge at 10 to 15 amperes and continue until the voltage reaches 28.5 volts—then reduce the current to 4 amperes and continue charging until the battery voltage reaches 28.5 volts, or until the battery temperature exceeds 100°F and the voltage begins to decline.

Never add electrolyte unless the battery is fully charged. Allow the fully charged battery to stand for a period of 3 or 4 hours before distilled water can be added to bring the electrolyte to the proper level. A hydrometer or syringe can be used for introduction of the distilled water—just enough to cover the top of the plates. The battery solution is then recycled to stir the water and prevent it from freezing during cold-weather operation.

SAFETY PRECAUTIONS

The electrolyte used in nickel-cadmium batteries is potassium hydroxide (KOH). This is a highly corrosive alkaline solution, and should be handled with the same degree of caution as sulfuric acid (H_2SO_4). Personnel should always wear rubber gloves, a rubber apron, and protective goggles when handling and servicing these batteries. If the electrolyte is spilled on the skin or clothing, the exposed area should be rinsed immediately with water, or if available, vinegar, lemon juice, or boric acid solution. If the face or eyes are affected, treat as above and report immediately for medical examination and treatment.

The battery shop used for nickel-cadmium batteries should be a separate isolated shop from the lead-acid battery shop.

SILVER-ZINC BATTERIES

Silver-zinc batteries are used largely in military applications and in some industrial applications where their unique characteristics are sufficiently important to justify their comparatively high cost.

The silver-zinc battery was developed for one major and one secondary purpose. The major purpose was to secure a large quantity of electrical power for emergency operations. The secondary purpose was to permit a design weight savings in new batteries. A lightweight, silver-zinc battery provides as much electrical capacity as a much larger lead-acid or nickel-cadmium battery.

Operational silver-zinc batteries have a nominal operating voltage of 24 volts, obtained with sixteen 1.5 volt cells. Cell electrolytic levels should be monitored and adjusted

NOTES:

periodically. The other required operations that might be considered maintenance are the normal recharging of the battery and keeping the top surfaces of the cells reasonably clean.

CHARACTERISTICS

Because of its extremely low internal resistance, the silver-zinc battery is capable of discharge rates of up to 30 times its ampere-hour rating. The low internal resistance (as low as 0.0003 ohms per cell) is due primarily to the excellent conductivity of its plates, the close plate spacing (possible because small amounts of electrolyte may be used successfully), and the fact that the composition (and therefore the conductivity) of the electrolyte does not change during discharge. The internal conductivity of the battery increases during discharge as the positive plates are changed from oxides of silver (fair conductors) to metallic silver.

The high electrical capacity per unit of space and weight is a result of the close plate spacing, the large degree to which the active plate materials are utilized, and the absence of heavy supporting grids in the plate. Silver-zinc batteries are capable of producing as much as six times more energy per unit of weight and volume than other types. Silver-zinc cells have been built with capacities ranging from tenths of ampere-hours to thousands of ampere-hours.

Good voltage regulation is provided by the relatively constant voltage discharge characteristic of the silver-zinc battery. Terminal voltage is essentially constant throughout most of the discharge when discharged at rates higher than 2- or 3-hour rate.

Silver-zinc batteries have a maximum service cycle life which is less than that of other types, but their life expectancy compares favorably with that of other types of batteries that are designed for maximum capacity per unit of space and weight such as nickel-cadmium batteries.

OPERATION

The construction and electrochemical reactions of the silver-zinc battery are somewhat similar to those of the nickel-cadmium type. When in the fully charged condition, the positive plates are composed of silver oxide and the negative plates of zinc. As the battery discharges, the positive plates are reduced to metallic silver and the negative plates are oxidized. Thus, when the battery is discharging, electrons are flowing out of the cathode (negative plates) and into the anode (positive plates) by way of the external circuit.

The electrolyte, potassium hydroxide in aqueous solution, exists as potassium (K) and hydroxide (OH) ions, which serve only to conduct the electric charge between the plates. Thus, the electronic or metallic conduction in the external circuit is balanced by the ionic or electrolytic conduction through the electrolyte, so as to maintain the net charge transfer into and out of each electrode the same.

As with other types of alkaline cells, and unlike lead-acid cells, the electrolyte does not take part in the chemical transformations and therefore its specific gravity does not change with the state of charge of the cell. As long as the plates are covered, the electrical capacity of the battery is independent of the amount of electrolyte present.

In general, silver-zinc batteries require maintenance which is similar in many respects to that which is required of the lead-acid batteries. Testing the open circuit voltage of the battery is the method by which its state of charge is determined.

A silver-zinc battery tester or a voltmeter which reads accurately to 0.1 volt should be used to test the open circuit voltage of the battery. If the reading is below 25.6 volts, remove the battery cover and inspect the top of the battery for corrosion or damaged cells. If any damage is evident, remove and replace the battery.

CHARGING

The silver-zinc battery is usually shipped in a dry condition. Only the special electrolyte furnished in the filling kit provided with each new battery should be used. Some battery types may use electrolytes containing special additives; other electrolytes, if used, may degrade the battery. The electrolyte must be kept in a closed alkali-resistant container, or it will absorb carbon dioxide from the air and deteriorate. The filling kit contains detailed instructions for filling and should be followed in detail. (NOTE: Batteries that will not be used within 30 days should be stored in the dry state.)

Silver-zinc batteries are sensitive to excessive voltage during charging and may be

NOTES:

damaged if the voltage exceeds 2.05 volts per cell. Precaution must therefore be taken to insure that the charging equipment is adjusted accurately to cut off the current at 28.7 volts.

Where charging is not monitored automatically or periodically, a voltage cut off system must be used which will interrupt the charging current when the voltage rises to 28.7 volts.

If possible, charging should be performed at an ambient temperature of 60° to 90°F, and the battery temperature during charging should not exceed 150°F as measured at the intercell connections.

While silver-zinc batteries do not generate any harmful gases during normal charge and discharge operations, they do generate both oxygen and hydrogen gases during excessive overcharging. All vent caps and spong-rubber plugs must be removed from the vent hole during charging operations. If electrolyte is forced from the vent holes or if excessive gassing is evident, it is an indication of overheating, and the charging should be interrupted for 8 hours to allow the battery to cool. After charging, the batteries should be allowed to stand idle at least 8 hours.

The level of the electrolyte of each cell of the battery should be checked after charging and the level adjusted by either removing any excess electrolyte or by adding distilled water if low.

The safety precautions relating to silver-zinc batteries are the same as those which have been presented for the nickel-cadmium batteries.

SILVER-CADMIUM BATTERY

One of the recent developments in storage batteries is the silver-cadmium battery. Generally, the most important requirements for evaluating and designing a battery are for high energy density, good voltage regulation, long shelf life, repeatable number of cycles, and long service life expectancy. The silver-cadmium battery is designed to offer the overall maximum performance in all of these expectations.

The silver-cadmium battery has more than twice the wet shelf life of the silver-zinc battery. The long shelf life plus the good voltage regulation makes the silver-cadmium battery a highly desirable addition to the family of electric storage batteries. Limitations include lower cell voltage than other rechargeable batteries and high initial cost.

NOTES:

CHAPTER 4

SERIES D-C CIRCUITS

SIMPLE ELECTRIC CIRCUIT

Whenever two unequal charges are connected by a conductor, a complete pathway for current flow exist. Current will flow from the negative to the positive charge. This was illustrated in chapter 2.

An electric circuit is a completed conducting pathway, consisting not only of the conductor, but including the path through the voltage source. Current flows from the positive terminal through the source, emerging at the negative terminal. As an example, a lamp connected by conductors across a dry cell forms a simple electric circuit. (Refer to fig. 4-1.)

Current flows from the negative (-) terminal of the battery through the lamp to the positive (+) battery terminal, and continues by going through the battery from the positive (+) terminal to the negative (-) terminal. As long as this pathway is unbroken, it is a closed circuit and current will flow. However, if the path is broken at ANY point, it is an open circuit and no current flows. (See fig. 4-1 (B).)

Current flow in the external circuit is the movement of electrons in the direction indicated by the arrows (from the negative terminal through the lamp to the positive terminal). (See fig. 4-1 (A).) Current flow in the internal battery circuit is the simultaneous movement in opposite directions of positive hydrogen ions toward the positive terminal of the battery and negative ions toward the negative terminal.

SCHEMATIC REPRESENTATION

A SCHEMATIC is a diagram in which symbols are used for the various components instead of pictures. These symbols are used in an effort to make the diagrams easier to draw and easier to understand. In this respect, schematic symbols aid the technician in the same way that shorthand aids the stenographer. In previous chapters the schematic symbols for cells and resistances were presented. These symbols will now be used to discuss the circuits of figure 4-1.

A schematic diagram of a basic circuit is shown in figure 4-2. The battery is designated by the letter symbols E_{bb}, the light bulb in the circuit is labeled R_1. Since in reality the light bulb element is nothing more than a wire wound resistor, the conventional resistor symbol is used for the bulb in this discussion. It should be noted, however, that the light bulb has its own specific schematic symbol and is not normally drawn as a resistor. The standard symbol used for a light bulb is discussed at a later time when need arises.

In studies of electricity and electronics many circuits are analyzed which consist mainly of specially designed resistive components. As previously stated, these components are called resistors. Throughout the remaining analysis of the basic circuit, the resistive component will be a physical resistor. However, the resistive component could be any one of several electrical devices.

A closed loop of wire (conductor) is not necessarily a circuit. A source of voltage must be included to make it an electric circuit. In any electric circuit where electrons move around a closed loop, current, voltage, and resistance are present. The physical pathway for current flow is actually the circuit. Its resistance controls the amount of current flow around the circuit. By knowing any two of the three quantities, such as voltage and current, the third (resistance) may be determined. This is done mathematically by the use of Ohm's LAW.

OHM'S LAW

In the early part of the 19th century Georg Simon Ohm proved by experiment that a precise

NOTES:

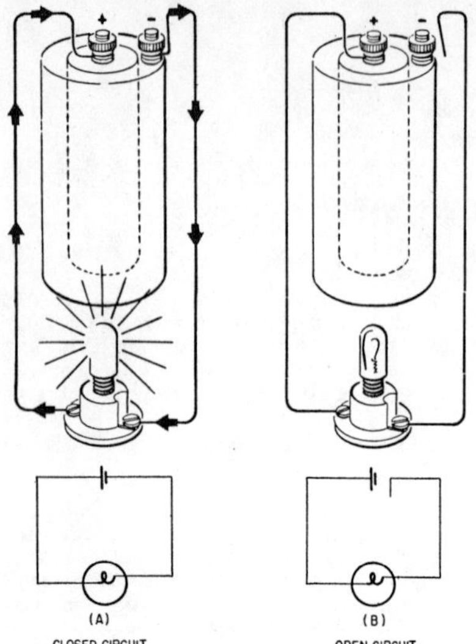

Figure 4-1.—(A) Simple electric circuit (closed); (B) simple electric circuit (open).

relationship exists between current, voltage, and resistance. This relationship is called Ohm's Law and is stated as follows:

The current in a circuit is DIRECTLY proportional to the applied voltage and INVERSELY proportional to the circuit resistance. Ohm's Law may be expressed as an equation:

$$I = \frac{E}{R} \qquad (4\text{-}1)$$

Where: I = current in amperes

E = voltage in volts

R = resistance in ohms

If any two of the quantities in equation (4-1) are known, the third may be easily found. For example, Figure 4-3 shows a circuit containing a resistance of 1.5 ohms and a source voltage

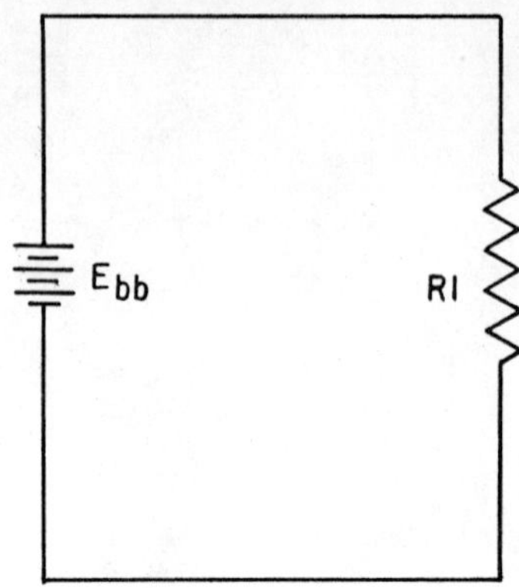

Figure 4-2.—Schematic diagram of a basic circuit.

of 1.5 volts. How much current flows in the circuit?

Given: E = 1.5 volts

R = 1.5 ohms

I = ?

Solution: $I = \frac{E}{R}$

$I = \frac{1.5}{1.5}$

I = 1 ampere

To observe the effect of source voltage on circuit current, the above problem will be solved again using double the previous source voltage.

Given: E = 3 volts

R = 1.5 ohms

I = ?

Solution: $I = \frac{E}{R}$

$I = \frac{3}{1.5}$

I = 2 amperes

NOTES:

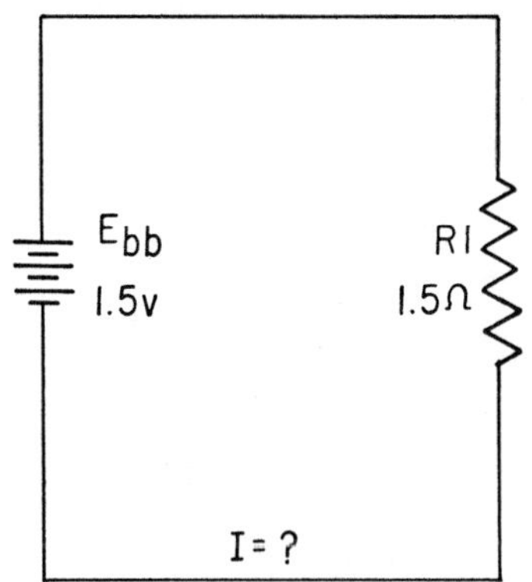

Figure 4-3.—Determining current in a basic circuit.

Notice that as the source voltage doubles, the circuit current also doubles.
Circuit current is directly proportional to applied voltage and will change by the same factor that the voltage changes.

To verify the statement that current is inversely proportional to resistance, assume the resistor in figure 4-3 to have a value of 3 ohms.

Given: $E = 1.5$ volts

$R = 3$ ohms

$I = ?$

Solution: $I = \frac{E}{R}$

$I = \frac{1.5}{3}$

$I = 0.5$ ampere

Comparing this current of 0.5 ampere for the 3 ohm resistor, to the 1 ampere of current obtained with the 1.5-ohm resistor, shows that doubling the resistance will reduce the current to one half the original value. Circuit current is inversely proportional to the circuit resistance.

NOTES:

In many circuit applications current is known and either the voltage or the resistance will be the unknown quantity. To solve a problem in which current and resistance are known, the basic formula for Ohm's Law must be transposed to solve for E as follows:

Basic equation: $I = \frac{E}{R}$ (4-1)

Multiply both sides of the equation by R.

$$IR = \frac{E}{\not{R}} \not{R}$$

$$IR = E$$

$$E = IR \qquad (4\text{-}2)$$

To transpose the basic formula when resistance is unknown:

Basic equation: $I = \frac{E}{R}$ (4-1)

Multiply both sides of the equation by R.

$$IR = \frac{E}{\not{R}} \not{R}$$

$$IR = E$$

Divide both sides of the equation by I.

$$\frac{\not{I}R}{\not{I}} = \frac{E}{I}$$

$$R = \frac{E}{I} \qquad (4\text{-}3)$$

Example: What voltage is required to properly light a lamp having a resistance of 10 ohms and a current rating of 1 ampere?

First draw a circuit like figure 4-4 including all the given information.

Given: $R = 10$ ohms

$I = 1$ ampere

$E = ?$

Solution: $E = IR$

$E = 1 \times 10$

$E = 10$ volts

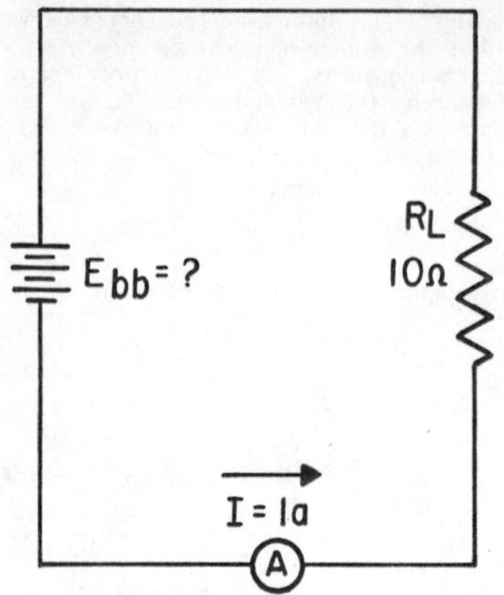

Figure 4-4.—Determining voltage in a basic circuit.

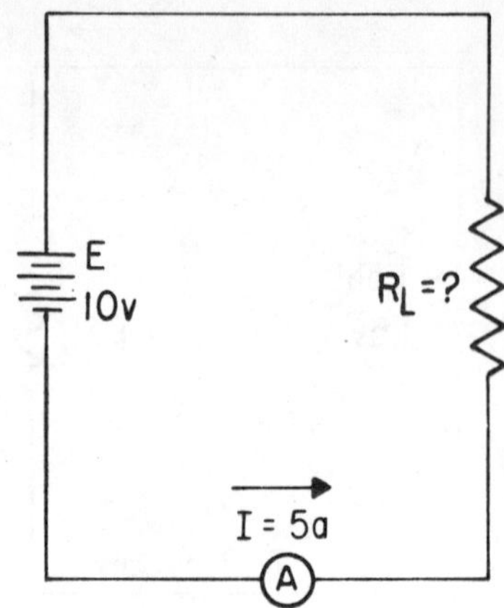

Figure 4-5.—Determining resistance in a basic circuit.

Example: When a 10 volt source is connected to a circuit, the circuit draws 5 amperes of current from the source. How much resistance is contained in the circuit?

Given: $E = 10$ volts

$I = 5$

$R = ?$

Draw and label a circuit like figure 4-5.

Solution:

$$R = \frac{E}{I} \qquad (4\text{-}3)$$

$$R = \frac{10}{5}$$

$$R = 2 \text{ ohms}$$

Although the three equations representing Ohm's Law are fairly simple, they are perhaps the most important of all electrical equations. These three equations and the law they represent must be thoroughly understood before continuing on to more advanced theory.

GRAPHICAL ANALYSIS

One of the most valuable methods of inquiry available to the technician is that of graphical analysis. No other method provides a more convenient or more rapid way to observe the characteristics of an electrical device.

The first step in constructing a graph consists of obtaining a table of data from which the graph will evolve. The information in the table can be obtained experimentally by taking laboratory measurements on the device under examination, or can be obtained theoretically through a series of computations. The latter method will be used here.

Let us assume that the characteristics of the circuit shown in figure 4-6 are to be investigated using Ohm's Law and graphical methods. Since there are three variables (E, I, and R) under consideration, there are three unique graphs that may be constructed.

In constructing any graph of electrical quantities, it is standard practice to vary one quantity in a specified way, and note the changes which occur in a second quantity. The quantity which is intentionally varied is called the independent variable and is plotted on the X-AXIS. The second quantity which changes as a result of changes in the first quantity is called the dependent variable and is plotted on the Y-AXIS. Any other quantities involved are held constant.

NOTES:

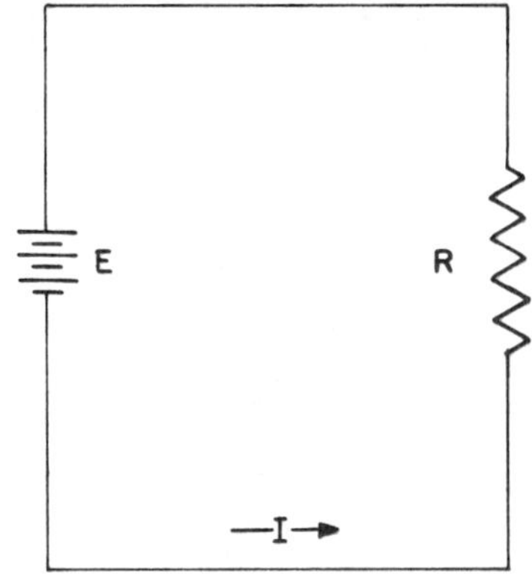

Figure 4-6.—Three variables in a series circuit.

In the circuit of figure 4-6 the resistance will remain fixed (constant) and the voltage (independent variable) will be varied. The resulting changes in current (dependent variable) will then be graphed.

To aid in compiling the data, a table of values is completed as shown in figure 4-7. This table shows R to be held constant at 10 ohms as E is varied from 0 to 20 volts in 5-volt steps. Through the use of Ohm's Law the value of current in column two of the table can be calculated for each value of voltage in column one. When the table is complete, the information it contains can be used to construct the graph in figure 4-7. For example, when the voltage applied to the 10-ohm resistor is 10 volts, the current is 1 ampere. These values of current and voltage determine a point on the graph. When all the points have been plotted, a smooth curve is drawn through the points. This curve is called the volt-ampere characteristic for the 10-ohm resistor.

Through the use of this curve the value of current through the resistor can be quickly determined for any value of voltage between 0 and 20 volts.

A very important characteristic of a fixed resistor is illustrated by the graph in figure 4-7. Since the volt-ampere characteristic curve is a straight line, it shows that equal changes of voltage across a resistor produce equal changes in current through the resistor. Because of this straight line characteristic, the fixed resistor is called a linear device.

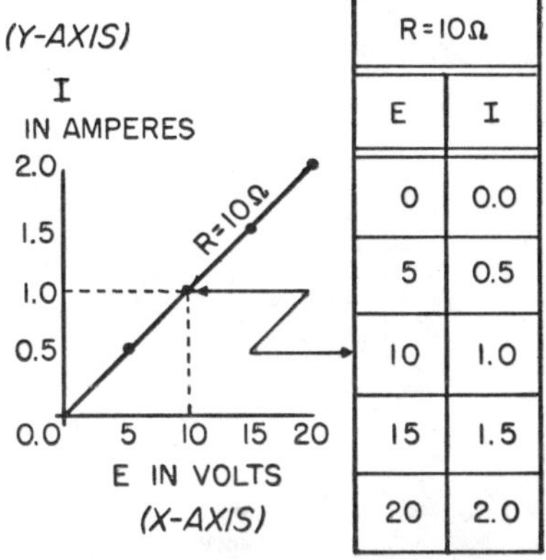

R=10Ω	
E	I
0	0.0
5	0.5
10	1.0
15	1.5
20	2.0

Figure 4-7.—Volt-ampere characteristic.

This graph illustrates an important characteristic of the basic law—namely, the current varies directly with the applied voltage if the resistance is constant.

If the voltage across a load is maintained at a constant value, the current through the load will depend solely upon the effective resistance of the load. For example, with a constant voltage of 12 volts and a resistance of 12 ohms, the current will be $\frac{12}{12}$, or 1 ampere. If the resistance is halved, the current will be doubled; if the resistance is doubled, the current will be halved. In other words, the current will vary inversely with the resistance.

If the resistance of the load is reduced in steps of 2 ohms starting at 12 ohms and continuing to 2 ohms, the current through the load becomes $\frac{12}{10} = 1.2$ amperes; $\frac{12}{8} = 1.5$ amperes; $\frac{12}{6} = 2$ amperes; and so forth. The relation between current and resistance in this example is expressed as a graph (fig. 4-8), whose equation is $I = \frac{12}{R}$. The numerator of the

NOTES:

fraction represents a constant value of 12 volts in this example. As R approaches a small value the current approaches a very large value. The example illustrates a second equally important relation in Ohm's law—namely, that the current varies inversely with the resistance.

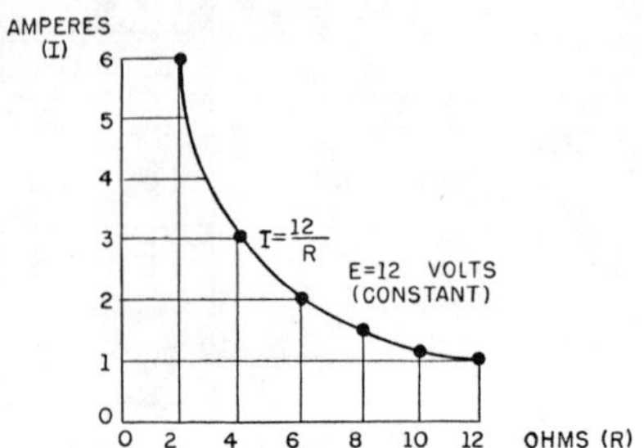

Figure 4-8.—Relation between current and resistance.

If the current through the load is maintained constant at 5 amperes, the voltage across the load will depend upon the resistance of the load and will vary directly with it. The relation between voltage and resistance is shown in the graph of figure 4-9. Values of resistance are plotted horizontally along the X-axis to the right of the origin, and corresponding values of voltage are plotted vertically along the Y-axis above the origin. The graph is a straight line having the equation E = 5R. The coefficient 5 represents the assumed current of 5 amperes which is constant in this example. Thus, a third important relation is illustrated—namely, that the voltage across a device varies directly with the effective resistance of the device provided the current through the device is maintained constant.

APPLYING OHM'S LAW

Equation (4-1) may be transposed to solve for the resistance if the current and voltage are known, or to solve for the voltage if the current and resistance are known. Thus, $R = \frac{E}{I}$ and $E = IR$. For example, if the voltage across a device is 50 volts and the current through it is 2 amperes, the resistance of the device will be

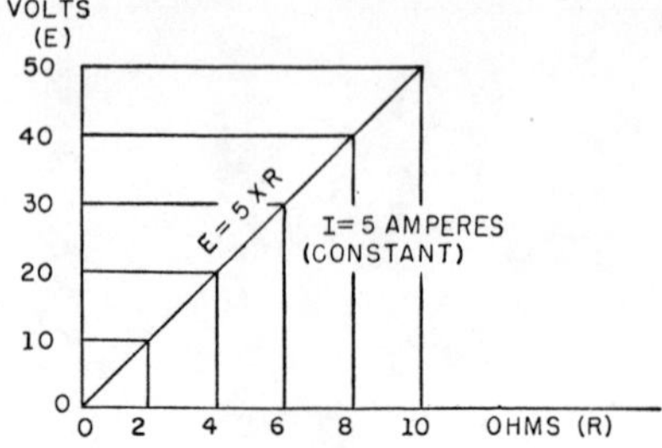

Figure 4-9.—Graph of voltage vs. resistance with constant current.

$\frac{50}{2}$, or 25 ohms. Also, if the current through a wire is 3 amperes and the resistance of the wire is 0.5 ohm, the voltage drop across the wire will be 3 x 0.5, or 1.5 volts.

Equation (4-1) and its transpositions may be obtained readily with the aid of figure 4-10. The circle containing E, I, and R is divided into two parts with E above the line and IR below it. To determine the unknown quantity, first cover that quantity with a finger. The location of the remaining uncovered letters in the circle will indicate the mathematical operation to be performed. For example, to find I, cover I with a finger. The uncovered letters indicate that E is to be divided by R, or $I = \frac{E}{R}$. To find E, cover E. The result indicates that I is to be multiplied by R, or $E = IR$. To find R, cover R. The result indicates that E is to be divided by I, or $R = \frac{E}{I}$.

The beginning student is cautioned not to rely wholly on the use of this diagram when transposing simple formulas but rather to use it to supplement his knowledge of the algebraic method. Algebra is a basic tool in the solution of electrical problems and the importance of knowing how to use it should not be underemphasized or bypassed after the student has learned a shortcut method such as the one indicated in this figure.

NOTES:

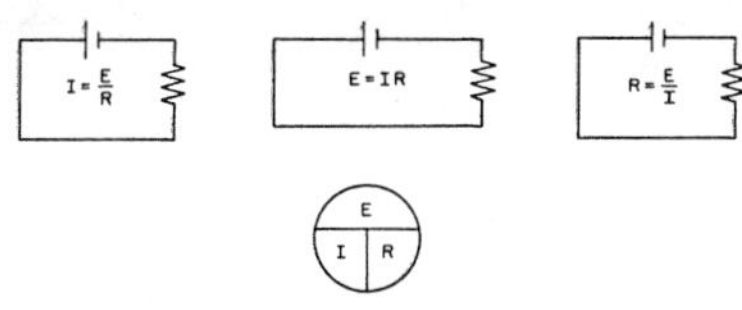

Figure 4-10.—Ohm's law in diagram form.

ELECTRIC POWER AND ENERGY

POWER

Power, whether electrical or mechanical, pertains to the rate at which work is being done. Work is done whenever a force causes motion. If a mechanical force is used to lift or move a weight, work is done. However, force exerted WITHOUT causing motion, such as the force of a compressed spring acting between two fixed objects, does not constitute work.

Previously, it was shown that voltage is electrical force, and that voltage forces current to flow in a closed circuit. However, when voltage exists between two points, but current cannot flow, no work is done. This is similar to the spring under tension that produced no motion. When voltage causes electrons to move, work is done. The instantaneous RATE at which this work is done is called the electric power rate, and its measure is the watt.

A total amount of work may be done in different lengths of time. For example, a given number of electrons may be moved from one point to another in 1 second or in 1 hour, depending on the RATE at which they are moved. In both cases, total work done is the same. However, when the work is done in a short time, the wattage, or INSTANTANEOUS POWER RATE is greater than when the same amount of work is done over a longer period of time.

As stated, the basic unit of power is the WATT, and it is equal to the voltage across a circuit multiplied by current through the circuit. This represents the rate at any given instant at which work is being done in moving electrons through the circuit. The symbol P indicates electrical power. Thus, the basic power formula is P = E x I. E is the voltage across and I is the current through the resistor or circuit whose power is being measured. The amount of power will change when either voltage or current, or both voltage and current change. This relation is shown with the graph and simple circuit in figure 4-11.

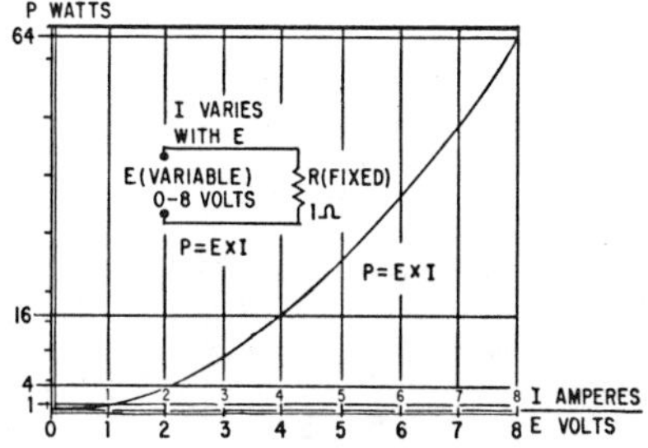

Figure 4-11.—Graph of power related to changing voltage and current.

The resistance is 1 ohm, and does not change. Voltage E is increased in steps of 1 volt from 0 to 8. By applying Ohm's law, the current I is determined for each step of voltage. For instance, when E is 1 volt, the current I is

$$I = \frac{E}{R} \tag{4-1}$$

$$I = \frac{1}{1}$$

$$I = 1 \text{ ampere}$$

Power P, in watts, is determined by applying the basic power formula P = E x I. When E is 1 volt, I is 1 ampere, so P is

$$P = E \times 1$$

$$P = 1 \times 1$$

$$P = 1 \text{ watt}$$

However, when E is 2 volts, and I is 2 amperes, P becomes:

$$P = E \times 1$$

$$P = 2 \times 2$$

$$P = 4 \text{ watts}$$

NOTES:

It is important to note that when voltage E was doubled from 1 volt to 2 volts, power P doubled TWICE from 1 watt to 4 watts. This occurred because the doubling of voltage caused a doubling of current, therefore power was doubled twice. This is shown as follows:

$$P = E \times I$$

$$P = E \times 2 \times I \times 2$$

$$P = (1 \times 2) \times (1 \times 2)$$

$$P = 2 \times 2$$

$$P = 4 \text{ watts.}$$

This shows that the power in a circuit of fixed resistance is caused to change at a SQUARE rate by changes in applied voltage. Thus, the basic power formula $P = E \times I$ may also be $P = \frac{E^2}{R}$. To further illustrate the square-rate relation between power and voltage, note on the graph that power is the square of voltage (when resistance is 1 ohm). For instance, when E is 2 volts, P is 4 watts. When E is doubled to 4 volts, P is 16 watts, and when E is redoubled to 8 volts, P becomes 64 watts. When resistance is any value other than 1 ohm, power will not be the exact square of voltage in quantity but it will still vary at a square rate. That is, regardless of the value of resistance, so long as it is fixed, when voltage doubles, power doubles twice. Also, when the voltage is halved, power is halved twice.

Another important relation may be seen by studying figure 4-11. Thus far power has been calculated with voltage and current ($P = E \times I$), and with voltage and resistance ($P = \frac{E^2}{R}$). Referring to figure 4-11, note that power also varies as the square of current just as it does with voltage. Thus, another formula for power, with current and resistance as its factors, is $P = I^2R$. Note that resistance R is a divisor in one formula ($P = \frac{E^2}{R}$), but is a multiplier in the other ($P = I^2R$). This is true because of substitutions in the original formula $P = E \times I$. That is, the Ohm's law equivalent of I is $\frac{E}{R}$. If this equivalent is substituted for I in the power formula $P = E \times I$, the results are as follows:

$$P = E \times I \quad (4\text{-}4)$$

$$P = E \times \frac{E}{R}$$

$$P = \frac{E^2}{R} \quad (4\text{-}5)$$

In addition, the Ohm's law equivalent of E is I x R. If this equivalent is substituted for E, the power formula becomes

$$P = E \times I$$

$$P = (I \times R) \times I$$

$$P = I^2R \quad (4\text{-}6)$$

In the foregoing discussion, and in figure 4-11, it was shown how variations of the voltage impressed across a fixed resistance caused variations in the circuit current and power. The following discussion refers to figure 4-12. In this circuit, the voltage E is fixed at 10 volts, and the resistance R is the variable factor. (The arrow through the resistance means it is variable).

When the resistance R is set at 1 ohm, the current I is 10 amperes, and the power is

$$P = I^2R \quad (4\text{-}6)$$

$$P = (10^2) \times R$$

$$P = 100 \times 1$$

$$P = 100 \text{ watts}$$

When the resistance is doubled to 2 ohms, this same calculation will show that power is halved to 50 watts, as the graph shows. Subsequent redoubling of the resistance to 4 and then to 8 ohms causes the power to be halved each time to 25 and 12-1/2 watts, respectively. Conversely, you should note the relation when starting with resistance at 10 ohms, with 10 watts of power. If resistance is halved to 5 ohms, power is doubled to 20 watts.

In figures 4-11 and 4-12, current and power were caused to vary as a function of voltage, in one case, and of resistance in the other. In figure 4-13, however, current is held constant.

NOTES:

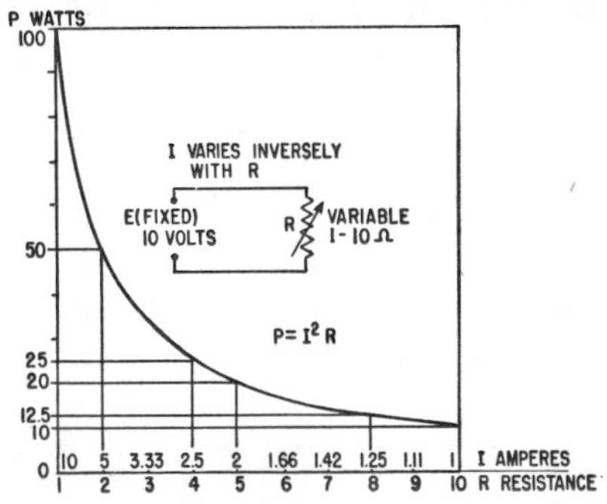

Figure 4-12.—Graph of power related to changing resistance and current.

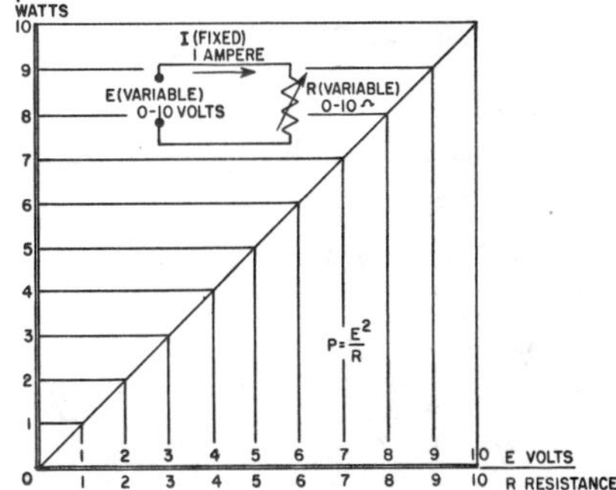

Figure 4-13.—Graph of power related to changing voltage and resistance.

This is done by raising or lowering voltage and resistance equally and directly with each other. The resulting variations in power are linear with those changes. That is, power is changed step-for-step with voltage. The changing resistance maintains a constant current despite changes of voltage. At any point on the graph, voltage divided by resistance is 1 ampere. Had the current been allowed to vary, power would have changed at a curved, or square rate, instead of linearly as on the graph.

Up to this point, four of the most important basic electrical quantities have been discussed. These are E, I, R, and P. It is of fundamental importance that you thoroughly understand the interrelation of these quantities. You should understand how any one of these quantities either controls or is controlled by the others in an electrical circuit. These relations are further explained in the treatment that follows. You should compare each statement carefully with its associated formula. Check each formula for correctness by applying it to the graphs in figures 4-11, 4-12, or 4-13. (The appropriate figure number is indicated after each of the following statements)

1. Power, as related to E and I: $P = EI$ (fig. 4-11).

   This formula states that P is the product of E multiplied by I, regardless of their individual values. If either E or I varies, P varies proportionally. If both E and I vary, P varies at a square rate.

2. Power, as related to I and R: $P = I^2R$ (fig. 4-12).

   This formula states that if R is held constant, and I is varied, P varies as the square of I, because I appears as a squared quantity (I^2) in the formula. Also, if I is held constant and R is varied, P varies directly and proportionally to R, because R is a multiplier in the formula.

3. Power, as related to E and R: $P=\frac{E^2}{R}$ (fig. 4-13).

   This formula states that if R is held constant as E is varied, P varies as the square of E, because E appears as a squared quantity (E^2) in the formula. Also, if E is held constant and R is varied, P varies inversely but proportionally to R, because R is a divisor in the formula.

In the preceding paragraph, P was expressed in terms of alternate pairs of the other three basic quantities E, I, and R. In practice, you should be able to express any one of the three basic quantities, as well as P, in terms of any two of the others. Figure 4-14 is a summary of twelve basic formulas you should know. The four quantities E, I, R, and P are at the center of the figure.

NOTES:

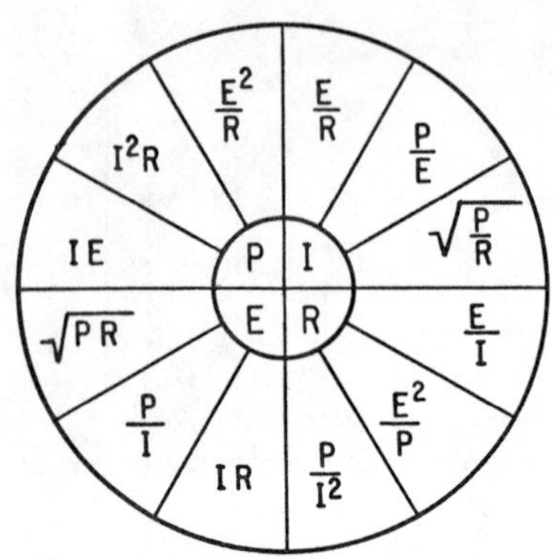

Figure 4-14.—Summary of basic formulas.

Adjacent to each quantity are three segments. Note that in each segment, the basic quantity is expressed in terms of two other basic quantities, and no two segments are alike.

RATING OF ELECTRICAL DEVICES BY POWER

Electrical lamps, soldering irons, and motors are examples of electrical devices that are rated in watts. The wattage rating of a device indicates the rate at which the device converts electrical energy into another form of energy, such as light, heat, or motion.

For example, a 100-watt lamp will produce a brighter light than a 75-watt lamp, because it converts more electrical energy into light.

Electric soldering irons are of various wattage ratings, with the high-wattage irons changing more electrical energy to heat than those of low-wattage ratings.

If the normal wattage rating is exceeded, the equipment or device will overheat and probably be damaged. For example, if a lamp is rated 100 watts at 110 volts and is connected to a source of 220 volts, the current through the lamp will double. This will cause the lamp to use four times the wattage for which it is rated, and it will burn out quickly.

NOTES:

POWER CAPACITY OF ELECTRICAL DEVICES

Rather than indicate a device's ability to do work, its wattage rating may indicate the device's operating limits. These power limits generally are given as the maximum or minimum safe voltages and currents to which a device may be subjected. However, in cases where a device is not limited to any specific operating voltage, its limits are given directly in watts.

Resistors

A resistor is an example of such a device. It may be used in circuits with widely different voltages, depending on the desired current. However, the resistor has a maximum current limitation for each voltage applied to it. The product of the resistor's voltage and current at any time must not exceed a certain wattage.

Thus, resistors are rated in watts, in addition to their ohmic resistance. Resistors of the same resistance value are available in different wattage values. Carbon resistors, for example, are commonly made in wattage ratings of 1/3, 1/2, 1, and 2 watts. (See fig. 4-15.) The larger the physical size of a carbon resistor, the higher its wattage rating, since a larger amount of material will absorb and give up heat more easily.

When resistors of wattage ratings greater than 2 watts are needed, wire-wound resistors

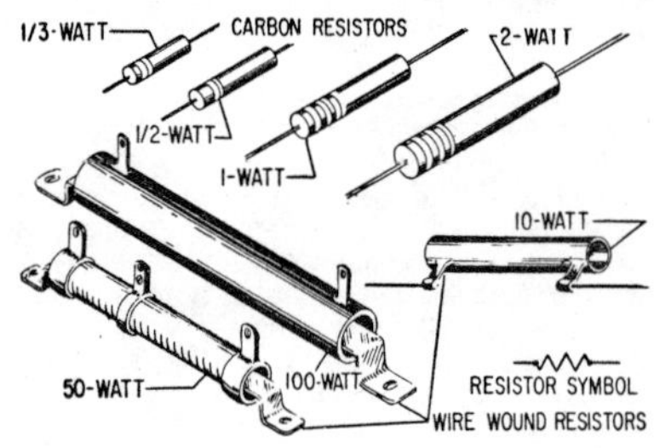

Figure 4-15.—Resistors of different wattage ratings.

are used. Such resistors are made in ranges between 5 and 200 watts, with special types being used for power in excess of 200 watts.

Fuses

When current passes through a resistor, electric energy is transformed into heat, which raises the temperature of the resistor. If the temperature becomes too high, the resistor may be damaged. The metal wire in a wound resistor may melt, opening the circuit and interrupting current flow. This effect is used to an advantage in fuses.

Fuses are actually metal resistors with very low resistance values. They are designed to "blow out" and thus open a circuit when current exceeds the fuse's rated value.

When the power consumed by the fuse raises the temperature of its metal too high, the metal melts, or "blows." In service, fuses are generally connected as shown in figure 4-16.

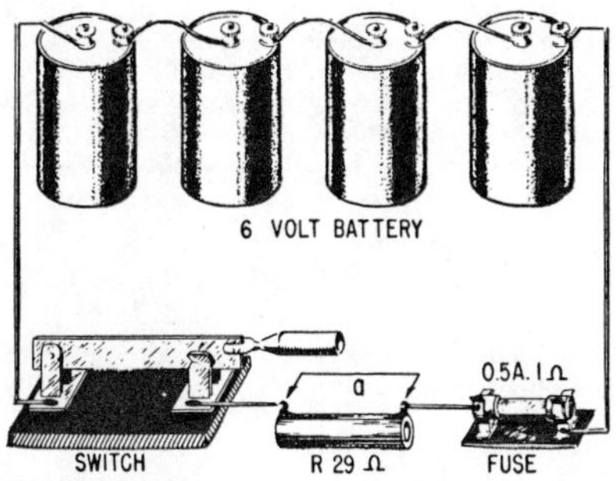

Figure 4-16.—Simple fused circuit.

Note that all current flowing through the load resistance of 29 ohms must also flow through the half-ampere fuse. Under normal conditions, the total resistance would be 29 + 1 = 30 ohms. If the switch were closed, the current I would be

$$I = \frac{E}{R}$$

$$I = \frac{6}{30}$$

$$I = 0.2 \text{ ampere}$$

This would be less than the rated current of the fuse, and it would not open. However, if the short conductor (a, fig. 4-16) were connected, the load resistance would be bypassed, or "shorted out." Only the fuse resistance of one ohm would remain in the circuit. Current would then be

$$I = \frac{E}{R}$$

$$I = \frac{6}{1}$$

$$I = 6 \text{ amperes}$$

Six amperes of current will cause the half-ampere fuse to open very quickly, because its wattage rating is greatly exceeded.

There are a great number of different types and sizes of fuses presently in use. Figure 4-17 shows three of the most common types. Fuses are discussed in greater detail in chapter 14 of this manual.

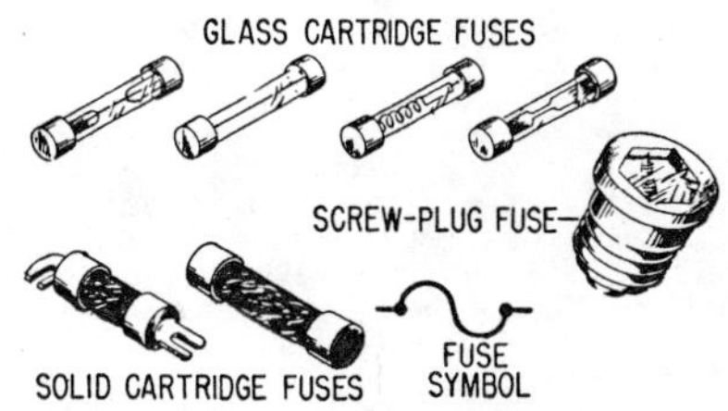

Figure 4-17.—Typical fuse types.

ENERGY

Energy is defined as the ability to do work. Energy is expended when work is done, because it takes energy to maintain a force when that force acts through a distance. The total energy expended to do a certain amount of work is

NOTES:

equal to the working force multiplied by the distance through which the force moved to do the work. This is the mechanical definition.

In electricity, total energy expended is equal to the rate at which work is done, multiplied by the length of time the rate is measured. Essentially, energy W is equal to power P times time t.

An equation for energy is derived by multiplying both sides of equation (4-4) by the common factor of time, t, and equating the expression to the energy, W, as

$$W = Pt$$

$$W = EIt. \qquad (4\text{-}7)$$

Similarly, both sides of equations (4-5) and (4-6) may be multiplied by the time factor, t, and equated to the energy, W, as

$$W = \frac{E^2}{R}\,t, \qquad (4\text{-}8)$$

and

$$W = I^2Rt. \qquad (4\text{-}9)$$

In the energy equations (4-7), (4-8), and (4-9), E is in volts and I in amperes. If t is expressed in hours, W will be in watt-hours.

If t is expressed in seconds, W will be in watt-seconds or joules (1 joule is equal to 1 watt-second). Since Q = It (where Q is in coulombs, I in amperes, and t in seconds), it is possible to substitute Q for It in equation (4-7) with the resulting expression for energy. Thus

$$W = QE \qquad (4\text{-}10)$$

where W is the energy in joules or watt-seconds, Q is the quantity in coulombs, and E is in volts. (As explained in chapter 2, Q is the symbol for coulombs. This is the measure of a quantity of electrons. The Q is to electricity as the gallon is to water.)

Electrical energy is bought and sold in units of kilowatt-hours ($3{,}600 \times 10^3$ joules), and is totalized in large central generating stations in terms of megawatt-hours ($3{,}600 \times 10^6$ joules). For example, if the average demand over a 10-hour period is 70 megawatts, the total energy delivered is 70 x 10, or 700 megawatt-hours. This amount of energy is equivalent to 700 x 1,000 = 700,000 kilowatt-hours, or $700 \times 3{,}600 \times 10^6 = 2{,}520{,}000 \times 10^6$ joules. The most practical unit to use depends in part upon the magnitude of the quantity of energy involved, and in this example the megawatt-hour is appropriate.

SERIES CIRCUIT CHARACTERISTICS

As previously mentioned, an electric circuit is a complete path through which electrons can flow from the negative terminal of the voltage source, through the connecting wires or conductors, through the load or loads, and back to the positive terminal of the voltage source. A circuit is thus made up of a voltage source, the necessary connecting conductors, and the effective load.

If the circuit is arranged so that the electrons have only ONE possible path, the circuit is called a SERIES CIRCUIT. Therefore, a series circuit is defined as a circuit that contains only one path for current flow. Figure 4-18 shows a series circuit having several lamps.

RESISTANCE

Referring to figure 4-18, the current in a series circuit, in completing its electrical path, must flow through each lamp inserted into the circuit. Thus, each additional lamp offers added resistance. In a series circuit, THE TOTAL CIRCUIT RESISTANCE (R_T) IS EQUAL TO THE SUM OF THE INDIVIDUAL RESISTANCES.

As an equation:

$$R_T = R_1 + R_2 + R_3 \cdots R_n \qquad (4\text{-}11)$$

NOTE: The subscript n denotes any number of additional resistances that might be in the equation.

Example: Three resistors of 10 ohms, 15 ohms, and 30 ohms are connected in series across a battery whose emf is 110 volts (fig. 4-19). What is the total resistance?

Given:

R_1 = 10 ohms

R_2 = 15 ohms

R_3 = 30 ohms

R_T = ?

NOTES:

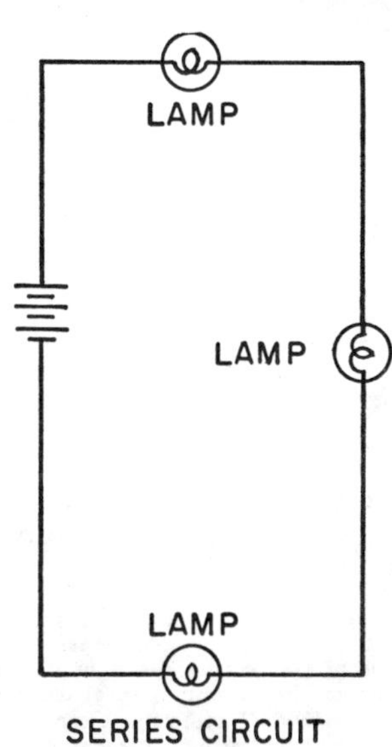

Figure 4-18.—Series circuit.

Solution:

$$R_T = R_1 + R_2 + R_3$$

$$R_T = 10 + 15 + 30$$

$$R_T = 55 \text{ ohms}$$

In some circuit applications, the total resistance is known and the value of a circuit resistor has to be determined. Equation (4-11) can be transposed to solve for the value of the unknown resistance.

Example: The total resistance of a circuit containing three resistors is 40 ohms (fig. 4-20). Two of the circuit resistors are 10 ohms each. Calculate the value of the third resistor.

NOTES:

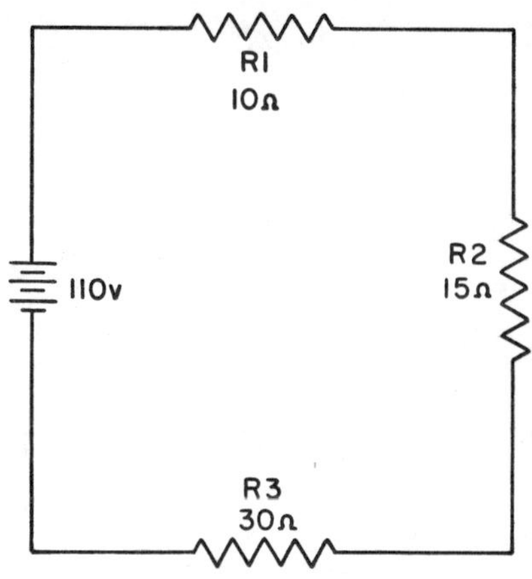

Figure 4-19.—Solving for total resistance in a series circuit.

Given:

$R_T = 40$ ohms

$R_1 = 10$ ohms

$R_2 = 10$ ohms

$R_3 = ?$

Solution: $R_T = R_1 + R_2 + R_3$ (4-11)

Subtracting $(R_1 + R_2)$ from both sides of the equation

$$R_3 = R_T - R_1 - R_2$$

$$R_3 = 40 - 10 - 10$$

$$R_3 = 40 - 20$$

$$R_3 = 20 \text{ ohms}$$

CURRENT

Since there is but one path for current in a series circuit, the same current must flow

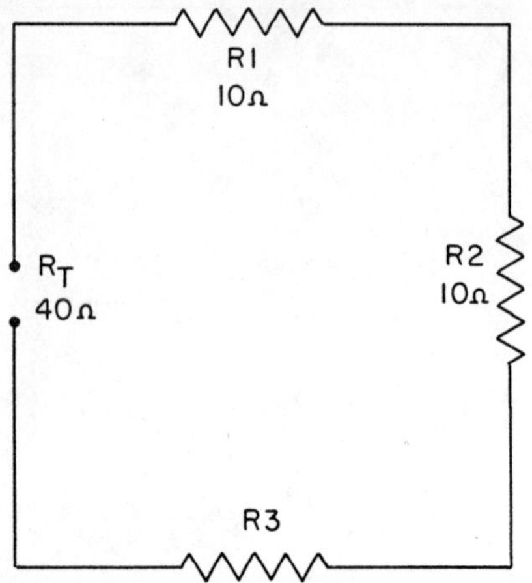

Figure 4-20.—Calculating the value of one resistance in a series circuit.

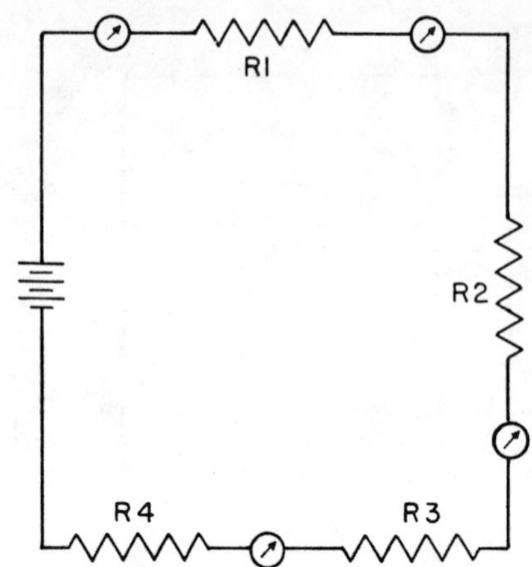

Figure 4-21.—Current in a series circuit.

through each part of the circuit. To determine the current throughout a series circuit, only the current through one of the parts need be known.

The fact that the same current flows through each part of a series circuit can be verified by inserting ammeters into the circuit at various points as shown in figure 4-21. If this were done, each meter would be found to indicate the same value of current.

VOLTAGE

As stated previously, the voltage drop across the resistor in the basic circuit is the total voltage across the circuit and is equal to the applied voltage. The total voltage across a series circuit is also equal to the applied voltage, but consists of the sum of two or more individual voltage drops. In any series circuit the SUM of the resistor voltage drops must equal the source voltage. This statement can be proven by an examination of the circuit shown in figure 4-22. In this circuit a source potential (E_T) of 20 volts is impressed across a series circuit consisting of two 5 ohm resistors. The total resistance of the circuit is equal to the sum of the two individual resistances, or 10 ohms. Using Ohm's Law the circuit current may be calculated as follows:

$$I = \frac{E_T}{R_T}$$

$$I = \frac{20}{10}$$

$$I = 2 \text{ amperes}$$

Knowing the size of the resistors to be 5 ohms each, and the current through the resistors to be 2 amperes, the voltage drops across the resistors can be calculated. The voltage (E_1) across R_1 is therefore:

$$E_1 = IR_1$$

$$E_1 = 2 \text{ amperes} \times 5 \text{ ohms}$$

$$E_1 = 10 \text{ volts}$$

Since R_2 is the same ohmic value as R_1 and carries the same current, the voltage drop

NOTES:

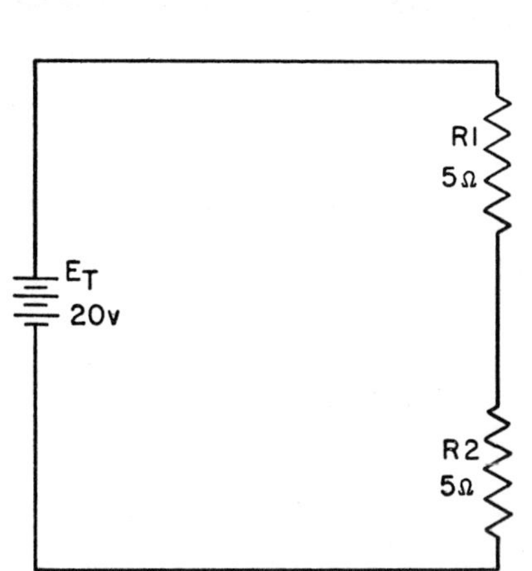

Figure 4-22.—Calculating total resistance in a series circuit.

across R_2 is also equal to 10 volts. Adding these two 10 volt drops together gives a total drop of 20 volts exactly equal to the applied voltage. For a series circuit then:

$$E_T = E_1 + E_2 + E_3 \ldots E_n \quad (4\text{-}12)$$

Example: A series circuit consists of three resistors having values of 20 ohms, 30 ohms, and 50 ohms respectively. Find the applied voltage if the current through the 30 ohm resistor is 2 amperes.

To solve the problem, a circuit diagram is first drawn and labeled as shown in figure 4-23.

Given: R_1 = 20 ohms

R_2 = 30 ohms

R_3 = 50 ohms

I = 2 amperes

Solution: Since the circuit involved is a series circuit, the same 2 amperes of current flows

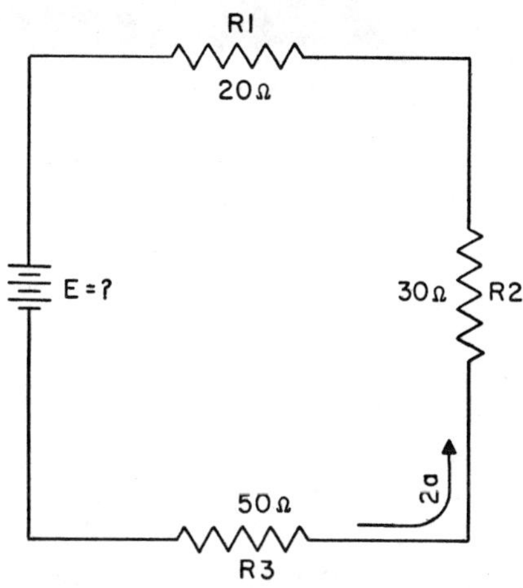

Figure 4-23.—Solving for applied voltage in a series circuit.

through each resistor. Using Ohm's Law, the voltage drops across each of the three resistors can be calculated and are:

E_1 = 40 volts

E_2 = 60 volts

E_3 = 100 volts

Once the individual drops are known they can be added to find the total or applied voltage:

$$E_T = E_1 + E_2 + E_3 \quad (4\text{-}12)$$

$$E_T = 40v + 60v + 100v$$

$$E_T = 200 \text{ volts}$$

NOTE: In using Ohm's Law, the quantities used in the equation MUST be taken from the SAME part of the circuit. In the above example the voltage across R_2 was computed using the current through R_2 and the resistance of R_2.

It must be emphasized that the potential difference across a resistor remains constant, for it is a measure of the amount of energy

NOTES:

required to move a unit charge from one point to another. As long as the source produces electric energy as rapidly as it is consumed in a resistance, the potential difference across the resistance will remain at a constant voltage. The value of this voltage is determined by the applied voltage and the proportional relationship of circuit resistances. The voltage drops that occur in a series circuit are in direct proportions to the resistance across which they appear. This is a result of having the same current flow through each resistor. Thus, the larger the resistor the larger will be the voltage drop across it.

POWER

Each of the resistors in a series circuit consumes power which is dissipated in the form of heat. Since this power must come from the source, the total power must be equal in amount to the power consumed by the circuit resistances. In a series circuit the total power is equal to the SUM of the powers dissipated by the individual resistors. Total power (P_T) is thus equal to:

$$P_T = P_1 + P_2 + P_3 + \ldots P_n \qquad (4\text{-}13)$$

Example: A series circuit consists of three resistors having values of 5 ohms, 10 ohms, and 15 ohms. Find the total power dissipation when 120 volts is applied to the circuit. (See figure 4-24.)

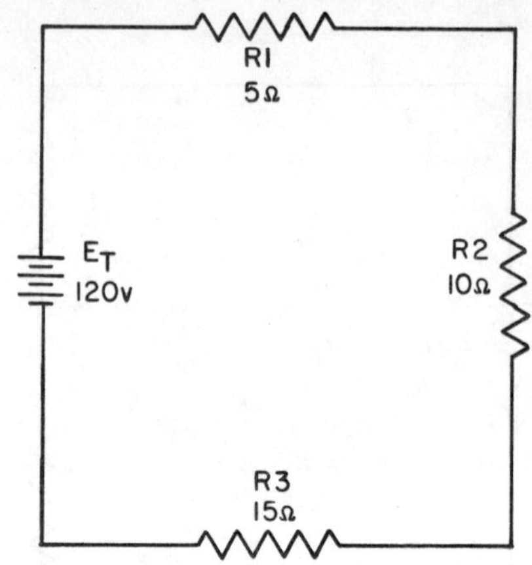

Figure 4-24.—Solving for total power in a series circuit.

given:

R_1 = 5 ohms

R_2 = 10 ohms

R_3 = 15 ohms

E = 120 volts

Solution: The total resistance is found first.

$$R_T = R_1 + R_2 + R_3 \qquad (4\text{-}11)$$

$$R_T = 5 + 10 + 15$$

$$R_T = 30 \text{ ohms}$$

Using total resistance and the applied voltage, the circuit current is calculated.

$$I = \frac{E_T}{R_T}$$

$$I = \frac{120}{30}$$

$$I = 4 \text{ amperes}$$

Using the power formulas, the individual power dissipations can be calculated. For resistor R_1:

$$P_1 = I^2R_1 \qquad (4\text{-}6)$$

$$P_1 = (4)^2 5$$

$$P_1 = 80 \text{ watts}$$

For R_2.

$$P_2 = I^2R_2 \qquad (4\text{-}6)$$

$$P_2 = (4)^2 10$$

$$P_2 = 160 \text{ watts}$$

NOTES:

For R_3:

$$P_3 = I^2R_3 \qquad (4\text{-}6)$$

$$P_3 = (4)^2 15$$

$$P_3 = 240 \text{ watts}$$

To obtain total power:

$$P_T = P_1 + P_2 + P_3 \qquad (4\text{-}13)$$

$$P_T = 80 + 160 + 240$$

$$P_T = 480 \text{ watts}$$

To check the answer the total power delivered by the source can be calculated:

$$P_{source} = I_{source} \times E_{source} \qquad (4\text{-}4)$$

$$P_{source} = 4 \text{ a.} \times 120 \text{ v.}$$

$$P_{source} = 480 \text{ watts}$$

Thus the total power is equal to the sum of the individual power dissipations.

RULES FOR SERIES D-C CIRCUITS

The important factors governing the operation of a series circuit are listed below. These factors have been set up as a group of rules so that they may be easily studied. These rules must be completely understood before the study of more advanced circuit theory is undertaken.

1. The same current flows through each part of a series circuit.
2. The total resistance of a series circuit is equal to the sum of the individual resistances.
3. The total voltage across a series circuit is equal to the sum of the individual voltage drops.
4. The voltage drop across a resistor in a series circuit is proportional to the size of the resistor.
5. The total power dissipated in a series circuit is equal to the sum of the individual power dissipations.

NOTES:

GENERAL CIRCUIT ANALYSIS

To establish a procedure for solving series circuits, the following sample problem will be solved.

Example: Three resistors of 5 ohms, 10 ohms, and 15 ohms are connected across a battery rated at 90 volts terminal voltage. Completely solve the circuit (fig. 4-25).

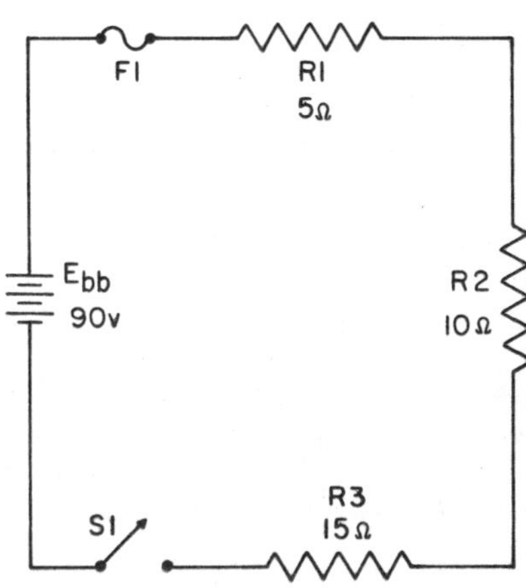

Figure 4-25.—Solving for various values in a series circuit.

In solving the circuit the total resistance will be found first. Next, the circuit current will be calculated. Once the current is known the voltage drops and power dissipations can be calculated.

The total resistance is:

$$R_T = R_1 + R_2 + R_3$$

$$R_T = 5 \text{ ohms} + 10 \text{ ohms} + 15 \text{ ohms}$$

$$R_T = 30 \text{ ohms}$$

By Ohm's Law the current is:

$$I = \frac{E_{bb}}{R_T}$$

$$I = \frac{90}{30}$$

$$I = 3 \text{ amperes}$$

The voltage (E_1) across R_1 is:

$$E_1 = IR_1$$

$$E_1 = 3 \text{ amperes} \times 5 \text{ ohms}$$

$$E_1 = 15 \text{ volts}$$

The voltage (E_2) across R_2 is:

$$E_2 = IR_2$$

$$E_2 = 3 \text{ amperes} \times 10 \text{ ohms}$$

$$E_2 = 30 \text{ volts}$$

The voltage (E_3) across R_3 is:

$$E_3 = IR_3$$

$$E_3 = 3 \text{ amperes} \times 15 \text{ ohms}$$

$$E_3 = 45 \text{ volts}$$

The power dissipated in R_1 is:

$$P_1 = I \times E_1$$

$$P_1 = 3 \text{ amperes} \times 15 \text{ volts}$$

$$P_1 = 45 \text{ watts}$$

The power dissipated in R_2 is:

$$P_2 = I \times E_2$$

$$P_2 = 3 \text{ amperes} \times 30 \text{ volts}$$

$$P_2 = 90 \text{ watts}$$

NOTES:

The power dissipated in R_3 is:

$$P_3 = I \times E_3$$

$$P_3 = 3 \text{ amperes} \times 45 \text{ volts}$$

$$P_3 = 135 \text{ watts}$$

The total power dissipated is:

$$P_T = E_T \times I$$

$$P_T = 90 \text{ volts} \times 3 \text{ amperes}$$

$$P_T = 270 \text{ watts}$$

Example: Four resistors $R_1 = 10$ ohms, $R_2 = 10$ ohms, $R_3 = 50$ ohms, and $R_4 = 30$ ohms are connected in series across a battery. The current through the circuit is 0.5 amperes. (See figure 4-26.)

a. What is the battery voltage?
b. What is the voltage across each resistor?
c. What is the power expended in each resistor?
d. What is the total power?

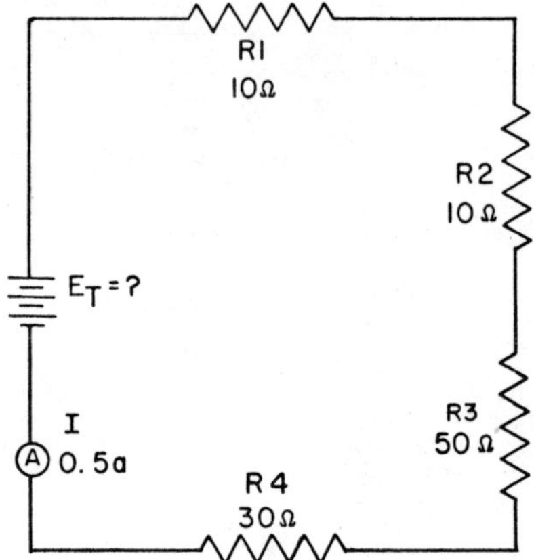

Figure 4-26.—Computing series circuit values.

Given:

R_1 = 10 ohms

R_2 = 10 ohms

R_3 = 50 ohms

R_4 = 30 ohms

I = 0.5 ampere

Find:

$E_1 = ?$ $P_1 = ?$

$E_2 = ?$ $P_2 = ?$

$E_3 = ?$ $P_3 = ?$

$E_4 = ?$ $P_4 = ?$

$E_T = ?$ $P_T = ?$

Solution:

(a) $R_T = R_1 + R_2 + R_3 + R_4$

$R_T = 10 + 10 + 50 + 30 = 100$ ohms

$E_T = IR_T = 0.5 \times 100 = 50$ volts

(b) $E_1 = IR_1 = 0.5 \times 10 = 5$ volts

$E_2 = IR_2 = 0.5 \times 10 = 5$ volts

$E_3 = IR_3 = 0.5 \times 50 = 25$ volts

$E_4 = IR_4 = 0.5 \times 30 = 15$ volts

Check:

$E_T = E_1 + E_2 + E_3 + E_4$

$E_T = 5 + 5 + 25 + 15 = 50$ volts

(c) Power consumed in R_1 is:

$P_1 = IE_1 = 0.5 \times 5 = 2.5$ watts

$P_2 = IE_2 = 0.5 \times 5 = 2.5$ watts

$P_3 = IE_3 = 0.5 \times 25 = 12.5$ watts

$P_4 = IE4 = 0.5 \times 15 = 7.5$ watts

(d) Total power

$P_T = P_1 + P_2 + P_3 + P_4$

$P_T = 2.5 + 2.5 + 12.5 + 7.5 = 25$ watts

Check:

$P_T = I_T{}^2R_T = 0.5^2 \times 100 = 25$ watts

or:

$P_T = I_TE_T = 0.5 \times 50 = 25$ watts

or:

$$P_T = \frac{E_T{}^2}{R_T} = \frac{50^2}{100} = \frac{2500}{100} = 25 \text{ watts}$$

An important fact to keep in mind when applying Ohm's Law to a series circuit is to consider whether the values used are component values or total values. When the information available enables the use of Ohm's Law to find total resistance, total voltage and total current, total values must be inserted into the formula.

To find total resistance:

$$R_T = \frac{E_T}{I_T}$$

To find total voltage:

$$E_T = I_T \times R_T$$

To find total current:

$$I_T = \frac{E_T}{R_T}$$

NOTE: I_T is equal to I in a series circuit. However, the distinction between I_T and I in the formula should be noted. The reason being that future circuits may have several currents, and it will be necessary to differentiate between I_T and other currents.

To compute any quantity (E, I, R, or P) associated with a single given resistor, the values used in the formula must be obtained from that

NOTES:

particular resistor. For exampler, to find the value of an unknown resistance, the voltage across and the current through that particular resistor must be used.

To find the value of a resistor:

$$R = \frac{E_R}{I_R}$$

To find the voltage drop across a resistor:

$$E_R = I_R \times R$$

To find current through a resistor:

$$I_R = \frac{E_R}{R}$$

KIRCHHOFF'S VOLTAGE LAW

In 1847 Kirchhoff extended the use of Ohm's Law by developing a simple concept concerning the voltages contained in a series circuit loop. Kirchhoff's Law is stated as follows.

> The algebraic sum of the instantaneous emf's and voltage drops around any closed circuit loop is zero.

Through the use of Kirchhoff's Law, circuit problems can be solved which would be difficult and often impossible with only a knowledge of Ohm's Law. When the law is properly applied, an equation can be set up for a closed loop and the unknown circuit values may be calculated.

POLARITY OF VOLTAGE

To apply Kirchhoff's Voltage Law, the meaning of voltage POLARITY must be understood. In the circuit shown in figure 4-27 the current is seen to be flowing in a counterclockwise direction due to the arrangement of the battery source E_{bb}. Notice that the end of resistor R_1 into which the current flows is marked NEGATIVE (-). The end of R_1 at which the current leaves is marked POSITIVE (+). These polarity markings are used to show that the end of R_1 into which the current flows is at a higher negative potential than is the end of the resistor at which the current leaves. Point A is thus more negative than point B.

Point C, which is at the same potential as point B, is labeled negative. This is to indicate that point C, though positive with respect to point A, is more negative than point D. To say a point is positive (or negative), without stating what it is positive IN RESPECT TO, has no meaning.

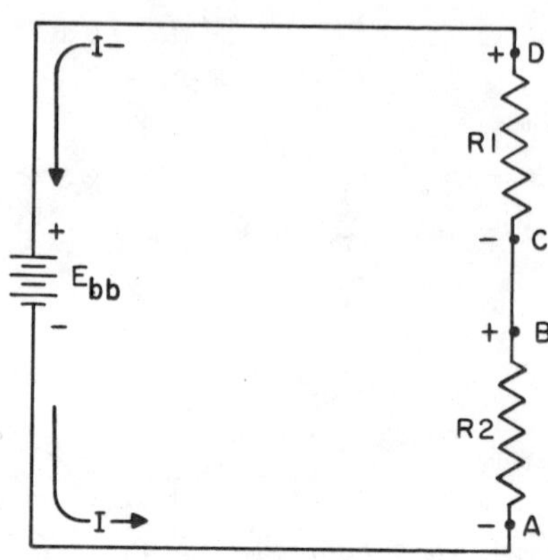

Figure 4-27.—Voltage polarities.

Kirchhoff's Voltage Law can be written as an equation as shown below:

$$E_a + E_b + E_c + \ldots E_n = 0 \qquad (4\text{-}14)$$

where E_a, E_b, etc., are the voltage drops and emf's around any closed circuit loop. To set up the equation for an actual circuit, the following procedure is used.

1. Assume a direction of current through the circuit. (Correct direction desirable but not necessary.)
2. Using assumed direction of current, assign polarities to all resistors through which the current flows.
3. Place correct polarities on any source included in the circuit.
4. Starting at any points in the circuit, trace around the circuit writing down the magnitude and polarity of the voltage across each component in succession. The polarity used is the sign AFTER the component is passed through. Stop

NOTES:

when reaching the point at which the trace was started.

5. Place these voltages with their polarities into equation (4-14) and solve for the desired quantity.

Example: Three resistors are connected across a 50 volt source. What is the voltage across the third resistor if the voltage drops across the first two resistors are 25 volts and 15 volts?

Solution: A diagram is first drawn as shown in figure 4-28. Next a direction of current is assumed as shown. Using this current, the polarity markings are placed at each end of each resistor and also on the terminals of the source. Starting at point A, trace around the circuit in the direction of current flow recording the voltage and polarity of each component. Starting at point A these voltages would be as follows:

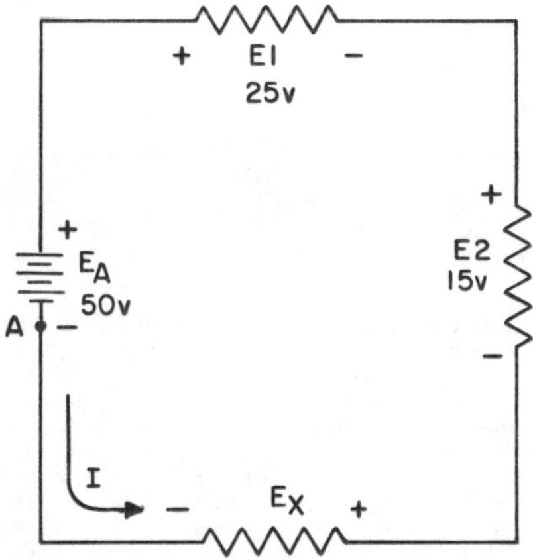

Figure 4-28.—Determining unknown voltage in a series circuit.

Basic formula:

$$E_a + E_b + E_c \ldots E_n = 0 \qquad (4\text{-}14)$$

From the circuit:

$$(+E_x) + (+E_2) + (+E_1) + (-E_A) = 0$$

Substituting values from circuit:

$$E_x + 15 + 25 + 50 = 0$$

$$E_x - 10 = 0$$

$$E_x = 10 \text{ volts}$$

Thus, the unknown voltage (E_x) is found to be 10 volts.

Using the same idea as above, a problem can be solved in which the current is the unknown quantity.

Example: A circuit having a source voltage of 60 volts contains three resistors of 5 ohms, 10 ohms, and 15 ohms. Find the circuit current.

Solution: Draw and label the circuit (fig. 4-29). Establish a direction of current flow and assign polarities. Next, starting at any point, point (A) will be chosen in this example; write out the loop equation.

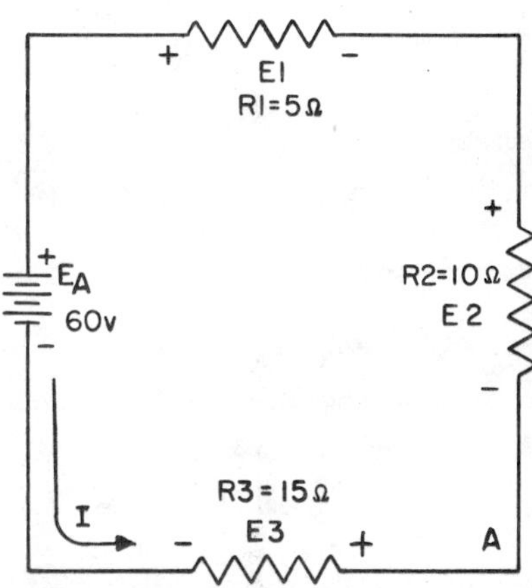

Figure 4-29.—Correct direction of assumed current.

Basic equation:

$$E_a + E_b + E_c + \ldots E_n = 0 \qquad (4\text{-}14)$$

$$+E_2 + E_1 - E_A + E_3 = 0$$

NOTES:

77

Since E = IR, by substitution:

$$IR_2 + IR_1 - E_A + IR_3 = 0$$

Substituting values:

$$10I = 5I - 60 + 15I = 0$$

Combining like terms:

$$30I - 60 = 0$$

$$30I = 60$$

$$I = 2 \text{ amperes}$$

Since the current obtained in the above calculations is a positive 2 amperes, the assumed direction of current was correct. To show what happens if the incorrect direction of current is assumed, the problem will be solved as before but with the opposite direction of current. The circuit is redrawn showing the new direction of current and new polarities in figure 4-30.

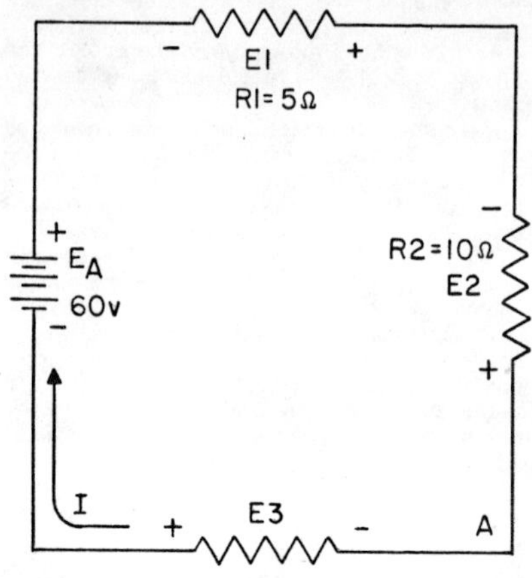

Figure 4-30.—Incorrect direction of assumed current.

Solution:

$$E_a + E_b + E_c + \ldots E_n = 0 \qquad (4\text{-}14)$$

Starting at point (A):

$$-E_2 - E_1 - E_A - E_3 = 0$$

$$-IR_2 - IR_1 - E_A - IR_3 = 0$$

$$10I - 5I - 60 - 15I = 0$$

$$-30I - 60 = 0$$

$$-30I = 60$$

$$I = -2 \text{ amperes}$$

Notice that the AMOUNT of current is the SAME as before. Its polarity, however, is NEGATIVE. The negative polarity simply indicates the wrong direction of current was assumed. Should it be necessary to use this current in further calculations on the circuit, the negative polarity should be retained in the calculations.

NOTES:

SERIES AIDING AND OPPOSING SOURCES

In many practical applications a circuit may contain more than one source. Sources of emf that cause current to flow in the same direction are considered to be SERIES AIDING and their voltages add. Sources of emf that would tend to force current in opposite directions are said to be SERIES OPPOSING, and the effective source voltage is the difference between the opposing voltages. When two opposing sources are inserted into a circuit, current flow would be in a direction determined by the larger source. Examples of series aiding and opposing sources are shown in figure 4-31.

MULTIPLE SOURCE SOLUTIONS

A simple solution may be obtained for a multiple source circuit through the use of Kirchhoff's Voltage Law. In applying this method, the exact same procedure is used for the multiple source as was used above for the single source circuit. This is demonstrated by the following problem.

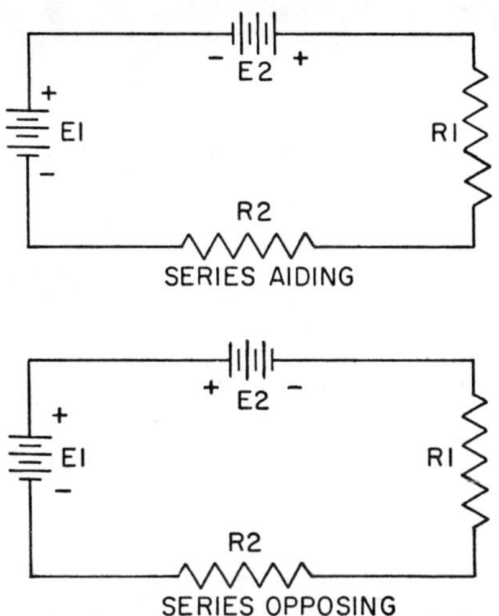

Figure 4-31. —Aiding and opposing sources.

Example: Using Kirchhoff's Voltage equation, find the amount of current in the circuit shown in figure 4-32.

Solution: As before, a direction of current flow is assumed and polarity signs are placed on the drawing. The loop equation will be starting at point A.

Basic equation:

$$E_a + E_b + E_c + \ldots . E_n = 0 \qquad (4\text{-}14)$$

From the circuit:

$$E_{bb2} + E_1 - E_{bb1} + E_{bb3} + E_2 = 0$$

$$+20 + 60I - 180 + 40 + 20I = 0$$

Combining like terms:

$$+80I - 120 = 0$$

$$80I = 120$$

$$I = 1.5 \text{ amperes}$$

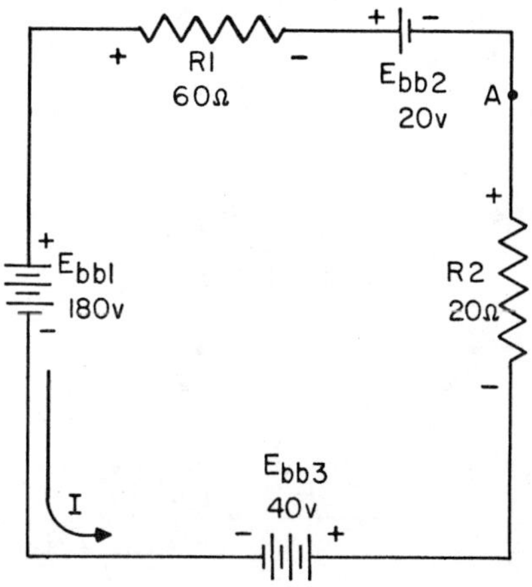

Figure 4-32.—Solving for circuit current using Kirchhoff's voltage equation.

VOLTAGE REFERENCES

REFERENCE POINT

A reference point is an arbitrarily chosen point to which all other points are compared. In series circuits, any point can be chosen as a reference and the electrical potential at all other points can be determined in reference to the initial point. In the example of figure 4-33 point A shall be considered as the reference. Each series resistor in the illustrated circuit is of equal value; therefore, the applied voltage is equally distributed across each resistor. The potential at point B is 25 volts more positive than A. Points C and D are 50 volts and 75 volts respectively more positive than point A.

NOTES:

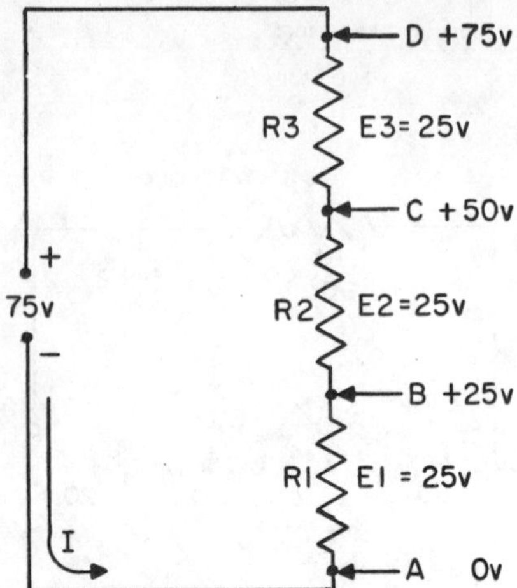

Figure 4-33.—Reference points in a series circuit.

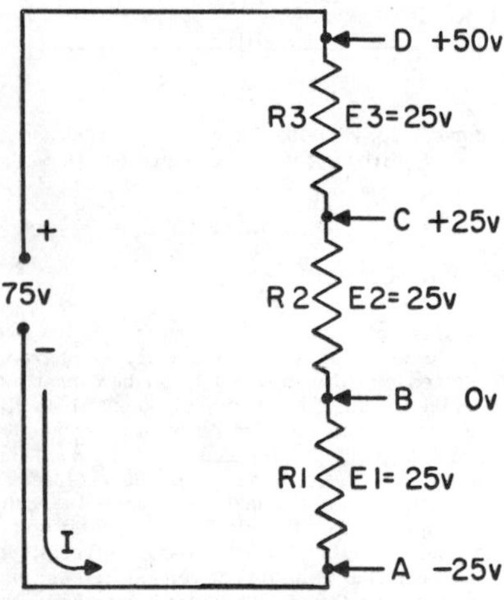

Figure 4-34.—Determining potentials with respect to a reference point.

If point B is used as the reference as in figure 4-34, point D would be positive 50 volts in respect to the new reference point B. The former reference point A is 25 volts negative in respect to point B.

GROUND

As in the previous circuit illustration, the reference point of a circuit is always considered to be at zero potential. Since the earth (ground) is said to be at a zero potential, the term GROUND is used to denote a common electrical point of zero potential. In figure 4-35, point A is the zero reference or ground and is symbolized as such.

Point C is 75 volts positive and point B is 25 volts positive in respect to ground.

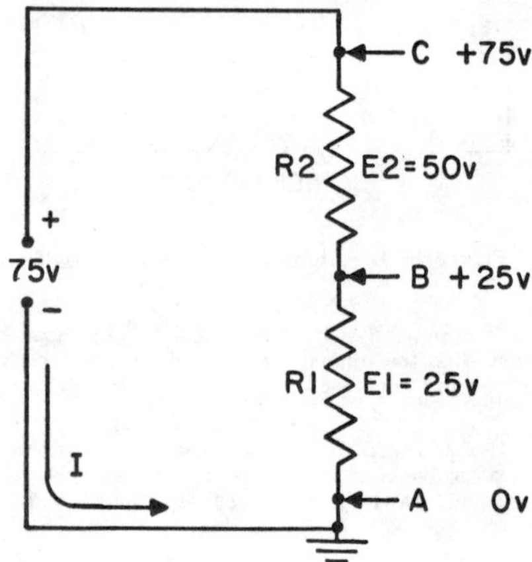

Figure 4-35.—Use of ground symbols.

In many electrical/electronic equipments, the metal chassis is the common ground for the many electrical circuits. The value of ground is noted when considering its contribution to economy, simplification of schematics, and ease of measurement. When completing each electrical circuit, common points of a circuit at zero potential are connected directly to the metal chassis thereby eliminating a large amount of connecting wire. The electrons pass through the metal chassis (conductor) to reach other points of the circuit. An example of a grounded circuit is illustrated in figure 4-36.

NOTES:

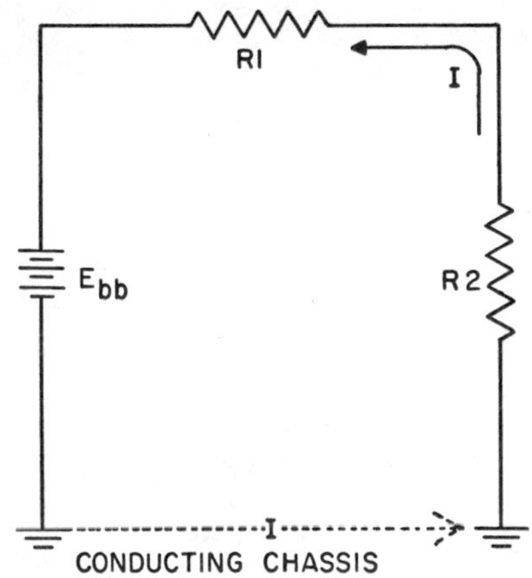

Figure 4-36.—Ground used as a conductor.

Most voltage measurements used to check proper circuit operation in electronic equipment are taken in respect to ground. One meter lead is attached to ground and the other meter lead is moved to various test points.

OPEN AND SHORT CIRCUITS

A circuit is said to be OPEN when a break exists in a complete conducting pathway. Although an open occurs any time a switch is thrown to deenergize a circuit, an open may also develop accidentally due to abnormal circuit conditions. To restore a circuit to proper operation, the open must be located and its cause determined.

Sometimes an open can be located visually by a close inspection of the circuit components. Defective components, such as burned out resistors and fuses can usually be discovered by this method. Others such as a break in wire covered by insulation, or the melted element of an enclosed fuse, are not visible to the eye. Under such conditions, the understanding of an open's effect on circuit conditions enables a technician to make use of a voltmeter or ohmmeter to locate the open component.

In figure 4-37, the series circuit consists of two resistors and a fuse. Notice the effects on circuit conditions when the fuse opens.

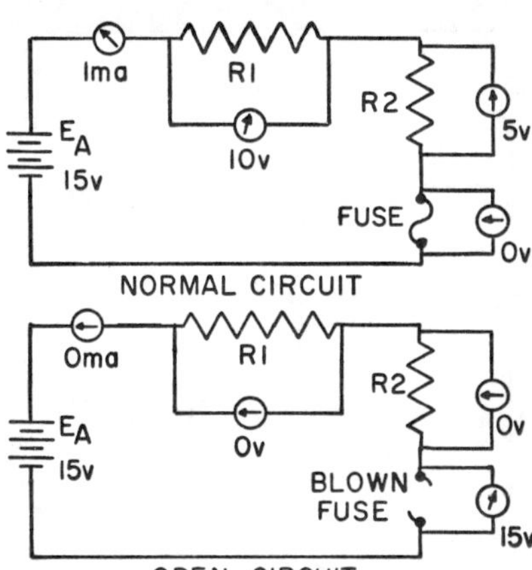

Figure 4-37.—Normal and open circuit conditions.

Current ceases to flow; therefore, there is no longer a voltage drop across the resistors. Each end of the open conducting path becomes an extension of the battery terminals and the voltage felt across the open is equal to the applied voltage.

An open circuit, such as found in figure 4-37 could also have been located with an ohmmeter. However, when using an ohmmeter to check a circuit, it is important to first deenergize the circuit. The reason being that an ohmmeter has its own power source and would be damaged if connected to an energized circuit.

The ohmmeter used to check a series circuit would indicate the ohmic value of each resistance it is connected across. The open circuit due to its almost infinite resistance would cause no deflection on the ohmmeter as indicated by the illustration, figure 4-38.

A SHORT CIRCUIT is an accidental path of low resistance which passes an abnormal amount of current. A short circuit exists whenever the resistance of the circuit or the resistance of a part of a circuit drops in value to almost zero ohms. A short often occurs as a result of improper wiring or broken insulation.

NOTES:

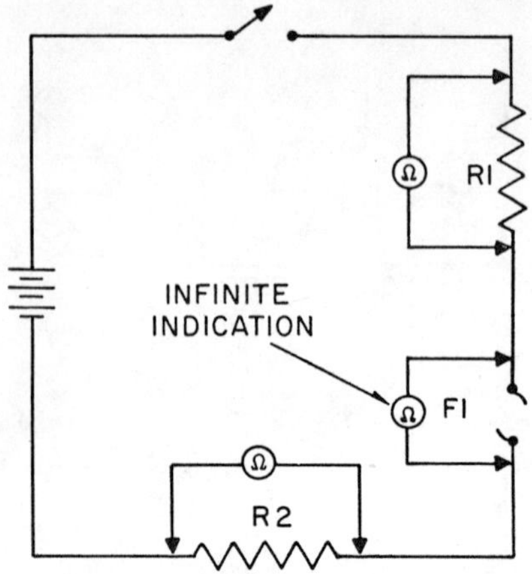

Figure 4-38.—Ohmmeter readings in a series circuit.

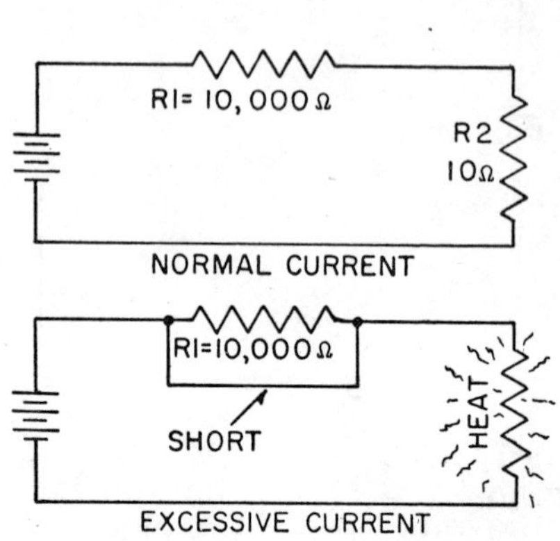

Figure 4-39.—Normal and short circuit conditions.

In figure 4-39 a short is caused by improper wiring. Note the effect on current flow. Since the resistor has in effect been replaced with a piece of wire, practically all the current flows through the short and very little current flows through the resistor. Electrons flow through the short, a path of almost zero resistance and complete the circuit by passing through the 10-ohm resistor and the battery. The amount of current flow increases greatly because its resistive path has decreased from 10,010 ohms to 10 ohms. Due to the excessive current through the 10-ohm resistor, the increased heat dissipated by the resistor will destroy the component.

EFFECT OF SOURCE RESISTANCE ON VOLTAGE, POWER, AND EFFICIENCY

All sources of emf have some internal resistance that acts in series with the load resistance. The source resistance is generally indicated in circuit diagrams as a separate resistor connected in series with the source. Both the voltage and power made available to the load may be increased if the resistance of the source is reduced.

NOTES:

The effects of source resistance, R_S, on load voltage may be illustrated by the use of figure 4-40. In figure 4-40 (A), the circuit is open, and therefore a voltmeter connected across the battery will read the open-circuit voltage. In the case of a dry cell, the open-circuit voltage is 1.5 volts. In figure 4-40 (B), the cell is short-circuited through the ammeter, and a current of 30 amperes flows from the source. In this case the voltage of the cell is developed across the internal resistance of the cell. The internal resistance of the cell is therefore,

$$R_S = \frac{E_S}{I} = \frac{1.5}{30} = 0.05 \text{ ohm}.$$

If a load, R_L, of 0.10 ohm is connected to the circuit, as shown in figure 4-40 (C), the current, I, becomes

$$I = \frac{E_S}{R_t} = \frac{1.5}{0.15} = 10 \text{ amperes}.$$

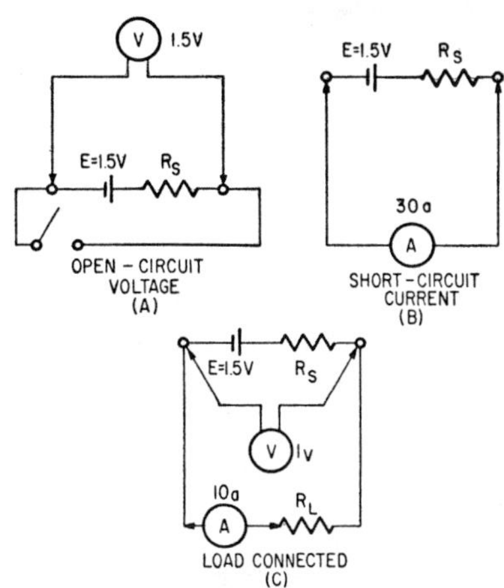

Figure 4-40.—Effect of source resistance on load voltage.

The voltage available at the load is

$$E_L = IR_L = 10 \times 0.1 = 1 \text{ volt.}$$

The voltage absorbed across the internal resistance of the cell is

$$IR_s = 10 \times 0.05 = 0.5 \text{ volt.}$$

Thus the effect of the internal resistance is to decrease the terminal voltage from 1.5 volts to 1 volt when the cell delivers 10 amperes to the load.

The effect of the source resistance on the power output of a d-c source may be shown by an analysis of the circuit in figure 4-41 (A). When the variable load-resistor, R_L, is set at the zero ohms position (equivalent to a short circuit) the current is limited only by the internal resistance, R_s, of the source. The short-circuit current, I, is determined as

$$I = \frac{E_s}{R_s} = \frac{100}{5} = 20 \text{ amperes}$$

This is the maximum current that may be drawn from the source. The terminal voltage across the short circuit is zero and all the voltage is absorbed within the terminal resistance of the source.

If the load resistance, R_L, is increased (the internal resistance remaining the same), the current drawn from the source will decrease. Consequently, the voltage drop across the internal resistance will decrease. At the same time, the terminal voltage applied across the load will increase and will approach a maximum as the current approaches zero.

The MAXIMUM POWER TRANSFER THEOREM says in effect that maximum power is transferred from the source to the load when the resistance of the load is equal to the internal resistance of the source. This theorem is illustrated in the tabular chart and the graph of figure 4-41 (B) and (C). When the load resistance is 5 ohms, thus matching the source resistance, the maximum power of 500 watts is developed in the load.

The efficiency of power transfer (ratio of output to input power) from the source to the load increases as the load resistance is increased. The efficiency approaches 100 percent as the load resistance approaches a relatively large value compared with that of the source, since less power is lost in the source. The efficiency of power transfer is only 50 percent at the maximum power transfer resistance of 5 ohms and approaches zero efficiency at relatively low values of load resistance compared with that of the source.

Thus the problem of high efficiency and maximum power transfer is resolved as a compromise somewhere between the low efficiency of maximum power transfer and the high efficiency of the high-resistance load. Where the amounts of power involved are large and the efficiency is important, the load resistance is made large relative to the source resistance so that the losses are kept small. In this case the efficiency will be high. Where the problem of matching a source to a load is of paramount importance, as in communications circuits, a strong signal may be more important than a high percentage of efficiency. In such cases, the efficiency of transmission will be only about 50 percent. However, the power of transmission will be the maximum of which the source is capable of supplying.

NOTES:

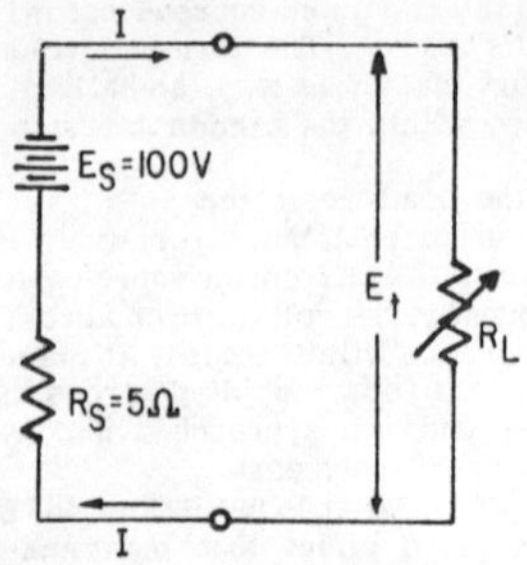

E_S = OPEN-CIRCUIT VOLTAGE OF SOURCE
R_S = INTERNAL RESISTANCE OF SOURCE
E_t = TERMINAL VOLTAGE
R_L = RESISTANCE OF LOAD
P_L = POWER USED IN LOAD
I = CURRENT FROM SOURCE
% EFF = PERCENTAGE OF EFFICIENCY

(A)
CIRCUIT AND SYMBOL DESIGNATIONS

R_L	E_t	I	P_L	%EFF
0	0	20	0	0
1	16.6	16.6	275.6	16.6
2	28.6	14.3	409	28.6
3	37.5	12.5	468.8	37.5
4	44.4	11.1	492.8	44.4
5	50	10	500	50
6	54.5	9.1	495.4	54.5
7	58.1	8.3	482.2	58.1
8	61.6	7.7	474.3	61.6
9	63.9	7.1	453.7	63.9
10	66	6.6	435.6	66
20	80	4	320	80
30	87	2.9	252	87
40	88	2.2	193.6	88
50	91	1.82	165	91

(B)
CHART

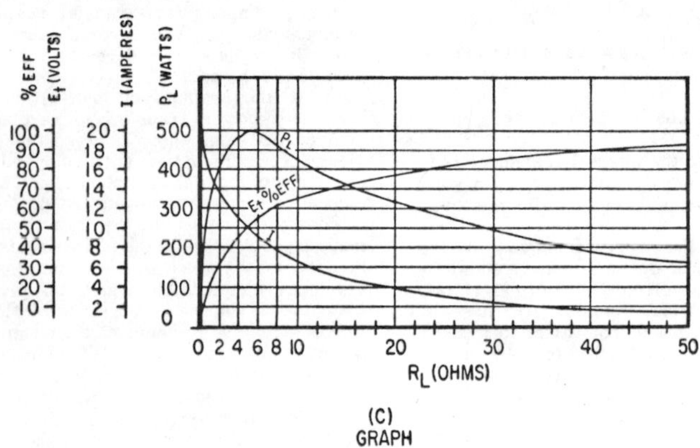

(C)
GRAPH

Figure 4-41.—Effect of source resistance on power output.

NOTES:

CHAPTER 5

PARALLEL D-C CIRCUITS

An adequate understanding of modern electrical equipment requires a progressive development in the study of typical electrical circuits. In stepping-stone fashion, the discussion of series d-c circuits will now be followed by a consideration of the characteristics of parallel d-c circuits. It will be shown how the principles applied to series circuits can be used to determine the reactions of such quantities as voltage, current, and resistance in parallel and series-parallel circuits.

Along with the progressive introduction of electrical theories and circuit characteristics comes a corresponding progression in the use of mathematical equations and problem solving methods. A basic knowledge of powers of ten, fractions, fractional equations, and the use of simultaneous equations is required for the comprehension of material presented in this chapter.

PARALLEL CIRCUIT CHARACTERISTICS

A parallel circuit is defined as one having more than one current path connected to a common voltage source. Parallel circuits, therefore, must contain two or more load resistances which are not connected in series. An example of a basic parallel circuit is shown in figure 5-1.

Commencing at the voltage source (E_{bb}) and tracing counterclockwise around the circuit, two complete and separate paths can be identified in which current can flow. One path is traced from the source through resistance R_1 and back to the source; the other, from the source through resistance R_2 and back to the source.

VOLTAGE

You have seen that the source voltage in a series circuit divides proportionately across each resistor in the circuit. In a parallel circuit (fig. 5-1), the same voltage is present across all the resistors of a parallel group. This voltage is equal to the applied voltage (R_{bb}). The foregoing statement can be expressed in equation form as

$$E_{bb} = E_{R1} = E_{R2} = E_{Rn} \qquad (5\text{-}1)$$

Voltage measurements taken across the resistors of a parallel circuit, as illustrated by figure 5-2, verify the above equation. Each voltmeter indicates the same amount of voltage. Notice that the voltage across each resistor is the same as the applied voltage.

Example. Assume that the current through a resistor of a parallel circuit is known to be 4.5 milliamperes (ma) and the value of the resistor is 30,000 ohms. Determine the potential across the resistor. The circuit is shown in figure 5-3.

Given:

$$R_2 = 30\text{K}$$

$$I_{R2} = 4.5 \text{ milliamperes}$$

Find:

$$E_{R2} = ?$$

$$E_{bb} = ?$$

Solution: Select proper equation.

$$E = IR \qquad (4\text{-}2)$$

Substitute known values:

$$E_{R2} = I_{R2} \times R_2$$

$$E_{R2} = 4.5 \text{ milliamperes} \times 30{,}000 \text{ ohms}$$

NOTES:

Express in powers of ten:

$$E_{R2} = (4.5 \times 10^{-3}) \times (30 \times 10^{3})$$

$$E_{R2} = 4.5 \times 30$$

Resultant:

$$E_{R2} = 135 \text{ v}$$

Therefore:

$$E_{bb} = 135 \text{ v}$$

Having determined the voltage across one resistor (R_2) in a parallel circuit, the value of the source voltage (E_{bb}) and the potentials across any other resistors that may be connected in parallel with it are known (equations 5-1).

CURRENT DIVISION

The current in a circuit . . . . is inversely proportional to the circuit resistance. This fact, obtained from Ohm's law, establishes the relationship upon which the following discussion is developed.

A single current flows in a series circuit. Its value is determined in part by the total resistance of the circuit. However, the source current in a parallel circuit divides among the available paths in relation to the value of the resistors in the circuit. Ohm's law remains unchanged. For a given voltage, current varies inversely with resistance.

The behavior of current in parallel circuits will be shown by a series of illustrations using example circuits with different values of resistance for a given value of applied voltage.

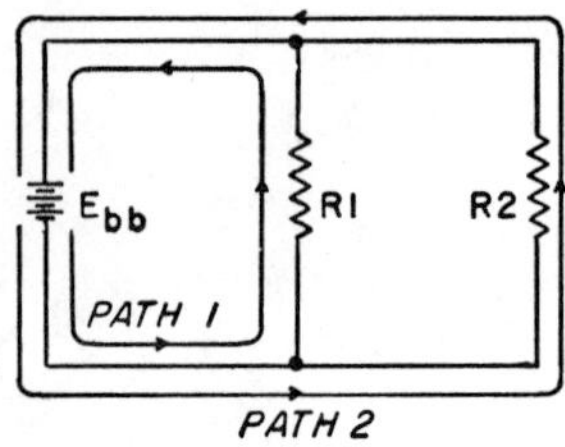

Figure 5-1.—Example of a basic parallel circuit.

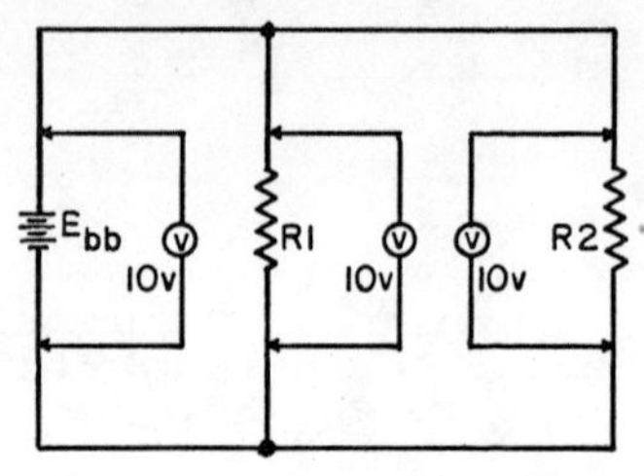

Figure 5-2.—Voltage comparison in a parallel circuit.

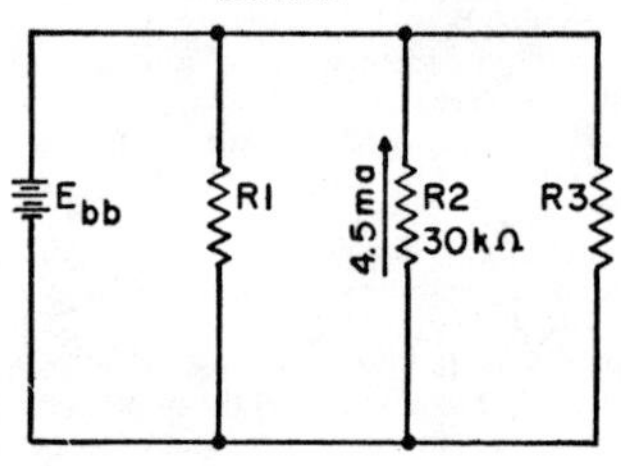

Figure 5-3.—Example problem parallel circuit.

Part (A) of figure 5-4 shows a basic series circuit. Here the total current must pass through the single resistor. The amount of current is determined as

$$I_t = \frac{E_{bb}}{R_1} = \frac{50}{10} = 5 \text{ amperes}$$

Part (B) of figure 5-4 shows the same resistor (R_1) with a second resistor (R_2) of equal value connected in parallel across the voltage source. Applying the proper equation from Ohm's law, the current flow through each resistor is seen to be the same as through the single resistor in part (A). These individual currents are determined as follows:

$$I_{R1} = \frac{E_{bb}}{R_1} = \frac{50}{10} = 5 \text{ amperes}$$

$$I_{R2} = \frac{E_{bb}}{R_2} = \frac{50}{10} = 5 \text{ amperes}$$

However, it is apparent that if 5 amperes of current flows through each of the two resistors,

NOTES:

there must be a total current of 10 amperes drawn from the source. The distribution of current in the simple parallel circuit shown in figure 5-4 (B) is as follows:

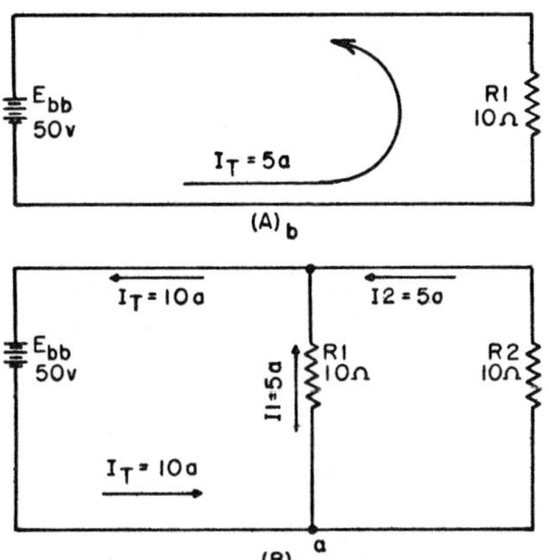

Figure 5-4.—Analysis of current in parallel circuit.

The total current of 10 amperes leaves the negative terminal of the battery and flows to point a. Since point a is a connecting point for the two resistors, it is called a junction. At junction a the total current divides into two smaller currents of 5 amperes each. These two currents flow through their respective resistors and rejoin at junction b. The total current then flows from junction b back to the positive terminal of the source. Thus, the source supplies a total current of 10 amperes and each of the two equal resistors carries one-half the total current.

Each individual current path in the circuit of figure 5-4 (B) is referred to as a branch. Each branch will carry a current that is a portion of the total current. Two or more branches form a network.

From the foregoing observations, the characteristics of current in a parallel circuit can be expressed in terms of the following general equation

$$I_t = I_1 + I_2 + \ldots . I_n \qquad (5\text{-}2)$$

NOTES:

The analysis of current in parallel circuits is continued with the use of the following example circuits.

Compare part (A) of figure 5-5 with part (B) of the preceding example circuit in figure 5-4. Notice that doubling the value of the second branch resistor (R_2) has no effect on the current in the first branch (I_{R1}), but does reduce its own branch current (I_{R2}) to one-half its original value. The total circuit current drops to a value equal to the sum of the branch currents. These facts are verified as follows:

$$I_1 = \frac{E_{bb}}{R_1} = \frac{50}{10} = 5 \text{ amperes}$$

$$I_2 = \frac{E_{bb}}{R_2} = \frac{50}{20} = 2.5 \text{ amperes}$$

$$I_t = I_1 + I_2$$

$$I_t = 5 + 2.5 = 7.5 \text{ amperes}$$

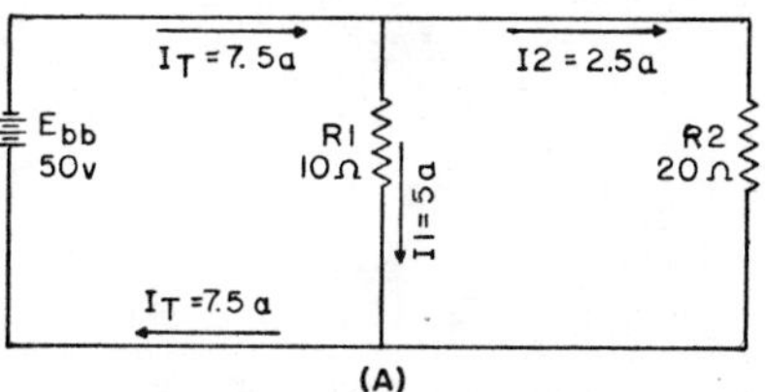

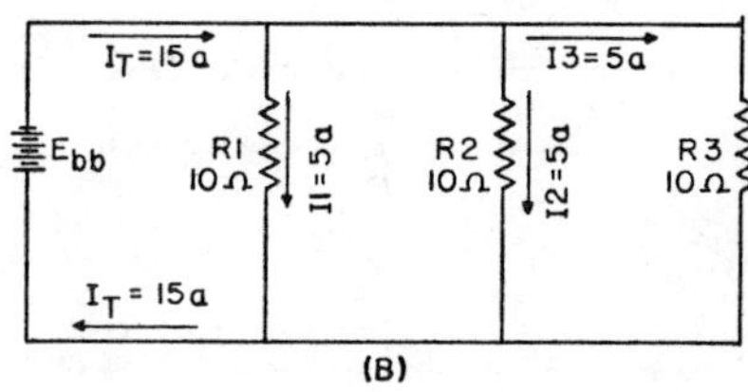

Figure 5-5.—Current behavior in parallel circuits.

Now compare the two circuits of figure 5-5. Notice that the sum of the ohmic values of the resistors in both circuits is equal and that the applied voltage is the same value. However, the total current in part (B) is twice the amount in part (A). It is apparent, therefore, that the manner in which resistors are connected in a circuit, as well as their actual ohmic value, affects the total current flow. This phenomenon will be illustrated in more detail in the discussion of resistance. The amount of current flow in the branch circuits and the total current in the circuit (fig. 5-5 (B)), are determined as follows:

$$I_1 = \frac{E_{bb}}{R_1} = \frac{50}{10} = 5 \text{ amperes}$$

$$I_2 = \frac{E_{bb}}{R_2} = \frac{50}{10} = 5 \text{ amperes}$$

$$I_3 = \frac{E_{bb}}{R_3} = \frac{50}{10} = 5 \text{ amperes}$$

$$I_t = I_1 + I_2 + 13$$

$$I_t = \frac{E_{bb}}{R_1} + \frac{E_{bb}}{R_2} + \frac{E_{bb}}{R_3}$$

$$I_t = \frac{50}{10} + \frac{50}{10} + \frac{50}{10} = 15 \text{ amperes}$$

The division of current in a parallel network follows a definite pattern. This pattern is described by Kirchhoff's current law which is stated as follows:

The algebraic sum of the currents entering and leaving any junction of conductors is equal to zero. This law can be stated mathematically as

$$I_a + I_b + \ldots\ldots + I_n = 0 \qquad (5\text{-}3)$$

where I_a, I_b, etc., are the currents entering and leaving the junction. Currents entering the junction are assumed to be positive, and currents leaving the junction are considered negative. When solving a problem using equation (5-3), the currents must be placed into the equation with the proper polarity signs attached.

NOTES:

Example. Solve for the value of I_3 in figure 5-6.

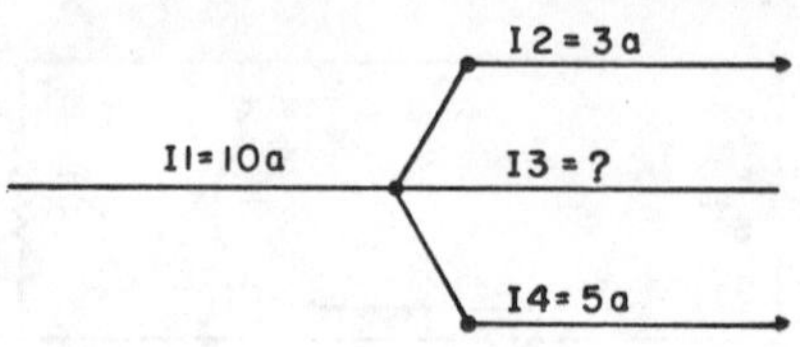

Figure 5-6.—Circuit for example problem.

Solution: First the currents are given proper signs.

$I_1 = +10$ amperes

$I_2 = -3$ amperes

$I_3 =$? amperes

$I_4 = -5$ amperes

these currents are placed into equation (5-3) with the proper signs as follows:

Basic equation:

$$I_a + I_b + \ldots\ldots I_n = 0 \qquad (5\text{-}3)$$

Substitution:

$$I_1 + I_2 + I_3 + I_4 = 0$$

$$(+10) + (-3) + (I_3) + (-5) = 0$$

Combining like terms:

$$I_3 + 2 = 0$$

$$I_3 = -2 \text{ amperes}$$

thus, I_3 has a value of 2 amperes, and the negative sign shows it to be a current leaving the junction.

Example. Using figure 5-7, solve for the magnitude and direction of I_3:

Solution:

$$I_a + I_b + \ldots\ldots I_n = 0 \qquad (5\text{-}3)$$

$$I_1 + I_2 + I_3 + I_4 = 0$$

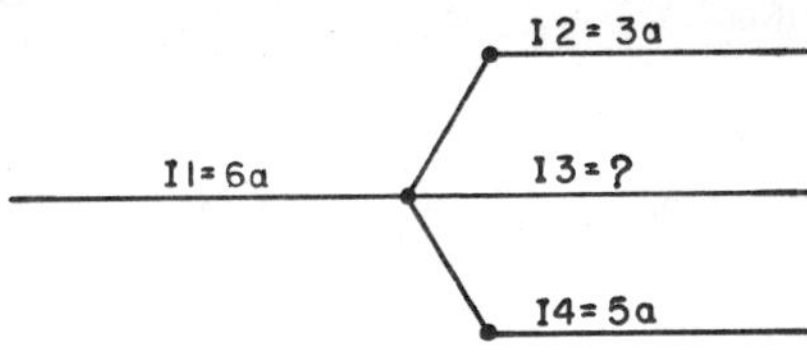

Figure 5-7.—Circuit for example problem.

$$(+6a) + (-3a) + (I_3) + (-5a) = 0$$

$$I_3 - 2a = 0$$

$$I_3 = 2 \text{ amperes}$$

Thus, I_3 is 2 amperes, and its positive sign shows it to be a current entering the junction.

PARALLEL RESISTANCE

The preceding discussion of current introduced certain principles involving the characteristics and effects of resistance in parallel circuits. A detailed explanation of the characteristics of parallel resistances will be considered in this section. The explanation will commence with a simple parallel circuit. Various methods used to determine the total resistance in parallel circuits will be described.

In the example diagram (fig. 5-8), two cylinders of conductive material having a resistance value of 10 ohms each are connected across a 5-volt battery. A complete circuit consisting of two parallel paths is formed and current will flow as shown.

Computing the individual currents shows that there is one-half an ampere of current flowing through each resistance. Accordingly, the total current flowing from the battery to the junction of the resistors, and returning from the resistors to the battery, is equal to 1 ampere. The total resistance of the circuit can be determined by substituting total values of voltage and current into the following equation. This equation is derived from Ohm's law.

$$R_t = \frac{E_t}{I_t}$$

$$R_t = \frac{5}{1} = 5 \text{ ohms}$$

This computation shows the total resistance to be 5 ohms, one-half the value of either of the two resistors.

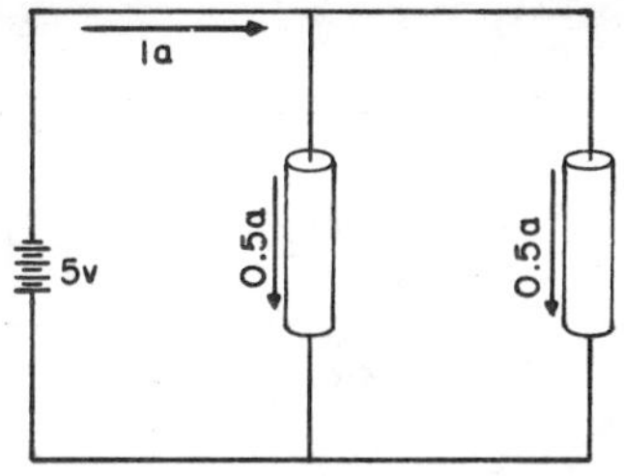

Figure 5-8.—Two equal resistors connected in parallel.

Since the total resistance of this parallel circuit is smaller than either of the two resistors, the term "total resistance" does not mean the sum of the individual resistor values. The total resistance of resistors in parallel is also referred to as equivalent resistance. In many texts the terms total resistance and equivalent resistances are used interchangeably.

There are several methods used to determine the equivalent resistance of parallel circuits. The most appropriate method for a particular circuit depends on the number and value of the resistors. For the circuit described above, the following simple equation is used:

$$R_{eq} = \frac{R}{N}$$

where

R_{eq} = equivalent parallel resistance

R = ohmic value of one resistor

N = number of resistors

This equation is valid for any number of equal value parallel resistors.

An understanding of why the equivalent resistance of two parallel resistors is smaller than the resistance of either of the two resistors can be gained by an examination of figure 5-8. The two 10-ohm cylinders have fixed equal volumes. If the cylinders were combined into one cylinder as shown in figure 5-9, the volume would double. If the same length is retained and the volume is doubled, the cross-sectional area will double. When the cross-sectional area of a material is increased, the resistance is decreased proportionately.

NOTES:

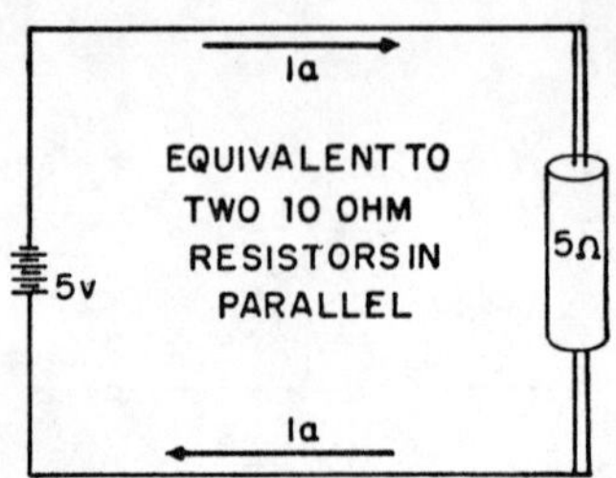

Figure 5-9.—Equivalent parallel circuit.

Since, in this case, the cross-sectional area is two times the original area, the resistance is one-half the former value. Therefore, when two equal value resistors are connected in parallel, they present a total resistance equivalent to a single resistor of one-half the value of either of the original resistors.

Example. Four 40-ohm resistors are connected in parallel. What is their equivalent resistance?

Solution:

$$R_{eq} = \frac{R}{N} = \frac{40}{4} = 10 \text{ ohms}$$

Circuits containing parallel resistance of unequal value will now be considered. Refer to example circuit in figure 5-10.

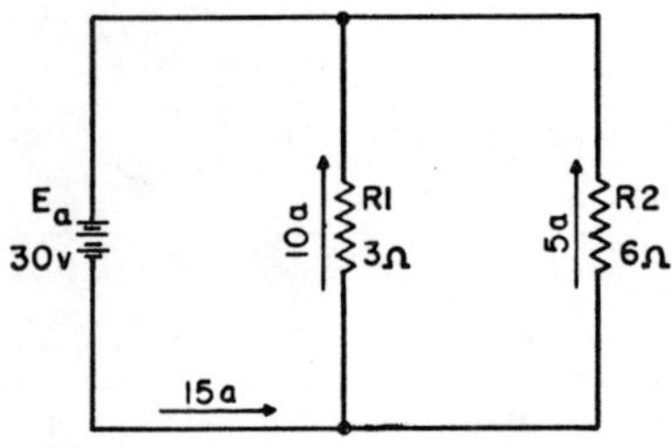

Figure 5-10.—Example circuit with unequal parallel resistors.

NOTES:

Given:

$$R_1 = 3\Omega, R_2 = 6\Omega, E_a = 30v$$

Known:

$I_1 = 10$ amperes, $I_2 = 5$ amperes, $I_t = 15$ amperes.

Determine:

$$R_{eq} = ?$$

Solution:

$$R_{eq} = \frac{E_a}{I_t}$$

$$R_{eq} = \frac{30}{15} = 2 \text{ ohms}$$

Notice that the equivalent resistance of two ohms is less than the value of either branch resistor. In parallel circuits the equivalent resistance will always be smaller than the resistance of any branch.

RECIPROCAL METHOD

Many circuits are encountered in which resistors of unequal value are connected in parallel. It is therefore desirable to develop a formula which can be used to compute the equivalent resistance of two or more unequal parallel resistors. This equation can be derived as follows:

Given:

$$I_t = I_1 + I_2 + \cdots\cdots I_n \qquad (5\text{-}2)$$

Substituting $\frac{E}{R}$ for I gives:

$$\frac{E_t}{R_t} = \frac{E_1}{R_1} + \frac{E_2}{R_2} + \cdots\cdots \frac{E_n}{R_n}$$

Since in a parallel circuit $E_t = E_1 = E_2 = E_n$

$$\frac{E}{R_t} = \frac{E}{R_1} + \frac{E}{R_2} + \cdots\cdots \frac{E}{R_n}$$

Dividing both sides by E:

$$\frac{E}{E\,R_t} = \frac{E}{E\,R_1} + \frac{E}{E\,R_2} + \cdots\cdots \frac{E}{E\,R_n}$$

$$\frac{1}{R_t} = \frac{1}{R_1} + \frac{1}{R_2} + \cdots\cdots \frac{1}{R_n}$$

Taking the reciprocal of both sides:

$$\frac{1}{\frac{1}{R_t}} = \frac{1}{\frac{1}{R_1} + \frac{1}{R_2} + \ldots \frac{1}{R_n}}$$

Simplifying:

$$R_t = \frac{1}{\frac{1}{R_1} + \frac{1}{R_2} + \ldots\ldots \frac{1}{R_n}}$$

This formula is called "the reciprocal of the sum of the reciprocals" and is the one normally used to solve for the equivalent resistance of a number of parallel resistors.

Example. Given three parallel resistors of 20 ohms, 30 ohms, and 40 ohms; find the equivalent resistance using the reciprocal equation. (See fig. 5-11.)

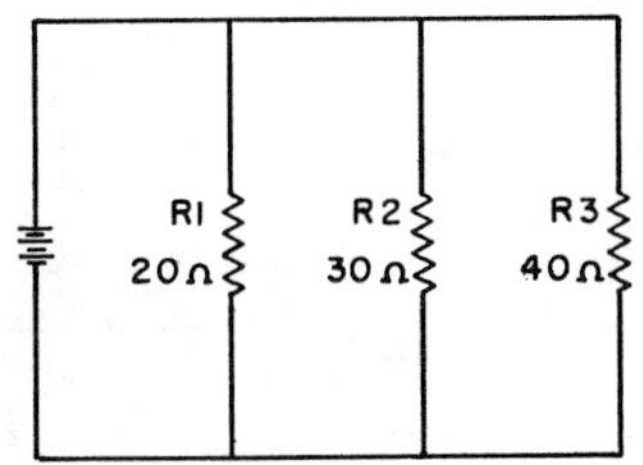

Figure 5-11.—Example parallel circuit with unequal branch resistors.

Solution:

Select the proper equation:

$$R_{eq} = \frac{1}{\frac{1}{R_1} + \frac{1}{R_2} + \frac{1}{R_3}}$$

Substitute:

$$R_{eq} = \frac{1}{\frac{1}{20} + \frac{1}{30} + \frac{1}{40}}$$

Find LCD:

$$R_{eq} = \frac{1}{\frac{6}{120} + \frac{4}{120} + \frac{3}{120}} = \frac{1}{\frac{13}{120}}$$

NOTES:

Invert:

$$R_{eq} = \frac{120}{13} = 9.23 \text{ ohms}$$

Some parallel circuit problems can be solved more conveniently by considering the ease with which current can flow. The degree to which a circuit permits or conducts current is called the conductance (G) of the circuit. The unit of conductance is the MHO, which is ohms spelled backwards. The conductance of a circuit is the reciprocal of the resistance. The conductance can therefore be found using the following formula:

$$G = \frac{1}{R}$$

also:

$$R = \frac{1}{G}$$

In a parallel circuit, the total conductance is equal to the sum of the individual branch conductances. As an equation:

$$G_t = G_1 + G_2 + \ldots\ldots G_n \qquad (5\text{-}6)$$

Example: Determine the equivalent (total) resistance of the circuit shown in the preceding example (fig. 5-11), using the conductance method.

Solution:

$$G_1 = \frac{1}{R_1} = \frac{1}{20} = 0.050 \text{ mho}$$

$$G_2 = \frac{1}{R_2} = \frac{1}{30} = 0.033 \text{ mho}$$

$$G_3 = \frac{1}{R_3} = \frac{1}{40} = 0.025 \text{ mho}$$

$$G_t = G_1 + G_2 + G_3 \qquad (5\text{-}6)$$

$$G_t = 0.050 + 0.033 + 0.025 = 0.108 \text{ mho}$$

Since:

$$R_t = \frac{1}{G_t}$$

$$R_t = \frac{1}{0.108} = 9.25 \text{ ohms}$$

The value of equivalent resistance determined by the conductance method is almost identical to the value determined by the reciprocal of the sum of the reciprocals methods.

PRODUCT OVER THE SUM

A convenient formula for finding the equivalent resistance of two parallel resistors can be derived from equation (5-5) as shown below:

$$R_t = \frac{1}{\frac{1}{R_1} + \frac{1}{R_2}} \qquad (5\text{-}5)$$

Finding the LCD:

$$R_t = \frac{1}{\frac{R_2 + R_1}{R_1 \times R_2}}$$

Taking the reciprocal:

$$R_t = \frac{R_1 \times R_2}{R_1 + R_2}$$

This equation, called the product over the sum formula, is used so frequently it should be committed to memory.

Example. What is the equivalent resistance of a 20 ohm and a 30 ohm resistor connected in parallel?

Given:

$$R_1 = 20$$

$$R_2 = 30$$

Find:

$$R_{eq} = ?$$

Solution:

$$R_t = \frac{R_1 \times R_2}{R_1 + R_2}$$

$$R_t = \frac{20 \times 30}{20 + 30}$$

$$R_t = 12 \text{ ohms}$$

PARALLEL CIRCUIT REDUCTION

In the study of electricity, it is often necessary to resolve a complex circuit into a simpler form. Any complex circuit consisting of resistances can be reduced to a basic equivalent circuit containing the source and total resistance. This process is called reduction to an equivalent circuit. An example of circuit reduction is shown in figure 5-12.

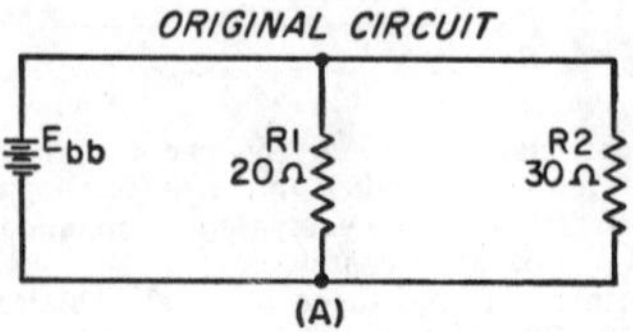

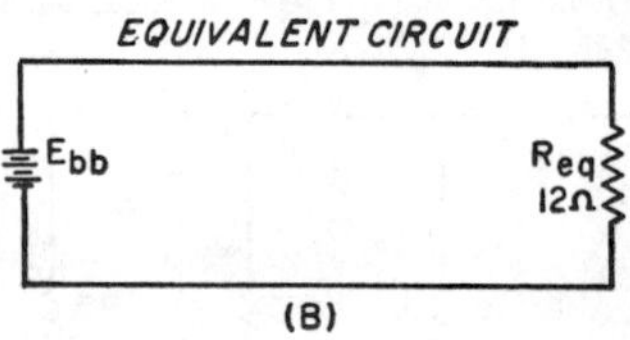

Figure 5-12.—Parallel circuit with equivalent circuit.

The circuit shown in figure 5-12 (A) is reduced to the simple circuit shown in (B).

COMPUTING TOTAL POWER

Power computations in a parallel circuit are essentially the same as those used for the series circuit. Since power dissipation in resistors consists of a heat loss, power dissipations are additive regardless of how the resistors are connected in the circuit. The total power dissipated is equal to the sum of the powers dissipated by the individual resistors. Like the series circuit, the total power consumed by the parallel circuit is

$$P_t = P1 + P2 + \ldots\ldots P_n \qquad (4\text{-}13)$$

Example. Find the total power consumed by the circuit in figure 5-13.

Solution:

$$P_{R1} = E_{bb} \times I_{R1}$$

$$P_{R1} = 50 \times 5$$

NOTES:

$P_{R1} = 250$ w

$P_{R2} = E_{bb} \times I_{R2}$

$P_{R2} = 50 \times 2$

$P_{R2} = 100$ w

$P_{R3} = E_{bb} \times I_{R3}$

$P_{R3} = 50 \times 1$

$P_{R3} = 50$ w

$P_t = P_1 + P_2 + P_3$

$P_t = 250 + 100 + 50$

$P_t = 400$ w

Note that the power dissipated in the branch circuits is determined in the same manner as the power dissipated by individual resistors in a series circuit. The total power (P_t) is then obtained by summing up the powers dissipated in the branch resistors using equation (4-13).

Since, in the example shown in figure 5-13, the total current is known, the total power could be determined by the following method:

$P_t = E_{bb} \times I_t$

$P_t = 50v \times 8a$

$P_t = 400$ w

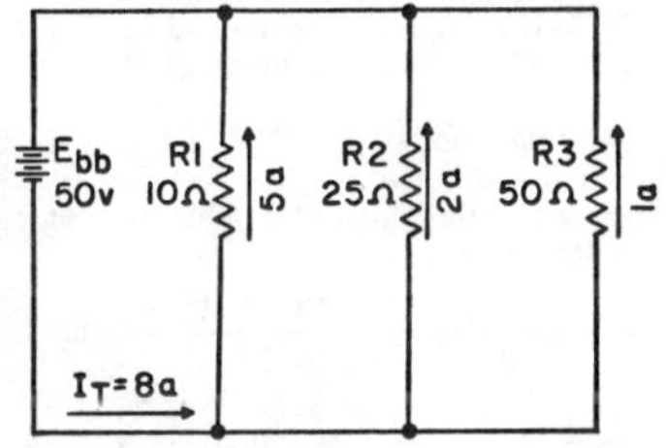

BE.94
Figure 5-13.—Example parallel circuit.

Rules for solving parallel d-c circuits are as follows:

1. The same voltage exists across each branch of a parallel circuit and is equal to the source voltage.
2. The current through a branch of a parallel network is inversely proportional to the amount of resistance of the branch.
3. The total current of a parallel circuit is equal to the sum of the currents of the individual branches of the circuit.
4. The total resistance of a parallel circuit is equal to the reciprocal of the sum of the reciprocals of the individual resistances of the circuit.
5. The total power consumed in a parallel circuit is equal to the sum of the power consumption of the individual resistances.

TYPICAL PROBLEMS IN PARALLEL CIRCUITS

Problems involving the determination of resistance, voltage, current, and power in a parallel circuit are solved as simply as in a series circuit. The procedure is the same—(1) draw a circuit diagram, (2) state the values given and the values to be found, (3) state the applicable equations, and (4) substitute the given values and solve for the unknown.

For example, the parallel circuit of figure 5-14 consists of 2 branches (a and b). Branch a consists of 3 lamps in parallel. Their ratings are L_1 = 50 watts, L_2 = 25 watts, and L_3 = 75 watts. Branch b also has 3 lamps in parallel with ratings of L_4 = 150 watts, L_5 = 200 watts, and L_6 = 250 watts. The source voltage is 100 volts.

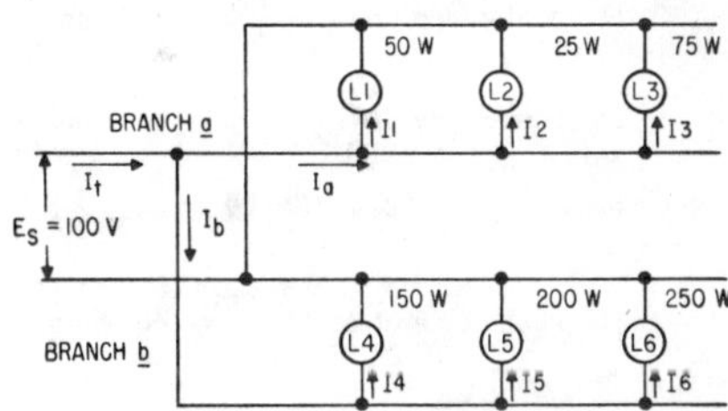

BE.95
Figure 5-14.—Typical parallel circuit.

NOTES:

Problem:

1. Find the current in each lamp.
2. Find the resistance of each lamp.
3. Find the current in branch a.
4. Find the current in branch b.
5. Find the total circuit current.
6. Find the total circuit resistance.
7. Find the total power supplied to the circuit.
8. Check 7 by a separate calculation.

Solution:

1. The current in L_1 is $I = \frac{P}{E} = \frac{50}{100} =$ 0.50 ampere.

   The current in L_2 is $\frac{25}{100} = 0.25$ ampere.

   The current in L_3 is $\frac{75}{100} = 0.75$ ampere.

   The current in L_4 is $\frac{150}{100} = 1.50$ amperes.

   The current in L_5 is $\frac{200}{100} = 2.00$ amperes.

   The current in L_6 is $\frac{250}{100} = 2.5$ amperes.

2. The resistance of L_1 is $R = \frac{E}{I} = \frac{100}{0.5} =$ 200 ohms.

   The resistance of L_2 is $\frac{100}{0.25} = 400$ ohms.

   The resistance of L_3 is $\frac{100}{0.75} = 133$ ohms.

   The resistance of L_4 is $\frac{100}{1.5} = 66.7$ ohms.

   The resistance of L_5 is $\frac{100}{2.0} = 50$ ohms.

   The resistance of L_6 is $\frac{100}{2.5} = 40$ ohms.

3. The current in branch a is

   $I_1 + I_2 + I_3 = 0.5 + 0.25 + 0.75 = 1.5$ amperes

4. The current in branch b is

   $I_4 + I_5 + I_6 = 1.5 + 2.0 + 2.5 = 6.0$ amperes

5. The total circuit current is

   $I_a + I_b = 1.5 + 6.0 = 7.5$ amperes

6. The total circuit resistance is

   $$R_t = \frac{E}{I_t} = \frac{100}{7.5} = 13.3 \text{ ohms}$$

7. The total power supplied to the circuit is:

   50w + 25w + 75w + 150w + 200w + 250w = 750 watts

8. The total power is also equal to

   $P_t = EI_t = 100 \times 7.5 = 750$ watts

SERIES-PARALLEL COMBINATIONS

In the preceding discussions, series and parallel d-c circuits have been considered separately. However, the technician will seldom encounter a circuit that consists solely of either type of circuit. Most circuits consist of both series and parallel elements. A circuit of this type will be referred to as a combination circuit. The solution of a combination circuit is simply a matter of application of the laws and rules discussed prior to this point.

SOLVING A COMBINATION CIRCUIT

At least three resistors are required to form a combination circuit. Two basic series-parallel circuits are shown in figure 5-15. In figure 5-15 (A), R_1 is connected in series with the parallel combination made up of R_2 and R_3.

The total resistance (R_t) of figure 5-15 (A) is determined in two steps. First, the equivalent resistance of the parallel combination of R_2 and R_3 is determined as follows:

$$R_{2,3} = \frac{R_2 R_3}{R_2 + R_3} = \frac{3 \times 6}{3 + 6} = \frac{18}{9} = 2 \text{ ohms}$$

The sum of $R_{2,3}$ and R_1 — that is, R_t — is

$$R_t = R_{2,3} + R_1 = 2 + 2 = 4 \text{ ohms}$$

NOTES:

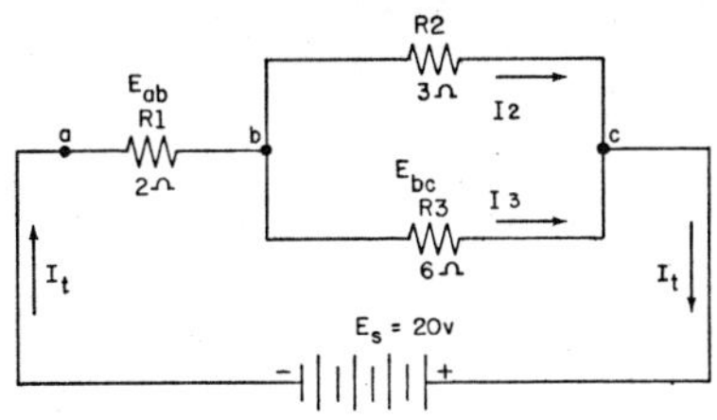

R1 IN SERIES WITH PARALLEL COMBINATION OF R2 AND R3
(A)

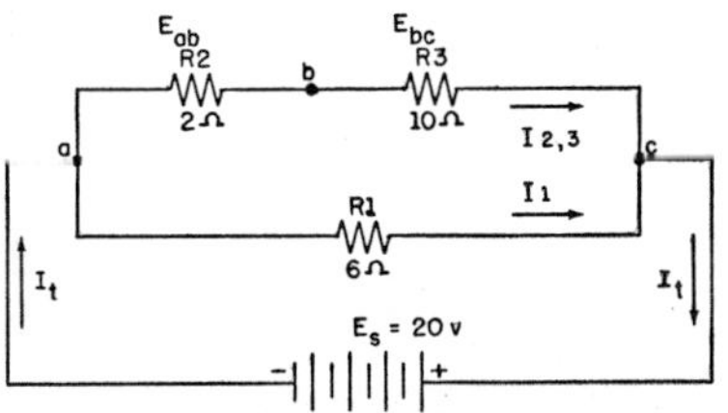

R1 IN PARALLEL WITH THE SERIES COMBINATION OF R2 & R3
(B)

Figure 5-15.—Compound circuits—series-parallel connections.

If the total resistance (R_t) and the source voltage (E_s) are known, the total current (I_t) may be determined by Ohm's law. Thus, in figure 5-15 (A),

$$I_t = \frac{E_s}{R_t} = \frac{20}{4} = 5 \text{ amperes}$$

If the values of the various resistors and the current through them are known, the voltage drops across the resistors may be determined by Ohm's law. Thus,

$$E_{ab} = I_t R_1 = 5 \times 2 = 10 \text{ volts}$$

and

$$E_{bc} = I_t R_{2,3} = 5 \times 2 = 10 \text{ volts}$$

According to Kirchhoff's voltage law, the sum of the voltage drops around the closed circuit is equal to the source voltage. Thus,

$$E_{ab} + E_{bc} = E_s$$

or

$$10 + 10 = 20 \text{ volts}$$

If the voltage drop (E_{bc}) across $R_{2,3}$—that is, the drop between points b and c—is known, the current through the individual branches may be determined as follows:

$$I_2 = \frac{E_{bc}}{R_2} = \frac{10}{3} = 3.333 \text{ amperes}$$

and

$$I_3 = \frac{E_{bc}}{R_3} = \frac{10}{6} = 1.666 \text{ amperes}$$

According to Kirchhoff's current law, the sum of the currents flowing in the individual parallel branches is equal to the total current. Thus,

$$I_2 + I_3 = I_t$$

or

$$3.333 + 1.666 = 5 \text{ amperes (approx)}$$

The total current flows through R_1; and at point b it divides between the two branches in inverse proportion to the resistance of each branch. Twice as much current goes through R_2 as through R_3 because R_2 has one-half the resistance of R_3. Thus, 3.333, or two-thirds of 5 amperes flows through R_2; and 1.666, or one-third of 5 amperes flows through R_3.

In figure 5-15 (B), R_1 is in parallel with series combination of R_2 and R_3. The total resistance (R_t) is determined in two steps. First, the sum of the resistance of R_2 and R_3—that is, $R_{2,3}$—is determined as follows:

$$R_{2,3} = R_2 + R_3 = 2 + 10 = 12 \text{ ohms}$$

Second, the total resistance (R_t) is the result of combining $R_{2,3}$ in parallel with R_1, or

$$R_t = \frac{R_{2,3}R_1}{R_{2,3} + R_1} = \frac{12 \times 6}{12 + 6} = 4 \text{ ohms}$$

If the total resistance (R_t) and the source voltage (E_s) are known, the total current (I_t) may be determined by Ohm's law. Thus, in figure 5-15 (B),

$$I_t = \frac{E_s}{R_t} = \frac{20}{4} = 5 \text{ amperes}$$

NOTES:

A portion of the total current flows through the series combination of R_2 and R_3 and the remainder flows through R_1. Because current varies inversely with the resistance, two-thirds of the total current flows through R_1 and one-third flows through the series combination of R_2 and R_3, since R_1 is one-half of $R_2 + R_3$.

The source voltage (E_s) is applied between points a and c, and therefore the current I_1 through R_1 is

$$I_1 = \frac{E_s}{R_1} = \frac{20}{6} = 3.333 \text{ amperes}$$

and the current, $I_{2,3}$, through $R_{2,3}$ is

$$I_{2,3} = \frac{E_s}{R_{2,3}} = \frac{20}{12} = 1.666 \text{ amperes}$$

According to Kirchhoff's current law the sum of the individual branch currents is equal to the total current, or

$$I_t = I_1 + I_{2,3}$$

$$5 = 3.333 + 1.666 \text{ (approx)}$$

Combination circuits may be made up of a number of resistors arranged in numerous series and parallel combinations. In more complicated circuits, special theorems, rules, and formulas are used. These are based on Ohm's Law and provide faster solutions for particular applications. Series formulas are applied to the series parts of the circuit, and parallel formulas are applied to the parallel parts. For example, in figure 5-16, the total resistance (R_t) may be obtained in three logical steps.

First, R_3, R_4, and R_5 in figure 5-16 (A), are in series (there is only one path for current) and may be combined in figure 5-16 (B), to give the resistance, R_s, of the three resistors. Thus,

$$R_s = R_3 + R_4 + R_5 = 5 + 9 + 10 = 24 \text{ ohms}$$

and it is now in parallel with R_2 (because they both receive the same voltage).

The combined resistance of R_s in parallel with R_2 is

$$R_{s,2} = \frac{R_2 R_s}{R_2 + R_s} = \frac{8 \times 24}{8 + 24} = 6 \text{ ohms}$$

as in the figure 5-16 (C).

NOTES:

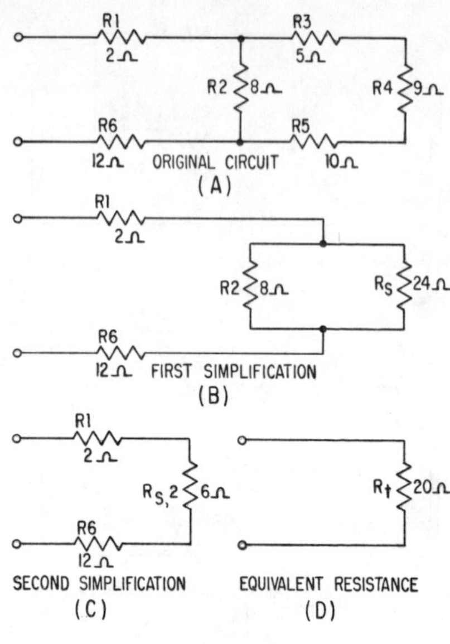

Figure 5-16.—Solving total resistance in a compound circuit.

Third, the total resistance (R_t) is determined by combining resistors R_1 and R_6 with $R_{s,2}$ as

$$R_t = R_1 + R_6 + R_{s,2} = 2 + 12 + 6 = 20 \text{ ohms}$$

Other compound circuits may be solved in a similar manner. For example, in figure 5-17, the total resistance (R_t) may be found by simplifying the circuit in successive steps beginning with the resistance, R_1 and R_2. Thus,

$$R_{1,2} = \frac{R_1 R_2}{R_1 + R_2} = \frac{3 \times 6}{3 + 6} = \frac{18}{9} = 2 \text{ ohms}$$

and it is in series with R_3.

The resistances, $R_{1,2}$ and R_3, are added to give the resultant resistance, $R_{1,2,3}$. Thus,

$$R_{1,2,3} = R_{1,2} + R_3 = 2 + 4 = 6 \text{ ohms}$$

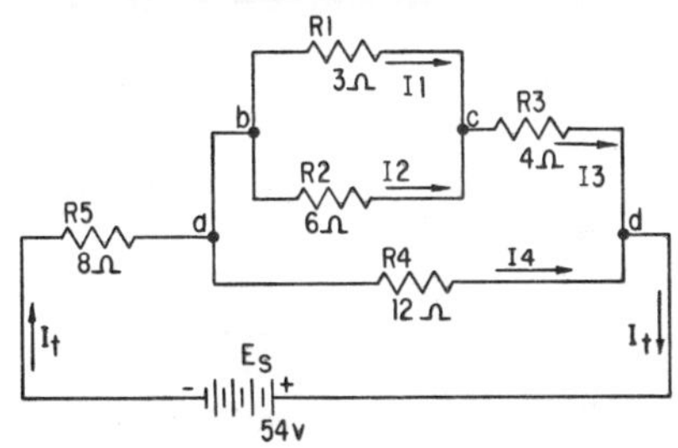

Figure 5-17.—Compound circuit for solving resistance, voltage, current, and power.

$R_{1,2,3}$ is in parallel with R_4. The combined resistance, $R_{1,2,3,4}$, is determined as follows:

$$R_{1,2,3,4} = \frac{R_{1,2,3}R_4}{R_{1,2,3} + R_4} = \frac{6 \times 12}{6 + 12} = \frac{72}{18} = 4 \text{ ohms}$$

This equivalent resistance is in series with R_5. Thus, the total resistance (R_t) of the circuit is

$$R_t = R_{1,2,3,4} + R_5 = 4 + 8 = 12 \text{ ohms}$$

By Ohm's Law, the line current (I_t) is

$$I_t = \frac{E_s}{R_t} = \frac{54}{12} = 4.5 \text{ amperes}$$

The line current flows through R_5 and therefore the voltage drop, E_5, across R_5 is

$$E_5 = I_tR_5 = 4.5 \times 8 = 36 \text{ volts}$$

According to Kirchhoff's voltage law, the sum of the voltage drops around the circuit is equal to the source voltage; accordingly, the voltage between points a and d is

$$E_{ad} = E_s - E_5 = 54 - 36 = 18 \text{ volts}$$

The current through R_4 is

$$I_4 = \frac{E_4}{R_4} = \frac{18}{12} = 1.5 \text{ amperes}$$

The resistance, $R_{1,2,3}$, of parallel resistors R_1 and R_2 in series with resistor R_3 is 6 ohms.

NOTES:

E_{ad} is applied across 6 ohms; therefore the current, I_3, through R_3 is

$$I_3 = \frac{E_{ad}}{R_{1,2,3}} = \frac{18}{6} = 3 \text{ amperes}$$

The voltage drop, E_3, across R_3 is

$$E_3 = I_3R_3 = 3 \times 4 = 12 \text{ volts}$$

and the voltage across the parallel combination of R_1 and R_2—that is, E_{bc}—is

$$E_{bc} = I_{1,2}R_{1,2} = 3 \times 2 = 6 \text{ volts}$$

where $I_{1,2}$ is the current through the parallel combinations of R_1 and R_2. By Kirchhoff's current law, $I_{1,2}$ is equal to I_3. The current, I_1 through R_1 is

$$I_1 = \frac{E_{bc}}{R_1} = \frac{6}{3} = 2 \text{ amperes}$$

and the current, I_2, through R_2 is

$$I_2 = \frac{E_{bc}}{R_2} = \frac{6}{6} = 1 \text{ ampere}$$

The preceding computations may be checked by the application of Kirchhoff's voltage and current law to the entire circuit. Briefly, the sum of the voltage drops around the circuit is equal to the source voltage. Voltage E_5 across R_5 is 36 volts and voltage E_{ad} across R_4 is 18 volts—that is,

$$E_s = E_5 + E_{ad}$$

or

$$54 = 36 + 18 \text{ volts}$$

Likewise, the voltage drop, E_{bc}, across the parallel combination of R_1 and R_2 plus the voltage drop, E_3, across R_3 should be equal to the voltage across points a and d. E_{bc} is 6 volts and E_3 is 12 volts. Therefore,

$$E_{ad} = E_{bc} + E_3 = 6 + 12 = 18 \text{ volts}$$

Kirchhoff's current law says in effect that the sum of the branch currents is equal to the line current, I_t. The line current is 4.5 amperes, and therefore the sum of I_4 and I_3 should be 4.5 amperes, or

$$I_t = I_4 + I_3 = 1.5 + 3 = 4.5 \text{ amperes}$$

The power consumed in a circuit element is determined by one of the three power formulas. For example, in figure 5-17 the power, P_1 consumed in R_1 is

$$P_1 = I_1E_{bc} = 2 \times 6 = 12 \text{ watts}$$

the power P_2 consumed in R_2 is

$$P_2 = I_2E_{bc} = 1 \times 6 = 6 \text{ watts}$$

the power P_3 consumed in R_3 is

$$P_3 = I_3E_3 = 3 \times 12 = 36 \text{ watts}$$

the power P_4 consumed in R_4 is

$$P_4 = I_4E_4 = 1.5 \times 18 = 27 \text{ watts}$$

and the power P_5 consumed in R_5 is

$$P_5 = I_5E_5 = 4.5 \times 36 = 162 \text{ watts}$$

The total power P_t, consumed is

$$P_t = P_1 + P_2 + P_3 + P_4 + P_5$$

$$= 12 + 6 + 36 + 27 + 162$$

$$= 243 \text{ watts}$$

The total power is also equal to the total current multiplied by the source voltage, or

$$P_t = I_tE_s = 4.5 \times 54 = 243 \text{ watts}$$

EFFECTS OF SOURCE RESISTANCE

The parallel circuits discussed up to this point have been explained and solved without considering the internal resistance of the source. Every known source possesses resistance. In a battery the resistance is partially due to the opposition offered to the movement of current through the electrolyte. A schematic representation of source resistance is shown in figure 5-18.

The internal resistance of the battery is labeled (R_i) and is always shown schematically connected in series with the source. Under load conditions this internal resistance will have a voltage drop across it and must be considered as part of the external circuit. The voltage at battery terminals A and B will always be less than the generated voltage of the battery since a portion of the generated voltage will be dropped across the internal resistance of the battery.

NOTES:

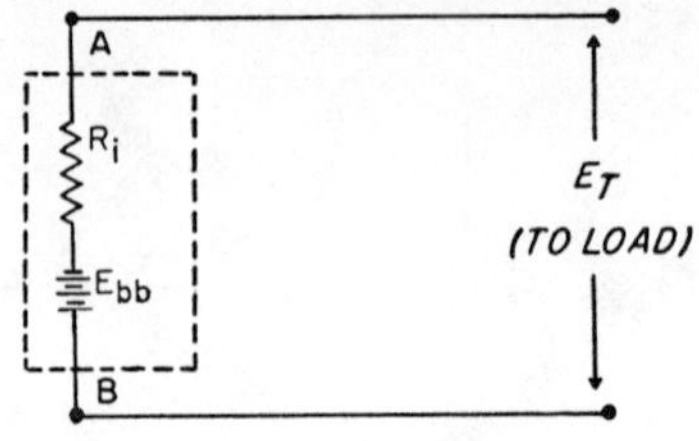

Figure 5-18.—Battery with internal resistance.

The presence of internal resistance results in (1) a diminished voltage supplied to the components that comprise the load, (2) a decrease in total current, and (3) an increase in total resistance. The power dissipated by the circuit is also affected. The effect of internal resistance on the circuit is analyzed using the example circuit shown in figure 5-19.

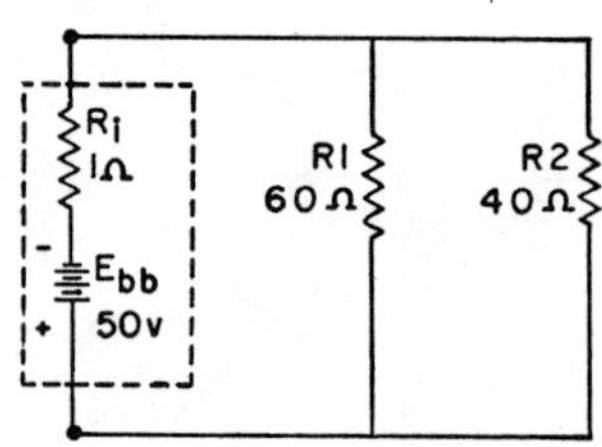

Figure 5-19.—Effect of source resistance on a parallel circuit.

The circuit shown in figure 5-19 can no longer be classified as a parallel circuit because there is a series resistance to be considered. The circuit is solved in the following manner.

Determine R_{eq} for the parallel network:

$$R_{eq} = \frac{R_1 \times R_2}{R_1 + R_2} = \frac{60 \times 40}{60 + 40} = \frac{2400}{100} = 24 \text{ ohms}$$

Reduce to an equivalent circuit (fig. 5-20).

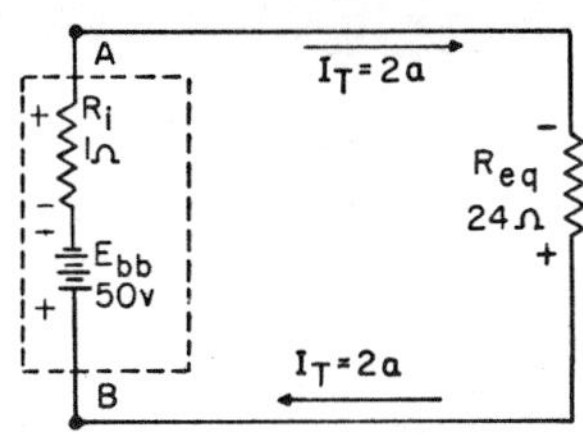

Figure 5-20.—Equivalent circuit.

Compute the total series resistance:

$$R_t = R_i + R_{eq}$$

$$R_t = 1 + 24 = 25 \text{ ohms}$$

Compute total current:

$$I_t = \frac{E_{bb}}{R_t} = \frac{50 \text{ v}}{25} = 2 \text{ amperes}$$

Determine voltage drop across R_{eq}:

$$E_{R_{eq}} = I_t \times R_{eq}$$

$$E_{R_{eq}} = 2 \times 24$$

$$E_{R_{eq}} = 48 \text{ volts}$$

Find voltage drop across R_i:

$$E_{R_i} = I_t \times R_i$$

$$E_{R_i} = 2\text{a} \times 1 \text{ ohm}$$

$$E_{R_i} = 2 \text{ volts}$$

Determine power dissipated by load resistors:

$$P_{R_{eq}} = I_t \times E_{R_{eq}}$$

$$P_{R_{eq}} = 2\text{a} \times 48 \text{ v}$$

$$P_{R_{eq}} = 96 \text{ w}$$

Determine power dissipated by source resistance:

$$P_{R_i} = I_t \times E_{R_i}$$

$$P_{R_i} = 2\text{a} \times 2 \text{ v}$$

$$P_{R_i} = 4 \text{ w}$$

Determine total power dissipation:

$$P_t = P_{R_{eq}} + P_{R_i}$$

$$P_t = 96\text{w} + 4\text{w}$$

$$P_t = 100 \text{ w}$$

Circuit efficiency is determined by the following formula:

$$\text{Percent Eff} = \frac{P_o}{P_{in}} \times 100$$

where

Percent Eff = percent of efficiency

P_o = power supplied to load device

P_{in} = power supplied by the source

For the circuit of figure 5-20 the percent Efficiency is:

$$\text{Percent Eff} = \frac{P_o}{P_{in}} \times 100$$

$$\text{Percent Eff} = \frac{96}{96 + 4} \times 100$$

$$\text{Percent Eff} = \frac{96\text{w}}{100\text{w}} \times 100 = 96 \text{ percent}$$

From this efficiency relationship, we may conclude that the source resistance does affect the total power dissipated by the equivalent (load) resistance. The source resistance also affects the transfer of power. As stated in the preceding chapter, maximum transfer of power occurs when the circuit is 50 percent efficient, or when there is an equal amount of voltage dropped across the load and the source resistance.

OPEN AND SHORT CIRCUITS

In comparing the effects of an open in series and parallel circuits, the major difference to be noted is that an open in a parallel circuit would not necessarily disable the entire circuit; i.e., the current flow would not be reduced to zero,

NOTES:

unless the open condition existed at some point electrically common to all other parts of the circuit.

A short circuit in a parallel network has an effect similar to a short in a series circuit. In general, the short will cause an increase in current and the possibility of component damage regardless of the type of circuit involved.

Opens and shorts, alike, if occurring in a branch circuit of a parallel network, will result in an overall change in the equivalent resistance. This can cause undesirable effects in other parts of the circuit due to the corresponding change in the total current flow.

To prevent damage to equipment due to a short circuit, a fuse or overload relay is normally placed in the circuit in series with the more sensitive components or in series with the source. The effects of a short circuit occurring in a fused network is shown in figure 5-21 and is explained as follows:

In figure 5-21, with the switch in position one (as shown), a value of current flows that does not exceed the rated current capacity of the fuse. If the switch is thrown to position two, the straight wire conductor will be in parallel with the load resistors. The equivalent resistance of the straight wire and the resistors, all connected in parallel, will be less than the resistance of the straight wire. This follows from the fact that the total resistance of a parallel circuit is always less than the smallest resistance in the branch. Since a complete path still exists to permit current flow, and the equivalent resistance is effectively zero, the current will rise rapidly until the current capacity of the fuse is reached. The fuse will then open the circuit causing the current to stop flowing. A short usually causes components to fail in a circuit which is not properly fused, or otherwise protected. The failure may take the form of a burned-out resistor, damaged source, or a fire in the circuit components and wiring.

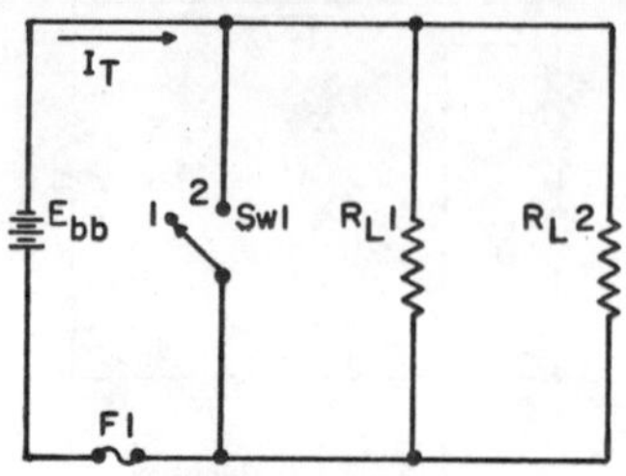

Figure 5-21.—Example of a circuit protected from shorts by a fuse.

VOLTAGE DIVIDER

In practically all electronic devices, such as radio receivers and transmitters, certain design requirements recur again and again. For instance, a typical radio receiver may require a number of different voltages at various points in its circuitry. In addition, all the various voltages must be derived from a single primary power supply. The most common method of meeting these requirements is by the use of a voltage-divider network. A typical voltage divider consists of two or more resistors connected in series across the primary power supply. The primary voltage E_S must be as high or higher than any of the individual voltages it is to supply. As the primary voltage is dropped by successive steps through the series resistors, any desired fraction of the original voltage may be "tapped off" to supply individual requirements. The values of the series resistors to be used is dictated by the voltage drops required.

If the total current flowing in the divider circuit is affected by the loads placed on it, then the voltage drops of each divider resistor will also be affected. When a voltage divider is being designed, the maximum current drawn by the loads will determine the value of the resistors that form the voltage divider. Normally, the resistance values chosen for the divider will permit a current equal to 10 percent of the total current drawn by the external loads. This current which does not flow through any of the load devices is called bleeder current.

A voltage divider circuit is shown in figure 5-22. The divider is connected across a 270-volt source and supplies three loads simultaneously—10 ma (1 milliampere is 0.001 ampere) at 90 volts, between terminal 1 and ground; 5 ma at 150 volts, between terminal 2 and ground; and 30 ma at 180 volts, between terminal 3 and ground. The current in resistor A is 15 ma. The current, voltage, resistance, and power of the 4 resistors are to be determined.

Kirchhoff's law of currents applied to terminal 1 indicates that the current in resistor B

NOTES:

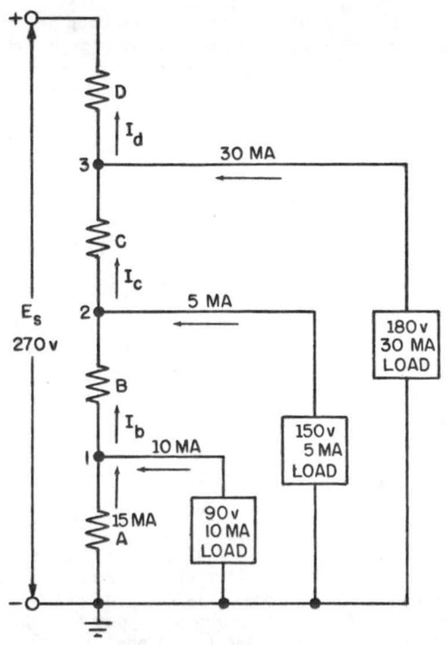

Figure 5-22.—Voltage divider, to determine R and P.

is equal to the sum of 15 ma from resistor A and 10 ma from the 90-volt load. Thus,

$$I_b = 15 + 10 = 25 \text{ ma}$$

Similarly,

$$I_c = 25 + 5 = 30 \text{ ma}$$

and

$$I_d = 30 + 30 = 60 \text{ ma}$$

Kirchhoff's voltage law indicates that the voltage across resistor A is 90 volts; the voltage across B is

$$E_b = 150 - 90 = 60 \text{ volts}$$

the voltage across C is

$$E_c = 180 - 150 = 30 \text{ volts}$$

NOTES:

and the voltage across D is

$$E_d = 270 - 180 = 90 \text{ volts}$$

Before solving for the various resistances, it should be recalled that in the formula, $R = \frac{E}{I}$, R will be in ohms if E is in volts and I is in amperes. In many electronic circuits, particularly those being considered, it is just as valid and considerably simpler to let R be in thousands of ohms (k-ohms), E in volts, and I in milliamperes. In the following formulas this convention will be followed.

Applying Ohm's law to determine the resistances—

resistance of A if $R_a = \frac{E_a}{I_a} = \frac{90}{15} = 6$ k-ohms

resistance of B if $R_b = \frac{E_b}{I_b} = \frac{60}{25} = 2.4$ k-ohms

resistance of C is $R_c = \frac{E_c}{I_c} = \frac{30}{30} = 1$ k-ohm

resistance of D is $R_d = \frac{E_d}{I_d} = \frac{90}{60} = 1.5$ k-ohms

The power absorbed by

resistor A is $P_a = E_a I_a = 90 \times 0.015 = 1.35$ watts

resistor B is $P_b = E_b I_b = 60 \times 0.025 = 1.50$ watts

resistor C is $P_c = E_c I_c = 30 \times 0.030 = 0.90$ watt

resistor D is $P_d = E_d I_d = 90 \times 0.060 = 5.40$ watts

The total power absorbed by the 4 resistors is

$$1.35 + 1.50 + 0.90 + 5.40 = 9.15 \text{ watts}$$

The power absorbed by the load connected to

terminal 1 is $P_1 = E_1 I_1 = 90 \times 0.010 = 0.90$ watt

terminal 2 is $P_2 = E_2 I_2 = 150 \times 0.005 = 0.75$ watt

terminal 3 is $P_3 = E_3 I_3 = 180 \times 0.030 = 5.4$ watts

The total power supplied to the 3 loads is

$$0.90 + 0.75 + 5.4 = 7.05 \text{ watts}$$

The total power supplied to the entire circuit including the voltage divider and the 3 loads is

$$9.15 + 7.05 = 16.2 \text{ watts}$$

This value is checked as

$$P_t = E \times I_t = 270 \times 0.060 = 16.2 \text{ watts}$$

In figure 5-23 the voltage divider resistances are given and the current in R_5 is to be found. The load current in R_1 is 6 ma; the current in R_2 is 4 ma; and the current in R_3 is 10 ma. The source voltage is 510 volts. Kirchhoff's current law may be applied at the junctions a, b, c, and d to determine expressions for the current in resistors R_4, R_5, R_6, and R_7. Accordingly, the current in R_4 is I + 6 + 4 + 10, or I + 20; the current in R_5 is I; the current in R_6 is I + 6; the current in R_7 is I + 6 + 4, or I + 10.

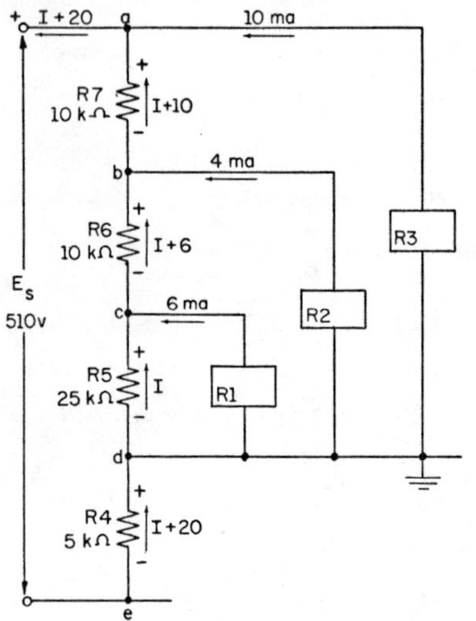

Figure 5-23.—Voltage divider, to determine E and R.

The voltage across R_4 may be expressed in terms of the resistance in k-ohms and the current in milliamperes as 5(I + 20) volts. Similarly, the voltage across R_5 is equal to 25I; the voltage across R_6 is 10(I + 6) and the voltage across R_7 is 10(I + 10). Kirchhoff's law of voltages may be applied to the voltage divider to solve for the unknown current, I, by expressing the source voltage in terms of the given values of voltage, resistance, and current (both known and unknown values). The sum of the voltages across R_4, R_5, R_6, and R_7 is equal to the source voltage as follows:

$$E_4 + E_5 + E_6 + E_7 = E_s$$

$$5(I + 20) + 25 I + 10(I + 6) + 10(I + 10) = 510$$

$$5 I + 100 + 25 I + 10 I + 60 + 10 I + 100 = 510$$

$$50 I + 260 = 510$$

$$50 I = 510 - 260$$
$$50 I = 250$$
$$I = 5 \text{ ma}$$

The current of 5 ma through R_5 produces a voltage drop across R_5 of 5 x 25, or 125 volts. Since R_1 is in parallel with R_5, the voltage across load R_1 is 125 volts. The current through R_4 is 5 + 20, or 25 ma and the corresponding voltage is 5 x 25, or 125 volts. Since point d is at ground potential, point c is 125 volts positive with respect to ground, whereas point e is 125 volts negative with respect to ground. The current in R_6 is 5 + 6 or 11 ma and the voltage drop across R_6 is 11 x 10, or 110 volts. The current in R_7 is 5 + 10, or 15 ma and the voltage drop is 15 x 10 or 150 volts. The total voltage is the sum of the voltages across the divider. Thus,

$$125 + 125 + 110 + 150 = 510$$

The power absorbed by each resistor in the voltage divider may be found by multiplying the voltage across the resistor by the current in the resistor. If the current is expressed in amperes and the emf in volts, the power will be expressed in watts. Thus the power in R_4 is

$$P_4 = E_4 I_4 = 125 \times 0.025 = 3.125 \text{ watts}$$

Similarly the power in R_5 is 125 x 0.005 = 0.625 watt; the power in R_6 is 110 x 0.011 = 1.21 watts; and in R_7 is 150 x 0.015 = 2.25 watts. The total power in the divider is

$$3.125 + 0.625 + 1.21 + 2.25 = 7.21 \text{ watts}$$

The voltage across load R_1 is the voltage across R_5, or 125 volts. The power in R_1 is

$$P_1 = E_1 I_1 = 125 \times 0.006 = 0.750 \text{ watts}$$

NOTES:

The voltage across load R_2 is equal to the sum of the voltages across R_5 and R_6. Thus,

$$E_2 = E_5 + E_6 = 125 + 110 = 235 \text{ volts}$$

The power in load R_2 is

$$P_2 = E_2 I_2 = 235 \times 0.004 = 0.940 \text{ watt}$$

The voltage across load R_3 is equal to the sum of the voltages across R_5, R_6, and R_7. Thus,

$$E_3 = E_5 + E_6 + E_7 = 125 + 110 + 150 = 385 \text{ volts}$$

The power in load R_3 is

$$P_3 = E_3 I_3 = 385 \times 0.010 = 3.85 \text{ watts}$$

The total power in the three loads is

$$0.75 + 0.94 + 3.85 = 5.54 \text{ watts}$$

and the total power supplied by the source is equal to the sum of the power absorbed by the voltage divider and the three loads, or

$$7.21 + 5.54 = 12.75 \text{ watts}$$

The total power may be checked by

$$P_t = E_t I_t = 510 \times 0.025 = 12.75 \text{ watts}$$

The resistances of load resistors R_1, R_2, and R_3 are determined by means of Ohm's law as follows:

$$R_1 = \frac{E_1}{I_1} = \frac{125}{6} = 20.83 \text{ k-ohms}$$

$$R_2 = \frac{E_2}{I_2} = \frac{235}{4} = 58.75 \text{ k-ohms}$$

and

$$R_3 = \frac{E_3}{I_3} = \frac{385}{10} = 38.5 \text{ k-ohms}$$

The variation of voltages and currents found in the previous examples are undesirable in a voltage divider. It must be designed to provide voltages that are as stable as possible. A voltage divider consisting of two resistors will be designed using the circuit configuration shown in figure 5-24. The supply voltage is 200 volts. It is desired to furnish voltages of 50 and 200 volts to two loads drawing 6 and 18 milliamperes respectively. Assume bleeder current to be 10 percent of the required load current.

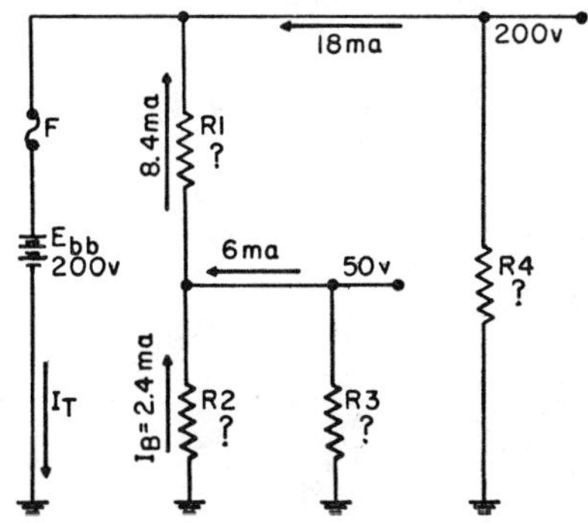

Figure 5-24.—Example circuit for proposed voltage divider.

Total load current is specified as 24 milliamperes. The bleeder current, therefore, should be

$$I_b = 10 \text{ percent } I_L$$

$$I_b = 10 \text{ percent} \times 24 \text{ ma}$$

$$I_b = 2.4 \text{ ma}$$

The bleeder current and the current through resistor R_3 combine and both currents flow through R_1. This current value may be computed

$$I_{R1} = I_b + I_{R3}$$

$$I_{R1} = 2.4 \text{ ma} + 6 \text{ ma}$$

$$I_{R1} = 8.4 \text{ ma}$$

The total current may also be determined

$$I_t = 8.4 \text{ ma} + 18 \text{ ma}$$

$$I_t = 26.4 \text{ ma}$$

The resistance values of R_3 and R_4 must be as follows:

$$R_3 = \frac{E_{R3}}{I_{R3}} = \frac{50}{6 \times 10^{-3}} = 8.33 \text{ k-ohms}$$

$$R_4 = \frac{E_{R4}}{I_{R4}} = \frac{200}{18 \times 10^{-3}} = 11.1 \text{ k-ohms}$$

NOTES:

Computing for R_1 and R_2

$$R_1 = \frac{E_{R1}}{I_{R1}} = \frac{150}{8.4 \times 10^{-3}} = 17.85 \text{ k-ohms}$$

$$R_2 = \frac{E_{R2}}{I_{R2}} = \frac{50}{2.4 \times 10^{-3}} = 20.82 \text{ k-ohms}$$

TYPICAL PROBLEMS IN SERIES-PARALLEL CIRCUITS

As seen by the preceding calculations, problems involving the determination of resistance, voltage, current, and power in a series-parallel circuit are relatively simple. The procedure is the same as for series and parallel circuits—(1) draw the circuit diagram, (2) state the values given and the values to be found, (3) state the applicable equations, and (4) substitute the given values and solve for the unknown. For an example refer to figure 5-25.

Problems:

1. Find the resistance of branch (a).
2. Find the resistance of branch (b).
3. Find the total circuit resistance.
4. Find the total circuit current.
5. Find the voltages E_{R1}, E_a, and E_b.
6. Find the current for branch (a) and (b).
7. Find the voltages E_{R2} and E_{R5}.
8. Find the currents I_1, I_2, I_3, and I_4.
9. Find the voltages E_{R3}, E_{R6}, and E_{R7}.
10. Find the power for R_8, branches (a) and (b), and R_1.
11. Find the total circuit power.

Solutions:

1. The resistance of branch (a) R_a is

$$R_a = \frac{R_3 \times R_4}{R_3 + R_4} + R_5$$

$$R_a = \frac{100 \times 100}{100 + 100} + 50$$

$$R_a = 50 + 50 = 100 \text{ ohms}$$

2. The resistance of branch (b) R_b is

$$R_b = R_2 + \frac{(R_7 + R_8)\, R6}{R_6 + R_7 + R_8}$$

$$R_b = 20 + \frac{(80 + 80)\ 160}{80 + 80 + 160}$$

$$R_b = 20 + 80 = 100 \text{ ohms}$$

3. The total circuit resistance R_T is

$$R_T = \frac{R_a \times R_b}{R_a + R_b} + R_1$$

$$R_T = \frac{100 \times 100}{100 + 100} + 50$$

$$R_T = 50 + 50 = 100 \text{ ohms}$$

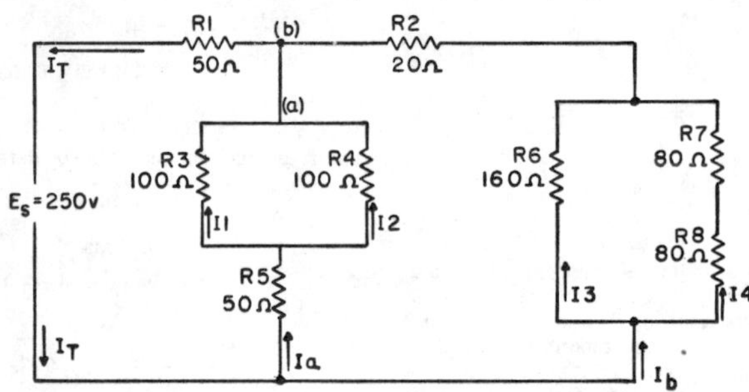

Figure 5-25.—Typical series-parallel circuit.

NOTES:

4. The total circuit current is $= I_T = \frac{E}{R_T}$

$$I_T = \frac{E}{R_T} = \frac{250}{100} = 2.5 \text{ amperes}$$

5. The voltage drop of R_1 is

$$E_{R1} = I_1 R_1 = 2.5 \times 50 = 125 \text{ volts}$$

The voltage for (a) is

$$E_a = E_S - E_{R1} = 250 - 125 = 125 \text{ volts}$$

The voltage for (b) is

$$E_b = E_S - E_{R1} = 250 - 125 = 125 \text{ volts}$$

6. The current for branch (a) is

$$I_a = \frac{E_a}{R_a} = \frac{125}{100} = 1.25 \text{ amperes}$$

The current for branch (b) is

$$I_b = \frac{E_b}{R_b} = \frac{125}{100} = 1.25 \text{ amperes}$$

7. The voltage drop across R_2 is

$$E_{R2} = I_b R_2 = 1.25 \times 20 = 25 \text{ volts}$$

The voltage drop across R_5 is

$$E_{R5} = I_a R_5 = 1.25 \times 50 = 62.5 \text{ volts}$$

8. The current I_1 is

$$I_1 = \frac{E_{R3}}{R_3} = \frac{62.5}{100} = 0.625 \text{ ampere}$$

The current I_2 is

$$I_2 = I_a - I_1 = 1.25 - 0.625 = 0.625 \text{ ampere}$$

The current I_3 is

$$I_3 = \frac{E_{R6}}{R_6} = \frac{100}{160} = 0.625 \text{ ampere}$$

The current for I_4 is

$$I_4 = I_b - I_3 = 1.25 - 0.625 = 0.625 \text{ ampere}$$

9. The voltage drop across R_3 is

$$E_{R3} = I_1 R_3 = 0.625 \times 100 = 62.5 \text{ volts}$$

The voltage drop across R_6 is

$$E_{R6} = I_3 R_6 = 0.625 \times 160 = 100 \text{ volts}$$

The voltage drop across R_8 is

$$E_{R8} = I_4 R_8 = 0.625 \times 80 = 50 \text{ volts}$$

10. The power consumed by R_8 is

$$P_{R8} = I_4 E_{R8} = 0.625 \times 50 = 31.25 \text{ watts}$$

The power consumed by branch (a) is

$$P_a = I_a E_a = 1.25 \times 125 = 156.25 \text{ watts}$$

The power consumed by branch (b) is

$$P_b = I_b E_b = 1.25 \times 125 = 156.25 \text{ watts}$$

The power consumed by R, is

$$P_{R1} = I_{R1} E_{R1} = 2.5 \times 125 = 312.5 \text{ watts}$$

11. The total power consumed by the circuit is

$$P_T = P_{R1} + P_a + P_b = 312.5 + 156.25 + 156.25 = 625 \text{ watts}$$

or $$P_T = E\, I_T = 250 \times 2.5 = 625 \text{ watts}$$

NOTES:

CHAPTER 6

NETWORK ANALYSIS OF D-C CIRCUITS

SPECIAL NETWORK TECHNIQUES

The circuit solutions studied up to this point have been accomplished mainly through the use of formulas derived from Ohm's law. Like many other fields of science, electronics has its share of special shortcut methods. These methods, however, must be reserved until enough background theory has been presented to make their use worthwhile. This chapter will therefore be devoted to methods of solution which either simplify circuit calculations or solve circuits which cannot be solved by ordinary methods.

LOOP ANALYSIS

Loop analysis is a valuable method of circuit analysis in which Kirchhoff's voltage law is the key to the solution. In this method, two or more equations are formed which are then solved simultaneously.

Figure 6-1 shows a network containing five resistors and a source. This network will not actually be solved but is included so that the terms and procedures can be defined. In solving this circuit by loop analysis, three currents are assumed as shown. Each current is arbitrarily assigned a counterclockwise direction (the true direction is unimportant at this time). Each of the three closed current paths is called a mesh. The individual circuit components (resistors) which form the meshes are called elements.

To solve the circuit in figure 6-1, three equations are formed, one for each mesh. The number of equations required is always equal to the number of meshes. These equations are then solved simultaneously.

TYPICAL SOLUTION

In this section the combination circuit shown in figure 6-2 will be solved using the loop method of analysis. Notice that the given circuit is a two mesh circuit and therefore two equations are required for the solution.

To begin the solution, a current circulating in a counterclockwise direction is assumed in each mesh. In addition, it is assumed that these currents cause voltage drops across the circuit resistors. Polarities are then assigned to each resistor, according to the direction of current flowing through the resistor. Notice that resistor R_2 carries two currents which flow in opposite directions. A separate set of polarity signs is used for the voltage drops caused by each of these currents. After assigning currents and polarities to the circuit, the voltage equations can be formed.

Recalling Kirchhoff's statement that the algebraic sum of the emf's and voltage drops around any closed loop is equal to zero, an equation can be formed for loop ABEF of figure 6-2. This equation is written starting at point A and tracing around the loop in the direction of assumed current. The polarities used for each voltage drop in the equation are those found following the component traced through. For example, in tracing through R_1 from E to F the positive polarity sign is used. (For those needing a review of Kirchhoff's voltage equations, refer to chapter 5.) Therefore, the equation for loop ABEF is

$$\text{ABEF:} \quad 20I_1 - 20I_2 + 8I_1 - 40 = 0 \qquad (1)$$

$$\text{Simpliying:} \quad 28I_1 - 20I_2 = 40 \qquad (2)$$

Notice that in passing through R_2, two voltage drops of opposite polarity are encountered. Both of these drops ($20I_1$ and $-20I_2$) must be included in the equation along with their proper polarities.

NOTES:

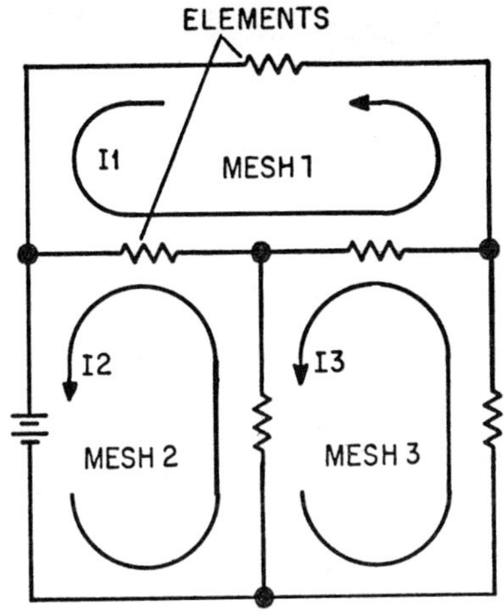

Figure 6-1.—A three mesh network.

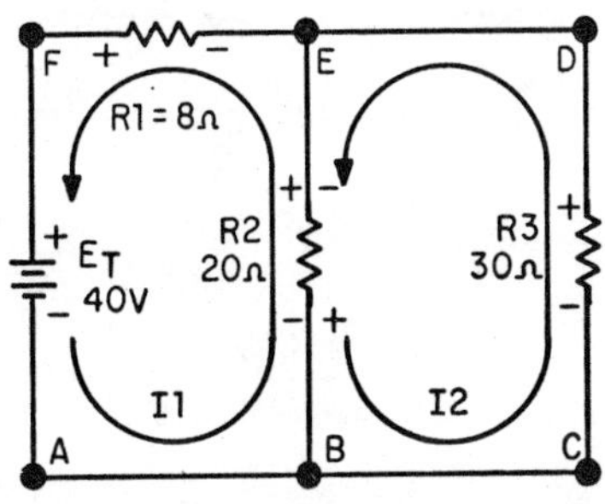

Figure 6-2.—Example circuit.

Next, an equation is written for loop BCDE.

This equation is

BCDE: $+30I_2 + 20I_2 - 20I_1 = 0$ (3)

Simplifying: $20I_1 + 50I_2 = 0$ (4)

NOTES:

Again notice that two opposing voltage drops ($+20I_2$ and $-20I_1$) were included for R_2.

Equations (2) and (4), repeated below, are now solved simultaneously to obtain I1 and I2.

$$28I_1 - 20I_2 = 40 \quad (2)$$

$$-20I_1 + 50I_2 = 0 \quad (4)$$

In order to eliminate I_2, equation (2) is multiplied by 5 and equation (4) is multiplied by 2. The resulting equations (5) and (6) are then added.

$$140I_1 - 100I_2 = 200 \quad (5)$$

$$-40I_1 + 100I_2 = 0 \quad (6)$$

$$100I_1 = 200$$

Dividing both sides by 100:

$$I_1 = \frac{200}{100}$$

$$I_1 = 2 \text{ amperes}$$

To obtain the value of I_2, 2 amperes is now substituted into equation (4) and in place of I_1.

$$-20 \times 2 + 50I_2 = 0$$

$$-40 + 50I_2 = 0$$

$$50I_2 = 40$$

$$I_2 = 0.8 \text{ ampere}$$

Thus, current I_1 is 2 amperes and current I_2 is 0.8 ampere. The voltage drops can now be evaluated using these currents and Ohm's law as follows:

$$E_{R1} = I_1R_1$$

$$E_{R1} = 2 \times 8$$

$$E_{R1} = 16 \text{ volts}$$

Notice that the actual current through R_2 is the algebraic sum of the two opposing currents. Therefore, the voltage across R_2 is

$$E_{R2} = (I_1 - I_2)\,R_2$$

$$E_{R2} = 1.2 \times 20$$

$$E_{R2} = 24 \text{ volts}$$

$$E_{R3} = I_2R_3$$

$$E_{R3} = 0.8 \times 30$$

$$E_{R3} = 24 \text{ volts}$$

MULTIPLE SOURCE CIRCUITS

Quite frequently, networks containing more than one source must be solved. Although a circuit of this type may look complicated, the solution is no more difficult than the one discussed previously. In fact, the same method of solution is used in both single and multiple source circuits.

Figure 6-3 shows a multiple source circuit which will be used in the example solution. In the diagram a counterclockwise current has been assumed in each mesh and polarities assigned accordingly. Note that R_2 has two opposing voltage drops, one for each current. The voltage equation for loop ABEF is

$$+5 + 20I_1 - 20I_2 + 20 + 5I_1 - 50 = 0 \quad (7)$$

Simplifying: $$25I_1 - 20I_2 = 25 \quad (8)$$

Loop BCDE:

$$-100 + 7I_2 + 20 + 10I_2 + 20I_2 - 20I_1 - 5 = 0 \quad (9)$$

Simplifying: $$-20I_1 + 37I_2 = 85 \quad (10)$$

Multiply (8) by 4: $$100I_1 - 80I_2 = 100 \quad (11)$$
Multiply (10) by 5: $$-100I_1 + 185I_2 = 425 \quad (12)$$
Add: $$+105I_2 = 525$$

$$I_2 = \frac{525}{105}$$

$$I_2 = 5 \text{ amperes}$$

Substitute (13) in (8)

$$25I_1 - 20(5) = 25$$

$$25I_1 - 100 = 25$$

$$25I_1 = 125$$

$$I_1 = 5 \text{ amperes} \quad (14)$$

Had one of the above currents been negative, this would have indicated an incorrectly assumed direction of current flow. The magnitude of current would be correct. Now that the currents are known, the voltage drops across the resistors can be calculated.

$$E_{R1} = I_1R_1$$

$$E_{R1} = 25 \text{ volts}$$

$$E_{R2} = (I_1 - I_2)R_2$$

$$E_{R2} = 0 \text{ volts}$$

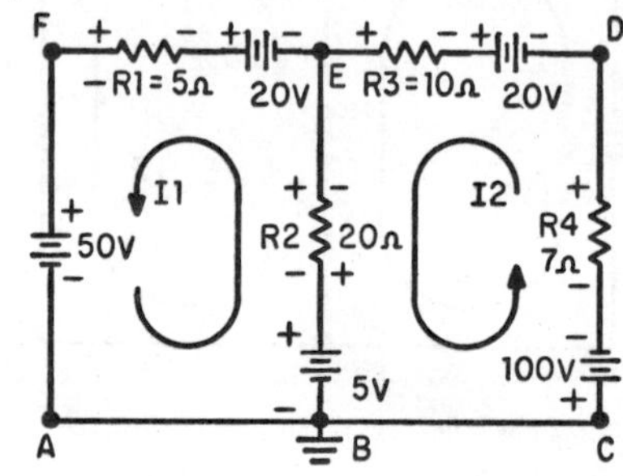

Figure 6-3.—Multiple source circuit.

Notice that equal and opposite currents flow through R_2 and no voltage drop occurs across it.

$$E_{R3} = I_2R_3$$

$$E_{R3} = 50 \text{ volts}$$

$$E_{R4} = I_2R_4$$

$$E_{R4} = 35 \text{ volts}$$

EQUIVALENT CIRCUITS

In the analysis of many circuits, the solution is primarily concerned with the computation of values for load current and load voltage. In most cases additional calculations are required as intermediate steps in the solution. The following discussion will show how many of these intermediate steps can be eliminated.

Let us now examine the load resistor (R_L) shown in figure 6-4. This resistor is connected

NOTES:

to a switch so that it can be connected across the terminals of either of the two circuits. When the switch is in the position illustrated, R_L is connected to circuit A.

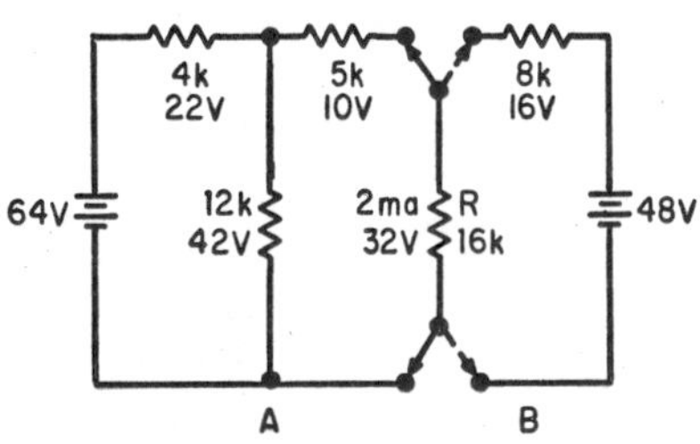

Figure 6-4.—Equivalent circuits.

By using a lengthy series of steps, the values of load current (2 ma) and load voltage (32 v) can be computed. Using the methods studied up to this time, computations of total resistance, total current, and various voltage drops would be required as intermediate steps of the solution. (An alternate method would be loop analysis.)

Keeping in mind the values of load voltage and current determined for circuit A, assume the switch to be in the position which places R_L across the terminals of circuit B. In this simple series circuit the load voltage and current can be easily determined. Notice that again the load voltage is 32 volts and the load current is 2 milliamperes. Since the load voltage and current are the same regardless of which circuit is used, as far as the load is concerned circuit B could be substituted for circuit A. Any circuit that may be substituted for another circuit is called an equivalent circuit. The purpose of this section is to show how a simple equivalent circuit can be developed for any complex circuit. This simple equivalent circuit is then used for the calculations instead of the original complex circuit. It should not be necessary to emphasize the hours of work that can be saved by utilizing an equivalent circuit.

THEVENIN'S THEOREM

One of the most valuable equivalent circuits is one known as Thevenin's equivalent circuit. This circuit is derived from Thevenin's Theorem, stated as follows:

Thevenin's Theorem: Any linear network of impedance and sources, if viewed from any two points in the network, can be replaced by an equivalent impedance Z_{th} in series with an equivalent voltage source E_{th}. (The term impedance means any opposition to current flow and in d-c circuits will be taken to mean resistance.)

According to this theorem, any linear d-c circuit regardless of its complexity can be replaced by a Thevenin's equivalent shown in figure 6-5.

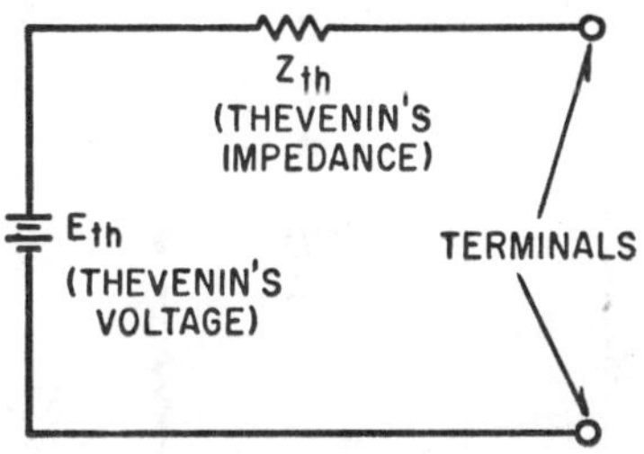

Figure 6-5.—Thevenin's equivalent circuit.

The process whereby a Thevenin's equivalent circuit is developed for a given network is best illustrated by an example.

Assume that the circuit in figure 6-6 (A) is to be used to develop a Thevenin's equivalent circuit. Basically the problem consists of finding values of E_{th} and Z_{th}. These quantities can be found using the following procedure.

APPLICATION OF THEVENIN'S THEOREM

1. Disconnect the section of the circuit considered as the load (R_L in fig. 6-6 (A)).
2. By measurement or calculation, determine the voltage that would appear between the load terminals with the load disconnected (terminals X and Y). This open circuit voltage is called Thevenin's voltage (E_{th}).
3. Replace each source within the circuit by its internal impedance. (A constant voltage

NOTES:

source such as a battery is replaced with a short, while a constant current source is replaced with an open. (See fig. 6-6 (B).)

4. By measurement or calculation, determine the impedance (resistance) the load would see, looking back into the network from the load terminals. This is Thevenin's impedance (Z_{th}). (See fig. 6-6 (B).)

5. Draw the equivalent circuit consisting of R_L and Z_{th} in series, connected across source E_{th} (fig. 6-6 (C). Solve for the load current and voltage.

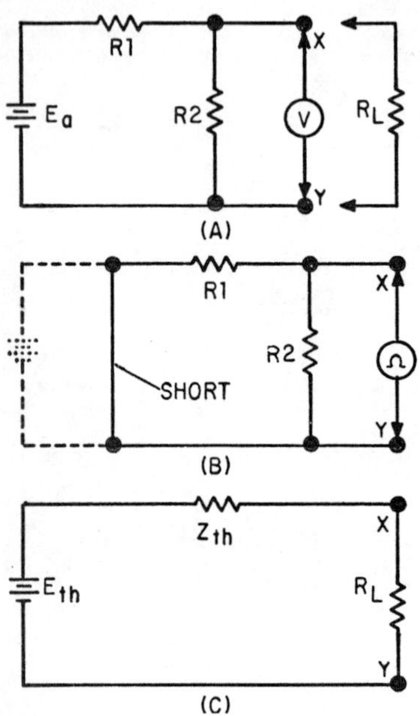

Figure 6-6.—Developing a Thevenin's equivalent.

A TYPICAL SOLUTION

In the following solution, the load voltage will be computed using loop analysis. The circuit will then be solved a second time using a Thevenin's equivalent circuit. The two methods can thus be compared as to results and ease of computation.

Figure 6-7 shows a three mesh circuit for which the output voltage E_L is to be determined. Currents are assumed in each mesh and appropriate polarities are assigned. The voltage equations are then written as follows:

ABGH: $$+10I_1 - 10I_2 - 110 = 0 \quad (15)$$

$$10I_1 - 10I_2 = 110 \quad (16)$$

BCFG: $$+30I_2 - 30I_3 + 20I_2 + 10I_2 - 10I_1 = 0 \quad (17)$$

$$-10I_1 + 60I_2 - 30I_3 = 0 \quad (18)$$

CDEF: $$+10I_3 + 30I_3 - 30I_2 = 0 \quad (19)$$

$$-30I_2 + 40I_3 = 0 \quad (20)$$

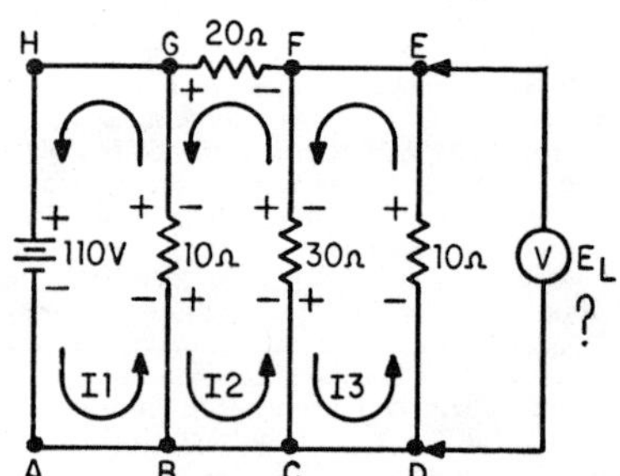

Figure 6-7.—Example circuit.

Adding (16) to (18):

$$\begin{aligned} 10I_1 - 10I_2 &= 110 \\ -10I_1 + 60I_2 - 30I_3 &= 0 \\ \hline 50I_2 - 30I_3 &= 110 \quad (21) \end{aligned}$$

Multiplying (20) by 3: $-90I_2 + 120I_3 = 0$

Multiplying (21) by 4: $200I_2 - 120I_3 = 440$

Adding: $$110I_2 = 440 \quad (22)$$

$$I_2 = 4 \text{ amperes} \quad (23)$$

NOTES:

Substituting (23) into (20)

$$-30(4) + 40I_3 = 0$$

$$-120 + 40I_3 = 0$$

$$40I_3 = 120$$

$$I_3 = 3 \text{ amperes} \qquad (24)$$

Substituting (23) into (16)

$$10I_1 - 10(4) = 110$$

$$10I_1 - 40 = 110$$

$$10I_1 = 150$$

$$I_1 = 15 \text{ amperes} \qquad (25)$$

Once all three currents are known, the voltages are computed and added for each loop as a check. If the computed voltages around each loop add up to zero, the computed currents are correct. The output voltage E_L is then

$$E_L = I_3 R_L$$

$$E_L = 30 \text{ volts}$$

Most people will agree that the above solution is long and tedious. The same circuit will now be solved through the use of a Thevenin's equivalent circuit. The procedure is as follows:

1. Disconnect the load as shown in (A) of figure 6-8.
2. Compute E_{th}, the no load (open circuit) voltage between terminals E and D. This voltage is the same as the voltage across R_3. Since 110 volts are applied to R_2 and R_3 in series, the voltage across R_3 is three-fifths of 110 volts or 66 volts.
3. Replace the source with its internal impedance. In this case the source is a battery (constant voltage source) and is replaced with a short as in (B) of figure 6-8. This shorts out both the battery terminals and R1, resulting in the circuit shown in (C) of figure 6-8.

NOTE: The internal resistance of a flashlight cell is approximately 0.005 ohm. Thus, the internal resistance of a battery is usually considered to be 0 ohms.

NOTES:

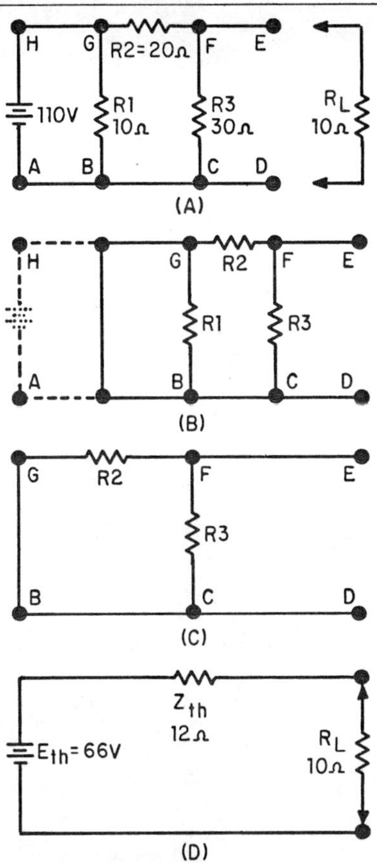

Figure 6-8.—Evolution of the equivalent circuit.

4. Determine the resistance (Z_{th}) the load would see "looking back" into the network from terminals E and D. Notice that in (C) of the figure, two separate paths exist between terminals E and D. One of these paths is through R_2 and the other is through R_3, indicating the two resistors to be in parallel. Since R_2 and R_3 are in parallel, this resistance Z_{th} is

$$Z_{th} = \frac{R_2R_3}{R_2+R_3}$$

$$Z_{th} = \frac{20x30}{20+30}$$

$$Z_{th} = 12 \text{ ohms}$$

5. Draw the equivalent circuit and connect the load resistance as in (D) of figure 6-8, including the values obtained for E_{th} and Z_{th}. Using Ohm's law, solve for the load current and voltage.

$$I_L = \frac{E_{th}}{Z_{th}+ R_L}$$

$$I_L = \frac{66}{22}$$

$$I_L = 3 \text{ amperes}$$

Notice that the load currents (I_3 in the loop analysis and I_L in Thevenin's equivalent) are identical.

$$E_L = I_L R_L$$

$$E_L = 3 \text{ x } 10$$

$$E_L = 30 \text{ volts}$$

This is the same value of voltage found by loop analysis. At this point one should stop and compare the labor required by each method in order to reach a solution. Once the steps used in applying Thevenin's theorem are learned, this method is by far the simpler of the two methods.

VOLTAGE DIVIDER EQUATION

As an aid to the application of Thevenin's equivalent circuit, an equation can be derived which will yield the load voltage in one simple calculation. This equation (the derivation will be left as a problem for the student) is:

$$E_L = \frac{E_{th} R_L}{Z_{th}+R_L} \qquad (26)$$

To illustrate the use of equation (26) the Thevenin equivalent will be developed for the circuit in figure 6-9. As before, the load circuit will be opened to determine E_{th}, the open circuit load voltage. With the switch in figure 6-9 open, no load current flows through R_L or R_3. The voltage between X and Y is therefore the same as the voltage across R_2. Applying the voltage divider formula this voltage is found to be:

$$E_{R2} = \frac{E_aR_2}{R_1+R_2}$$

$$E_{R2} = \frac{100 \text{ x } 30}{50}$$

$$E_{R2} = 60 \text{ volts}$$

Since this is the open circuit load voltage, this voltage is E_{th}.

$$E_{th} = 60 \text{ volts}$$

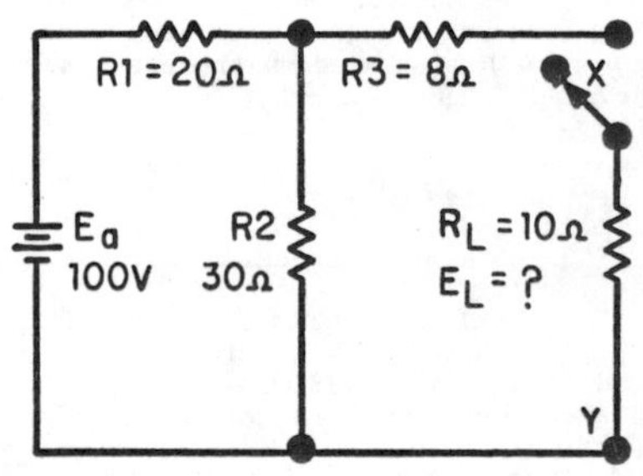

Figure 6-9.—Example circuit.

Next the source is replaced with a short circuit. This places R_1 and R_2 in parallel and the impedance Z_{th} looking back into the network is:

$$Z_{th} = R_3 + \frac{R_1R_2}{R_1+R_2}$$

$$Z_{th} = 8 + \frac{20 \text{ x } 30}{50}$$

$$Z_{th} = 20 \text{ ohms}$$

NOTES:

The Thevenin's equivalent circuit is drawn and the load connected as in figure 6-10. In one step the output voltage across the load can be found as follows:

$$E_L = \frac{E_{th} R_L}{Z_{th} + R_L}$$

$$E_L = \frac{60 \times 10}{30}$$

$$E_L = 20 \text{ volts}$$

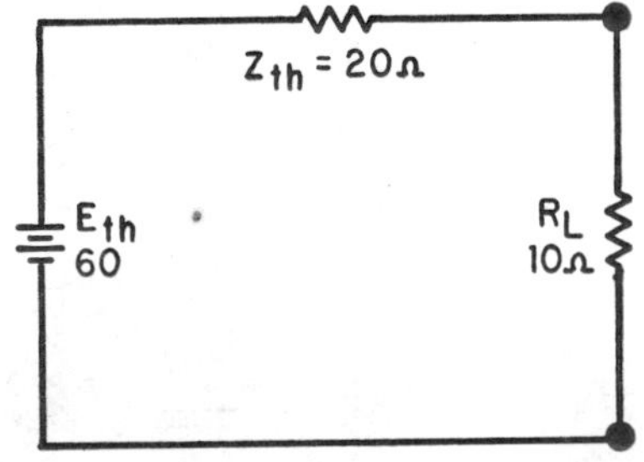

Figure 6-10.—Thevenin's equivalent for figure 6-9.

Using Thevenin's equivalent circuit and the voltage divider equation as tools, certain complex circuits can be quickly solved. This technique should not be forgotten as it will be a time saving aid in future chapters.

NORTON'S THEOREM

Another important theorem that can be used as an aid in solving complex circuits is called Norton's Theorem. This theorem is similar to Thevenin's and is stated as follows:

Norton's Theorem: Any linear network of impedance and sources, if viewed from any two points in the network, can be replaced by an equivalent impedance Z_{th} in shunt with an equivalent current source I_n.

This equivalent circuit is shown in figure 6-11. Notice that the impedance Z_{th} is placed in shunt (parallel) with a constant current source. The impedance Z_{th} is the same impedance used in Thevenin's equivalent circuit.

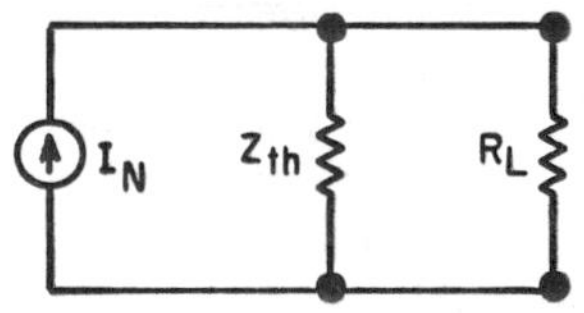

Figure 6-11.—Norton's equivalent circuit.

To illustrate the application of Norton's theorem, a Norton's equivalent circuit will be developed for the network shown in figure 6-12. The quantities Z_{th} and I_n can be found as follows:

1. Disconnect the section of the circuit considered as the load (R_L in fig. 6-12 (A)).

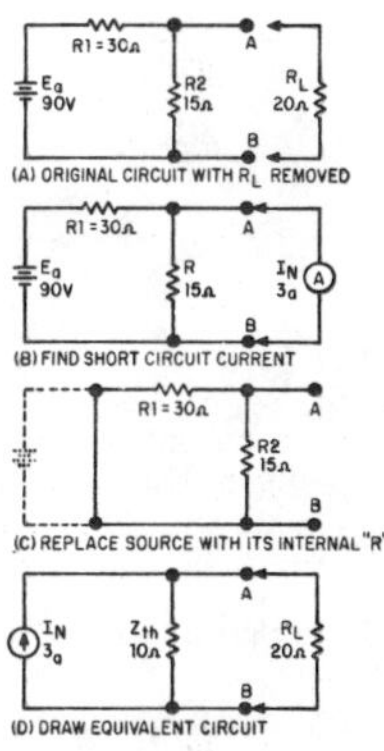

Figure 6-12.—Evolution of Norton's equivalent circuit.

2. By measurement or calculation determine the current that would flow through a wire connected between the load terminals (A and B of fig. 6-12 (B)). This short-circuit load current is Norton's current (I_n)

3. Remove short from load terminals. Replace each source within the network with its internal impedance (the battery in fig. 6-12 (C) is replaced with a short).

NOTES:

4. By measurement or calculation, determine the impedance Z_{th} looking back into the network. (Same as for Thevenin's equivalent.)

5. Draw the equivalent circuit consisting of R_L, Z_{th}, and source I_n all in parallel (fig. 6-12 (C)) and then solve for the desired quantities.

As an aid to the application of Norton's theorem, a current divider equation can be derived which will yield the load current in one calculation. Using the notation from Norton's equivalent circuit this equation is

$$I_L = \frac{I_n Z_{th}}{Z_{th} + R_L} \tag{27}$$

The current through the short (I_n) in figure 6-12 (B) is 3 amperes, since the short effectively places R_1 directly across the source. With I_n equal to 3 amperes and Z_{th} equal to 10 ohms, the load current I_L is

$$I_L = \frac{I_n Z_{th}}{Z_{th} + R_L} \tag{27}$$

$$I_L = \frac{3 \times 10}{10 + 20}$$

$$I_L = 1 \text{ ampere}$$

Should it be desired to develop the Thevenin's equivalent circuit, E_{th} can be easily determined. The Thevenin's and Norton's equivalent circuits are closely related such that

$$E_{th} = I_n Z_{th} \tag{28}$$

Therefore, E_{th} for figure 6-12 is

$$E_{th} = I_n Z_{th}$$

$$E_{th} = 3 \times 10$$

$$E_{th} = 30 \text{ volts}$$

Thus, if either equivalent circuit is known it is a simple matter to convert to the other. Usually a Norton's equivalent circuit is used when the load current is desired, and a Thevenin's equivalent circuit is used when the load voltage is required.

NOTES:

BRIDGE CIRCUITS

SIMPLE RESISTANCE BRIDGE

A resistance bridge circuit in its simplest form is shown in figure 6-13. Two identical resistors, R_1 and R_2, are connected in parallel across a 20-volt power source. The network becomes a bridge circuit when a cross connection or "bridge" is placed between the two resistors.

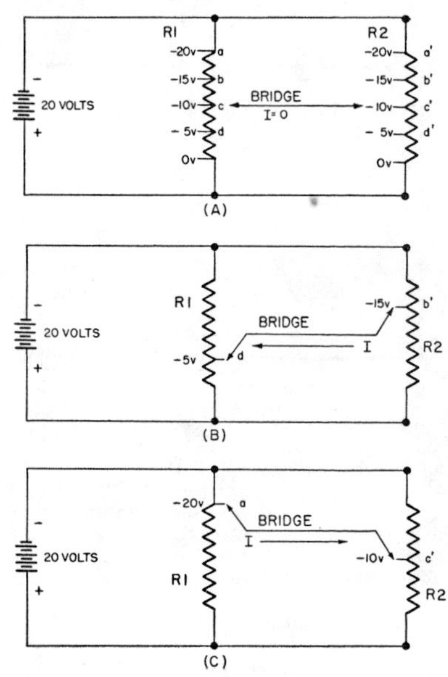

Figure 6-13.—Simple resistance bridges.

The voltage across both resistors is dropped at the same rate, because the resistors are identical. Therefore, points a-a', b-b', c-c', and d-d' are at equal potentials. If the bridge is connected between points of equal potential, as in figure 6-13 (A), no current will flow through the bridge. However, if the bridge is connected between points of unequal potential, current will flow from the more negative to the less negative end, as shown in figure 6-13 (B). In (B), current flows right-to-left from b' to d. In (C), current flows left-to-right from a to c'. Thus, it can be seen that the direction of bridge current is controlled by the relative potential of the two ends of the bridge resistor. When the bridge is across equal potentials, and no current flows, it is said to be balanced. When it is across unequal potentials, and current flows, it is said to be unbalanced. The bridge may be unbalanced in either or both of two ways—(1) by connecting the bridge to unequal potentials, or (2) by using resistors of unequal values, thus causing an unbalanced condition.

UNBALANCED RESISTANCE BRIDGE

Figure 6-14 shows an unbalanced bridge using unequal resistors. The two parallel legs are R_1-R_4 and R_3-R_5. R_2 is the bridge. The current and voltage drop of each resistor, and the source voltage E_S, are to be determined.

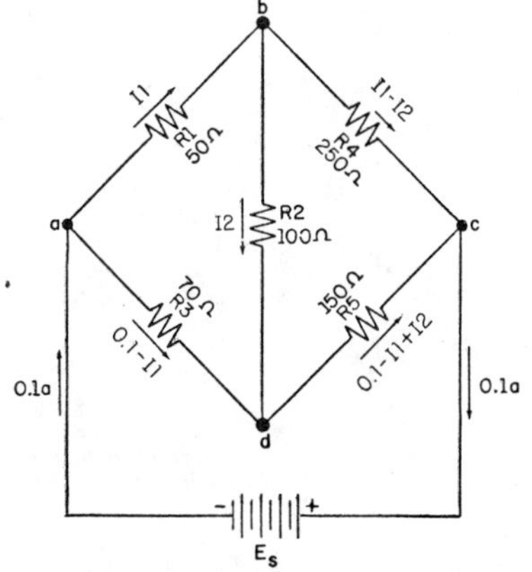

Figure 6-14.—Unbalanced resistance bridge.

The current of 0.1 ampere flowing into junction a divides into two parts. The part flowing through R1 is indicated as I_1 and the part through R3 is 0.1 - I1. Similarly, at junction b, I_1 divides, part flowing through R_2 and the remainder through R_4. The part through R_2 is designated I_2, and the part through R_4 is $I_1 - I_2$. The direction of current through R_2 may be assumed arbitrarily.

If the solution indicates a positive value for I_2, the assumed direction is proved to be correct. The current through R_4 is $I_1 - I_2$. At junction d the currents may be analyzed in a similar manner. Current I_2 through R_2 joins current $0.1 - I_1$, from R_3; and the current through R_5 is $0.1 - I_1 + I_2$.

The unknown currents, I_1 and I_2, may be determined by establishing two voltage equations in which they both appear. These equations are solved for I_1 and I_2 in terms of the given values of current and resistance. The first voltage equation is developed by tracing clockwise around the closed circuit containing resistors R_1, R_2, and R_3. The trace starts at junction a, and proceeds to b, to d, and back to a. The algebraic sum of the voltages around this circuit is zero. These voltages are expressed in terms of resistance and current. Going from a to b, the voltage drop is in the direction of the arrow and is equal to $-50I_1$; the drop across R_2, going from b to d, is $-100I_2$; and the voltage from d to a (in opposite direction to the arrow) is $+ 70(0.1 - I_1)$. Thus,

$$-50I_1 - 100I_2 + 70(0.1 - I_1) = 0$$

Multiplying both sides by -1,

$$50I_2 + 100I_2 - 70(0.1 - I_1) = 0$$

and transposing and simplifying,

$$120I_1 + 100I_2 = 7 \quad (29)$$

The second voltage equation is established by tracing clockwise around the circuit, which includes resistors R_4, R_5, and R_2. Starting at junction b, the trace proceeds to c, to d, and returns to b. The voltage across R_4, from b to c, is $-250(I_1 - I_2)$; the voltage across R_5, from c to d, is $+ 150(0.1 - I_1 + I_2)$; and the voltage across R_2, from d to b, is $+ 100I_2$. Thus,

$$-250(I_1 - I_2) + 150(0.1 - I_1 + I_2) + 100I_2 = 0$$

from which,

$$400I_1 - 500I_2 = 15 \quad (30)$$

Equations (29) and (30) may be solved simultaneously by multiplying equation (29) by the

NOTES:

factor 5 and then adding the equations to eliminate I_2 as follows:

$$400I_1 - 500I_2 = 15$$

$$600I_1 + 500I_2 = 35$$

$$1{,}000I_1 = 50$$

$$I_1 = 0.05 \text{ ampere}$$

Substituting the value of 0.05 for I_1 in equation (29) and solving for I_2,

$$120(0.05) + 100I_2 = 7$$

$$100I_2 = 1$$

$$I_2 = 0.01 \text{ ampere}$$

Thus the current in R_1 is $I_1 = 0.05$ ampere. The current in R_2 is $I_2 = 0.01$ ampere. The current in R_3 is $0.1 - I_1 = 0.1 - 0.05$, or 0.05 ampere. The current in R_4 is $I_1 - I_2 = 0.05 - 0.01$, or 0.04 ampere. The current in R_5 is $0.1 - I_1 + I_2$ or $0.1 - 0.05 + 0.01 = 0.06$ ampere. The voltages E_1, E_2, E_3, E_4, and E_5 are as follows:

E_1 across R_1 is $I_1R_1 = 0.05 \times 50 = 2.5$ volts.

E_2 across R_2 is $I_2R_2 = 0.01 \times 100 = 1.0$ volt.

E_3 across R_3 is $(0.1 - I_1)R_3 = 0.05 \times 70 = 3.5$ volts.

E_4 across R_4 is $(I_1 - I_2)R_4 = 0.04 \times 250 = 10$ volts.

E_5 across R_5 is $(0.1 - I_1 + I_2)R_5 = 0.06 \times 150 =$ 9.0 volts.

The source voltage E_s is equal to the sum of voltages across R_3 and R_5 or R_1 and R_4. Thus,

$$E_s = E_1 + E_4$$

$$= 2.5 + 10 = 12.5 \text{ volts}$$

and

$$E_s = E_3 + E_5$$

$$= 3.5 + 9 = 12.5 \text{ volts}$$

The voltage across R_2 is the difference in the voltages across R_3 and R_1. It is also the difference in the voltages across R_4 and R_5.

WHEATSTONE BRIDGE

A type of circuit that is widely used for precision measurements of resistance is the Wheatstone bridge. The circuit diagram of a Wheatstone bridge is shown in figure 6-15 (A). R_1, R_2, and R_3 are precision variable resistors, and R_x is the resistor whose unknown value of resistance is to be determined. After the bridge has been properly balanced, the unknown resistance may be determined by means of a simple formula. The galvanometer, G, is inserted across terminals b and d to indicate the condition of balance. When the bridge is properly balanced there is no difference in potential across terminals bd, and the galvanometer deflection, when the switch is closed, will be zero.

The operation of the bridge is explained in a few logical steps. When the switch to the battery is closed, electrons flow from the negative terminal of the battery to point a. Here the current divides, as it would in any parallel circuit, a part of it passing through R_1 and R_2 and the remainder passing through R_3 and R_x. The two currents, labeled I_1 and I_2, unite at point c and return to the positive terminal of the battery. The value of I_1 depends on the sum of resistances R_1 and R_2, and the value of I_2 depends on the sum of resistances R_3 and R_x. In each case, according to Ohm's law, the current is inversely proportional to the resistance.

R_1, R_2, and R_3 are adjusted so that when the galvanometer switch is closed there will be no deflection of the needle. When the galvanometer shows no deflection there is no difference of potential between points b and d. This means that the voltage drop (E_1) across R_1, between points a and b, is the same as the voltage drop (E_3) across R_3, between points a and d. By similar reasoning, the voltage drops across R_2 and R_x—that is, E_2 and E_x—are also equal. Expressed algebraically,

$$E_1 = E_3$$

or

$$I_1R_1 = I_2R_3$$

and

$$E_2 = E_x$$

or

$$I_1R_2 = I_2R_x$$

NOTES:

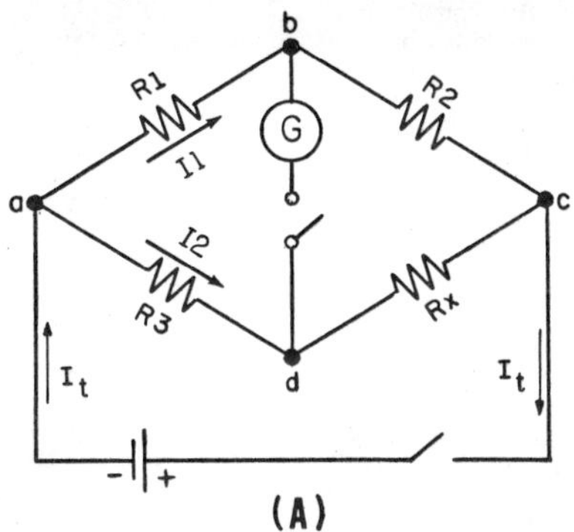

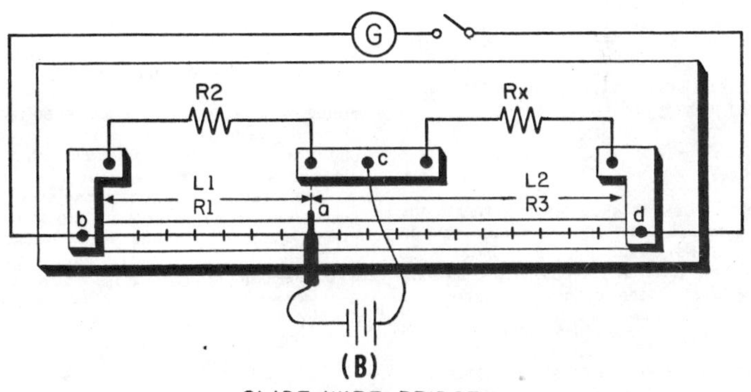

Figure 6-15.—Wheatstone bridge circuit.

Dividing the voltage drops across R_1 and R_3 by the respective voltage drops across R_2 and R_x,

$$\frac{I_1R_1}{I_1R_2} = \frac{I_2R_3}{I_2R_x}$$

Simplifying,

$$\frac{R_1}{R_2} = \frac{R_3}{R_x}$$

Therefore,

$$R_x = \frac{R_2R_3}{R_1}$$

NOTES:

The resistance values of R_1, R_2, and R_3 are readily determined from the markings on the standard resistors, or from the calibrated dials if a dial-type bridge is used.

The Wheatstone bridge may be of the slide-wire type, as shown in figure 6-15 (B). In the slide-wire circuit, the slide-wire (b to d), corresponds to R_1 and R_3 of figure 6-15 (A). The wire may be an alloy of uniform cross section; for example German silver or nichrome, having a resistance of about 100 ohms. Point a is established where the slider contacts the wire. The bridge is balanced by moving the slider along the wire.

The equation for solving for R_x in the slide-wire bridge of figure 6-15 (B), is similar to the one used for solving for R_x in figure 6-15 (A). However, in the slide-wire bridge the length L_1 corresponds to the resistance R_1, and the length L_2 corresponds to the resistance R_3. Therefore, L_1 and L_2 may be substituted for R_1 and R_3 in the equation. The resistance of L_1 and L_2 varies uniformly with slider movement because in a wire of uniform cross section the resistance varies directly with the length; therefore the ratio of the resistances equals to the corresponding ratio of the lengths. Substituting L_1 and L_2 for R_1 and R_3

$$R_x = \frac{L_2 R_2}{L_1}$$

A meter stick is mounted underneath the slide-wire and L_1 and L_2 are easily read in centimeters. For example, if a balance is obtained when R_2 = 150 ohms, L_1 = 25 cm. and L_2 = 75 cm., the unknown resistance is

$$R_x = \frac{75}{25} \times 150 = 450 \text{ ohms}$$

PARALLEL SOURCES SUPPLYING A COMMON LOAD

The circuit shown in figure 6-16 illustrates two sources of emf, E_{s1} and E_{s2}, having internal resistances of 2 and 2.5 ohms respectively, connected in parallel, and supplying a 5-ohm load. Neglecting the resistance of the lead wires, it is desired to determine the current delivered by each source to the load, the load current, and the load voltage.

The problem may be solved by establishing two voltage equations in which the voltages are expressed in terms of the unknown currents I_1 and I_2, the known resistances, and the known

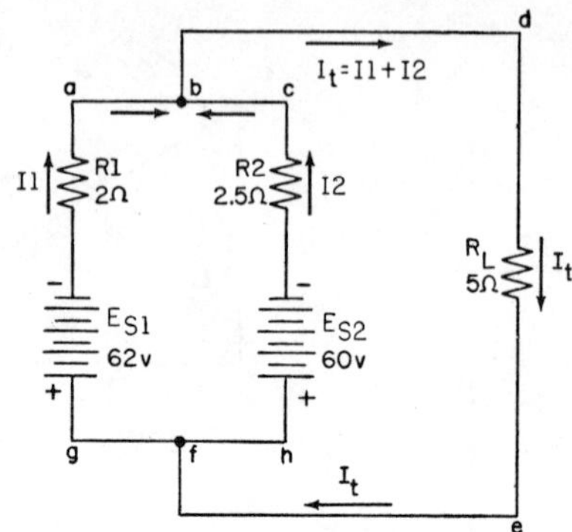

Figure 6-16.—Parallel sources supplying a common load.

voltages. The equations are then solved simultaneously as in previous examples, to eliminate one of the unknown currents. The other unknown current is solved by substitution.

The first voltage equation is established by starting at point g and tracing clockwise around circuit gabdefg. The total load current is equal to the sum of the source currents, $I_1 + I_2$. The first voltage equation is,

$$62 - 2I_1 - 5(I_1 + I_2) = 0$$

from which,

$$7I_1 + 5I_2 = 62 \qquad (31)$$

The second voltage equation is established by starting at point h and tracing around circuit hcbdefh. Thus,

$$60 - 2.5I_2 - 5(I_1 + I_2) = 0$$

from which,

$$5I_1 + 7.5I_2 = 60 \qquad (32)$$

NOTES:

I_2 is eliminated by multiplying equation (31) by 1.5 and subtracting equation (32) from the result, as follows:

$$10.5I_1 + 7.5I_2 = 93$$
$$5.0I_1 + 7.5I_2 = 60$$
$$5.5I_1 = 33$$
$$I_1 = 6 \text{ amperes}$$

Substituting this value in equation (31),

$$7 \times 6 + 5I_2 = 62$$

from which

$$I_2 = 4 \text{ amperes}$$

The load current is $I_1 + I_2 = 6 + 4 = 10$ amperes. Thus source E_{s1} supplies 6 amperes and source E_{s2} supplies 4 amperes.

The load voltage is equal to the voltage developed across terminals f and b and is equal to the difference in a given source voltage and the internal voltage absorbed across the coresponding source resistance.

The statement applies equally to either source since both are in parallel with the load. In terms of source E_{s1}

$$E_{fb} = E_{s1} - I_1R_1$$
$$= 62 - 6 \times 2$$
$$= 50 \text{ volts}$$

and in terms of source E_{s2}

$$E_{fb} = E_{s2} - I_2R_2$$
$$= 60 - 4 \times 2.5$$
$$= 50 \text{ volts}$$

A further check on the load voltage is to express this value in terms of the load current and the load resistance as follows:

$$E_{fb} = (I_1 + I_2)\ R_L$$
$$= (6 + 4)\ (5)$$
$$= 50 \text{ volts}$$

NOTES:

DISTRIBUTION CIRCUITS

TWO-WIRE DISTRIBUTION CIRCUITS

Up to this point the voltage drop and the power lost in the line wires connecting the load and the source have been neglected. When the load is located at some distance from the source, the line resistance becomes an appreciable part of the total circuit resistance and the voltage and power lost in the line become significant even with moderate loads.

In figure 6-17, load M draws 7 amperes through terminals b and e, and the parallel group of lamps draws 5 amperes through terminals c and d. The current in line wires bc and de is 5 amperes. The line current in wires ab and ef is 5 + 7 = 12 amperes.

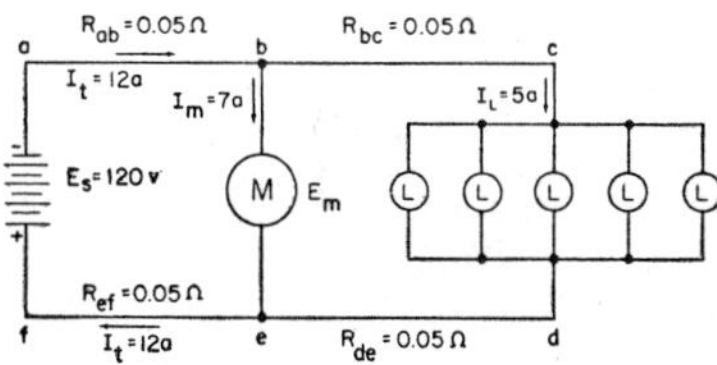

Figure 6-17.—Simple two-wire distribution circuit.

The voltage source supplies a constant potential of 120 volts between points a and f. The resistance of line wires ab and ef is 2 x 0.05 = 0.1 ohm. The voltage drop across line wires ab and ef is 12 x 0.1 = 1.2 volts. The voltage drop across line wires bc and de is 5 x 0.1 = 0.5 volt. The voltage across M is 120 - 1.2 = 118.8 volts and the voltage across the five lamps is 118.8 - 0.5 = 118.3 volts.

The power dissipated in line wires ab and ef is equal to $(12)^2 \times 0.1 = 14.4$ watts. The power absorbed by line wires bc and de is $(5)^2 \times 0.1 = 2.5$ watts. The total power absorbed by the line wires is 14.4 + 2.5 = 16.9 watts.

The power delivered to load M is 118.8 x 7 = 831.6 watts, and to the 5 lamps is 118.3 x 5 = 591.5 watts. The total power supplied to the entire circuit is equal to 16.9 + 831.6 + 591.5 = 1,440 watts and is equal to the product of the total applied voltage and the total current. Thus,

$$P_t = E_t I_t = 120 \times 12 = 1{,}440 \text{ watts}$$

THREE-WIRE DISTRIBUTION CIRCUITS

Three-wire distribution circuits transmit power at 240 volts and utilize it at 120 volts. The direct-current 3-wire system includes a positive feeder, a negative feeder, and a neutral wire, as shown in figure 6-18 (A). The loads are connected between the negative feeder and the neutral, and between the positive feeder and the neutral. When the loads are unbalanced (unequal), the neutral wire carries a current equal to the difference in the currents in the negative and positive feeders.

In the example of figure 6-18 (A), load L_1 draws 10 amperes, load L_2 draws 4 amperes, and the neutral wire carries a current of 10 - 4 = 6 amperes. The direction of flow of the current in the neutral wire is always the same as that of the smaller of the currents in the positive and negative feeders. Thus the flow is to the left in the lower (positive) wire and also in the neutral. The current in the upper (negative) wire is 10 amperes and in the lower wire it is 4 amperes. The algebraic sum of the currents entering and leaving junction c is equal to zero. Thus,

$$+10 - 6 - 4 = 0$$

To find load voltage, E_1, a voltage equation is established in which E_1 is expressed in terms of the source voltage, E_{s1}, and the IR drops in the negative feeder and neutral wire. The algebraic sum of the voltages around the circuit fabcf, is equal to zero. Starting at f and proceeding clockwise,

$$+120 - 10 \times 0.5 - E_1 - 6 \times 0.5 = 0$$

$$E_1 = 112 \text{ volts}$$

Thus the voltage across load L_1 is 112 volts. This voltage is less than the source voltage by an amount equal to the sum of the voltage drops in the negative (5 volts) and the neutral (3 volts) wires.

To find load voltage E_2 a voltage equation is established in which E_2 is expressed in terms of the source voltage, E_{s2}, and the IR drops in the positive feeder and the neutral wire. The algebraic sum of the voltages around the circuit efcde is zero. Starting at e and proceeding clockwise,

$$+120 + (6 \times 0.5) - E_2 - (4 \times 0.5) = 0$$

$$E_2 = 121 \text{ volts}$$

In tracing the circuit from f to c, note that the direction is against the arrow representing current flow, and therefore that the IR drop of (6 x 0.5) volts is preceded by a plus sign. The load voltage, E_2, is 121 volts, which is 1 volt higher than the source voltage, E_{s2}. The total source voltage ($E_{s1} + E_{s2}$) is 240 volts and the total load voltage ($E_1 + E_2$) is $112 + 121 = 233$ volts. This value is also equal to the difference between the total source voltage and the sum of the voltage drops in the positive and negative feeders, or $240 - (2 + 5) = 233$ volts.

When the loads are balanced on the positive and negative sides of the 3-wire system, the neutral current is zero and the currents in the outside wires (positive and negative) are equal. When the loads are unbalanced the neutral wire carries the unbalanced current. The voltage on the heavily loaded side falls while the voltage on the lightly loaded side rises. The lower the resistance of the neutral wire, the less imbalance in voltage there will be for a given unbalanced load.

A more complicated 3-wire circuit is shown in figure 6-18 (B). The source voltage is 120 volts between each outside wire and the neutral, or center, wire. Load currents in the upper side of the system are indicated as 10, 4, and 8 amperes respectively for loads 1, 2, and 3. In the lower side of the system, the load currents are 12 and 6 amperes respectively for loads 4 and 5. In order to determine the various load voltages it is necessary to find the currents in each outside wire and in the neutral wire. The resistances of these wires are indicated, and therefore the voltage drops and the load voltages may be calculated after the currents are determined.

To find the currents in the various sections of the wires, it is best to start at the load farthest removed from the source. The polarities of the sources are such that electrons flow out of the negative terminal at n and return to the positive terminal at b.

Currents flowing toward a junction are assumed to be positive, and those flowing away from a junction are assumed to be negative. Applying Kirchhoff's current law at junction h, the

NOTES:

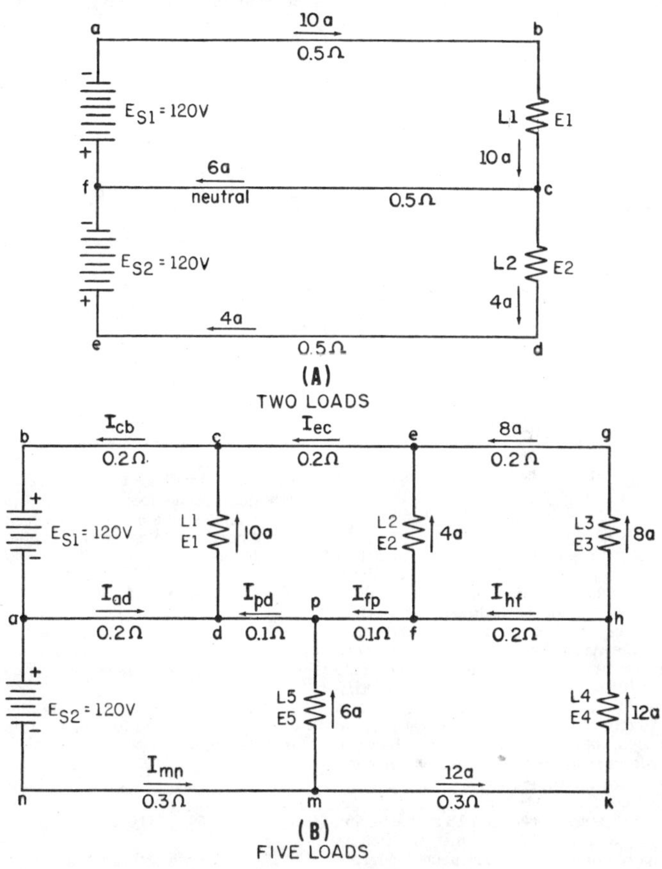

Figure 6-18.—Three-wire distribution circuits.

neutral current, I_n (flowing from h to f) is determined as

$$12 - 8 - I_{hf} = 0$$

$$I_{hf} = 4 \text{ amperes}$$

Applying the same rule successively to junctions f, e, p, m, d, and c, it follows that at junction f,

$$4 - 4 - I_{fp} = 0$$

$$I_{fp} = 0 \text{ amperes}$$

NOTES:

at junction e,

$$4 + 8 - I_{ec} = 0$$

$$I_{ec} = 12 \text{ amperes}$$

at junction p,

$$6 + 0 - I_{pd} = 0$$

$$I_{pd} = 6 \text{ amperes}$$

at junction m,

$$+ I_{mn} - 6 - 12 = 0$$

$$I_{mn} = 18 \text{ amperes}$$

at junction d,

$$I_{ad} + 6 - 10 = 0$$

$$I_{ad} = 4 \text{ amperes}$$

at junction c,

$$- I_{cb} + 10 + 12 = 0$$

$$I_{cb} = 22 \text{ amperes}$$

Thus, E_{s1} supplies 22 amperes and E_{s2} supplies 18 amperes. The electron flow in all parts of the lower wire is outward from the source, and the electron flow in all parts of the upper wire is back toward the source. The current in the neutral wire is always equal to the difference in the currents in the two outside wires, and the electron flow is in the direction of the smaller of these two currents. Thus in figure 6-18 (B), the neutral current in section ad is 4 amperes, which is the difference between 18 amperes and 22 amperes; and it is in the direction of the smaller current in section nm. The neutral current in section pd is 6 amperes, which is the difference between 18 amperes and 12 amperes; and it is in the same direction as the 12 amperes in section ec. The neutral current in section fp is zero because the current in each outside wire in that section is 12 amperes. The neutral current in section hf is 4 amperes, which is the difference between 12 amperes and 8 amperes, and it is in the direction of the smaller outside current in section ge.

In order fo find the voltages across the loads in figure 6-18 (B), Kirchhoff's voltage law is applied to the various individual circuits. Thus to find the voltage, E_1, across load L_1, the algebraic sum of the voltages around the circuit abcda is equated to zero, and E1 is then readily determined. Starting at a,

$$-120 + (22 \times 0.2) + E_1 + (4 \times 0.2) = 0$$

$$E_1 = 114.8 \text{ volts}$$

To find load voltage E_2, the algebraic sum of the voltages around circuit dcefpd is set equal to zero. Starting at d,

$$-114.8 + (12 \times 0.2) + E_2$$
$$+ (0 \times 0.1) - (6 \times 0.1) = 0$$

$$E_2 = 113 \text{ volts}$$

To find load voltage E_3, the algebraic sum of the voltages around loop feghf is set equal to zero. Starting at f,

$$-113 + (8 \times 0.2) + E_3 - (4 \times 0.2) = 0$$

$$E_3 = 112.2 \text{ volts}$$

To find load voltage E_4, the algebraic sum of the voltages around loop nadpfhkmn is set equal to zero. Loop mpfhkm cannot be used because it would contain two unknown voltages, E_5 and E_4. Starting at n,

$$-120 - (4 \times 0.2) + (6 \times 0.1) + (0 \times 0.1)$$
$$+ (4 \times 0.2) + E_4 + (12 \times 0.3) + (18 \times 0.3) = 0$$

$$E_4 = 110.4 \text{ volts}$$

To find load voltage E_5, the algebraic sum of the voltages around loop nadpmn is set equal to zero. Starting at n,

$$-120 - (4 \times 0.2) + (6 \times 0.1)$$
$$+ E_5 + (18 \times 0.3) = 0$$

$$E_5 = 114.8 \text{ volts}$$

In each case, the equations used contain one unknown; and thus a simple solution is quickly obtained. It is necessary that the path traced include a completely closed loop and that all but one of the voltages within that loop be known. Simple transposition of the resulting equation gives the desired voltage.

NOTES:

CHAPTER 7

ELECTRICAL CONDUCTORS AND WIRING TECHNIQUES

Since all electrical circuits utilize conductors of one type of another, it is essential that you know the basic physical features and electrical characteristics of the most common types of conductors.

As stated previously, any substance that permits the free motion of a large number of electrons is classed as a conductor. A conductor may be made from many different types of metals, but only the most commonly used types of materials will be discussed in this chapter.

To compare the resistance and size of one conductor with that of another, a standard or unit size of conductor must be established. A convenient unit of linear measurement, as far as the diameter of a piece of wire is concerned, is the mil (0.001 of an inch); and a convenient unit of wire length is the foot. The standard unit of size in most cases is the MIL-FOOT; that is, a wire will have unit size if it has diameter of 1 mil and a length of 1 foot. The resistance in ohms of a unit conductor of a given substance is called the specific resistance, or specific resistivity, of the substance.

Gage numbers are a further convenience in comparing the diameter of wires. The gage commonly used is the American wire gage (AWG), formerly the Brown and Sharpe (B and S) gage.

MIL

SQUARE MIL

The square mil is a convenient unit of cross-sectional area for square or rectangular conductors. A square mil is the area of a square, the sides of which are 1 mil, as shown in figure 7-1 (A). To obtain the cross-sectional area in square mils of a square conductor, square one side measured in mils. To obtain the cross-sectional area in square mils of a rectangular conductor, multiply the length of one side by that of the other, each length being expressed in mils.

For example, find the cross-sectional area of a large rectangular conductor 3/8 inch thick and 4 inches wide. The thickness may be expressed in mils as 0.375 x 1,000 = 375 mils, and the width as 4 x 1,000, or 4,000 mils. The cross-sectional area is 375 x 4,000, or 1,500,000 square mils.

CIRCULAR MIL

The circular mil is the standard unit of wire cross-sectional area used in American and English wire tables. Because the diameters of round conductors, or wires, used to conduct electricity may be only a small fraction of an inch, it is convenient to express these diameters in mils, to avoid the use of decimals. For example, the diameter of a wire is expressed as 25 mils instead of 0.025 inch. A circular mil is the area of a circle having a diameter of 1 mil, as shown in figure 7-1 (B). The area in circular mils of a round conductor is obtained by squaring the diameter measured in mils. Thus, a wire having a diameter of 25 mils has an area of 25^2 or 625 circular mils. By way of comparison, the basic formula for the area of a circle is $A = \pi R^2$ and in this example the area in square inches is

$$A = \pi R^2 = 3.14(0.0125)^2 = 0.00049 \text{ sq. in.}$$

If D is the diameter of a wire in mils, the area in square mils is

$$A = \pi\left(\frac{D}{2}\right)^2 = \frac{3.1416}{4} D^2 = 0.7854D^2 \text{ sq. mils.}$$

NOTES:

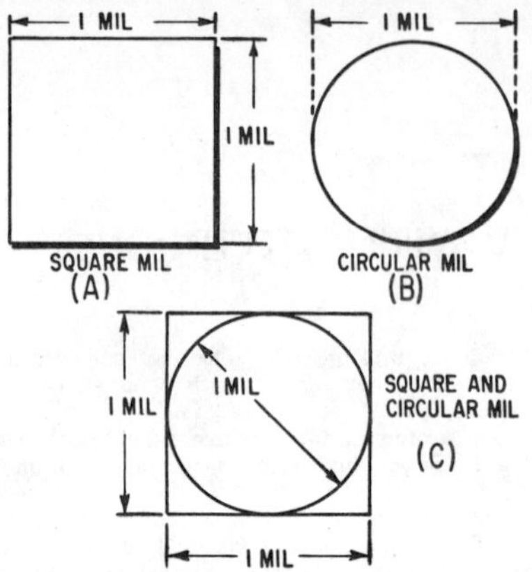

Figure 7-1.—(A) Square mil; (B) circular mil; (C) comparison of circular to square mil.

Therefore, a wire 1 mil in diameter has an area of

$$A = 0.7854 \times 1^2 = 0.7854 \text{ sq. mils},$$

which is equivalent to 1 circular mil. The cross-sectional area of a wire in circular mils is therefore determined as

$$A = \frac{0.7854D^2}{0.7854} = D^2 \text{ circular mils},$$

where D is the diameter in mils. Thus, the constant $\frac{\pi}{4}$ is eliminated from the calculation.

In comparing square and round conductors it should be noted that the circular mil is a smaller unit of area than the square mil, and therefore there are more circular mils than square mils in any given area. The comparison is shown in figure 7-1 (C). The area of a circular mil is equal to 0.7854 of a square mil. Therefore, to determine the circular-mil area when the square-mil area is given, divide the area in square mils by 0.7854. Conversely, to determine the square-mil area when the circular-mil area is given, multiply the area in circular mils by 0.7854.

NOTES:

For example, a No. 12 wire has a diameter of 80.81 mils. What is (1) its area in circular mils and (2) its area in square mils?

Solution:

(1) $A = D^2 = 80.81^2 = 6{,}530$ circular mils

(2) $A = 0.7854 \times 6{,}530 = 5{,}128.7$ square mils

A rectangular conductor is 1.5 inches wide and 0.25 inch thick. (1) What is its area in square mils? (2) What size of round conductor in circular mils is necessary to carry the same current as the rectangular bar?

Solution:

(1) $1.5'' = 1.5 \times 1{,}000 = 1{,}500$ mils

$0.25'' = 0.25 \times 1{,}000 = 250$ mils

$A = 1{,}500 \times 250 = 375{,}000$ sq. mils

(2) To carry the same current, the cross-sectional area of the rectangular bar and the cross-sectional area of the round conductor must be equal. There are more circular mils than square mils in this area, and therefore

$$A = \frac{375{,}000}{0.7854} = 477{,}000 \text{ circular mils}$$

A wire in its usual form is a slender rod or filament of drawn metal. In large sizes, wire becomes difficult to handle, and its flexibility is increased by stranding. The strands are usually single wires twisted together in sufficient numbers to make up the necessary cross-sectional area of the cable. The total area in circular mils is determined by multiplying the area of one strand in circular mils by the number of strands in the cable.

CIRCULAR-MIL-FOOT

A circular-mil-foot, as shown in figure 7-2, is actually a unit of volume. It is a unit conductor 1 foot in length and having a cross-sectional area of 1 circular mil. Because it is considered a unit conductor, the circular-mil-foot is useful in making comparisons between wires that are made of different metals. For example, a basis of comparison of the RESISTIVITY (to be treated later) of various substances may be made by determining the resistance of a circular-mil-foot of each of the substances.

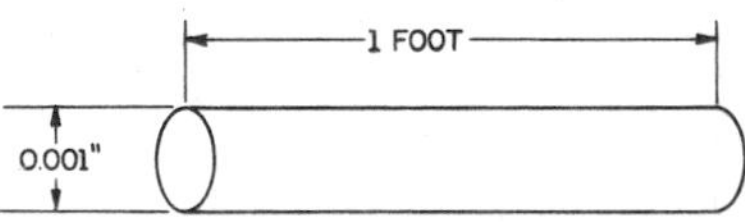

Figure 7-2.—Circular-mil-foot.

In working with certain substances it is sometimes more convenient to employ a different unit volume. Accordingly, unit volume may also be taken as the centimeter cube; and specific resistance becomes the resistance offered by a cube-shaped conductor 1 cm. long and 1 sq. cm. in cross-sectional area. The inch cube may also be used. The unit of volume employed is given in tables of specific resistances.

SPECIFIC RESISTANCE OR RESISTIVITY

Specific resistance, or resistivity, is the resistance in ohms offered by unit volume (the circular-mil-foot) of a substance to the flow of electric current. Resistivity is the reciprocal of conductivity. A substance that has a high resistivity will have a low conductivity, and vice versa.

Thus, the specific resistance of a substance is the resistance of a unit volume of that substance. Many tables of specific resistance are based on the resistance in ohms of a volume of the substance 1 foot long and 1 circular mil in cross-sectional area. The temperature at which the resistance measurement is made is also specified. If the kind of metal of which a conductor is made is known, the specific resistance of the metal may be obtained from a table. The specific resistances of some common substances are given in table 7-1.

The resistance of a conductor of uniform cross section varies directly as the product of the length and the specific resistance of the conductor and inversely as the cross-sectional area of the conductor. Therefore the resistance of a conductor may be calculated if the length, cross-sectional area, and specific resistance of the substance are known. Expressed as an equation, the resistance, R in ohms, of a conductor is

Table 7-1.—Specific resistance.

Substance	Specific resistance at 20° C.	
	Centimeter cube (microhms)	Circular-mil-foot (ohms)
Silver	1.629	9.8
Copper (drawn).	1.724	10.37
Gold	2.44	14.7
Aluminum . . .	2.828	17.02
Carbon (amorphous.)	3.8 to 4.1	. . . .
Tungsten	5.51	33.2
Brass	7.0	42.1
Steel (soft) . . .	15.9	95.8
Nichrome . . .	109.0	660.0

$$R = \rho \frac{L}{A}$$

where ρ (Greek rho) is the specific resistance in ohms per circular-mil-foot, L the length in feet and A the cross-sectional area in circular mils.

For example, what is the resistance of 1,000 feet of copper wire having a cross-sectional area of 10,400 circular mils (No. 10 wire), the wire temperature being 20° C?

Solution:

The specific resistance, from table 7-1, is 10.37. Substituting the known values in the preceding equation, the resistance, R, is determined as

$$R = \rho \frac{L}{A} = 10.37 \times \frac{1{,}000}{10{,}400} = 1 \text{ ohm, approximately}$$

If R, ρ, and A are known, the length may be determined by a simple mathematical transposition. This is of value in many applications. For example, when it is desired to locate a ground in a telephone line, special test equipment is used that operates on the principle that the resistance of a line varies directly with its length. Thus, the distance between the test point and a fault can be computed accurately.

As has been mentioned in preceding chapters, conductance (G) is the reciprocal of resistance.

NOTES:

When R is in ohms, the conductance is expressed in mhos. Where resistance is opposition to flow, conductance is the ease with which the current flows. Conductance is mhos is equivalent to the number of amperes flowing in a conductor per volt of applied emf. Expressed in terms of the specific resistance, length, and cross section of a conductor,

$$G = \frac{A}{\rho L}$$

The conductance, G, varies directly as the cross-sectional area, A, and inversely as the specific resistance, ρ and the length, L. When A is in circular mils, ρ is in ohms per circular-mil-foot, L is in feet, and G is in mhos.

The relative conductance of several substances is given in table 7-2.

Table 7-2.—Relative conductance.

Substance	Relative conductance (Silver = 100%)
Silver	100
Copper	98
Gold	78
Aluminum	61
Tungsten	32
Zinc	30
Platinum	17
Iron	16
Lead	15
Tin	9
Nickel	7
Mercury	1
Carbon	0.05

WIRE MEASURE

RELATION BETWEEN WIRE SIZES

Wires are manufactured in sizes numbered according to a table known as the American wire gage (AWG). As may be seen in table 7-3, the wire diameters become smaller as the gage numbers become larger. The largest wire size shown in the table is 0000 (read "4 naught"), and the smallest is number 40. Larger and smaller sizes are manufactured but are not commonly used by the Navy. The ratio of the diameter of one gage number to the diameter of the next higher gage number is a constant, 1.123. The cross-sectional area varies as the square of the diameter. Therefore, the ratio of the cross section of one gage number to that of the next higher gage number is the square of 1.123, or 1.261. Because the cube of 1.261 is very nearly 2, the cross-sectional area is approximately halved, or doubled, every three gage numbers. Also because 1.261 raised to the 10th power is very nearly equal to 10, the cross-sectional area is increased or decreased 10 times every 10 gage numbers.

A No. 10 wire has a diameter of approximately 102 mils, a cross-sectional area of approximately 10,400 circular mils, and a resistance of approximately 1 ohm per 1,000 feet. From these facts it is possible to estimate quickly the cross-sectional area and the resistance of any size copper wire without referring directly to a wire table.

For example, to estimate the cross-sectional area and the resistance of 1,000 feet of No. 17 wire, the following reasoning might be employed. A No. 17 wire is 3 sizes removed from a No. 20 wire and therefore has twice the cross-sectional area of a No. 20 wire. A No. 20 wire is 10 sizes removed from a No. 10 wire and therefore has one-tenth the cross section of a No. 10 wire. Therefore, the cross-sectional area of a No. 17 wire is 2 x 0.1 x 10,000 = 2,000 circular mils. Since resistance varies inversely with the cross-sectional area, the resistance of a No. 17 wire is 10 x 1 x 0.5 = 5 ohms per 1,000 feet.

A wire gage is shown in figure 7-3. It will measure wires ranging in size from number 0 to number 36. The wire whose size is to be measured is inserted in the smallest slot that will just accommodate the bare wire. The gage number corresponding to that slot indicates the wire size. The slot has parallel sides and should not be confused with the semicircular opening at the end of the slot. The opening simply permits the free movement of the wire all the way through the slot.

STRANDED WIRES AND CABLES

A WIRE is a slender rod or filament of drawn metal. The definition restricts the term to what would ordinarily be understood as "solid wire." The word "slender" is used because the length of a wire is usually large in comparison with the diameter. If a wire is covered with insulation, it is properly called an insulated wire. Although the term "wire" properly refers to the metal, it is generally understood to include the insulation.

NOTES:

Table 7-3.—Standard annealed solid copper wire.

(American wire gage--B & S)

Gage number	Diameter (mils)	Cross section		Ohms per 1,000 ft.		Ohms per mile 25° C. (=77° F.)	Pounds per 1,000 ft.
		Circular mils	Square inches	25° C. (=77° F.)	65° C. (=149° F.)		
0000	460.0	212,000.0	0.166	0.0500	0.0577	0.264	641.0
000	410.0	168,000.0	.132	.0630	.0727	.333	508.0
00	365.0	133,000.0	.105	.0795	.0917	.420	403.0
0	325.0	106,000.0	.0829	.100	.116	.528	319.0
1	289.0	83,700.0	.0657	.126	.146	.665	253.0
2	258.0	66,400.0	.0521	.159	.184	.839	201.0
3	229.0	52,600.0	.0413	.201	.232	1.061	159.0
4	204.0	41,700.0	.0328	.253	.292	1.335	126.0
5	182.0	33,100.0	.0260	.319	.369	1.685	100.0
6	162.0	26,300.0	.0206	.403	.465	2.13	79.5
7	144.0	20,800.0	.0164	.508	.586	2.68	63.0
8	128.0	16,500.0	.0130	.641	.739	3.38	50.0
9	114.0	13,100.0	.0103	.808	.932	4.27	39.6
10	102.0	10,400.0	.00815	1.02	1.18	5.38	31.4
11	91.0	8,230.0	.00647	1.28	1.48	6.75	24.9
12	81.0	6,530.0	.00513	1.62	1.87	8.55	19.8
13	72.0	5,180.0	.00407	2.04	2.36	10.77	15.7
14	64.0	4,110.0	.00323	2.58	2.97	13.62	12.4
15	57.0	3,260.0	.00256	3.25	3.75	17.16	9.86
16	51.0	2,580.0	.00203	4.09	4.73	21.6	7.82
17	45.0	2,050.0	.00161	5.16	5.96	27.2	6.20
18	40.0	1,620.0	.00128	6.51	7.51	34.4	4.92
19	36.0	1,290.0	.00101	8.21	9.48	43.3	3.90
20	32.0	1,020.0	.000802	10.4	11.9	54.9	3.09
21	28.5	810.0	.000636	13.1	15.1	69.1	2.45
22	25.3	642.0	.000505	16.5	19.0	87.1	1.94
23	22.6	509.0	.000400	20.8	24.0	109.8	1.54
24	20.1	404.0	.000317	26.2	30.2	138.3	1.22
25	17.9	320.0	.000252	33.0	38.1	174.1	0.970
26	15.9	254.0	.000200	41.6	48.0	220.0	0.769
27	14.2	202.0	.000158	52.5	60.6	277.0	0.610
28	12.6	160.0	.000126	66.2	76.4	350.0	0.484
29	11.3	127.0	.0000995	83.4	96.3	440.0	0.384
30	10.0	101.0	.0000789	105.0	121.0	554.0	0.304
31	8.9	79.7	.0000626	133.0	153.0	702.0	0.241
32	8.0	63.2	.0000496	167.0	193.0	882.0	0.191
33	7.1	50.1	.0000394	211.0	243.0	1,114.0	0.152
34	6.3	39.8	.0000312	266.0	307.0	1,404.0	0.120
35	5.6	31.5	.0000248	335.0	387.0	1,769.0	0.0954
36	5.0	25.0	.0000196	423.0	488.0	2,230.0	0.0757
37	4.5	19.8	.0000156	533.0	616.0	2,810.0	0.0600
38	4.0	15.7	.0000123	673.0	776.0	3,550.0	0.0476
39	3.5	12.5	.0000098	848.0	979.0	4,480.0	0.0377
40	3.1	9.9	.0000078	1,070.0	1,230.0	5,650.0	0.0299

A CONDUCTOR is a wire or combination of wires not insulated from one another, suitable for carrying an electric current.

A STRANDED CONDUCTOR is a conductor composed of a group of wires or of any combination of groups of wires. The wires in a stranded conductor are usually twisted together.

A CABLE is either a stranded conductor (single-conductor cable) or a combination of conductors insulated from one another

NOTES:

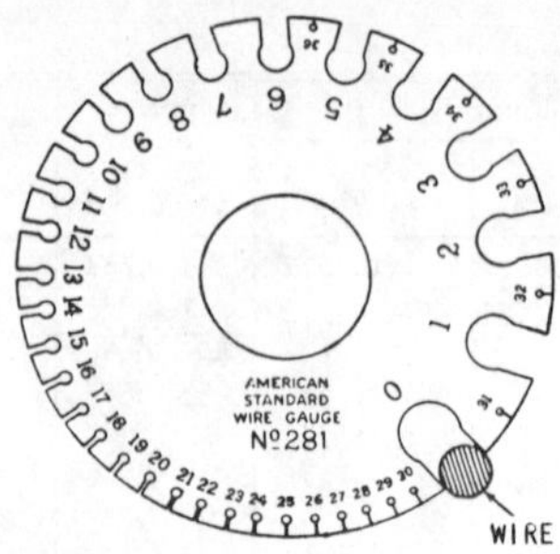

Figure 7-3.—Wire gage.

(multiple-conductor cable). The term cable is a general one, and in practice it is usually applied only to the larger sizes of conductors. A small cable is more often called a stranded wire or cord. Cables may be bare or insulated. The insulated cables may be sheathed (covered) with lead, or protective armor.

Figure 7-4 shows some of the different types of wire and cables used.

Conductors are stranded mainly to increase their flexibility. The arrangement of the wire strands in concentric-layer cables is as follows:

The first layer of strands around the center is made up of 6 conductors; the second layer is made up of 12 conductors; the third layer is made up of 18 conductors; and so on. Thus, standard cables are composed of 1, 7, 19, 37, and so forth strands.

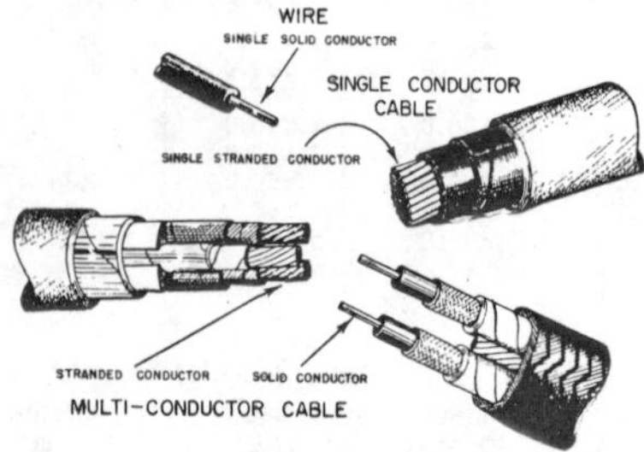

Figure 7-4.—Conductors.

NOTES:

The overall flexibility may be increased by further stranding of the individual strands.

Figure 7-5 shows a typical 37-strand cable. It also shows how the total circular-mil cross-sectional area of a stranded cable is determined.

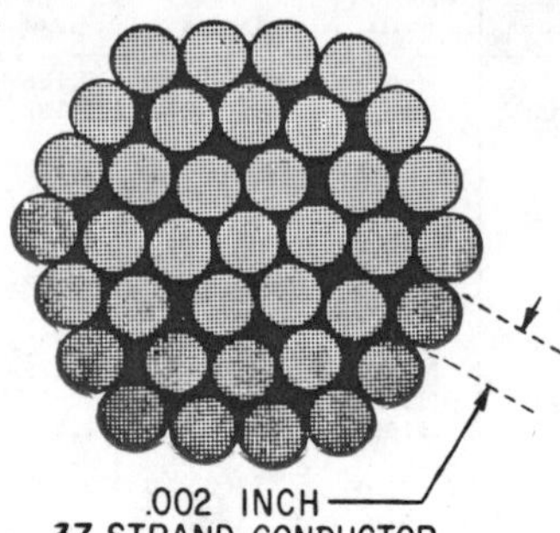

DIAMETER OF EACH STRAND = .002 INCH
DIAMETER OF EACH STRAND, MILS = 2 MILS
CIRCULAR MIL AREA OF
EACH STRAND = D^2 = 4 CM
TOTAL CM AREA OF
CONDUCTOR = 4 x 37 = 148 CM

Figure 7-5.—Stranded conductor.

FACTORS GOVERNING THE SELECTION OF WIRE SIZE

Several factors must be considered in selecting the size of wire to be used for transmitting and distributing electric power.

One factor is the allowable power loss (I^2R loss) in the line. This loss represents electrical energy converted into heat. The use of large conductors will reduce the resistance and therefore the I^2R loss. However, large conductors are more expensive initially than small ones; they are heavier and require more substantial supports.

A second factor is the permissible voltage drop (IR drop) in the line. If the source maintains a constant voltage at the input to the line, any variation in the load on the line will cause a variation in line current, and a consequent variation in the IR drop in the line. A wide variation in the IR drop in the line causes poor voltage regulation at the load. The obvious remedy is to reduce either *I* or *R*. A reduction in load current lowers the amount of power being transmitted, whereas a reduction in line resistance increases the size and weight of conductors required. A compromise is generally reached whereby the voltage variation at the load is within tolerable limits and the weight of line conductors is not excessive.

A third factor is the current-carrying ability of the line. When current is drawn through the line, heat is generated. The temperature of the line will rise until the heat radiated, or otherwise dissipated, is equal to the heat generated by the passage of current through the line. If the conductor is insulated, the heat generated in the conductor is not so readily removed as it would be if the conductor were not insulated. Thus, the protect the insulation from too much heat, the current through the conductor must be maintained below a certain value. Rubber insulation will begin to deteriorate at relatively low temperatures. Varnished cloth insulation retains its insulating properties at higher temperatures; and other insulation—for example, asbestos or silicon—is effective at still higher temperatures.

Electrical conductors may be installed in locations where the ambient (surrounding) temperature is relatively high; in which case the heat generated by external sources constitutes an appreciable part of the total conductor heating. Due allowance must be made for the influence of external heating on the allowable conductor current and each case has its own specific limitations. The maximum allowable operating temperature of insulated conductors is specified in tables and varies with the type of conductor insulation being used.

Tables have been prepared by the National Board of Fire Underwriters giving the safe current ratings for various sizes and types of conductors covered with various types of insulation. For example, the allowable current-carrying capacities of copper conductors at not over 30°C (86° F) room temperature for single conductors in free air are given in table 7-4.

COPPER VS. ALUMINUM CONDUCTORS

Although silver is the best conductor, its cost limits its use to special circuits where a substance with high conductivity is needed.

The two most generally used conductors are copper and aluminum. Each has characteristics that make its use advantageous under certain circumstances. Likewise, each has certain disadvantages.

Copper has a higher conductivity; it is more ductile (can be drawn out), has relatively high tensile strength, and can be easily soldered. It is more expensive and heavier than aluminum.

Table 7-4. —Current-carrying capacities (in amperes) of single copper conductors at ambient temperature of below 30°C

Size	Rubber or thermoplastic	Thermoplastic asbestos, var-cam, or asbestos var-cam	Impregnated asbestos	Asbestos	Heat-Resistant or Moisture Resistant
0000	300	385	475	510	370
000	260	330	410	430	320
00	225	285	355	370	275
0	195	245	305	325	235
1	165	210	265	280	205
2	140	180	225	240	175
3	120	155	195	210	150
4	105	135	170	180	130
6	80	100	125	135	100
8	55	70	90	100	70
10	40	55	70	75	55
12	25	40	50	55	40
14	20	30	40	45	30

NOTES:

Although aluminum has only about 60 percent of the conductivity of copper, its lightness makes possible long spans, and its relatively large diameter for a given conductivity reduces corona—that is, the discharge of electricity from the wire when it has a high potential. The discharge is greater when smaller diameter wire is used than when larger diameter wire is used. However, aluminum conductors are not easily soldered, and aluminum's relatively large size for a given conductance does not permit the economical use of an insulation covering.

A comparison of some of the characteristcs of copper and aluminum is given in table 7-5.

Table 7-5.—Characteristics of copper and aluminum.

Characteristics	Copper	Aluminum
Tensile strength (lb/in^2).	55,000	25,000
Tensile strength for same conductivity (lb).	55,000	40,000
Weight for same conductivity (lb).	100	48
Cross section for same conductivity (C.M.).	100	160
Specific resistance (Ω/mil ft).	10.6	17

TEMPERATURE COEFFICIENT

The resistance of pure metals—such as silver, copper, and aluminum—increases as the temperature increases. However, the resistance of some alloys—such as constantan and manganin—changes very little as the temperature changes. Measuring instruments use these alloys because the resistance of the circuits must remain constant if accurate measurements are to be achieved.

In table 7-1 the resistance of a circular-mil-foot of wire (the specific resistance) is given at a specific temperature, 20°C in this case. It is necessary to establish a standard temperature because, as has been stated, the resistance of pure metals increase with an increase in temperature; and a true basis of comparison cannot be made unless the resistances of all the substances being compared are measured at the same temperature. The amount of increase in the resistance of a 1-ohm sample of the conductor per degree rise in temperature above 0° C is called the TEMPERATURE COEFFICIENT OF RESISTANCE. For copper, the value is approximately 0.00427 ohm. For pure metals, the temperature coefficient of resistance ranges between 0.003 and 0.006 ohm.

Thus, a copper wire having a resistance of 50 ohms at an initial temperature of 0°C will have an increase in resistance of 50 x 0.00427, or 0.214 ohms for the entire length of wire for each degree of temperature rise above 0° C. At 20°C the increase in resistance is approximately 20 x 0.214, or 4.28 ohms. The total resistance at 20°C is 50 + 4.28, or 54.28 ohms.

CONDUCTOR INSULATION

To be useful and safe, electric current must be forced to flow only where it is needed. It must be "channeled" from the power source to a useful load. In general, current-carrying conductors must not be allowed to come in contact with one another, their supporting hardware, or personnel working near them. To accomplish this, conductors are coated or wrapped with various materials. These materials have such a high resistance that they are, for all practical purposes, nonconductors. They are generally referred to as insulators or insulating material.

Because of the expense of insulation and its stiffening effect, together with the great variety of physical and electrical conditions under which the conductors are operated, only the necessary minimum of insulation is applied for any particular type of cable designed to do a specific job. Therefore there is a wide variety of insulated conductors available to meet the requirements of any job.

Two fundamental properties of insulation materials (for example, rubber, glass, asbestos, and plastic) are insulation resistance and dielectric strength. These are entirely different and distinct properties.

INSULATION RESISTANCE is the resistance to current leakage through and over the surface of insulation materials. Insulation resistance can be measured by means of a megger without damaging the insulation, and information so obtained serves as a useful guide in appraising the general condition of insulation. However, the data obtained in this manner may not give a true picture of the condition of the insulation. Clean-dry insulation having cracks or other faults may show a high value of insulation resistance but would not be suitable for use.

NOTES:

DIELECTRIC STRENGTH is the ability of the insulator to withstand potential difference and is usually expressed in terms of the voltage at which the insulation fails because of the electrostatic stress. Maximum dielectric strength values can be measured by raising the voltage of a TEST SAMPLE until the insulation breaks down.

RUBBER

One of the most common types of insulation is rubber. The voltage that may be applied to a rubber-covered pair of conductors (twisted pair) is dependent on the thickness and the quality of the rubber covering. Other factors being equal, the thicker the insulation the higher may be the applied voltage. Figure 7-6 shows two types of rubber-covered wire. One is a single, solid conductor, and the other is a 2-conductor cable in which each stranded conductor is covered with rubber insulation. In each case the rubber serves the same purpose—to confine the current to its conductor.

It may be seen from the enlarged cross-sectional view that a thin coating of tin separates the copper conductor from the rubber insulation. If the thin coating of tin were not used, chemical action would take place and the rubber would become soft and gummy where it makes contact with the copper. When small, solid, or stranded conductors are used, a winding of cotton threads is applied between the conductors and the rubber insulation.

PLASTICS

Plastic has become one of the more common types of material used as insulation for electrical conductors. It has good insulating, flexibility, and moisture resistant qualities under various conditions. There are various types of plastics used as insulating material, thermoplastic being one of the most common. With the use of thermoplastic the conductor temperature can be higher than with some other types of insulating materials without damage to the insulating quality of the material.

VARNISHED CAMBRIC

Heat is developed when current flows through a wire, and when a large amount of current flows, considerable heat may be developed. The heat can be dissipated if air is circulated freely around the wire. If a cover of insulation is used, the heat is not removed so readily and the temperature may reach a high value.

Rubber is a good insulator at relatively low voltage as long as the temperature remains low. Too much heat will cause even the best grade of rubber insulation to become brittle and crack. VARNISHED CAMBRIC insulation will stand much higher temperatures than rubber insulation. Varnished cambric is cotton cloth that has been coated with an insulating varnish. Figure 7-7 shows some of the detail of a cable covered with varnished cambric insulation. The varnished cambric is in tape form and is wound around the conductor in layers. An oily compound is applied between each layer of the tape. This compound prevents water from seeping through the insulation. It also acts as a lubricant between the layers of tape, so they will slide over each other when the cable is bent.

This type of insulation is used on high-voltage conductors associated with switch gear in substations and power houses and other locations subjected to high temperatures. It is also used on high-voltage generator coils and leads, and also on transformer leads because it is unaffected by oils or grease and because it has a high dielectric strength. Varnished cambric and paper insulation for cables are the two types of insulating materials most widely used at voltages above 15,000 volts, but such cables are always lead-covered to keep the moisture out.

ASBESTOS

Even varnished cambric may break down when the temperature goes above 85°C (185° F). When the combined effects of a high ambient (surrounding) temperature and a high internal temperature due to large current flow through the wire makes the total temperature of the wire go above 85°C, ASBESTOS insulation is used.

Asbestos is a good insulation for wires and cables used under very high temperature conditions. It is fire resistant and does not change with age. One type of asbestos-covered wire is shown in figure 7-8. It consists of a stranded copper conductor covered with felted asbestos, which is, in turn, covered with asbestos braid. This type of wire is used in motion-picture projectors, arc lamps, spotlights, heating element leads, and so forth.

NOTES:

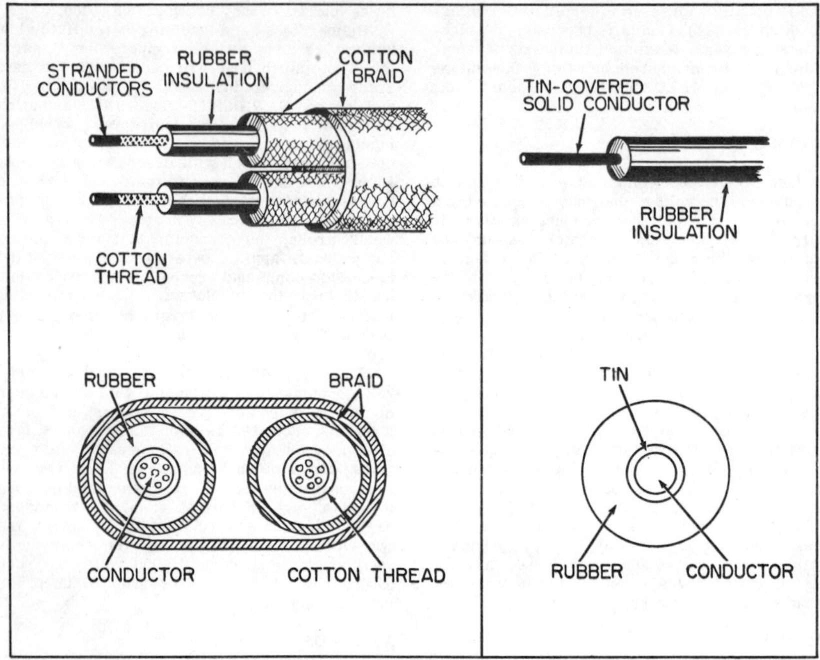

Figure 7-6.—Rubber insulation.

Another type of asbestos-covered cable is shown in figure 7-9. It serves as leads for motors and transformers that sometimes must operate in hot, wet locations. The varnished cambric covers the inner layer of felted asbestos and prevents moisture from reaching the innermost layer of asbestos. Asbestos loses its insulating properties when it becomes wet, and will in fact become a conductor. The varnished cambric prevents this from happening because it resists moisture. Although this insulation will withstand some moisture, it should not be used on conductors that may at times be partly immersed in water, unless the insulation is protected with an outer lead sheath.

PAPER

Paper has little insulation value alone, but when impregnated with a high grade of mineral oil, it serves as a satisfactory insulation for high-voltage cables. The oil has a high dielectric strength, and tends to prevent breakdown of paper insulation when the paper is thoroughly saturated with it. The thin paper tape is wrapped in many layers around the conductors, and it is then soaked with oil.

The 3-conductor cable shown in figure 7-10 consists of paper insulation on each conductor with a spirally wrapped nonmagnetic metallic tape over the insulation. The space between

NOTES:

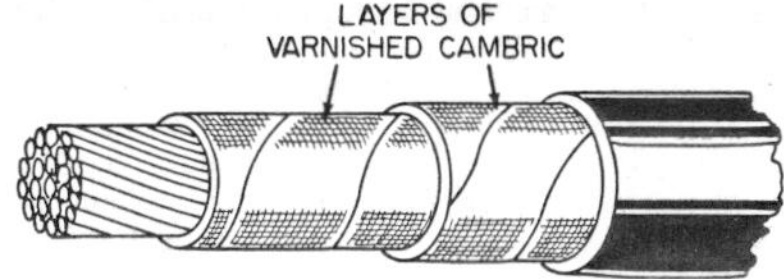

Figure 7-7.—Varnished cambric insulation.

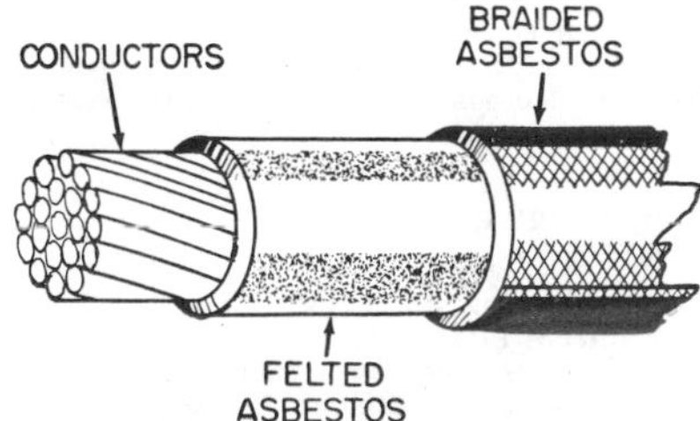

Figure 7-8.—Asbestos insulation.

conductors is filled with a suitable spacer to round out the cable and another nonmagnetic metal tape is used to secure the entire cable, and then a lead sheath is applied over all. This type of cable is used on voltages from 10,000 volts to 35,000 volts.

SILK AND COTTON

In certain types of circuits—for example communications circuits—a large number of conductors are needed, perhaps as many as several hundred. Figure 7-11 shows a cable containing many conductors, each insulated from the others by silk and cotton threads.

The use of silk and cotton as insulation keeps the size of the cable small enough to be handled easily. The silk and cotton threads are wrapped around the individual conductors in reverse directions, and the covering is then impregnated with a special wax compound.

Because the insulation in this type of cable is not subjected to high voltage, thin layers of silk and cotton are used.

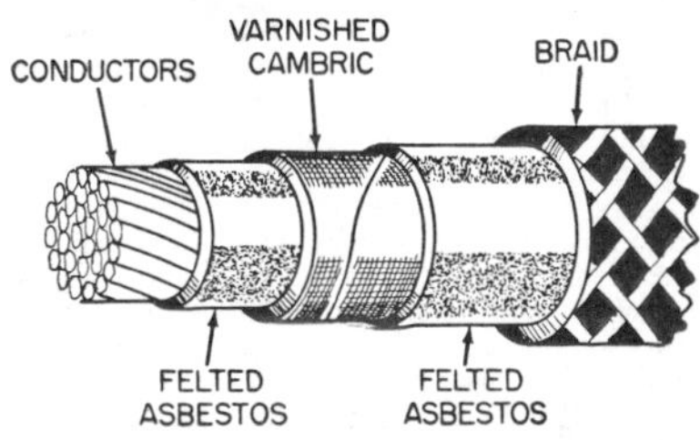

Figure 7-9.—Asbestos and varnished cambric insulation.

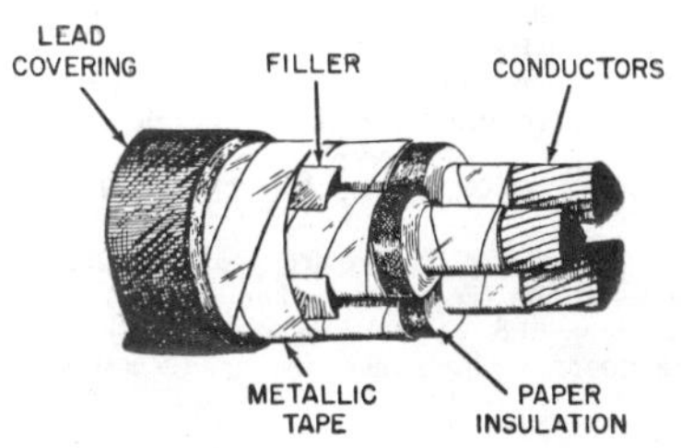

Figure 7-10.—Paper-insulated power cables.

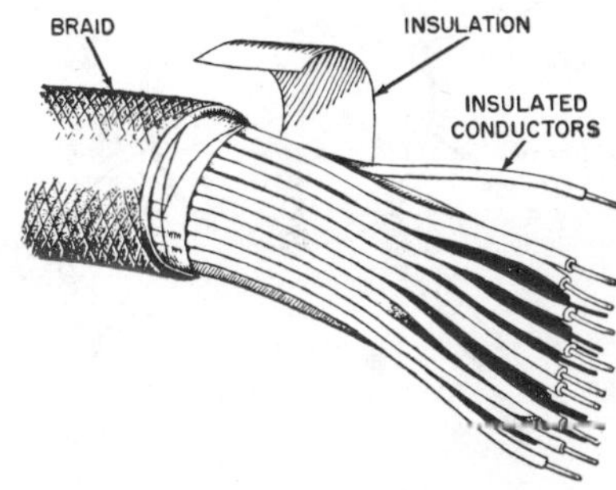

Figure 7-11.—Silk and cotton insulation.

NOTES:

ENAMEL

The wire used on the coils of meters, relays, small transformers, and so forth, is called MAGNET WIRE. This wire is insulated with an enamel coating. The enamel is a synthetic compound of cellulose acetate (wood pulp and magnesium). In the manufacture, the bare wire is passed through a solution of the hot enamel and then cooled. This process is repeated until the wire acquires from 6 to 10 coatings. Enamel has a higher dielectric strength than rubber for equal thickness. It is not practical for large wires because of the expense and because the insulation is readily fractured when large wires are bent.

Figure 7-12 shows an enamel-coated wire. Enamel is the thinnest insulating coating that can be applied to wires. Hence, enamel-insulated magnet wire makes smaller coils. Enameled wire is sometimes covered with one or more layers of cotton covering to protect the enamel from nicks, cuts, or abrasions.

CONDUCTOR PROTECTION

Wires and cables are generally subject to abuse. The type and amount of abuse depends on how and where they are installed and the manner in which they are used. Cables buried directly in the ground must resist moisture, chemical action, and abrasion. Wires installed in buildings must be protected against mechanical injury and overloading. Wires strung on crossarms on poles are kept far enough apart so that they do not touch; but snow, ice, and strong winds necessitate the use of conductors having high tensile strength and substantial supporting frame structures.

Generally, except for overload transmission lines, wires or cables are protected by some form of covering. The covering may be some type of insulator like rubber or plastic. Over this an outer covering of fibrous braid may be applied. If conditions require, a metallic outer covering may be used. The type of outer covering used depends on how and where the wire or cable is to be used.

FIBROUS BRAID

Cotton, linen, silk, rayon, and jute are types of FIBROUS BRAIDS. They are used for outer covering under conditions where the wires or cables are not exposed to heavy mechanical injury. Interior wiring for lights or power is usually done with impregnated cotton, braid-covered, rubber-insulated wire. Generally, the wire will be further protected by a flame-resistant nonmetallic outer covering or by a flexible or rigid conduit.

Figure 7-13 shows a typical building wire. In this instance two braid coverings are used for extra protection. The outer braid is soaked with a compound that resists moisture and flame.

Impregnated cotton braid is used as a covering for outdoor overhead conductors to afford protection against abrasion. For example,

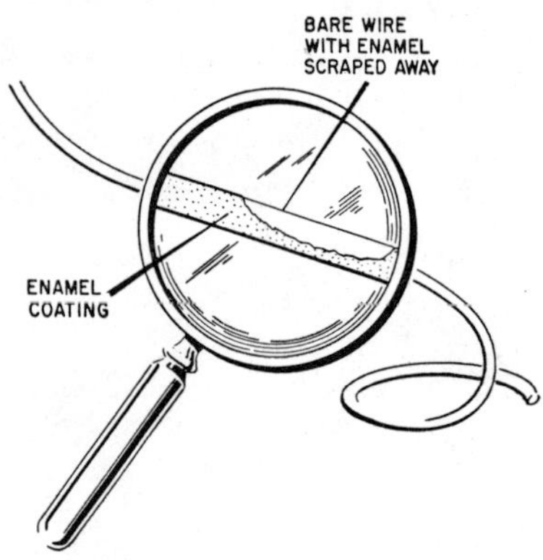

Figure 7-12.—Enamel insulation.

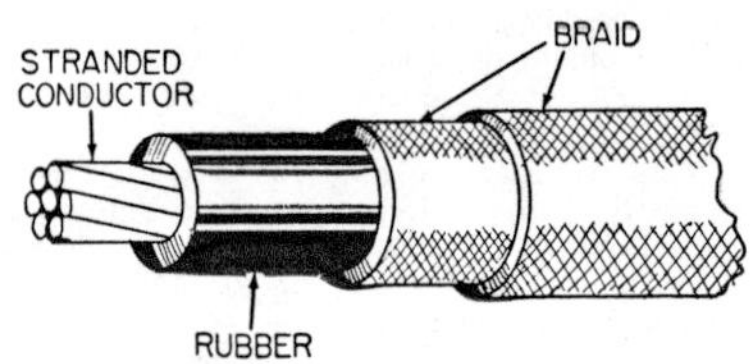

Figure 7-13.—Fibrous braid covering.

NOTES:

the service wires from the transformer secondary mains to the service entrance and also the high-voltage primary mains to the transformer are protected in this manner.

LEAD SHEATH

Subway-type cables or wires that are continually subjected to water must be protected by a watertight cover. This watertight cover is made either of a continuous lead jacket or a rubber sheath molded around the cable.

Figure 7-14 is an example of a lead-sheathed cable used in power work. The cable shown is a stranded 3-conductor type. Each conductor is insulated and then wrapped with a layer of rubberized tape. The conductors are twisted together and fillers or rope are added to form a rounded core. Over this is wrapped a second layer of tape called the SERVING, and finally the lead sheath is molded around the cable.

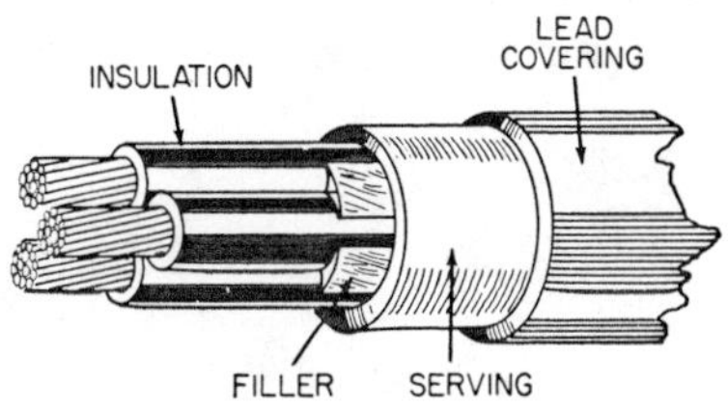

Figure 7-14.—Lead sheathed cable.

METALLIC ARMOR

Metallic armor provides a tough protective covering for wires or cables. The type, thickness, and kind of metal used to make the armor depend on the use of the conductors, the circumstances under which the conductors are to be used, and on the amount of rough treatment that is to be expected.

Four types of metallic armor cables are shown in figure 7-15.

Wire braid armor is used wherever light, flexible protection is needed. This type of armor is used almost exclusively aboard ship. The individual wires that are woven together to form the metal braid may be made of steel, copper, bronze, or aluminum. Besides mechanical protection, the wire braid also presents a static shield. This is important in radio work aboard ship to prevent interference from stray fields.

When cables are buried directly in the ground, they might be injured from two sources—moisture, and abrasion. They are protected from moisture by a lead sheath, and from abrasion by steel tape or interlocking armor covers. The steel tape covering, as shown in figure 7-15, is wrapped around the cable and then covered with a serving of jute. It is known as Parkway cable. The interlocking armor covering can withstand impacts better than steel tape. Interlocking armor has other uses besides underground work. In wiring the interior of buildings, interlocking armor-covered wire (BX cable) without the lead sheath is frequently used.

Armor wire is the best type of covering to withstand severe wear and tear. Underwater leaded cable usually has an outer armor wire cover.

All wires and cables do not have the same type of protective covering. Some coverings are designed to withstand moisture, others to withstand mechanical strain, and so forth. A cable may have a combination of each type, each doing its own job.

CONDUCTOR SPLICES AND TERMINAL CONNECTIONS

Conductor splices and connections are an essential part of any electric circuit. When conductors join each other, or connect to a load, splices or terminals must be used. It is important that they be properly made, since any electric circuit is only as good as its weakest link. The basic requirement of any splice or connection is that it be both mechanically and electrically as strong as the conductor or device with which it is used. High-quality workmanship and materials must be employed to insure lasting electrical contact, physical strength, and insulation (if required). The most common methods of making splices and connections in electric cables will now be discussed.

The first step in making a splice is preparing the wires or conductor. Insulation must be removed from the end of the conductor and the exposed metal cleaned. In removing the insulation from the wire, a sharp knife is used in much the same manner as in sharpening a pencil. That is, the knife blade is moved at a

NOTES:

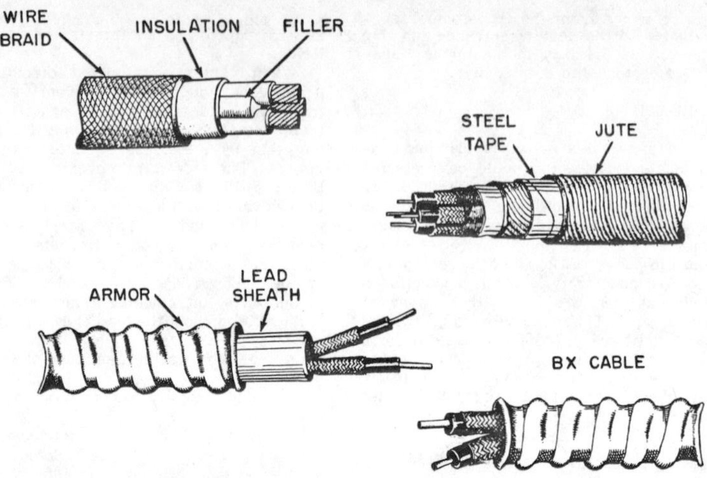

Figure 7-15.—Metallic armor.

small angle with the wire to avoid "nicking" the wire. This produces a taper on the cut insulation, as shown in figure 7-16. The insulation may also be removed by using a plier-like hand-operated wire stripper. After the insulation is removed, the bare wire ends should then be scraped bright with the back of a knife blade or rubbed clean with fine sandpaper.

WESTERN UNION SPLICE

Small, solid conductors may be joined together by a simple connection known as the WESTERN UNION SPLICE. In most instances the wires may be twisted together with the fingers and the ends clamped into position with a pair of pliers.

Figure 7-17 shows the steps in making a Western Union splice. First, the wires are prepared for splicing by removing sufficient insulation and cleaning the conductor. Next, the wires are brought to a crossed position and a long twist or bend is made in each wire. Then one of the wire ends is wrapped four of five times around the straight portion of the wire. The other end wire is wrapped in a similar manner. Finally, the ends of the wires should be pressed down as close as possible to the straight portion of the wire this prevents the sharp ends from puncturing the tape covering that is wrapped over the splice.

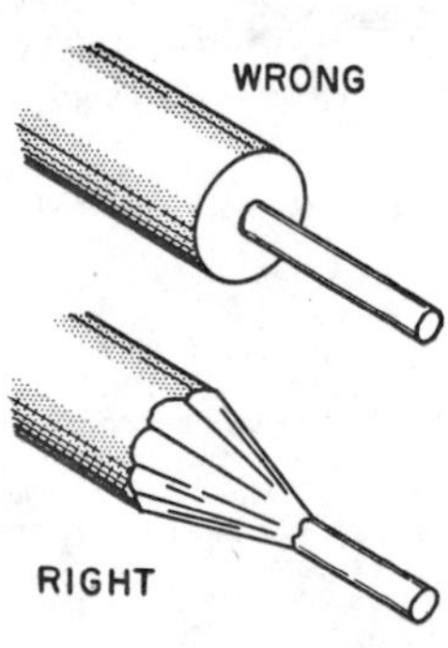

Figure 7-16.—Removing insulation from a wire.

NOTES:

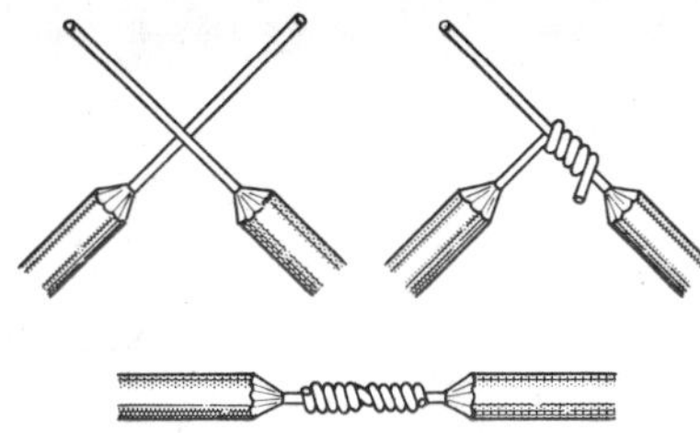

Figure 7-17.—Western Union splice.

STAGGERED SPLICE

Joining small, multiconductor cables together presents somewhat of a problem. Each conductor must be spliced and taped; and if the splices are directly opposite each other, the overall size of the joint becomes large and bulky. A smoother and less bulky joint may be made by staggering the splices.

Figure 7-18 shows how a 2-conductor cable is joined to a similar cable by means of the staggered splice. Care should be exercised to ensure that a short wire is connected to a long wire, and that the sharp ends are clamped firmly down on the conductor.

RATTAIL JOINT

Wiring that is installed in buildings is usually placed inside long lengths of steel pipe (conduit). Whenever branch circuits are required, junction or pull boxes are inserted in the conduit. One type of splice that is used for branch circuits is the rettail joint shown in figure 7-19.

The ends of the conductors to be joined are stripped of insulation. The wires are then twisted to form the rattail effect.

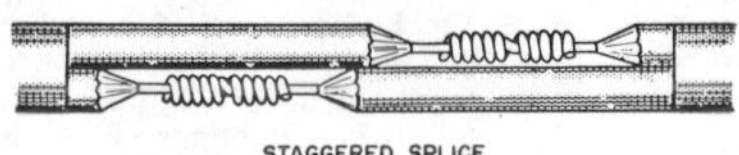

Figure 7-18.—Staggered splice.

NOTES:

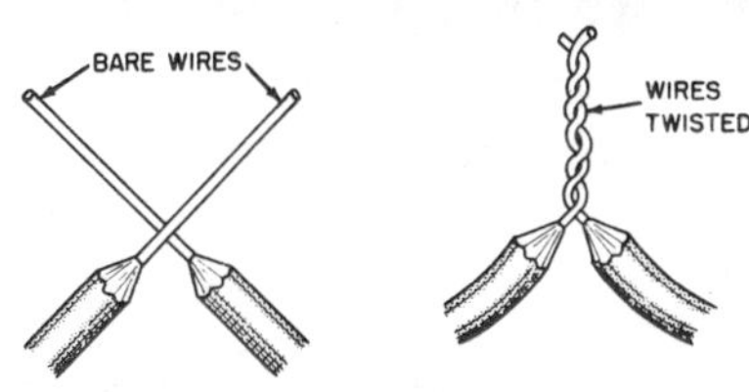

Figure 7-19.—Rattail joint.

FIXTURE JOINT

A fixture joint is used to connect a light fixture to the branch circuit of an electrical system where the fixture wire is smaller in diameter than the branch wire. Like the rattail joint, it will not stand much mechanical strain.

The first step is to remove the insulation from the wires to be joined. Figure 7-20 shows the steps in making a fixture joint.

After the wires are prepared, the fixture wire is wrapped a few times around the branch wire, as shown in the figure. The wires are not twisted, as in the rattail joint. The end of the branch wire is then bent over the completed turns. The remainder of the bare fixture wire

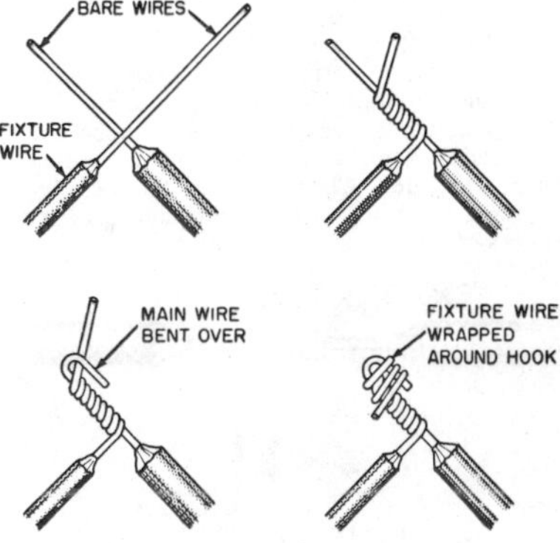

Figure 7-20.—Fixture joint.

is then wrapped over the bent branch wire. Soldering and taping completes the job.

KNOTTED TAP JOINT

All of the splices considered up to this point are known as BUTTED splices. Each was made by joining the FREE ends of the conductors together. Sometimes, however, it is necessary to join a conductor to a CONTINUOUS wire, and such a junction is called a TAP joint.

The main wire, to which the branch wire is to be tapped, has about one inch of insulation removed. The branch wire is stripped of about three inches of insulation. The steps in making the tap are shown in figure 7-21.

The branch wire is crossed over the main wire, as shown in figure, with about three-fourths of the bare portion of the branch wire extending above the main wire. The end of the branch wire is bent over the main wire, brought under the main wire, around the branch wire, and then over the main wire to form a knot. It is then wrapped around the main conductor in short, tight turns and the end is trimmed off.

The knotted tap is used where the splice is subject to strain or slip. When there is no mechanical strain, the knot may be eliminated.

SOLDERING EQUIPMENT

Soldering operations are a vital part of electrical/electronics maintenance procedures. It is a manual skill which can and must be learned by all personnel who work in the field of electricity. Practice is required to develop proficiency in the techniques of soldering; however, practice serves no useful purpose unless it is founded on a thorough understanding of basic principles. This discussion is devoted to providing information regarding some important aspects of soldering operations.

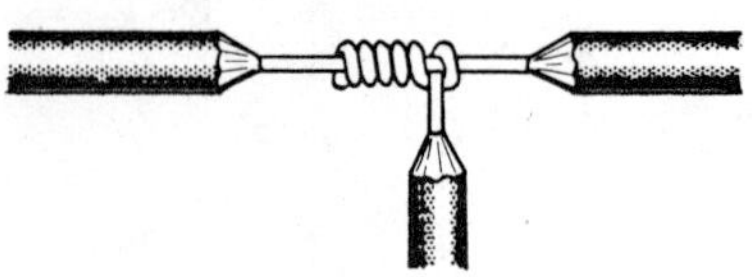

KNOTTED TAP JOINT

Figure 7-21.—Knotted tap joint.

NOTES:

Both the solder and the material to be soldered must be heated to a temperature which allows the solder to flow. If either is heated inadequately, "cold" solder joints result. Such joints do not provide either the physical strength or the electrical conductivity required. Appreciably exceeding the flow point temperature, however, is likely to cause damage to the parts being soldered. Various types of solder flow at different temperatures. In soldering operations it is necessary to select a solder that will flow at a temperature low enough to avoid damage to the part being soldered, or to any other part or material in the immediate vicinity.

The duration of high heat conditions is almost as important as the temperature. Insulation and many other materials in electrical equipment are susceptible to damage from heat. They are damaged if exposed to excessively high temperatures, or deteriorate if exposed to less drastically elevated temperatures for prolonged periods. The time and temperature limitations depend on many factors—the kind and amount of metal involved, the degree of cleanliness, the ability of the material to withstand heat, and the heat transfer and dissipation characteristics of the surroundings.

SOLDER

The three grades of solder generally used for electrical work are 40-60, 50-50, and 60-40 solder. The first figure is the percentage of tin, while the other is the percentage of lead. The higher the percentage of tin content, the lower the temperature required for melting. Also, the higher the tin content, the easier the flow, less the time required to harden, and generally the easier it is to do a good soldering job.

In addition to the solder, there must be flux to remove any oxide film on the metals being joined, otherwise they cannot fuse. The flux enables the molten solder to wet the metals so the solder can stick. The two types of flux are acid flux and rosin flux. Acid flux is more active in cleaning metals but is corrosive. Rosin is always used for the light soldering work in making wire connections. Generally, the rosin is in the hollow core of solder intended for electrical work, so that a separate flux is unnecessary. Such rosin-core solder is the type generally used. It should be noted, though, that the flux is not a substitute for cleaning the metals to be soldered. The metal must be shiny clean for the solder to stick.

SOLDERING PROCESS

Cleanliness is a prime perequisite for efficient, effective soldering. Solder will not adhere to dirty, greasy, or oxidized surfaces. Heated metals tend to oxidize rapidly, and the oxide must be removed prior to soldering. Oxides, scale, and dirt can be removed by mechanical means (such as scraping or cutting with an abrasive) or by chemical means. Grease or oil films can be removed by a suitable solvent. Cleaning should be accomplished immediately prior to the actual soldering operation.

Items to be soldered should normally be tinned before making mechanical connection. When the surface has been properly cleaned, a thin, even coating of flux may be placed over the surface to be tinned to prevent oxidation while the part is being heated to soldering temperature. Rosin core solder is usually preferred in electrical work, but a separate rosin flux may be used instead. Separate rosin flux is frequently used when tinning wires in cable fabrication. Tinning is the coating of the material to be soldered with a light coat of solder.

The tinning on a wire should extend only far enough to take advantage of the depth of the terminal or receptacle. Tinning or solder on wires subject to flexing causes stiffness, and may result in breakage.

The tinned surfaces to be joined should be shaped and fitted, then mechanically joined to make good mechanical and electrical contact. They must be held still with no relative movement of the parts. Any motion between parts will likely result in a poor solder connection.

SOLDERING TOOLS

Soldering Irons

All high quality irons operate in the temperature range of 500° to 600° F. Even the little 25-watt midget irons produce this temperature. The important difference in iron sizes is not temperature, but thermal inertia (the capacity of the iron to generate and maintain a satisfactory soldering temperature while giving up heat to the joint to be soldered). Although it is not practical to try to solder a heavy metal box with the 25-watt iron, that iron is quite suitable for replacing a half-watt resistor in a printed circuit. An iron with a rating as large as 150 watts would be satisfactory for use on a printed circuit, provided that suitable soldering techniques are used. One advantage of using a small iron for small work is that it is light and easy to handle and has a small tip which is easily inserted into close places. Also, even though its temperature is high, it does not have the capacity to transfer large quantities of heat.

Some irons have built-in thermostats. Others are provided with thermostatically controlled stands. These devices control the temperature of the soldering iron, but are a source of trouble. A well-designed iron is self-regulating by virtue of the fact that the resistance of its element increases with rising temperature, thus limiting the flow of current. For critical work, it is convenient to have a variable transformer for fine adjustment of heat; but for general-purpose work, no temperature regulation is needed.

Soldering Gun

The soldering gun has gained great popularity in recent years because it heats and cools rapidly. It is especially well adapted to maintenance and troubleshooting work where only a small part of the technician's time is spent actually soldering. A soldering iron, if kept hot constantly, oxidizes rapidly and is therefore difficult to keep clean.

A transformer in the soldering gun supplies approximately 1 volt at high current to a loop of copper which acts as the tip. It heats to soldering temperature in 3 to 5 seconds, but may overheat to the point of incandescence if left on over 30 seconds. The gun is operated with a finger switch so that the gun heats only while the switch is depressed.

Since the gun normally operates only for short periods at a time, it is comparatively easy to keep clean and well tinned; thus, little oxidation is allowed to form. However, the tip is made of pure copper, and is susceptible to pitting which results from the dissolving action of the solder.

Tinning of the tip is always desirable unless it has already been done. The gun or iron should always be kept tinned in order to permit proper heat transfer to the work to be soldered. Tinning also provides adequate control of the heat to prevent thermal spillover to nearby materials. Tinning of the tip of a gun may be somewhat more difficult than tinning the tip of an iron. Maintaining the proper tining on either type, however, may be made easier by tinning with silver solder. The temperature at which the

NOTES:

bond is formed between the copper tip and the silver solder is considerably higher than with lead-tin solder. This tends to decrease the pitting action of the solder on the copper tip.

Pitting of the tip indicates the need for retinning, after first filing away a portion of the tip. Retinning too often results in using up the tip too fast.

Overheating can easily occur when using the gun to solder delicate wiring. With practice, however, the heat can be accurately controlled by pulsing the gun on and off with its trigger. For most jobs, even the LOW position of the trigger overheats the soldering gun after 10 seconds; the HIGH position is used only for fast heating and for soldering heavy connections.

Heating and cooling cycles tend to loosen the nuts or screws which retain the replaceable tips on soldering irons or guns. When the nut on a gun is loosened, the resistance of the tip connection increases, and the temperature of the connection is increased. Continued loosening may eventually cause an open circuit. Therefore, the nut should be tightened periodically.

Resistance Soldering

A time-controlled resistance soldering set is now available. The set consists of a transformer that supplies 3 or 6 volts at high current to stainless steel or carbon tips. The transformer is turned ON by a foot switch and OFF by an electronic timer. The timer can be adjusted for as long as 3 seconds soldering time. This set is especially useful for soldering cables to plugs and similar connectors—even the smallest types available.

In use, the double-tip probes of the soldering unit are adjusted to straddle the connector cup to be soldered. One pulse of current heats it for tinning and, after the wire is inserted, a second pulse of current completes the job. Since the soldering tips are hot only during the brief period of actual soldering, burning of wire insulation and melting of connector inserts are greatly minimized.

The greatest difficulty with this device is keeping the probe tips free of rosin and corrosion. A cleaning block is mounted on the transformer case for this purpose. Some technicians prefer fine sandpaper for cleaning the double tips. CAUTION: Do not use steel wool. It is dangerous when used around electrical equipment.

Pencil Iron and Special Tips

An almost indispensable item is the pencil type soldering iron with an assortment of tips (fig. 7-22). Miniature soldering irons, with wattage ratings of less than 40 watts are easy to used and are recommended. In an emergency, larger irons can be converted and used on subminiature equipment as described later in this section.

One type of iron is equipped with several different tips that range from one-fourth to one-half inch in size (diameter) and are of various shapes. This feature makes it adaptable to a variety of jobs. Unlike most tips which are held in place by setscrews, these tips have threads and screw into the barrel. This feature provides excellent contact with the heating element, thus improving heat transfer efficiency. A pad of "antiseize" compound is supplied with each iron. This compound is applied to the threads each time a tip is installed in the iron, thereby enabling the tip to be easily removed when another is to be inserted.

A special feature of this iron is the soldering pot that screws in like a tip and holds about a thimbleful of solder. It is useful for tinning the ends of large numbers of wires.

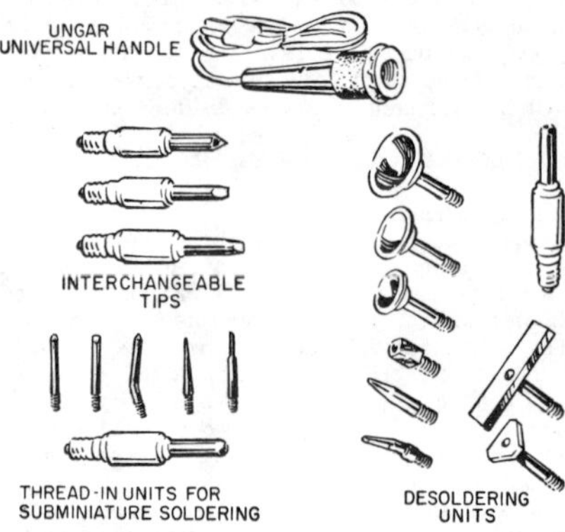

Figure 7-22.—Pencil iron kit with special tips.

NOTES:

The interchangeable tips are of various sizes and shapes for specific applications. Extra tips may be obtained and shaped to serve special purposes. The thread-in units are useful in soldering subminiature items. The desoldering units are specifically designed for performing special and individual functions.

Another advantage of the pencil soldering iron is its possible use as an improvised light source for inspections. Simply remove the soldering tip and insert a 120-volt, 6-watt, type 6S6, candelabra screwbase lamp bulb into the socket.

If leads, tabs, or small wires are bent against a board or terminal, slotted tips may be used to simultaneously melt the solder and straighten the leads.

A hollow tip, which fits over a pin terminal, may be used to desolder and resolder wiring at cables or feed-through terminals.

Many miniature components have multiple connections, all of which must be desoldered to permit removal of the component in one operation. These connections may be desoldered individually by heating each connection and brushing away the solder. With this method, particular care must be taken to insure that loose solder does not stick to other parts of become lodged where it may cause a short circuit. A more efficient method is to use the specially shaped desoldering units (fig. 7-22). Select the proper size and shape tip that will contact all terminals to be desoldered—and nothing else. Do not permit the tip to remain in contact with the terminals too long at one time.

If no suitable tip is available for a particular operation, an improvised tip may be made. Wrap a length of copper wire around one of the regular tips and bend the wire into the proper shape for the purpose. This method also serves to reduce tip temperature when a larger iron must be used on miniature components. (See fig. 7-23.)

In connection with the discussion of soldering tools and devices, the selection of solder and flux is also critical. A small diameter rosin core solder with a high tin-lead ratio (60/40) is normally preferred in miniature circuits where heat is critical.

Soldering Aids

Several devices other than the soldering iron and its tips are required in soldering miniature circuits. Several of these (brushes, probes, scrapers, knives, etc.) have been mentioned previously.

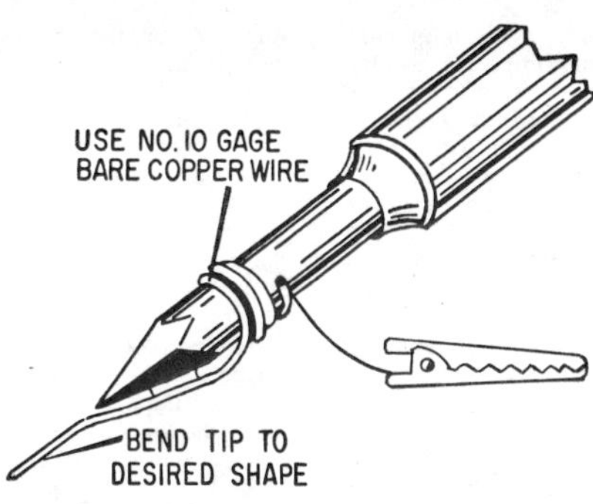

Figure 7-23.—Improvised tip to reduce tip temperature.

Some type of thermal shunt is essential in all soldering operations which involve heat-sensitive components. Pliers, tweezers, or hemostats may be used for some applications, but their effectiveness is limited. A superior heat shunt, as shown in figure 7-24, permits soldering the leads of component parts without overheating the part itself.

For maximum effectiveness, any protective coating should be removed before applying the heat shunt. The shunt should be attached carefully to prevent damage to the leads, terminals, or component parts. The shunt should be clipped to the lead, between the joint and the part being protected. As the joint is heated, the shunt absorbs the excess heat before it can reach the part and cause damage.

A small piece of beeswax may be placed between the protected unit and the heat shunt. When the beeswax begins to melt, the temperature limit has been reached. The heat source

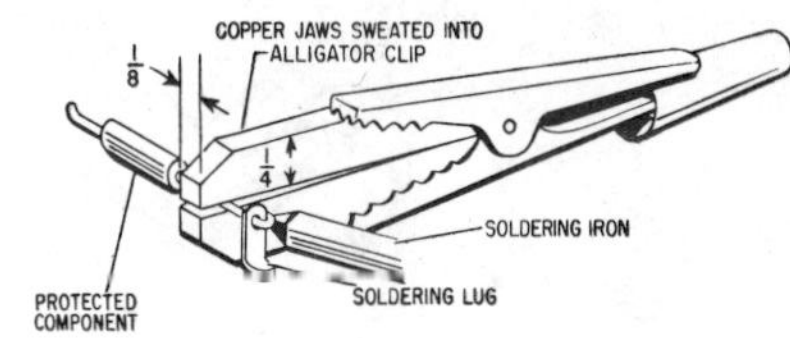

Figure 7-24.—Heat shunt.

NOTES:

should be removed immediately, but the shunt should be left in place.

Premature removal of the heat shunt permits the unrestricted flow of heat from the melted solder into the component. The shunt should be allowed to remain in place until it cools to room temperature. A clip-on type shunt is preferred because it requires positive action to remove the shunt, but does not require that the technician maintain pressure to hold it in place.

Another invaluable soldering aid is the "solder sucker" syringe. One type is shown in figure 7-25. Its purpose is to "suck up" excess solder (and incidentally the excess heat) from a joint. The only requirements of an efficient solder sucker are a controllable source of vacuum (squeeze bulb), a solder receiver, and a tip. The tip must be able to withstand the heat of molten solder. Teflon is ideal, but may be difficult to acquire. A silicon rubber-covered Fiberglas sleeving with an inner diameter of 0.162 inch and the bulb from a medicine dropper makes a suitable syringe. (The glass or plastic tip of the medicine dropper cannot withstand the heat.)

SOLDER CONNECTIONS

Frequent arguments occur in electrical shops concerning the proper method of making soldered connections to terminals and binding posts. For many years it was considered necessary to wrap the lead tightly around the terminal, so as to provide maximum mechanical support and strength. General Specification for Soldering Process states that the parts to be joined shall be held together in such a manner that the parts shall not move in relation to one another during the soldering process. The joint must not be disturbed until the solder has completely solidified.

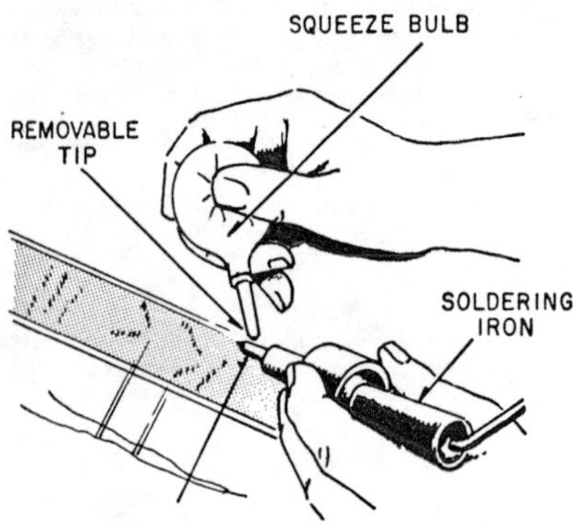

Figure 7-25.—Solder sucker.

Electronics Laboratories tested many standard capacitors and resistors soldered to terminals of various types. The joints were then subjected to vibrations far in excess of those normally encountered in electrical and electronic equipment. The connections were made with various degrees of wrapping around the terminals, with main reliance for physical strength being placed on the solder. As a result of these tests and others conducted by other organizations, the joints illustrated in figure 7-26 are recommended. Wrappings of three-eighths to three-fourths turn are usually recommended so that the joint need not be held during the application and cooling of the solder.

Excessive wrappings of leads results in increased heat requirements, more strain on parts, greater difficulty of inspection, greater difficulty of assembly and disassembly of the joints, and increased danger of breaking the parts or terminals during desoldering operations. Insufficient wrapping may result in poor solder joints due to movement of the lead during the soldering operation.

The areas to be joined must be heated to or slightly above the flow temperature of the solder. The application of heat must be carefully controlled to prevent damage to components of the assembly, insulation, or nearby materials. Solder is then applied to the heated area. Only enough solder should be used to make a satisfactory joint. Heavy fillets or beads must be avoided.

Solder should not be melted with the soldering tip and allowed to flow onto the joint. The joint should be heated and the solder applied to the joint. When the joint is adequately heated, the solder will flow evenly. Excessive temperature tends to carbonize flux, thus hindering the soldering operation.

No liquid should be used to cool a solder joint. By using the proper tools and soldering technique, a joint should not become so hot that rapid cooling is needed.

NOTES:

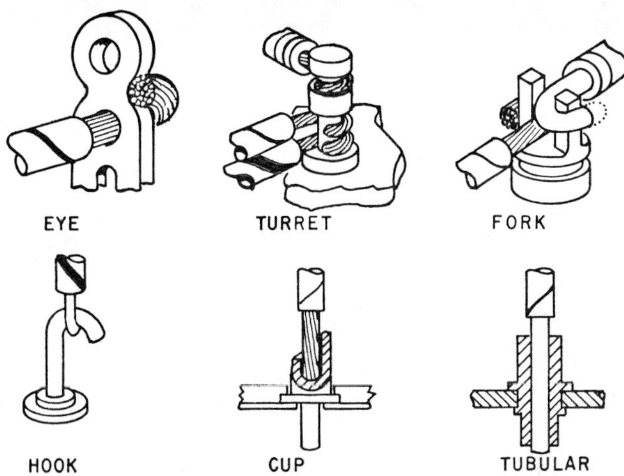

Figure 7-26.—Wrapping of terminals for soldering.

If, for any reason, a satisfactory joint is not initially obtained, the joint must be taken apart, the surfaces cleaned, excess solder removed, and the entire soldering operation (except tinning) repeated.

After the joint has cooled, all flux residues should be removed. Any flux residue remaining on the surface of electrical contacts may collect dirt and promote arcing at a later time. This cleaning is necessary even when rosin-core solder is used.

Connections should never be soldered or desoldered while equipment power is on or while the circuit is under test. Always discharge any capacitors in the circuit prior to any soldering operation.

SOLDER SPLICERS

The solder-type splicer is essentially a short piece of metal tube. Its inside diameter is just large enough to allow the tip of a stranded conductor to be inserted in either end, after the conductor tip has been stripped of insulation. This type of splicer is shown in figure 7-27.

The splicer is first heated and filled with solder. While still molten, the solder is then poured out, leaving the inner surfaces tinned. When the conductor tips are stripped, the length of exposed strands should be long enough so that the insulation butts against the splicer when the conductors are tinned and fully inserted. (See fig. 7-27 (B).) When heat is applied to the connection and the solder melts, excess solder will be squeezed out through the vents. This must be cleaned away. After the splice has cooled, insulating material must be wrapped or tied over the joint.

SOLDER TERMINAL LUGS

In addition to being joined or spliced to one another, conductors are often connected to other objects, such as motors and switches. Since this is where a length of conductor ends (terminates), such connections are referred to as terminal points. In some cases, it is allowable to bend the end of the conductor into a small "eye" and put it around a terminal binding post. Where a mounting screw is used, the screw is passed through the eye. The conductor tip which forms the eye should be bent as

NOTES:

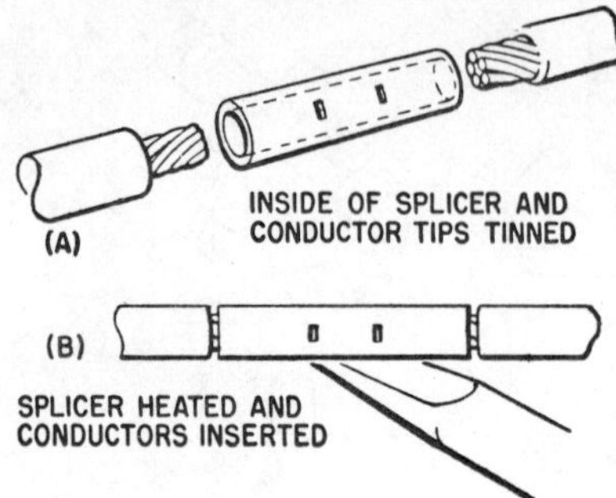

Figure 7-27.–Steps in using solder splicer.

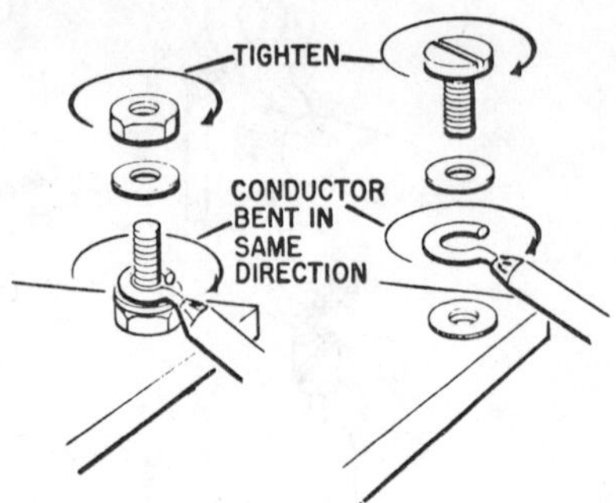

Figure 7-28.—Conductor terminal connection.

shown in figure 7-28. Note that when the screw or binding nut is tightened, it also tends to tighten the conductor eye.

This method of connection is sometimes not desirable. When design requirements are more rigid, terminal connections are made by using special hardware devices called terminal lugs. There are terminal lugs of many different sizes and shapes, but all are essentially the same as the type shown in figure 7-29.

Each type of lug has a barrel (sleeve) which is wedged, crimped, or soldered to its conductor. There is also a tongue with a hold or slot in it to receive the terminal post or screw. When mounting a solder-type terminal lug to a conductor, first tin the inside of the barrel. The conductor tip is stripped and also tinned, then inserted in the preheated lug. When mounted, the conductor insulation should butt against the lug barrel, so that there is no exposed conductor.

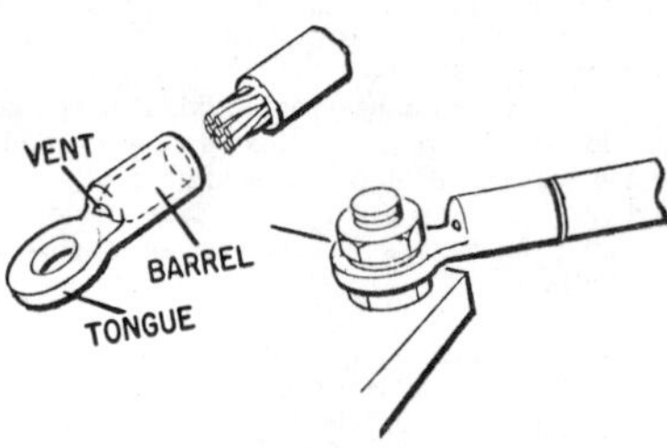

Figure 7-29.—Solder-type terminal lug.

SOLDERLESS CONNECTORS

Splicers and terminal lugs which do not require solder are more widely used than those which do require solder. Solderless connectors are attached to their conductors by means of several different devices, but the principle of each is essentially the same. They are all squeezed (crimped) tightly onto their conductors. They afford adequate electrical contact, plus great mechanical strength. In addition, solderless connectors are easier to mount correctly because they are free from the most common problems of solder connector mounting; namely, cold solder joints, burned insulation, and so forth.

Solderless connectors are made in a great variety of sizes and shapes, and for many different purposes. Only a few are discussed here.

NOTES:

SOLDERLESS SPLICERS

Three of the most common types solderless splicers, classified according to their methods of mounting, are the split-sleeve, split-tapered-sleeve, and crimp-on splicers.

Split-Sleeve Splicer

A split-sleeve splicer is shown in figure 7-30. To connect this splicer to its conductor, the stripped conductor tip is first inserted between the split-sleeve jaws. Using a tool designed for that purpose, the slide ring is forced toward the end of the sleeve. The sleeve jaws are closed tightly on the conductor, and the slide ring holds them securely.

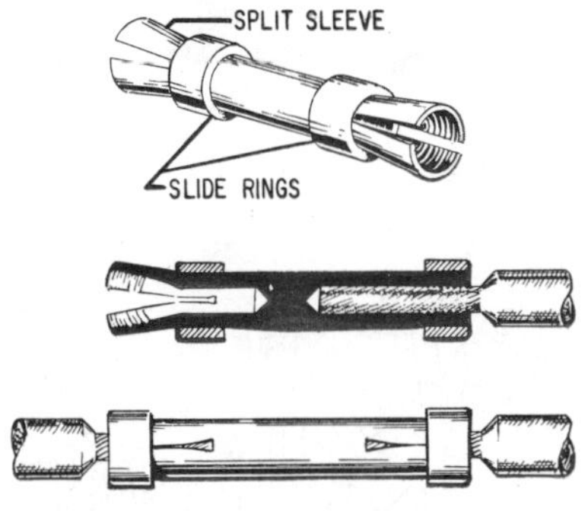

Figure 7-30.—Split-sleeve splicer.

Split-Tapered-Sleeve Splicer

A cross-sectional view of a split-tapered-sleeve splicer is shown in figure 7-31 (A). To mount this type of splicer, the conductor is stripped and inserted in the split-tapered sleeve. The threaded sleeve is turned or screwed into the tapered bore of the body. As the sleeve is turned in, the split segments are squeezed tightly around the conductor by the narrowing bore. The finished splice (fig. 7-31 (B)), must be covered with insulation.

Crimp-On Splicer

The crimp-on splicer (fig. 7-32) is the simplest of the splicers discussed. The type

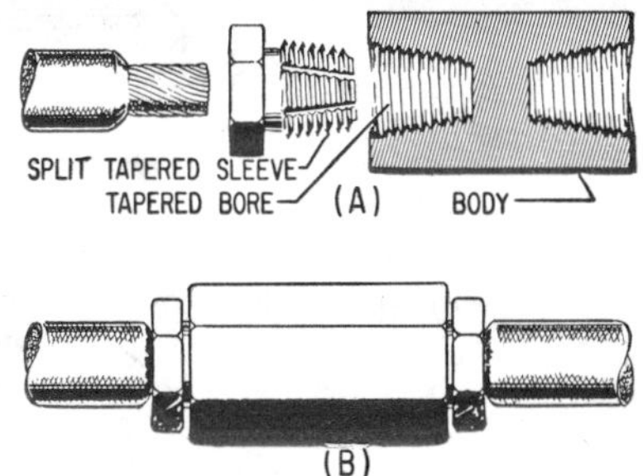

Figure 7-31.—Split-tapered-sleeve splice.

shown is preinsulated, though uninsulated types are manufactured. These splicers are mounted with a special plier-like hand-crimping tool designed for that purpose. The stripped conductor tips are inserted in the splicer, which is then squeezed tightly closed. The insulating sleeve grips the outer insulated conductor, and the metallic internal splicer grips the bare conductor strands.

SOLDERLESS TERMINAL LUGS

Solderless terminal lugs are used more widely than solder terminal lugs. They afford adequate electrical contact, plus great mechanical strength. In addition, solderless lugs are easier to attach correctly, because they are free from the most common problems of solder terminal lugs; namely, cold solder joints, burned insulation, and so forth. There are many sizes and shapes of these lugs, each intended for a different type of service of conductor size. Only a few are discussed here.

These are classified according to their method of mounting. They are the split-tapered-sleeve (wedge-type), split-tapered-sleeve (threaded-type), and crimp-on.

Split-Tapered-Sleeve Terminal Lug (Wedge)

This type lug is shown in figure 7-33. It is commonly referred to as a "wedge-on," because of the manner in which it is secured to a

NOTES:

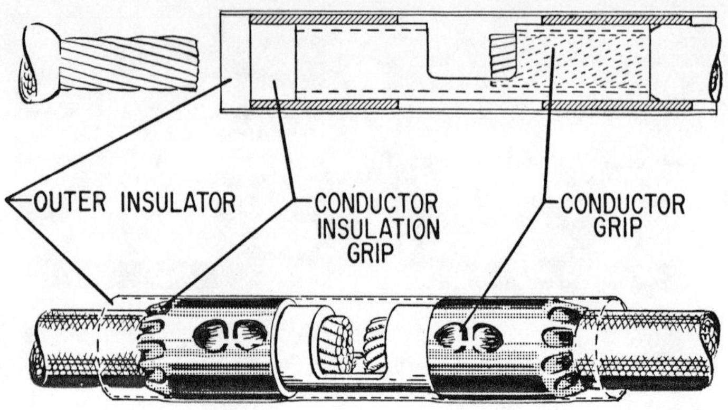

Figure 7-32.—Crimp-on splicer.

conductor. The stripped conductor is inserted through the hole in the split sleeve. When the sleeve is forced or "wedged" down into the barrel, its tapered segments are squeezed tightly around the conductor.

Split-Tapered-Sleeve
Terminal Lug (Threaded)

This lug (fig. 7-34) is attached to a conductor in exactly the same manner as a split-sleeve splicer. The segments of the threaded split sleeve squeezes tightly around the conductor as it is turned into the tapered bore of the barrel. For this reason, the lug is commonly referred to as a "screw-wedge."

Crimp-On Terminal Lug

The crimp-on lug is shown in figure 7-35. This lug is simply squeezed or "crimped" tightly onto a conductor. This is done by using the same tool used with the crimp-on splicer. The lug shown is preinsulated, but uninsulated types are manufactured. When mounted, both the conductor and its insulation are gripped by the lug.

TAPING A SPLICE

The final step in completing a splice or joint is the placing of insulation over the bare

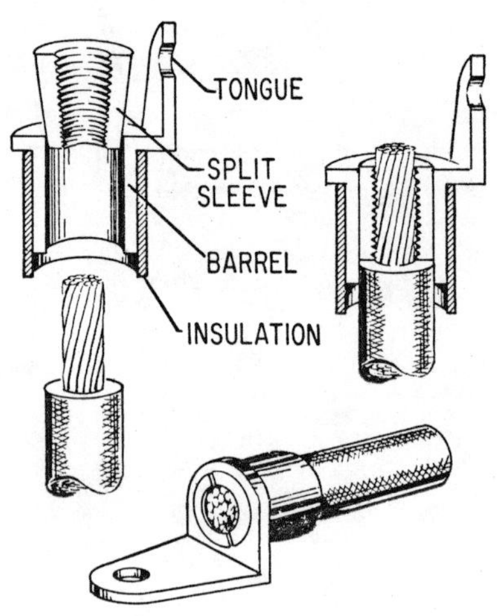

Figure 7-33.—Split-tapered-sleeve terminal lug (wedge type).

NOTES:

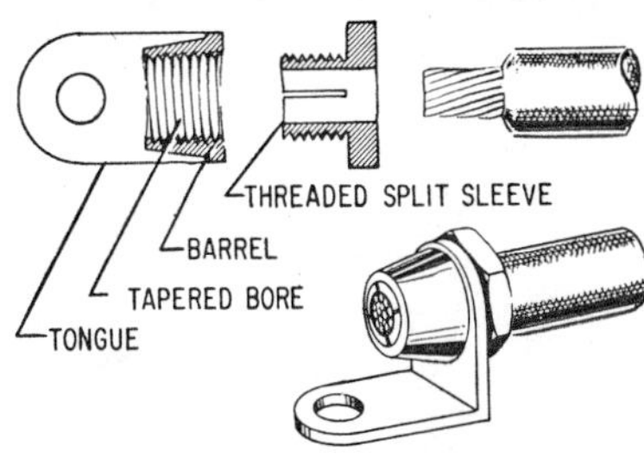

Figure 7-34.—Split-tapered-sleeve terminal lug (threaded).

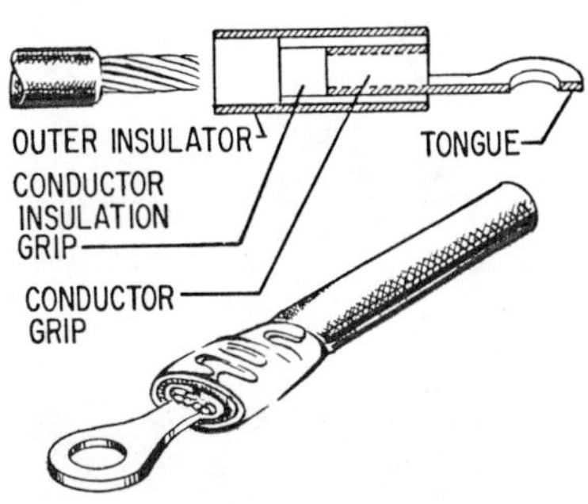

Figure 7-35.—Crimp-on terminal lug.

wire. The insulation should be of the same basic substance as the original insulation. Usually a rubber splicing compound is used.

RUBBER TAPE

Latex (rubber) tape is a splicing compound. It is used where the original insulation was rubber. The tape is applied to the splice with a light tension so that each layer presses tightly against the one underneath it. This pressure causes the rubber tape to blend into a solid mass. When the application is completed, an insulation similar to the original has been restored.

Between each layer of latex tape, when it is in roll form, there is a layer of paper or treated cloth. This layer prevents the latex from fusing while still on the roll. The paper or cloth is peeled off and discarded before the tape is applied to the splice.

Figure 7-36 shows the correct way to cover a splice with rubber insulation. The rubber splicing tape should be applied smoothly and under tension so that there will be no air spaces between the layers. In putting on the first layer, start near the middle of the joint instead of the end. The diameter of the completed insulated joint should be somewhat greater that the overall diameter of the original cable, including the insulation.

FRICTION TAPE

Putting rubber tape over the splice means that the insulation has been restored to a great degree. It is also necessary to restore the protective covering. Friction tape is used for this purpose; it also affords a minor degree of electrical insulation.

Friction tape is a cotton cloth that has been treated with a sticky rubber compound. It comes in rolls similar to rubber tape except that no paper or cloth separator is used. Friction tape is applied like rubber tape; however, it does not stretch.

The friction tape should be started slightly back on the original braid covering. Wind the tape so that each turn overlaps the one before it; and extend the tape over onto the braid covering at the other end of the splice. From this point a second layer is wound back along the splice until the original starting point is reached. Cutting the tape and firmly pressing down the end complete the job. When proper care is taken, the splice can take as much abuse as the rest of the wire.

Weatherproof wire has no rubber insulation, just a braid covering. In that case, no rubber tape is necessary, only friction tape need be used.

NOTES:

PLASTIC ELECTRICAL TAPE

Plastic electrical tape has come into wide use in recent years. It has certain advantages over rubber and friction tape. For example, it will withstand higher voltages for a given thickness. Single thin layers of certain commercially available plastic tape will stand several thousand volts without breaking down. However, to provide an extra margin of safety, several layers are usually wound over the splice. Because the tape is very thin, the extra layers add only a very small amount of bulk; but at the same time the added protection, normally furnished by friction tape, is provided by the additional layers of plastic tape. In the choice of plastic tape, the factor of expense must be balanced against the other factors involved.

Plastic electric tape normally has a certain amount of stretch so that it easily conforms to the contour of the splice without adding unnecessary bulk. The lack of bulkiness is especially important in some junction boxes where space is at a premium.

For high temperatures—for example, above 175° F.—a special type of tape backed with glass cloth is used.

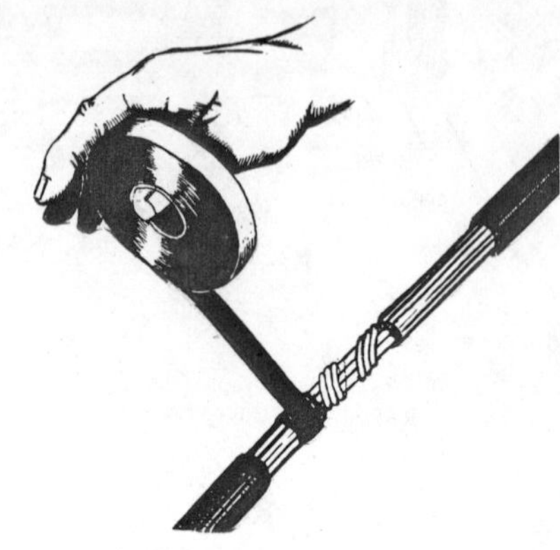

Figure 7-36.—Applying rubber tape.

NOTES:

CHAPTER 8

ELECTROMAGNETISM AND MAGNETIC CIRCUITS

The fundamental theories concerning simple magnets and magnetism were discussed in chapter 2 of this manual. Those discussions dealt mainly with forms of magnetism that were not related directly to electricity—permanent magnets for instance. Only brief mention was made of those forms of magnetism having direct relation to electricity (such as "producing electricity with magnetism"). This chapter resumes the study of magnetism where chapter 2 left off. Therefore, it may be necessary for you to review parts of chapter 2 from time to time. This chapter begins the more advanced study of magnetism as it is affected by electric current flow, and the closely related study of electricity as it is affected by magnetism. This general subject area is most often referred to as electromagnetism.

Magnetism and basic electricity are so closely related that one cannot be studied at length without involving the other. This close fundamental relationship will be continually borne out in other chapters of this manual, such as in the study of generators, transformers, and motors. The technician, to be proficient in electricity, must become familiar with such general relationships that exist between magnetism and electricity as follows:

1. Electric current flow will always produce some form of magnetism.
2. Magnetism is by far the most commonly used means for producing or using electricity.
3. The peculiar behavior of electricity under certain conditions is caused by magnetic influences.

MAGNETIC FIELD AROUND A CURRENT CARRYING CONDUCTOR

In 1819 Hans Christian Oersted, a Danish physicist, found that a definite relation exists between magnetism and electricity. He discovered that an electric current is accompanied by certain magnetic effects and that these effects obey definite laws. If a compass is place in the vicinity of a current-carrying conductor, the needle alines itself at right angles to the conductor, thus indicating the presence of a magnetic force. The presence of this force can be demonstrated by passing an electric current through a vertical conductor which passes through a horizontal piece of cardboard, as illustrated in figure 8-1. The magnitude and direction of the force are determined by setting a compass at various points on the cardboard and noting the deflection.

The direction of the force is assumed to be the direction the north pole of the compass points. These deflections show that a magnetic field exists in circular form around the conductor. When the current flows upward, the field direction is clockwise, as viewed from the top, but if the polarity of the supply is reversed so that the current flows downward, the direction of the field is counterclockwise.

The relation between the direction of the magnetic lines of force around a conductor and the direction of current flow along the conductor may be determined by means of the left-hand rule for a conductor. If the conductor is grasped in the left hand with the thumb extended in the direction of electron flow (- to +), the fingers will point in the direction of the magnetic lines of force. This is the same direction in which the north pole of a compass would point if the compass was placed in the magnetic field.

Arrows generally are used in electric diagrams to denote the direction of current flow along the length of wire. Where cross sections of wire are shown, a special view of the arrow is used. A cross-sectional view of a conductor

NOTES:

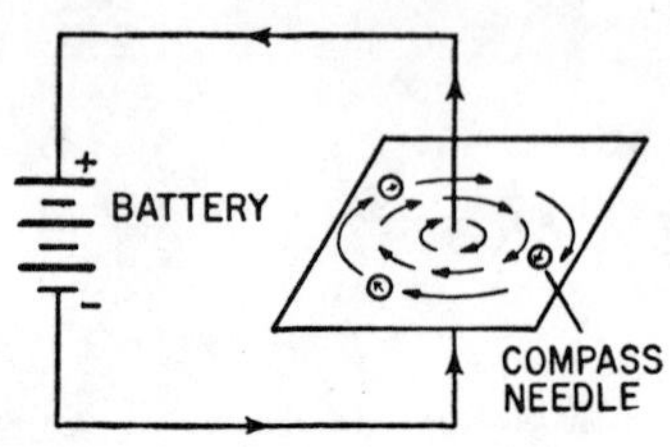

Figure 8-1.—Magnetic field around a current-carrying conductor.

that is carrying current toward the observer is illustrated in figure 8-2 (A). The direction of current is indicated by a dot, which represents the head of the arrow. A conductor that is carrying current away from the observer is illustrated in figure 8-2 (B). The direction of current is indicated by a cross, which represents the tail of the arrow.

When two parallel conductors carry current in the same direction, the magnetic fields tend to encircle both conductors, drawing them together with a force of attraction, as shown in figure 8-3 (A). Two parallel conductors carrying currents in opposite directions are shown in figure 8-3 (B). The field around one conductor is opposite in direction to the field around the other conductor. The resulting lines of force are crowded together in the space between the wires, and tend to push the wires apart. Therefore, two parallel adjacent conductors carrying currents in the same direction attract each other and two parallel conductors carrying currents in opposite directions repel each other.

MAGNETIC FIELD OF A COIL

The magnetic field around a current-carrying wire exists at all points along its length. The field consists of concentric circles in a plane perpendicular to the wire. (See fig. 8-1.) When

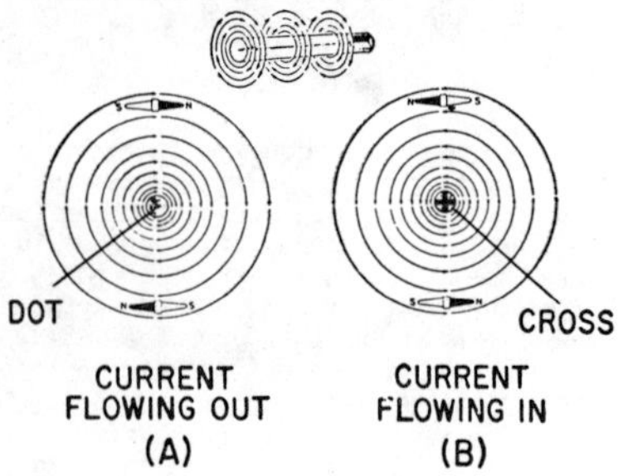

Figure 8-2.—Magnetic field around a current-carrying conductor, detailed view.

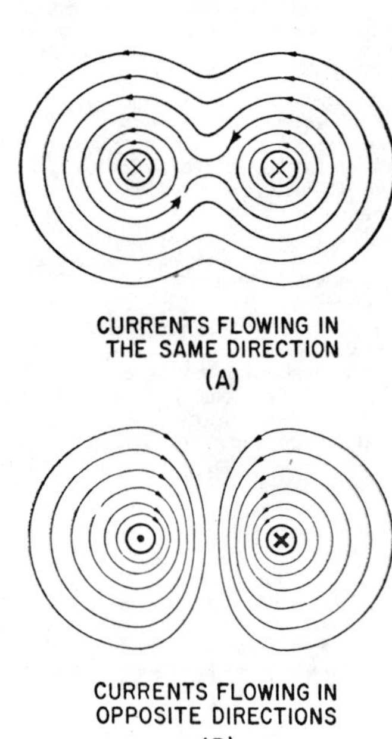

Figure 8-3.—Magnetic field around two parallel conductors.

NOTES:

this straight wire is wound around a core, as shown in figure 8-4 (A), it becomes a coil and the magnetic field assumes a different shape. Part (A) is a partial cutaway view which shows the construction of a simple coil. Part (B) is a complete cross-sectional view of the same coil. The two ends of the coil are identified as a and b. When current is passed through the coiled conductor, as indicated, the magnetic field of each turn of wire links with the fields of adjacent turns, as explained in connection with figure 8-3 (A). The combined influence of all the turns produces a two-pole field similar to that of a simple bar magnet. One end of the coil will be a north pole and the other end will be a south pole.

POLARITY OF AN ELECTROMAGNETIC COIL

In figure 8-2, it was shown that the direction of the magnetic field around a straight conductor depends on the direction of current flow through that conductor. Thus, a reversal of current flow through a conductor causes a reversal in the direction of the magnetic field that is produced. It follows that a reversal of the current flow through a coil also causes a reversal of its two-pole field. This is true because that field is the product of the linkage between the individual turns of wire on the coil. Therefore, if the field of each turn is reversed, it follows that the total field (coil's field) is also reversed.

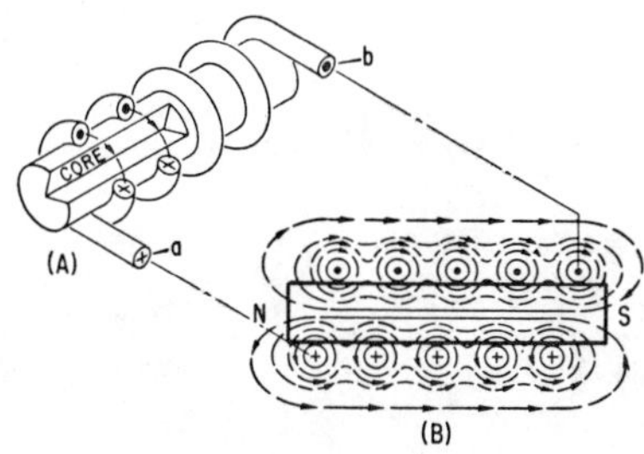

Figure 8-4.—Magnetic field produced by a current-carrying coil.

When the direction of electron flow through a coil is known, its polarity may be determined by use of the left-hand rule for coils. This rule is illustrated in figure 8-5, and is stated as follows: Grasping the coil in the left hand, with the fingers "wrapped around" in the direction of electron flow, the thumb will point toward the north pole.

STRENGTH OF AN ELECTROMAGNETIC FIELD

The strength, or intensity, of a coil's field depends on a number of factors. The major factors are listed below. All of these factors are discussed under headings that follow.

1. The number of turns of conductor.
2. The amount of current flow through the coil.
3. The ratio of the coil's length to its width.
4. The type of material in the core.

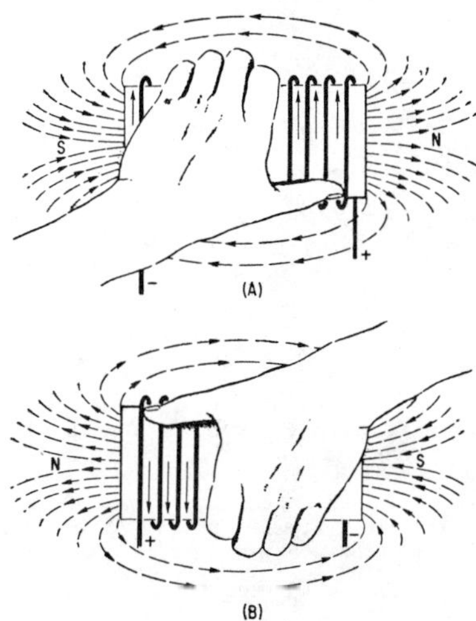

Figure 8-5.—Left-hand rule for coil polarity.

NOTES:

MAGNETIC CIRCUITS

Many electrical devices depend upon magnetism in one or more forms for their operation. To have these devices function efficiently, engineers work out intricate designs for the required magnetic conditions. The magnets designed to do a particular job must have the required strength, and must be provided with paths, or circuits, of suitable shapes and materials.

A magnetic circuit is defined as the path (or paths) taken by the magnetic lines of force leaving a north pole, passing through the entire circuit and returning to the south pole. A magnetic circuit may be a series or parallel circuit or any combination.

OHM'S LAW EQUIVALENT FOR MAGNETIC CIRCUITS

The law of current flow in the electric circuit is similar to the law for the establishing of flux in the magnetic circuit.

Ohm's law for electric circuits states that the current is directly proportional to the applied voltage and inversely proportional to the resistance offered by the circuit. Expressed mathematically,

$$I = \frac{E}{R}$$

Rowland's law for magnetic circuits states, in effect, that the number of lines of magnetic flux in maxwells (Φ) is directly proportional to the magnetomotive force in gilberts (F) and inversely proportional to the reluctance ($\mathcal{R}$) offered by the circuit. The unit of reluctance sometimes used is the REL, $\mathcal{R}$. Expressed mathematically.

$$\Phi = \frac{F}{\mathcal{R}}$$

The similarity of Ohm's law and Rowland's law is apparent. However, the units used in the expression for Rowland's law need to be explained.

The magnetic flux, Φ, (phi) is similar to current in the Ohm's law formula, and comprises the total number of lines of force existing in the magnetic circuit. The maxwell is the unit of flux—that is, 1 line of force is equal to 1 maxwell. However, the maxwell is often referred to as simply a line of force, line of induction, or line.

The magnetomotive force, **F**, or mmf, comparable to electromotive force in the Ohm's law formula, is the force that produces the flux in the magnetic circuit. The practical unit of magnetomotive force is the ampere-turn. Another unit of magnetomotive force sometimes used is the gilbert, designated by the capital letter, **F**. The gilbert is the magnetomotive force required to establish 1 maxwell in a magnetic circuit having 1 unit of reluctance (1 rel). The magnetomotive force in gilberts is expressed in terms of ampere-turns as

$$F = 1.257\ IN$$

where **F** is in gilberts, I is in amperes, and N is the number of complete turns of wire encircling the circuit.

The unit of intensity of magnetizing force per unit of length is designated as H, and is sometimes expressed as gilberts per centimeter of length. Expressed mathematically,

$$H = \frac{1.257\ IN}{l}$$

where l is the length in centimeters.

The reluctance, $\mathcal{R}$, similar to resistance in the Ohm's law formula, is the opposition offered by the magnetic circuit to the passage of magnetic flux. The unit of reluctance, symbol $\mathcal{R}$ pronounced REL, is the reluctance of 1 centimeter-cube of air. The reluctance of a magnetic substance varies directly as the length of the flux path and inversely as the cross-sectional area and the permeability, μ, of the substance. Expressed mathematically,

$$\mathcal{R} = \frac{l}{\mu A}$$

where l is the length in centimeters, and A is the cross-sectional area in square centimeters.

Permeability, designated by the Greek letter mu, μ, is treated under a separate heading. However, it is defined here to permit a fuller interpretation of Rowland's law and also a practical application of this law. Permeability is a measure of the relative ability of a substance to conduct magnetic lines of force as compared with air. The permeability of air is taken as 1. Permeability is indicated as the ratio of the flux density in lines per square centimeter (gauss, B) to the intensity of the magnetizing force in gilberts per centimeter of length, indicated by H. Expressed mathematically,

NOTES:

$$\mu = \frac{B}{H}$$

Another term used in magnetic circuits is permeance. Permeance, indicated by the symbol P, is the reciprocal of reluctance—that is,

$$P = \frac{1}{\mathcal{R}}$$

Values of B, H, and μ for common magnetic substances are given in table 8-1.

Flux density B is expressed as

$$B = \frac{\Phi}{A} = \frac{20{,}000}{4} = 5{,}000 \text{ lines/cm.}^2$$

and from table 8-1 the corresponding value of H for cast steel is 3.9. The formula for H has previously been given as

$$H = \frac{1.257\ IN}{l}$$

Table 8-1.—B, H, and μ for common magnetic material.

B	Sheet steel		Cast steel		Wrought iron		Cast iron	
	H	μ	H	μ	H	μ	H	μ
3,000	1.3	2,310	2.8	1,070	2.0	1,500	5.0	600
4,000	1.6	2,500	3.4	1,177	2.5	1,600	8.5	471
5,000	1.9	2,630	3.9	1,281	3.0	1,666	14.5	347
6,000	2.3	2,605	4.5	1,332	3.5	1,716	24.0	250
7,000	2.6	2,700	5.1	1,371	4.0	1,750	38.5	182
8,000	3.0	2,666	5.8	1,380	4.5	1,778	60.0	133
9,000	3.5	2,570	6.5	1,382	5.0	1,800	89.0	101
10,000	3.9	2,560	7.5	1,332	5.6	1,782	124.0	80.6
11,000	4.4	2,500	9.0	1,222	6.5	1,692	166.0	66.4
12,000	5.0	2,400	11.5	1,042	7.9	1,520	222.0	54.1
13,000	6.0	2,166	16.0	813	10.0	1,300	290.0	44.8
14,000	9.0	1,558	21.5	651	15.0	934	369.0	38.0
15,000	15.5	970	32.0	469	25.0	600	....	...
16,000	27.0	594	49.0	327	49.0	327	....	...
17,000	52.5	324	74.0	230	93.0	183	....	...
18,000	92.0	196	115.0	156	152.0	118	....	...
19,000	149.0	127	175.0	108	229.0	83	....	...
20,000	232.0	86	285.0	70	....	....	....	...

*B = flux density in lines per square centimeter; H = gilberts per centimeter of length; μ = permeability; $\mu = \frac{B}{H}$.

Permeance is like conductance in electric circuits, and is defined as the property of a magnetic circuit that permits lines of magnetic flux to pass through the circuit.

A comparison of the units, symbols, and equations used in applying Ohm's law to electric circuits and Rowland's law to magnetic circuits is given in table 8-2.

As a practical application of Rowland's law, let it be required to find the ampere-turns (IN) necessary to produce 20,000 lines of flux in a cast steel ring having a cross-sectional area of 4 square centimeters and an average length of 20 centimeters, as shown in figure 8-6.

from which

$$IN = \frac{Hl}{1.257}$$

Substituting 3.9 for H and 20 for l in the preceding equation,

$$IN = \frac{3.9 \times 20}{1.257} = 62 \text{ ampere-turns}$$

NOTES:

Table 8-2.—Comparison of electric and magnetic circuits.

	Electric circuit	Magnetic circuit
Force.	Volt, E, or e.m.f.	Gilberts, F, or m.m.f.
Flow	Ampere, I	Flux, Φ, in maxwells
Opposition.	Ohms, R	Reluctance, $\mathfrak{R}$, or rels
Law.	Ohm's law, $I = \frac{E}{R}$	Rowland's law, $\Phi = \frac{F}{\mathfrak{R}}$
Intensity of force .	Volts per cm. of length	$H = \frac{1.257IN}{\ell}$, gilberts per centimeter of length.
Density.	Current density—for example, amperes per cm.^{2}.	Flux density, B.—for example, lines per cm.^{2}, or gausses.

PROPERTIES OF MAGNETIC MATERIALS

PERMEABILITY

When an annealed sheet steel core is used in an electromagnet it produces a stronger magnet than if a cast iron core is used. This is true because annealed sheet steel is more readily acted upon by the magnetizing force of the coil than is hard cast iron. Therefore, soft sheet steel is said to have greater permeability because the magnetic lines are established more easily in it than in cast iron. The ratio of the flux produced by a coil when the core is iron (or some other substance) to the flux produced when the core is air is called the permeability of the iron (or whatever substance is used), the current in the coil being the same in each case.

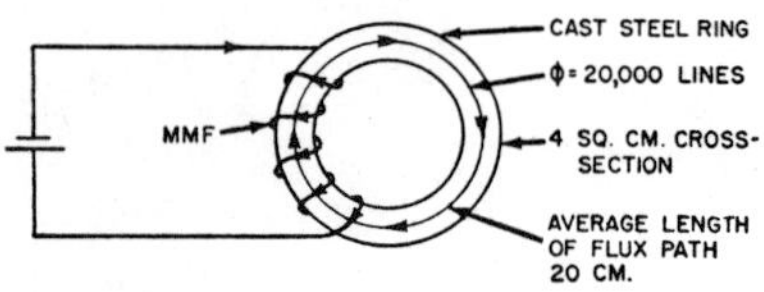

Figure 8-6.—Determining ampere-turns in a magnetic circuit.

NOTES:

The permeability of a substance is thus a measure of the relative ability to conduct magnetic lines of force, or its magnetic conductivity. The permeability of air is 1. The permeability of nonmagnetic materials, such as wood, aluminum, copper, and brass is essentially unity, or the same as for air.

Magnetization curves for the four magnetic materials listed in table 8-1 are given in figure 8-7.

The permeability of magnetic materials varies with the degree of magnetization, being smaller for high values of flux density, as is indicated in table 8-1 and figure 8-8.

HYSTERESIS

The simplest method of illustrating the property of hysteresis is by graphical means such as the hysteresis loop shown in figure 8-9.

In this figure the magnetizing force is indicated in gilberts per centimeter of length along the plus and minus H axis, and the flux density is indicated in gausses along the plus and minus B axis. The intensity of the magnetizing force, H, applied by means of a current-carrying coil of wire around the sample of magnetic material, is varied uniformly through one cycle of operation, starting at zero. The force, H, is increased in the positive direction (current flowing in a given direction through the coil) to 11 gilberts per centimeter. During this time the flux density, B, increases from zero to 14,000 at point A. If H is decreased to zero, the descending curve of flux density does not return to zero via its rise path. Instead, it returns to point

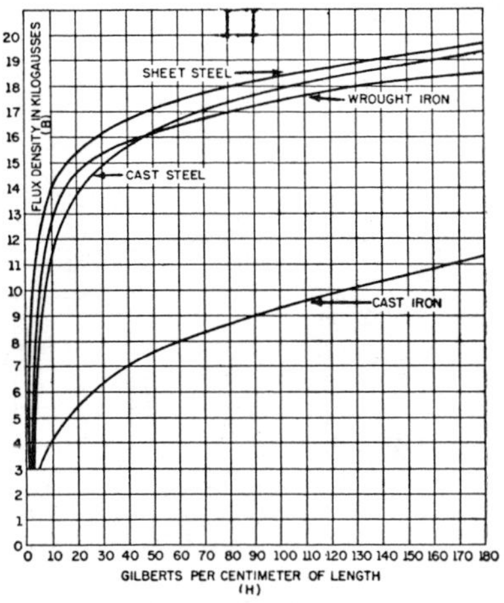

Figure 8-7.—Magnetization curves for four magnetic materials.

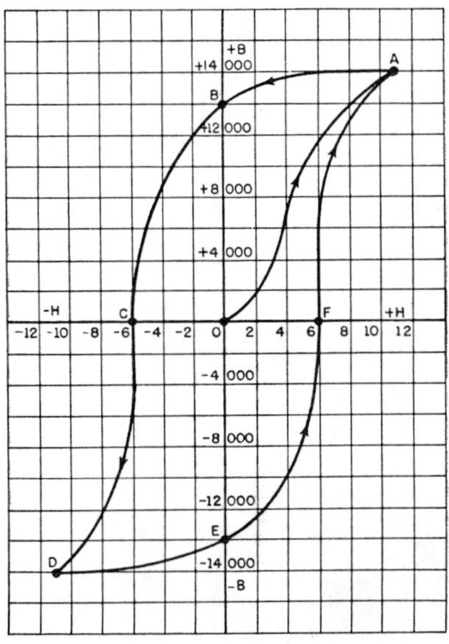

Figure 8-9.—Hysteresis loop.

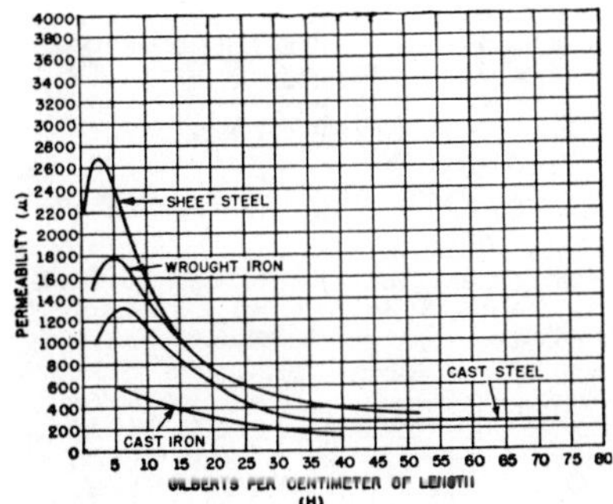

Figure 8-8.—Permeability curves.

B, where the flux density is 13,000. The magnetic flux indicated by the length of line OB represents the retentivity of the magnetic substance.

Retentivity is the ability of a magnetic substance to retain its magnetism after the magnetizing force has been removed. Retentivity is most apparent in hard steel and is least apparent in soft iron.

The value of the residual, or remaining, magnetism when H has been reduced to zero, depends on the substance used and the degree of flux density attained. In this example the residual magnetism is 13,000 gausses.

If current is now sent through the coil in the opposite direction, so that the intensity of the magnetizing force becomes -H, the force will have to be increased to point C before the residual magnetism is reduced to zero. The magnetizing force, OC, necessary to reduce the

NOTES:

residual magnetism to zero is called the coercive force. In this example the coercive force is 6 gilberts per centimeter.

If the magnetizing force is continued to -11 gilberts per centimeter, the curve descends from C to D, magnetizing the sample of magnetic material with the opposite polarity. If the magnetizing force is reduced again to zero, the flux density is reduced to point E. The magnetic flux indicated by the length of line OE represents the retentivity of the magnetic substance, as did line OB. The residual magnetism is again 13,000 gausses.

If the current through the coil is again reversed (sent through in the original direction), the magnetization curve moves to zero when the magnetising force is increased to point F.

Thus, when the magnetizing force goes through a complete cycle, the resulting magnetization likewise goes through a complete cycle.

From the foregoing analysis it is apparent that hysteresis is the property of a magnetic substance that causes the magnetization to lag behind the force that produces it. The lag of magnetization behind the force that produces it is caused by molecular friction. Energy is needed to move the molecules (or domains) through a cycle of magnetization. If the magnetization is reversed slowly, the energy loss may be negligible. However, if the magnetization is reversed rapidly, as when commercial alternating current is used, considerable energy may be disipated. If the molecular friction is great, as when hard steel is used, the losses may be very great. Another factor that determines hysteresis loss is the maximum density of the flux established in the magnetic material.

A comparison of the hysteresis loops for annealed steel and hard steel is shown in figure 8-10. The area within each loop is a measure of the hysteresis energy loss per cycle of operation. Thus, as shown in the figure, more energy is dissipated in molecular friction in hard steel than in annealed steel. It is therefore important that substances having low hysteresis loss be used for transformer cores and similar a-c applications.

ELECTROMAGNETS

An electromagnet is composed of a coil of wire wound around a core of soft iron. When direct current flows through the coil the core will become magnetized with the same polarity that the coil (solenoid) would have without the core. If the current is reversed, the polarity of both the coil and the soft-iron core is reversed.

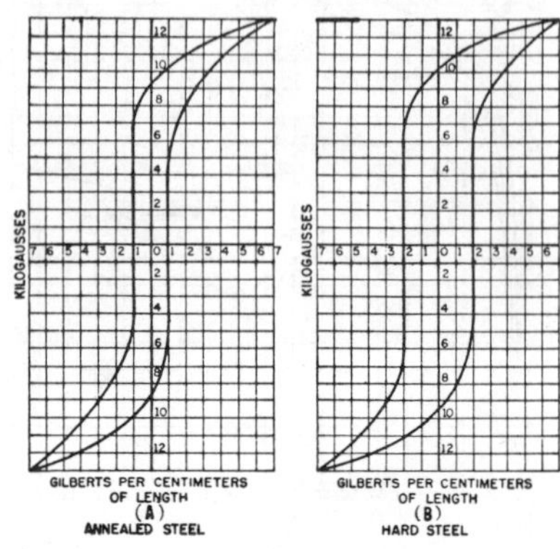

Figure 8-10.—Comparison of hysteresis loops.

The polarity of the electromagnet is determined by the left-hand rule in the same manner that the polarity of the solenoid in figure 8-5 was determined. If the coil is grasped in the left hand in such a way that the fingers curve around the coil in the direction of electron flow (- to +), the thumb will point in the direction of the north pole.

The addition of the soft-iron core does two things for the current-carrying coil, or solenoid. First, the magnetic flux is increased because the soft-iron core is more permeable than the air core; second, the flux is more highly concentrated. The permeability of soft iron is many times that of air, and therefore the flux density is increased considerably when a soft-iron core is inserted in the coil.

The magnetic field around the turns of wire making up the coil influences the molecules in the iron bar causing, in effect, the individual molecular magnets, or domains, to line up in the direction of the field established by the coil. Essentially the same effect is produced in a soft-iron bar when it is under the influence of a permanent magnet.

NOTES:

The magnetomotive force resulting from the current flow around the coil does not increase the magnetism that is inherent in the iron core, it merely reorientates the "atomic" magnets that were present before the magnetizing force was applied. If substantial numbers of the tiny magnets are orientated in the same direction, the core is said to be magnetized.

When soft iron is used, most of the atomic magnets return to what amounts to a miscellaneous orientation upon removal of the magnetizing current, and the iron is said to be demagnetized, if hard steel is used, more of them will remain in alinement with the direction of the flux produced by the flow of current through the coil, and the metal is said to be a permanent magnet. Soft iron and other magnetic materials having high permeability and low retentivity are generally used in electromagnets.

It is known from experience that a piece of soft iron is attracted to either pole of a permanent magnet. A soft-iron bar is likewise attracted by a current-carrying coil, if the coil and bar are orientated as in figure 8-11. As shown in the figure, the lines of force extend through the soft iron and magnetize it. Because unlike poles attract, the iron bar is pulled toward the coil. If the bar is free to move, it will be drawn into the coil to a position near the center where the field is the strongest.

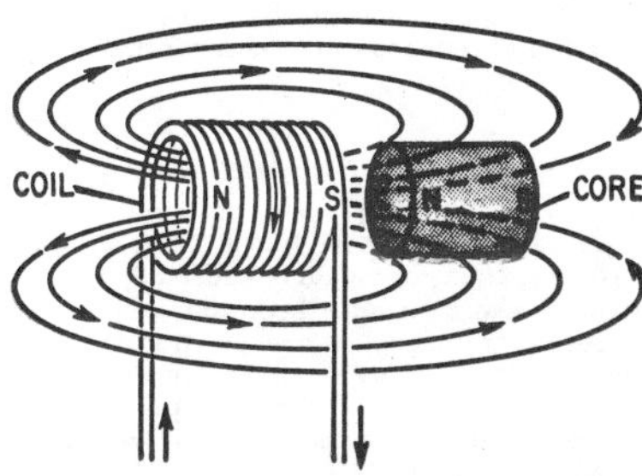

Figure 8-11.—Solenoid with iron core.

The solenoid-and-plunger type of magnet in various forms is employed extensively aboard ships and aircraft. These are used to operate the feeding mechanism of carbon-arc searchlights; to open circuit-breakers automatically when the load current becomes excessive; to close switches for motorboat starting; to fire guns; and to operate flood valves, magnetic brakes, and many other devices.

The armature-type of electromagnet also has extensive applications. In this type of magnet the coil is wound on and insulated from the iron core. The core is not movable. When current flows through the coil the iron core becomes magnetized and causes a pivoted soft-iron armature located near the electromagnet, to be attracted toward it. This type of magnet is used in door bells, relays, circuit breakers, telephone receivers, and so forth.

APPLICATIONS

Electric Bell

The electric bell is one of the most common devices employing the electromagnet. A simple electric bell is shown in figure 8-12. Its operation is explained as follows:

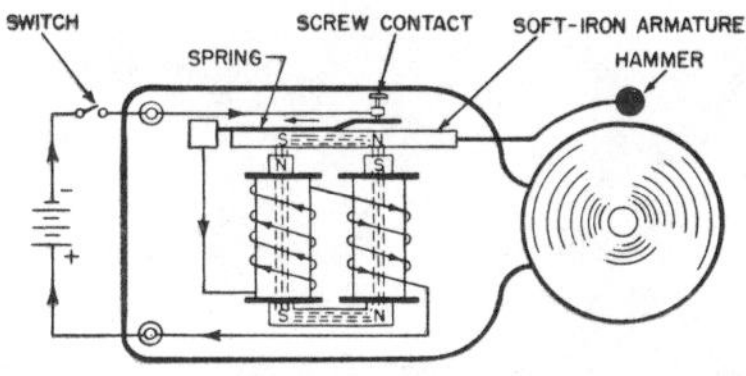

Figure 8-12.—Electric bell.

1. When the switch is closed, current flows from the negative terminal of the battery, through the contact points, the spring, the two coils, and back to the positive terminal of the battery.
2. The cores are magnetized, and the soft-iron armature (magnetized by induction) is pulled down, thus causing the hammer to strike the bell.
3. At the instant the armature is pulled down, the contact is broken, and the electromagnet loses its magnetism. The spring pulls the armature up so that contact is reestablished, and the operation is repeated. The speed with which the hammer is moved up and down depends on the

NOTES:

stiffness of the spring and the mass of the moving element.

The magnetomotive forces of the two coils are in series aiding and therefore the magnetization of the core is increased over that produced by one coil alone.

Circuit Breaker

A circuit breaker, like a fuse, protects a circuit against overloading caused by excessive loads or short circuits. In this device the winding of an electromagnet is connected in series with the load circuit to be protected and with the switch contact points. The principle of operation is shown in figure 8-13. Excessive current through the magnet winding causes the switch to be tripped, and the circuit to both breaker and load is opened by a spring. When the circuit fault has been cleared, the circuit is closed again by manually resetting the circuit breaker. Circuit breakers and their applications are discussed in greater detail in chapter 14 of this manual.

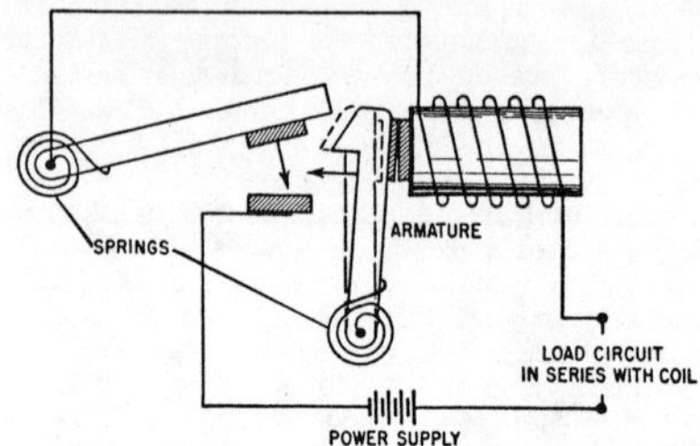

Figure 8-13.—Magnetic circuit breaker.

Many more applications of electromagnets are discussed throughout this manual. Their applications to generators, motors, voltage regulators, reverse current relays, and servomechanisms will be covered.

NOTES

CHAPTER 9

INTRODUCTION TO ALTERNATING-CURRENT ELECTRICITY

The first faltering steps of scientific achievement in the field of electricity were performed with crude, and for the most part, homemade apparatus. Great men, such as George Simon Ohm, had to fabricate nearly all the laboratory equipment used in their experiments. The only convenient source of electrical energy available to these early scientists was the voltaic cell, invented some years earlier. Due to the fact that cells and batteries were the only sources of power available, some of the early electrical devices were designed to operate from DIRECT CURRENT.

When the use of electricity became widespread, certain disadvantages in the use of direct current became apparent. In a direct-current system the supply voltage must be generated at the level required by the load. To operate a 240-volt lamp for example, the generator must deliver 240 volts. A 120-volt lamp could not be operated from this generator by any convenient means. A resistor could be placed in series with the 120-volt lamp to drop the extra 120 volts, but the resistor would waste an amount of power equal to that consumed by the lamp.

Another disadvantage of direct-current systems is the large amount of power lost due to the resistance of the transmission wires used to carry current from the generating station to the consumer. This loss could be greatly reduced by operating the transmission line at very high voltage and low current. This is not a practical solution in a d-c system, however, since the load would also have to operate at high voltage. As a result of the difficulties encountered with direct current, practically all modern power distribution systems use a type of current known as ALTERNATING CURRENT (a.c.). In an alternating-current system, the current flows first in one direction then reverses and flows in the opposite direction.

Unlike d-c voltage, a-c voltage can be stepped up or down by a device called a TRANSFORMER. This permits the transmission lines to be operated at high voltage and low current for maximum efficiency. Then at the consumer end the voltage is stepped down to whatever value the load requires by using a transformer. Due to its inherent advantages and versatility, alternating current has replaced direct current in all but a few commercial power distribution systems.

Many other types of current and voltage exist in addition to direct current and voltage. If a graph is constructed showing the magnitude of d-c voltage across the terminals of a battery with respect to time it would appear as in figure 9-1 (A). The d-c voltage is shown to have a constant amplitude. Some voltages go through periodic changes in amplitude like those shown in figure 9-1 (B). The pattern which results when these changes in amplitude with respect to time are plotted on graph paper is known as a WAVEFORM. Figure 9-1 (B) shows some of the common electrical waveforms. Of those illustrated, the sine wave will be dealt with most often.

BASIC A-C GENERATOR

An alternating-current generator converts mechanical energy into electrical energy. It does this by utilizing the principle of electromagnetic induction. In the study of magnetism, it was shown that a current-carrying conductor produces a magnetic field around itself. It is also true that a changing magnetic field may produce an emf in a conductor. If a conductor lies in a magnetic field, and either the field or conductor moves, an emf is induced in the conductor. This effect is called electromagnetic induction.

CYCLE

Figure 9-2 shows a suspended loop of wire (conductor) being rotated (moved) in a counter-

NOTES:

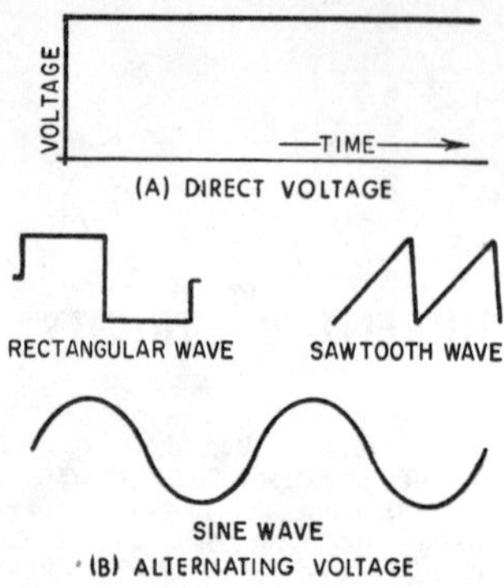

Figure 9-1.—Voltage waveforms.

clockwise direction through the magnetic field between the poles of a permanent magnet. For ease of explanation, the loop has been divided into a dark and a light half. Notice that in part (A), the dark half is moving along (parallel to) the lines of force. Consequently, it is cutting none of these lines. The same is true of the light half, moving in the opposite direction. Since the conductors are cutting no lines of force, no emf is induced. As the loop rotates toward the position shown in part (B), it cuts more and more lines of force per second because it is cutting more directly across the field (lines of force) as it approaches the position shown in (B). At position (B) the induced voltage is greatest because the conductor is cutting directly across the field.

As the loop continues to be rotated toward the position shown in part (C), it cuts fewer and fewer lines of force per second. The induced voltage decreases from its peak value. Eventually, the loop is once again moving in a plane parallel to the magnetic field, and no voltage (zero voltage) is induced. The loop has now been rotated through half a circle (one alternation, or 180°). The sine curve shown in the lower part of the figure shows the induced voltage at every instant of rotation of the loop. Notice that this curve contains 360°, or two alternations. Two alternations represent one complete cicle of rotation.

The direction of current flow during the rotation from (B) to (C), when a closed path is provided across the ends of the conductor loop, can be determined by using the LEFT-HAND RULE FOR GENERATORS. The left-hand rule is applied as follows: Extend the left hand so that the THUMB points in the direction of conductor movement, and the FOREFINGER points in the direction of magnetic flux (north to south). By pointing the MIDDLE FINGER 90 degrees from the forefinger, it will point in the direction of current flow within the conductor.

Applying the left-hand rule to the dark half of the loop in part (B), the direction of current flow can be determined and is depicted by the heavy arrow. Similarly, direction of current flow through the light half of the loop can be determined. The two induced voltages add together to form one total emf. When the loop is further rotated to the position shown in part (D), the action is reversed. The dark half is moving up instead of down, and the light half is moving down instead of up. By applying the left-hand rule once again, it is readily apparent that the direction of the induced emf and its resulting current have reversed. The voltage builds up to maximum in this new direction, as shown by the sine-wave tracing. The loop finally returns to it original position (part E), at which point voltage is again zero. The wave of induced voltage has gone through one complete cycle.

(As mentioned previously, the cycle is two complete alternations in a period of time. Recently, the hertz (Hz) has been designated to be used in lieu of cycles per second. While it may seem confusing to the reader that in one place a cycle is used to designate two alternations per period of time and in another instance a hertz is used to designate two alternations per second, the key to determine which is used is the time factor. One hertz is one cycle per second. Therefore, throughout this manual a cycle is used when no specific time element is involved and a hertz (Hz) is used when the time element is measured in seconds.)

If the loop is rotated at a steady rate, and if the strength of the magnetic field is uniform, the number of hertz and the voltage will remain at fixed values. Continuous rotation will produce a series of sine-wave voltage cycles, or, in other words, an a-c voltage. In this way mechanical energy is converted into electrical energy.

The rotating loop in figure 9-2 is called an armature. The armature may have any number of loops or coils.

NOTES:

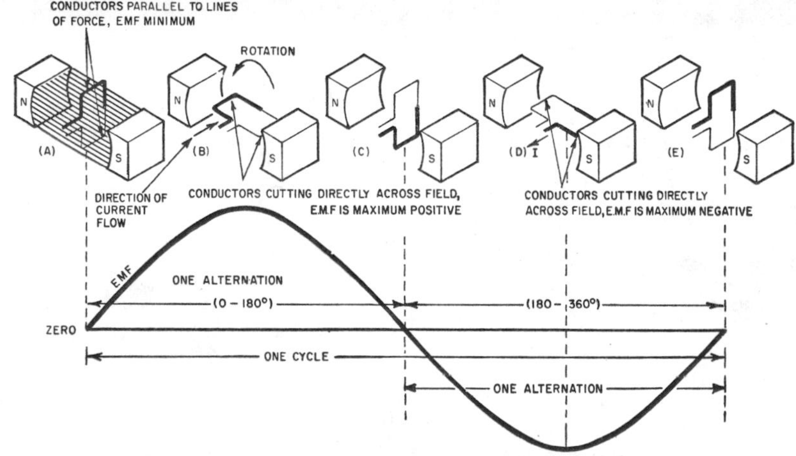

Figure 9-2.—Basic alternating-current generator.

FREQUENCY

The frequency (f) of an alternating current or voltage is the number of complete cycles occurring in each second of time. Hence, the speed of rotation of the loop determines the frequency. For a single loop rotating in a two-pole field you can see that each time the loop makes one complete revolution the current reverses direction twice. A single hertz will result if the loop makes one revolution each second. If it makes two revolutions per second, the output frequency will be 2 hertz. In other words, the frequency of a two-pole generator happens to be the same as the number of revolutions per second. As the speed is increased, the frequency is increased.

If an alternating-current generator has four pole pieces, as in figure 9-3, every complete mechanical revolution of the armature will produce 2 a-c cycles. When the dark half of the loop passes between poles S1 and N2, a voltage is induced which causes current to flow into the slipring attached to the dark end of the loop.

When the dark half passes between N2 and S2, the induced voltage reverses direction. Another reversal occurs when it passes between S2 and N1. The voltage at the sliprings reverses direction FOUR TIMES during each revolution. In other words, 2 cycles of a a-c voltage are generated for each mechanical revolution. If each revolution lasts 1 second, the frequency of the output is 2 hertz. The more poles that are added, the higher the frequency per revolution becomes. To find the output frequency of any a-c generator, the following formula can be used:

$$f = \frac{P \times rpm}{120}$$

where f is frequency in hertz, rpm is revolutions per minute, and P is the number of poles.

A generator made to deliver 60 hertz, having two field poles, would need an armature designed to rotate at 3,600 rpm. If it had four field poles, it would need an armature designed to rotate at 1,800 rpm. In either case, frequency would be the same. In actual practice, a generator designed for low-speed operation generally has a greater number of pole pieces, while a high-speed machine will have relatively fewer pole pieces, if both are to deliver power at the same frequency.

NOTES:

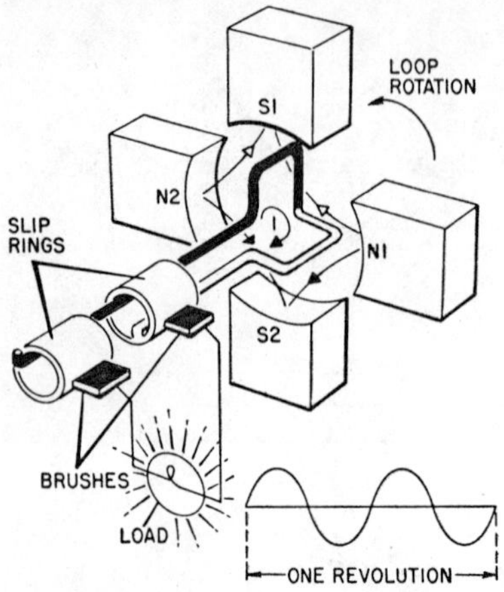

Figure 9-3.—Four-pole basic a-c generator.

PERIOD

An individual cycle of any sine wave represents a finite amount of TIME. Figure 9-4 shows 2 cycles of a sine wave which has a frequency of 2 hertz (Hz). Since 2 cycles occur each second, 1 cycle must require one-half second of time. The time required to complete 1 cycle of a waveform is called the PERIOD of the wave. In this example the period is one-half second.

Each cycle of the waveform in figure 9-4 is seen to consist of two pulse shaped variations in voltage. The pulse which occurs during the time the voltage is postive is called the POSITIVE ALTERNATION. The pulse which occurs during the time the voltage is negative is called the NEGATIVE ALTERNATION. For a sine wave these two alternations will be identical in size and shape, and opposite in polarity.

The period of a wave is inversely proportional to its frequency. Thus, the higher the frequency (greater number of Hz), the shorter the period. In terms of an equation:

$$t = \frac{1}{f}$$

where t = period in seconds
f = frequency in hertz

NOTES:

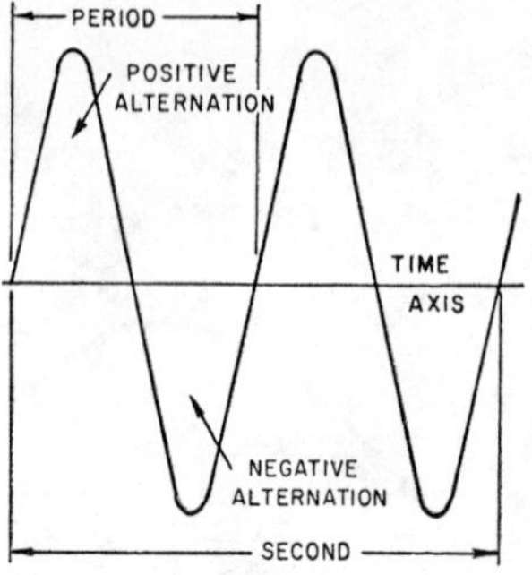

Figure 9-4.—Period of a sine wave.

GENERATED VOLTAGE

Alternating-current generators are usually constant-potential machines because they are driven at constant speed and have fixed magnetic field strength for a given load. The effective voltage, E, generated by a single-winding (phase) a-c generator is related to the total field strength per pole, Φ ; the frequency f; and the total number of active conductors, N, in the armature winding; as indicated in the following equation:

$$E = 2.22\Phi fN10^{-8}$$

For example, if $\Phi = 2.5 \times 10^6$, f = 60 hertz, and N = 96 conductors, the voltage generated is

$$E = 2.22 \times 2.5 \times 10^6 \times 60 \times 96 \times 10^{-8}$$

$$E = 320 \text{ volts}$$

The factors of poles (P) and speed (S in rpm), which appear in the equation for generated voltage in multipolar series-wound d-c generators, do not appear in the formula for the voltage generated in each phase of the a-c generator. This is because they are replaced by the equivalent factor of frequency ($f = \frac{PS}{120}$).

The length of active conductor extending under a pole does not appear in the equation directly because it is included in the factor of total magnetic flux per pole. The longer the active conductor, the more flux there will be for each pole, since the pole length and conductor length are assumed to be the same. For example, if an active conductor length is doubled, the pole length is doubled, the flux per pole is doubled, and the generated voltage is doubled.

ANALYSIS OF A SINE WAVE

VECTORS DEFINED

An alternating current or voltage is one in which the direction changes periodically. The electron movement is first in one direction, then in the other. The variation is of sine waveform. Straight lines drawn to scale, called VECTORS, are used in solving problems involving sine-wave currents and voltages.

A simple vector is a straight line used to denote the magnitude and direction of a given quantity. Magnitude is denoted by the length of the line, drawn to scale, and direction is indicated by an arrow at one end of the line, together with the angle that the vector makes with a horizontal reference vector.

For example, if a certain point B (fig. 9-5) lies 1 mile east of point A, the direction and distance from A to B can be shown as vector e by using a scale of approximately 1/2 inch = 1 mile.

Vectors may be rotated like the spokes of a wheel to generate angles. Positive rotation is counterclockwise and generates positive angles. Negative rotation is clockwise and generates negative angles.

The vertical projection (dotted line in fig. 9-6) of a ROTATING VECTOR may be used to represent the voltage at any instant. Vector E_m represents the maximum voltage induced in a conductor rotating at uniform speed in a 2-pole field (points 3 and 9). The vector is rotated counterclockwise through one complete revolution (360°). The point of the vector describes a circle. A line drawn from the point of the vector perpendicular to the horizontal diameter of the circle is the vertical projection of the vector.

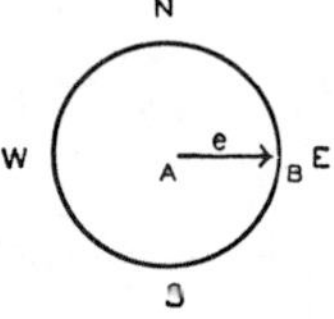

Figure 9-5.—Designating direction and distance by vectors.

NOTES:

The circle also describes the path of the conductor rotating in the bi-polar field. The vertical projection of the vector represents the voltage generated in the conductor at any instant corresponding to the position of the rotation vector as indicated by angle θ. Angle θ represents selected instants at which the generated voltage is plotted. The sine curve plotted at the right of the figure represents successive values of the a-c voltage induced in the conductor as it moves at uniform speed through the 2-pole field because the instantaneous values of rotationally induced voltage are proportional to the sine of the angle θ that the rotating vector makes with the horizontal.

EQUATION OF SINE WAVE OF VOLTAGE

The sine curve is a graph of the equation

$$e = E_m \sin \theta,$$

where e is the instantaneous voltage, E_m is the maximum voltage, θ is the angle of the generator armature, and "sine" is one of the trigonometric functions shown in figure 9-6.

The instantaneous voltage, e, depends on the sine of the angle. It rises to a maximum positive value as the angle reaches 90°. This occurs because the conductor cuts directly across the flux at 90°. It falls to zero at 180°, because the conductor cuts no lines of flux at 180°. It reaches a negative peak at 270°, and becomes zero again at 360°. For example, when $\theta=60°$ and $E_m = 100$ volts, $e = 100 \times \sin 60°$ 86.6 volts. When $\theta = 240°$, $e = 100 \sin 240° = -86.6$ volts.

Another form of the trigonometric equation for a sine wave of voltage involves the angular velocity of its rotating vector. The term "angular velocity" refers to the number of degrees (angles) through which a voltage vector rotates per second. Angular velocity is symbolized by the Greek letter omega (ω). In practice, ω is generally given in terms of RADIANS per second, rather than degrees per second. A radian is a segment of the circumference of a circle. This segment is always exactly equal in length to the RADIUS of that circle. There are π (3.14)

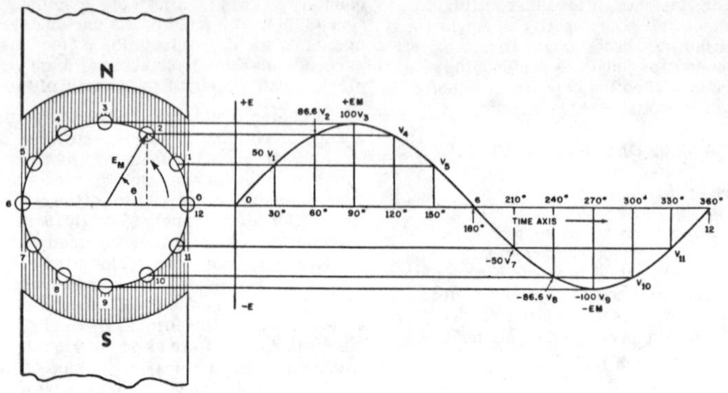

Figure 9-6.—Generation of sine-wave voltage.

radians in half a circle, and 2π (6.28) radians in the circumference of a complete circle. Therefore, when a voltage vector makes one complete revolution, describing one complete circle, it traverses 2π or 6.28 radians. In terms of degrees of rotation, one radian is 360°/6.28 or 57.32°.

The number of radians per second traversed by an alternating-voltage vector is closely related to its frequency, because either may be used to express the other. Since one vector revolution equals one complete cycle, then each cycle equals 6.28 radians of vector travel; that is, an alternating voltage whose frequency is 60 hertz would be said to have an angular velocity of 6.28 x 60, or roughly 377 radians per second. Written as a formula involving angular velocity, the term $2\pi f$ may be replaced by the simpler symbol ω, if convenient. As previously stated, another form of the trigonometric equation for a sine wave of voltage involves the angular velocity of the generating vector.

The equation is

$$e = E_m \times \sin \omega t.$$

It is used to determine the voltage of a rotating vector at some given instant of time. The starting reference, or "time zero," is usually when the voltage vector is at zero. The equation time factor t is the elapsed time from time zero, and is the exact instant at which the voltage is to be determined. To determine the exact angular position at any instant, multiply the angular velocity (ω) of the vector by the time elapsed (t). By dividing elapsed time t by the period for 1 cycle, and multiplying by 360°, the angular position of the voltage vector may be determined for any instant. Multiplying E_m by the sine of that instantaneous angle will yield the instantaneous voltage e.

For example, consider a 60-Hz voltage whose peak value is 100 volts. To determine the voltage 0.00139 second from the zero point the equation would be:

$$\sin \omega t = \sin 2\pi ft = \sin 360(60)(0.00139) = \sin 30°$$

$$= 0.5$$

The instantaneous voltage e is

$$e = E_m \times \sin 30°$$

$$e = 100 \times 0.5$$

$$e = 50 \text{ volts.}$$

NOTES:

There are four important values associated with sine waves of voltage or current:

Instantaneous-designated as e or i; maximum-designated as E_m or I_m; average-designated as E_{avg} or I_{avg}; and effective-designated as E or I.

The INSTANTANEOUS value may be any value between zero and maximum depending on the instant chosen, as indicated by the equation $e = E_m \sin \omega t$.

The MAXIMUM value of voltage is reached twice each cycle and is the greatest value of instantaneous voltage generated during each cycle.

The ratio of the instantaneous value of voltage to the maximum value is equal to the sine of the angle corresponding to that instant.

PEAK AMPLITUDE

One of the most frequently measured characteristics of a sine wave is its amplitude. Unlike d-c measurement, the amount of alternating current or voltage present in a circuit can be measured in various ways. In one method of measurement, the maximum amplitude of either the positive or the negative alternation is measured. The value of current or voltage obtained is called the PEAK VOLTAGE or the PEAK CURRENT. To measure the peak value of current or voltage, an oscilloscope or a special meter (peak reading meter) must be used. The peak value of a sine wave is illustrated in figure 9-7.

PEAK-TO-PEAK AMPLITUDE

A second method of indicating the amplitude of a sine wave consists of determining the total voltage or current between the positive and negative peaks. This value of current or voltage is called the PEAK-TO-PEAK VALUE (fig. 9-7). Since both alternations of a pure sine wave are identical, the peak-to-peak value is twice the peak value. Peak-to-peak voltage is usually measured with an oscilloscope, although some voltmeters have a special scale calibrated in peak-to-peak volts.

INSTANTANEOUS AMPLITUDE

The instantaneous value of a sine wave of voltage for any angle of rotation is expressed by the formula:

$$e = E_m \times \sin \theta$$

where e = the instantaneous voltage

E_m = the maximum or peak voltage

$\sin \theta$ = the sine of angle at which e is desired.

Similarly the equation for the instantaneous value of a sine wave of current would be:

$$i = I_m \times \sin \theta$$

where i = the instantaneous current

I_m = the maximum or peak current

$\sin\theta$ = the sin of the angle at which i desired

EFFECTIVE OR RMS VALUE

As the use of alternating current gained popularity, it became increasingly apparent that some common basis was needed on which a.c. and d.c. could be compared. A 100-watt light bulb, for example, should work just as well on 120 volts a.c. as it does on 120 volts d.c. It can be seen, however, that a sine wave of voltage having a peak value of 120 volts would not supply the

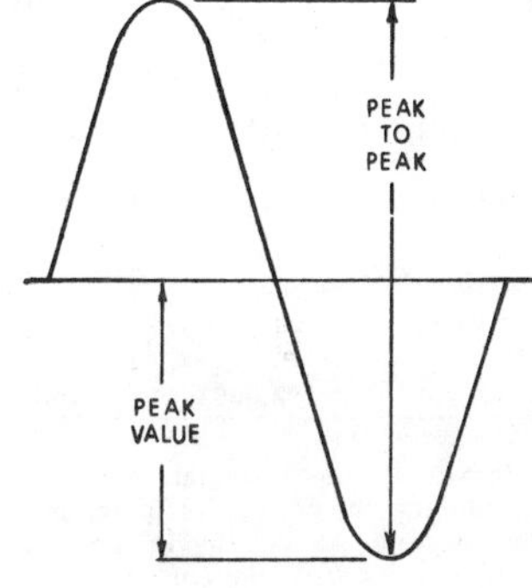

Figure 9-7.—Peak and peak-to-peak values.

NOTES:

lamp with as much power as a steady value of 120 volts d.c.

Since the power dissipated by the lamp is a result of current flow through the lamp, the problem resolves to one of finding a MEAN alternating current ampere which is equivalent to a steady ampere of direct current.

Figure 9-8 shows a circuit in which the peak alternating current through the 10 ohm resistor is 1.414 amperes. Since the current through the resistor is changing continuously the power dissipated by the resistor will also vary. It will be minimum when the current is zero.

The variations in power throughout the cycle can best be analyzed by plotting a curve showing the instantaneous power at each point in the cycle. In the procedure to follow, the instantaneous current, the square of the instantaneous current, and the instantaneous power will be calculated in 10° steps for the first quarter of the cycle. These values are shown in table 9-1.

Table 9-1.—Instantaneous values of current and Power.

Degrees	i	i^2	P
0°	.000	.000	.00
10°	.245	.060	.60
20°	.486	.236	2.36
30°	.707	.500	5.00
40°	.909	.826	8.26
50°	1.083	1.173	11.73
60°	1.225	1.500	15.00
70°	1.329	1.766	17.66
80°	1.393	1.940	19.40
90°	1.414	2.000	20.00

Notice that at 0° the instantaneous current (i) is zero causing the power dissipated by the resistor to be zero. At 10° the instantaneous current is 0.245 amperes, the current squared is 0.060 and the power is 0.60 watt. At 90° the current has reached its maximum value of 1.414 amperes, the square of the current is 2.000 and the power dissipated is 20.00 watts.

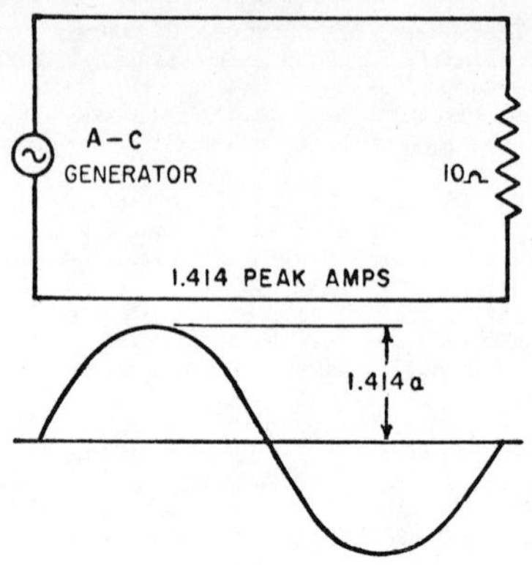

Figure 9-8.—Basic a-c circuit.

During the part of the sine wave of current from 90° to 180° the same values could be used as before but in a reverse order. Thus, at 100° the values of current and power would be identical to those at 80°.

Using the values of i and P from table 9-1, a graph can be constructed showing the way in which power varies throughout the cycle. This graph is shown in figure 9-9.

In this graph a sine wave of current is plotted first, using the instantaneous values from table 9-1. Next the curve representing i^2 and power is constructed.

Notice that the power curve has twice the frequency of the current curve, and that ALL POWER IS POSITIVE. This is due to the fact that heat is dissipated regardless of which way the current flows through the resistor.

Since all the alternations of the power curve are identical, the MEAN or AVERAGE POWER is the value HALF-WAY between the maximum and minimum values of power. Thus, the average power dissipated by the 10 ohm resistor is 10 watts, one-half the peak power. Since the curve representing power also represents current squared (i^2), the average or mean of the curve also lies half-way between the maximum and minimum values of i^2. As power is proportional to i^2, a d-c current having a value equal to the square root of the mean of the i^2 values would produce the same average power as the original sine wave of current. This mean current is called the ROOT MEAN SQUARE (RMS)

NOTES:

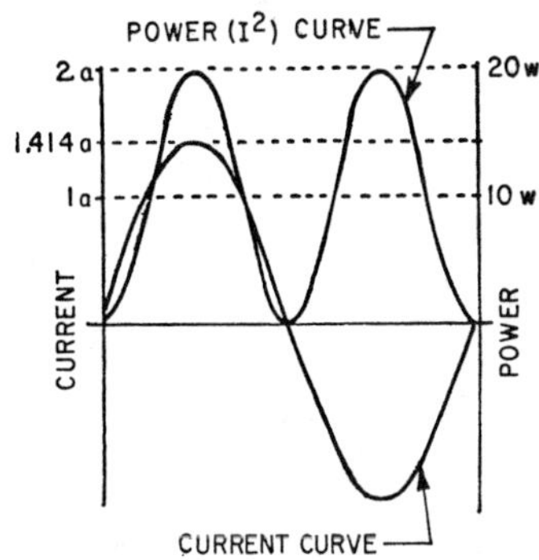

Figure 9-9.—Current and power curves.

current. One RMS ampere of alternating current is as effective in producing heat as one steady ampere of direct current. For this reason an RMS ampere is also callen an EFFECTIVE ampere. In figure 9-9 the peak current of 1.414 amperes produces the same amount of average power as 1 ampere of effective (RMS) current.

ANYTIME AN ALTERNATING VOLTAGE OR CURRENT IS STATED WITHOUT ANY QUALIFICATIONS, IT IS ASSUMED TO BE AN EFFECTIVE VALUE. Since effective values of a.c. are the ones generally used, most meters are calibrated to indicate effective values of voltage and current.

In many instances it is necessary to convert from effective to peak or vice-versa. Figure 9-9 shows that the peak value of a sine wave is 1.414 times the effective value and therefore:

$$E_m = E \times 1.414 \qquad (9\text{-}1)$$

where E_m = maximum or peak voltage

E = effective or RMS voltage

and

$$I_m = I \times 1.414$$

where

I_m = maximum or peak current

I = effective or RMS current.

NOTES:

Upon occasion it is necessary to convert a peak value of current or voltage to an effective value. The conversion factor may be derived as follows:

$$E_m = E \times 1.414 \qquad (9\text{-}1)$$

Multiplying both sides of the equation by 1/1.414

$$E_m \times \frac{1}{1.414} = E \times 1.414 \times \frac{1}{1.414}$$

$$E_m \times \frac{1}{1.414} = E$$

Dividing 1 by 1.414

$$E = E_m \times 0.707$$

where E = the effective voltage

E_m = the maximum or peak voltage

Similarly for current

$$I = I_m \times 0.707$$

where I = the effective current

I_m = the maximum or peak current.

AVERAGE VALUE

The average value of a complete cycle of a sine wave is zero, since the positive alternation is idential to the negative alternation. In certain types of circuits however, it is necessary to compute the average value of one alternation. This could be accomplished by adding together a series of instantaneous values of the wave between 0° and 180°, and then dividing the sum by the number of instantaneous values used. Such a computation would show one alternation of a sine wave to have an average value equal to 0.637 of the peak value. In terms of an equation:

equation:

$$E_{avg} = E_m \times 0.637$$

where

E_{avg} = the average voltage of one alternation

E_m = the maximum or peak voltage

similarly

$$I_{avg} = I_m \times 0.637$$

where

I_{avg} = the average current in one alternation

I_m = the maximum or peak current.

Figure 9-10 shows a comparison between the various values that are used to indicate the amplitude of a sine wave.

SINE WAVES IN PHASE

If a sine wave of voltage is applied to a resistance, the resulting current will also be a sine wave. This follows Ohm's law which states that the current is directly proportional to the applied voltage. Figure 9-11 shows a sine wave of voltage and the resulting sine wave of current superimposed on the same time axis. Notice that as the voltage increases in a positive direction the current increases along with it. When the voltage reverses direction, the current reverses direction. At all times the voltage and current pass through the same relative parts of their respective cycles at the same time. When two waves, such as those in figure 9-11, are precisely in step with one another they are said to be IN PHASE. To be in phase, the two waves must go through their maximum and minimum points at the same time and in the same direction.

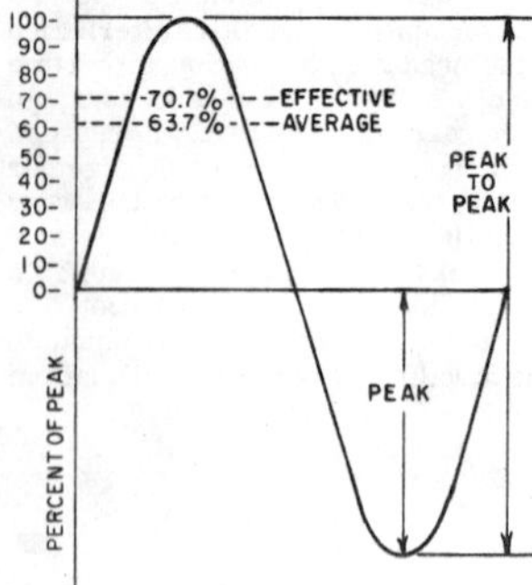

Figure 9-10.—Various values used to indicate sine wave amplitude.

NOTES:

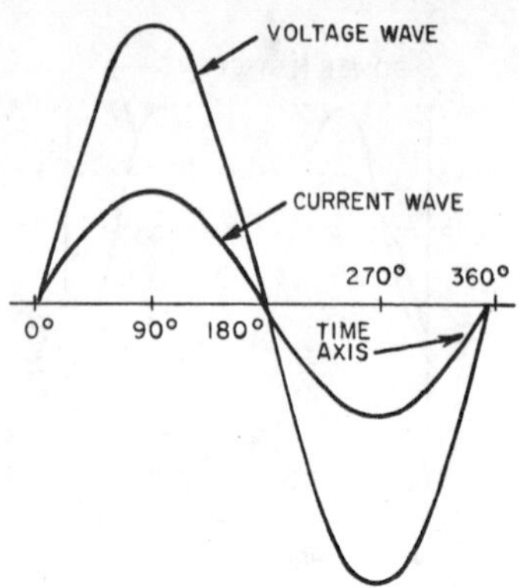

Figure 9-11.—Voltage and current waves in phase.

In some circuits, several sine waves can be in phase with each other. Thus, it is possible to have two or more voltage drops in phase with each other and also in phase with the circuit current.

SINE WAVES OUT OF PHASE

Figure 9-12 shows a voltage wave E_1 considered to start at 0° (time 1). As voltage wave E_1 reaches its positive peak, a second voltage wave E_2 starts its rise (time 2). Since these waves do not go through their maximum and minimum points at the same instant of time, a PHASE DIFFERENCE exists between the two waves. The two waves are said to be out of phase. For the two waves in figure 9-12 this phase difference is 90°.

To further describe the phase relationship between two waves the terms LEAD and LAG are used. The amount by which one wave leads or lags another is measured in degrees. Referring again to figure 9-12, wave E_2 is seen to start 90° later in time than wave E_1, thus wave E_2 lags wave E_1 by 90°. This relationship could also be

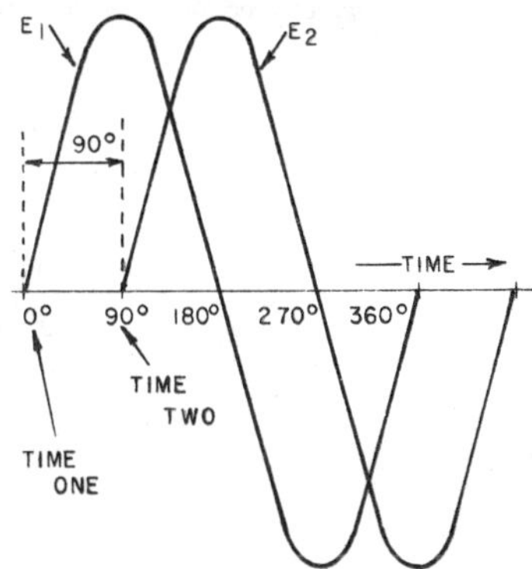

Figure 9-12.—Voltage waves 90° out of phase.

described by stating that wave E_1 leads wave E_2 by 90°.

It is possible for one wave to lead or lag another by any number of degrees, except 0 or 360°, in which condition the two waves are in phase. Thus, two waves may differ in phase by 45°, but two waves differing by 360° would be considered as in phase.

A phase relationship that is quite common is the one shown in figure 9-13. The two waves illustrated have a phase difference of 180°. Notice that although the waves pass through their maximum and minimum values at the same time, their instantaneous voltages are always of opposite polarity. If two such waves existed across the same component they would have a cancelling effect on each other. If the two waves are equal in amplitude the resultant wave would be zero. However, if they have different amplitudes the resultant wave would have the polarity of the larger and be the difference of the two.

To determine the phase difference between two sine waves, locate the points on the time axis where the two waves cross the time axis traveling in the same direction. The number of degrees between the crossing points is the phase difference. The wave that crosses the axis at the later time (to the right on the time axis) lags the other.

NOTES:

COMBINING A-C VOLTAGES

Vectors may be used to combine a-c voltages of sine waveform and of the same frequency. The angle between the vectors indicates the time difference between their positive maximum values. The length of the vectors represents either the effective value or the positive maximum values as desired.

The sine wave voltages generated in coils a and b of the simple generators (fig. 9-14 (A)) are 90° out of phase because the coils are located 90° apart on the two-pole armatures. The armatures are on a common shaft. When coil a is cutting squarely across the field, coil b is moving parallel to the field and not cutting through it. Thus, the voltage in coil a is maximum when the voltage in coil b is zero. If the frequency is 60 hertz, the time difference between the positive maximum values of these voltages is $\frac{90}{360} \times \frac{1}{60} = 0.00416$ second.

Vector E_a leads vector E_b by 90° in figure 9-14 (B) and sine wave a leads sine wave b by 90° in figure 9-14 (C). The curves are shown on separate axes to identify them with their respective generators; they are also projected on a common axis to show the relation between their instantaneous values. If coils a and b are connected in series and the maximum voltage generated in each coil is 10 volts, the total

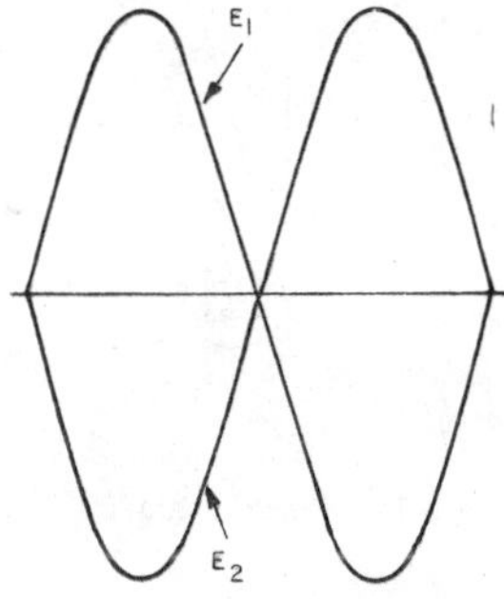

Figure 9-13.—Two waves 180° out of phase.

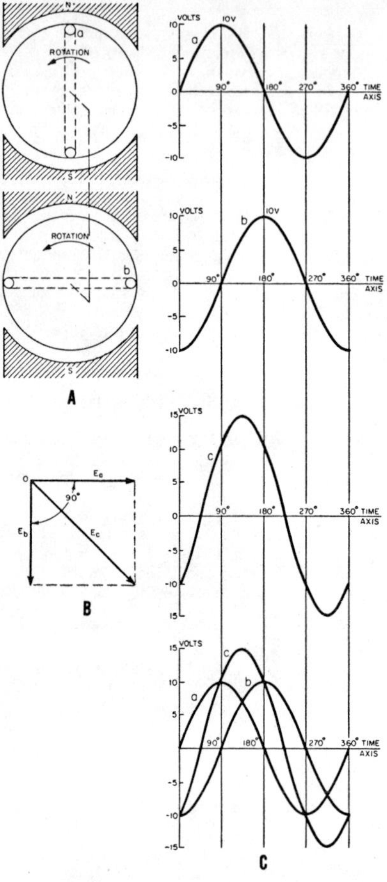

Figure 9-14.—Combining a-c voltages.

voltage is not 20 volts because the two maximum values of voltage do not occur at the same instant, but are separated one-fourth of a cycle.

The voltages cannot be added arithmetically because they are out of phase. These values, however, can be added vectorially.

E_c in figure 9-14 (B) is the vector sum of E_a and E_b, and is the diagonal of the parallelogram, the sides of which are E_a and E_b. The effective voltage in each coil is 0.707 x 10 or 7.07 volts (where 10 volts is the maximum) and represents the length of the sides of the parallelogram. The effective voltage of the series combination is 7.07 x$\sqrt{2}$ or 10 volts and represents the length of the diagonal of the parallelogram.

Curve c in figure 9-14 (C) represents the sine wave variations of the total voltage E_c developed in the series circuit connecting coils a and b. Voltage E_a leads E_b by 90°. Voltage E_c lags E_a by 45°, and leads E_b by 45°.

Counterclockwise rotation of the vectors is considered positive rotation thus giving the sense of lead or lag. Thus, if E_a and E_b (fig. 9-14 (B)) are rotated counterclockwise, and their movement observed from a fixed position, E_a will pass this position first, then 90° later E_b will pass the position. Thus E_b lags E_a by 90°. If the maximum voltage in each coil is 10 volts, the maximum value of the combined voltage will be 10 x$\sqrt{2}$ or 14.4 volts. (See fig. 9-14 (C)).

The important point to be made in this discussion is that two out-of-phase voltages may be resolved into a single resultant value by the use of vectors. Though generated voltages are used to illustrate the point in this case, the same rules apply to VOLTAGE DROPS as well. Any number of out-of-phase voltages may be combined vectorially, as long as they all have the same frequency; that is, as long as they remain a fixed number of degrees apart, like the two generator loops. Their magnitudes (LENGTH of their vectors) may be different.

NOTES:

CHAPTER 10

INDUCTANCE

The study of inductance presents a very challenging but rewarding segment of electricity. It is challenging in the sense that, at first, it will seem that new concepts are being introduced. The reader will realize as this chapter progresses that these "new concepts" are merely extensions and enlargements of fundamental principles that have been acquired previously in the study of magnetism and electron physics. The study of inductance is rewarding in the sense that a thorough understanding of it will enable the reader to acquire a working knowledge of electrical circuits more rapidly and with more surety of purpose than would otherwise be possible.

Inductance is the characteristic of an electrical circuit that makes itself evident by opposing the starting, stopping, or changing of current flow. The above statement is of such importance to the study of inductance that it bears repeating in a simplified form. Inductance is the characteristic of an electrical conductor which opposes a CHANGE in current flow.

One does not have to look far to find a physical analogy of inductance. Anyone who has ever had to push a heavy load (wheelbarrow, car, etc.) is aware that it takes more work to start the load moving than it does to keep it moving. This is because the load possess the property of inertia. Inertia is the characteristic of mass which opposes a CHANGE in velocity. Therefore, inertia can hinder us in some ways and help us in others. Inductance exhibits the same effect on current in an electric circuit as inertia does on velocity of a mechanical object. The effects of inductance are sometimes desirable—sometimes undesirable.

On September 22, 1791 in Newington Butts, London, a man was born who was destined to play a great part in the laying of the foundation of the growing science of electricity. This man, Michael Faraday, started to experiment with electricity around the year 1805 while working as an apprentice bookbinder. It was in 1831 that Faraday performed experiments on magnetically coupled coils. A voltage was induced in one of the coils due to a magnetic field created by current flow in the other coil. From this experiment came the induction coil, the theory of which eventually made possible many of our modern conveniences such as the automobile, doorbell, auto radio, etc. In performing this experiment Faraday also invented the first transformer, but since alternating current had not yet been discovered the transformer had few practical applications. Two months later, based on these experiments, Faraday constructed the first direct current generator. At the same time Faraday was doing his work in England, Joseph Henry was working independently along the same lines in New York. The discovery of the property of self-induction of a coil was actually made by Henry a little in advance of Faraday and it is in honor of Joseph Henry that the unit of inductance is called the HENRY.

It was from the experiments performed by these, and many other men that the laws and theories of inductance grew.

UNIT OF INDUCTANCE

The unit for measuring inductance, L, is the HENRY, h. An inductor has an inductance of 1 henry if an emf of 1 volt is induced in the inductor when the current through the inductor is changing at the rate of 1 ampere per second. The relation between the induced voltage, inductance, and rate of change of current with respect to time is stated mathematically as

$$E = L \frac{\Delta I}{\Delta t}$$

where E is the induced emf in volts, L is the inductance in henrys, and ΔI is the change

NOTES:

in current in amperes occurring in Δt seconds. (Delta, symbol Δ, means "a change in......")

The henry is a large unit of inductance and is used with relatively large inductors. The unit employed with small inductors is the millihenry, mh. For still smaller inductors the unit of inductance is the microhenry, μh.

SELF-INDUCTANCE

Even a perfectly straight length of conductor has some inductance. As previously explained, current in a conductor always produces a magnetic field surrounding, or linking with, the conductor. When the current changes, the magnetic field changes, and an emf is induced in the conductor. This emf is called a SELF-INDUCED EMF because it is induced in the conductor carrying the current. The direction of the induced emf has a definite relation to the direction in which the field that induces the emf varies. When the current in a circuit is increasing, the flux linking with the circuit is increasing. This flux cuts across the conductor and induces an emf in the conductor in such a direction as to oppose the increase in current and flux. This emf is sometimes referred to as counterelectromotive force (cemf). The two terms are used synonymically throughout this course. Likewise, when the current is decreasing, an emf is induced in the opposite direction and opposes the decrease in current. These effects are summarized by Lenz's law, which states that THE INDUCED EMF IN ANY CIRCUIT IS ALWAYS IN A DIRECTION TO OPPOSE THE EFFECT THAT PRODUCED IT.

The inductance is increased by shaping a conductor so that the electromagnetic field around each portion of the conductor cuts across some other portion of the same conductor. This is shown in its simplest form in figure 10-1 (A). A length of conductor is looped so that two portions of the conductor lie adjacent and parallel to one another. These portions are labeled conductor 1 and conductor 2. When the switch is closed, electron flow through the conductor establishes a typical concentric field around ALL portions of the conductor. For simplicity, however, the field is shown in a single plane that is perpendicular to both conductors. Although the field originates simultaneously in both conductors it is considered as originating in conductor 1 and its effect on conductor 2 will be noted. With increasing current, the field expands outward, cutting across a portion of conductor 2. The resultant induced emf in conductor 2 is shown by the dashed arrow. Note that it is in OPPOSITION to the battery current and voltage, according to Lenz's law.

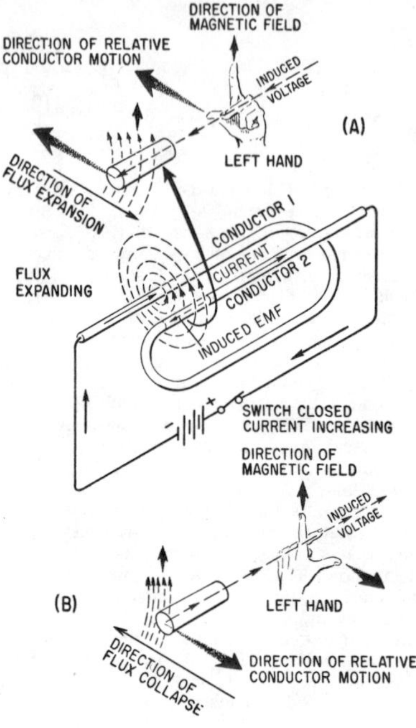

Figure 10-1.—Self-inductance.

The direction of this induced voltage may be determined by applying the Left-Hand Rule for Generators. This rule is applied to a portion of conductor 2 that is "lifted" and enlarged for the purpose in figure 10-1 (A). In applying this rule, the thumb of the left hand points in the direction that a conductor is moved through a field. (In this case, the field is moving, or expanding, in one direction, which is the same as the conductor moving in the opposite direction.) The index finger points in the direction of the magnetic field. The middle finger, extended as shown,

NOTES:

will now indicate the direction of induced (generated) voltage.

In figure 10-1 (B), the same section of conductor 2 is shown, but with the switch opened and the flux collapsing. Applying the left-hand rule in this case shows that the reversal of flux MOVEMENT has caused a reversal in the direction of the induced voltage. The most important thing to note, however, is that the voltage of self-induction opposes both CHANGES in current. It delays the initial buildup of current by opposing the battery voltage, and delays the breakdown of current by exerting an induced voltage in the same direction that the battery voltage acted.

FACTORS AFFECTING SELF-INDUCTANCE

Many things affect the self-inductance of a circuit. An important factor is the degree of linkage between the circuit conductors and its electromagnetic flux. In a straight length of conductor, there is very little flux linkage between one part of the conductor and another. Therefore, its inductance is extremely small. Conductors become much more inductive when they are wound into coils, as shown in figure 10-2. This is true because there is maximum flux linkage between the conductor turns, which lie side by side in the coil.

Inductance is further affected by the manner in which a coil is wound. The coil in figure 10-2 (A) is a poor inductor if compared to the other in the figure, because its turns are widely spaced, thus decreasing the flux linkage between its turns. Also, its lateral flux movement, indicated by the dashed arrows, does not link effectively, because there is only one layer of turns. A more inductive coil is shown in figure 10-2 (B). The turns are closely spaced, and the two layers link each other with a greater number of flux loops during all lateral flux movements. Note that nearly all turns, such as (a) are directly adjacent to four other turns (shaded), thus affording increased flux linkage.

The coil is made still more inductive by winding it in three layers, and providing a highly permeable core, as in figure 10-2 (C). The increased number of layers (cross-sectional area) improves lateral flux linkage. Note that some turns, such as (b) lie directly adjacent to six other turns (shaded). The magnetic properties of the iron core increase the total coil flux strength many times that of an air core coil of the same number of turns.

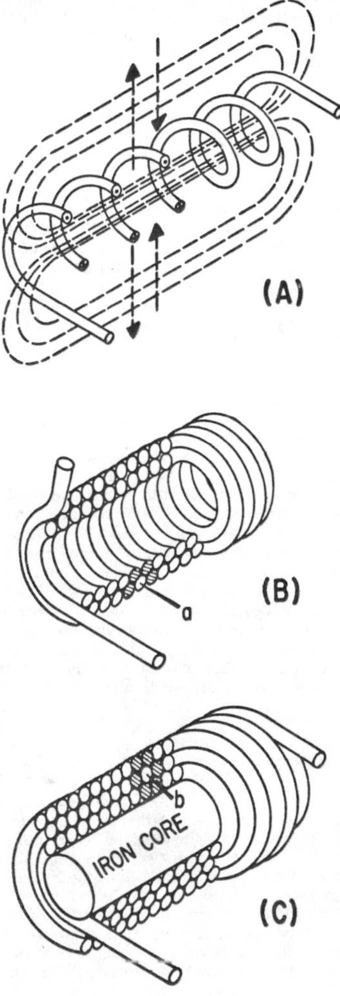

Figure 10-2.—Coils of various inductances.

From the foregoing, it can be seen that the primary factors controlling the inductance of a coil are (1) the number of turns of conductor, (2) the ratio of the cross-sectional area of the coil to its length, and (3) the permeability of its core material. The inductance of a coil is affected by the magnitude of current when the core is a magnetic material. When the core is air, the inductance is independent of the current. This is discussed in greater detail in the following paragraphs.

To summarize in an electric circuit, the cemf (counter electromotive force) is an induced emf or voltage. This voltage is induced in conductors of the circuit not by means of an external magnetic field, as in the case of the simple two-pole generator, but by means of the magnetic field already surrounding any conductor carrying a current. Any change in current changes the intensity of this magnetic field, and the resultant emf induced, the counter emf, is a self induced voltage. Thus, the property of a circuit which produces such an emf is called self-inductance. Actually, all elements in a circuit, including connecting wires, show some self inductance, but for all practical purposes only those elements designed to make use of this property to advantage are known as inductances or inductors. Moreover, it may be said that counter emf is present in any a-c circuit, but its effect is negligible in a circuit of moderate power, such as an electric lamp, which uses almost pure resistance as a load. But the effect of counter emf is considerable in circuits (even of very low power) which use an inductance as part of the load, such as the primary of the power transformer in an ordinary radio receiver.

GENERATION OF COUNTER EMF, LENZ'S LAW

It will be remembered that in the case of the simple generator, motion either of the conductor or of the magnet was necessary to an induced emf. In self inductance, the equivalent of motion, that is—a change in flux density of the magnetic lines of force about a conductor, is caused by the rise or fall of the current, since, as was seen in the study of electromagnetism, the intensity of a magnetic field about a conductor is directly proportional to the current through the conductor. The force setting up the flux lines is equal to $0.4\pi NI$. Since 0.4π is a constant, the factor NI is called the ampere turns. Any change in current changes the factor NI or ampere turns and, accordingly, the flux density. Thus, self inductance is present constantly in an a-c circuit because current is constantly changing, but it is present in a d-c circuit only at the moments of closing or opening the circuit.

That the emf self-induced in a conductor carrying current is a counter emf was deduced by H.F.E. Lenz from the principle of the conservation of energy. If the emf self-induced was not a counter emf, then an increase of current would aid the applied voltage, and this increase in applied voltage would in turn tend to increase the current. This process would continue, of course, until current reached an infinite amount, a condition not possible in the physical universe. As previously mentioned, Lenz' Law states: An induced emf always has such a direction as to oppose the action that produced it. Thus, when a current flowing through a circuit is varying in magnitude, it produces a varying magnetic field which sets up an induced emf that opposes the current change producing it. Or, it may be said that when the current in a circuit is increasing, the induced emf opposes the applied voltage and tends to keep the current from increasing; and when the current is decreasing, the induced emf aids the line voltage and tends to keep the current from decreasing.

The effect of counter emf may be observed experimentally in that an alternating current through an inductor is opposed by a force much greater than its simple d-c resistance. For example, the d-c resistance of the primary of an ordinary power transformer used in a typical radio receiver is approximately 6 ohms. As in figure 10-3, this primary is connected directly to the 120-volt, a-c outlet in the home. From Ohm's Law:

$$I = \frac{E}{R} = \frac{120}{6}$$

$$I = 20 \text{ amperes}$$

Thus, the current is calculated to be 20 amperes, but when actually measured the current is found to be approximately 1 ampere. It may be seen that some opposition other than the 6-ohm resistance is present in an a-c circuit. This opposition is the counter emf. If by mistake such a radio set is connected to a 120-volt, d-c line, the current through the primary of the transformer is 20 amperes, and the transformer burns up. Hence, Ohm's Law as stated for d-c circuits

NOTES:

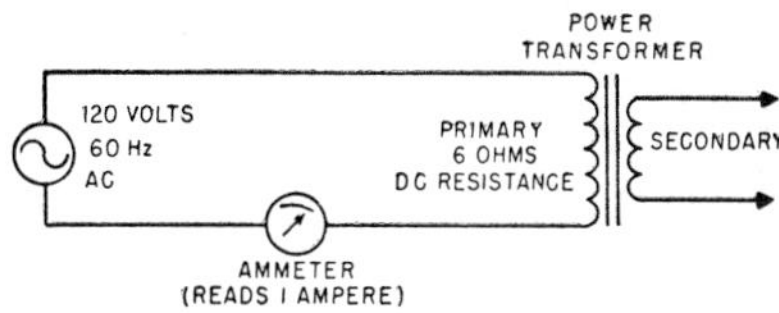

Figure 10-3.—Effect of cemf on current flow.

must be modified to include this effect of electrical inertia present in a-c circuits.

MAGNITUDE OF A COUNTER EMF

The magnitude of a counter emf depends upon the same factors that govern any induced emf. In the analysis of an induced emf it was shown that the magnitude of the emf induced in a conductor of unit length depended on the number of flux lines cut per second.
This principle may be restated as Faraday's Law of electromagnetic induction: THE EMF INDUCED IN ANY CIRCUIT IS DEPENDENT UPON THE RATE OF CHANGE OF THE FLUX LINKING THE CIRCUIT. Since there is no physical movement of the conductor or of the lines of force in self inductance, the rate of change of the flux density is equivalent to movement. But, as was shown above and in the study of electromagnetism, the flux density about a conductor is directly proportional to the current in the conductor. Therefore, the magnitude of self-induced emf depends directly upon the rate of change of the current in the circuit. Thus, a rapidly changing current induces a greater counter emf than a slowly changing current. But for any a.c. the rate of change of current depends on the number of hertz or the frequency. The counter emf then depends directly upon frequency.

As was mentioned previously, the total magnitude of an induced emf depends also on the length of the conductor, since in the simple generator the length of the conductor and the number of conductors in a coil side, determine the total emf induced. Thus, a long conductor has greater counter emf induced, or has more self inductance, than a short one. If, however, a long conductor is wound on itself in the form of a coil, its self inductance is increased because of the increase in total flux density. Such a coil, or inductance, is a solenoid, and, as was shown in the study of electromagnetism, the flux density about it may also be increased by the addition of a core material of high permeability, such as soft iron. Figure 10-4 illustrates this type of inductance.

MEASUREMENT OF INDUCTANCE

As previously mentioned, the unit of measurement of inductance is the henry, named after Joseph Henry, the co-discoverer with Faraday of the principle of electromagnetic induction. A henry is defined as the inductance of a circuit in which a current change of 1 ampere per second causes a counter emf of 1 volt. Since the henry is defined in terms of practical units, the factor 10^{-8} must be used if the cemf is to be read in volts and the rate of change of the current in amperes per second. Then

$$L = \frac{0.4\pi N^2 \mu A}{l} \times 10^{-8} \text{ (henrys)}$$

where L = self inductance of solenoid in henrys

N = number of turns of coil

μ = permeability of core in electromagnetic units

A = cross-sectional area of core in cm^2

l = mean length of core in cm.

This formula reveals the following important relationships:

1. The inductance of a coil is proportional to the square of the number of turns.
2. The inductance of a coil increases directly as the permeability of the material making up the core increases.
3. The inductance of coil increases directly as the cross-sectional area of the core increases.
4. The inductance of a coil decreases as its length increases. Figure 10-5 (A) shows two coils of a fixed number of turns with different cross-sectional areas. The larger coil has a greater total flux, or less reluctance, and therefore greater inductance. Figure 10-5 (B) shows two coils of a fixed number of turns and the

NOTES:

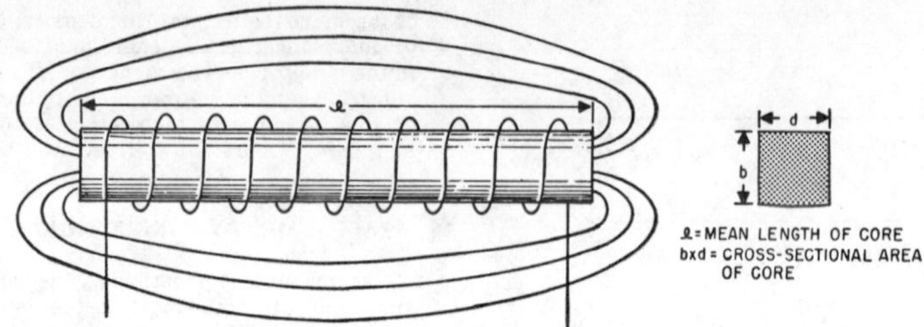

Figure 10-4.—Inductance wound on iron core.

same cross-sectional area, but of different lengths. The longer coil has less total flux, or greater reluctance, and therefore, less inductance.

It should also be noted that for ferromagnetic (iron-core) inductances, the permeability, μ, of the core material is not a constant, but depends on the magnitude of the magnetizing current. In a-c circuits, the current is constantly changing in magnitude and periodically in direction, and accordingly, an error is introduced in calculations of the magnitude of the inductance. In figure 10-6 (A) the relationship between the flux density B and the field intensity H is shown in the form of a hysteresis loop graph. But the ratio B/H is the definition of permeability. Therefore, it may be seen from this graph that the value of μ varies as the ratio of B to H varies for different points on the loop.

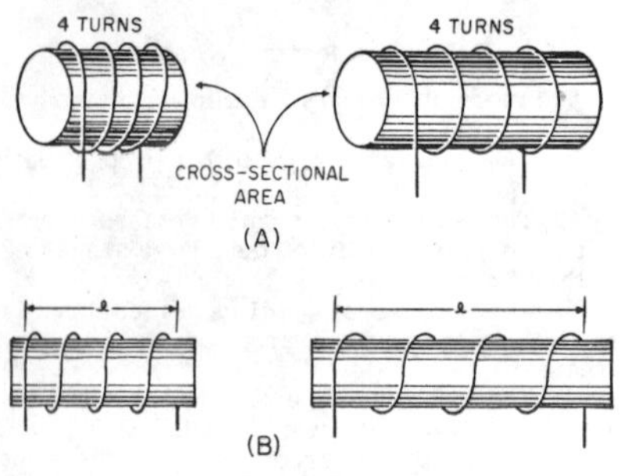

Figure 10-5.—Variation of inductance with size of solenoid.

In figure 10-6 (B) a permeability curve for cast steel is shown. Permeability for this material increases to a maximum value at approximately 7,000 lines per square centimeter and then falls off as the flux density increases. This variation of μ invalidates calculations of inductance based on the formula given previously.

MEASUREMENT OF COUNTER EMF: VOLTAGE

The formula for the magnitude of a cemf is as follows:

$$\text{cemf} = -L\frac{\Delta i}{\Delta t}$$

An examination of this formula reveals that the greater the inductance, or the faster the rate of change of the current, the greater is the cemf induced in the circuit. For example, a coil of 1-henry inductance has a current of 1 ampere flowing through it. If this current changes to 2 amperes in 1 second, the cemf will be:

$$\text{cemf} = -1 \times \frac{(2-1)}{1} = -1 \text{ volt}$$

If the current change remains the same but the coil used has an inductance of 10 henrys, then:

$$\text{cemf} = -10 \times \frac{(2-1)}{1} = -10 \text{ volts}$$

If the inductance remains, as in the first instance, at 1 henry, and the current change from 1

NOTES:

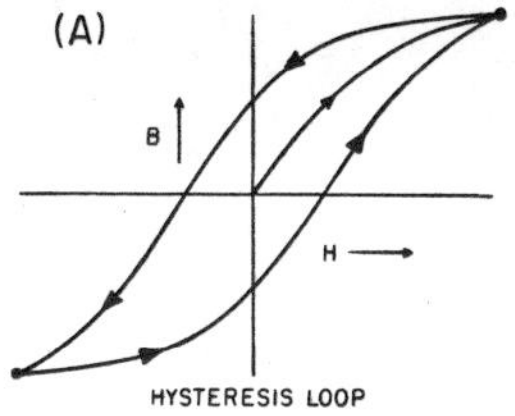

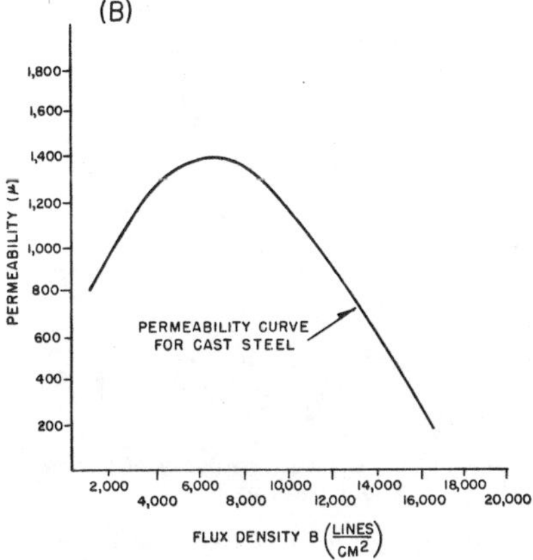

Figure 10-6.—Variation of μ for iron-core materials.

ampere to 2 amperes takes place in one-tenth of a second, then:

$$\text{cemf} = -1 \times \frac{(2 - 1)}{1/10} = -10 \text{ volts}$$

From these examples it may be seen that a high value of opposition to the flow of current may be obtained either by increasing the inductance or the speed of the change of current in a circuit, or both. Thus, low frequency a-c circuits, because of the slow speed of change of the current, generally employ high values of inductance (iron cores) to obtain a high cemf. High-frequency a-c circuits, because of the great speed of the change of the current, often may generate sufficient cemf with small air core inductances. Table 10-1 illustrates the rise in cemf as the rate of change of the current increases:

where Δi (called delta i) = change in current

Δt (called delta t) = time required for change in current.

Table 10-1.—Relationship of cemf and frequency.

L henrys	Δi amperes	Δt seconds	cemf in volts
1	1	1	-1
1	1	1/2	-2
1	1	1/4	-4
1	1	1/10	-10
1	1	1/20	-20
1	1	1/50	-50
1	1	1/100	-100
1	1	1/500	-500
1	1	1/1,000	-1,000
1	1	1/1,000,000	-1,000,000

From table 10-1 it is apparent that if a change of 1 ampere were to take place instantaneously, that is, if $\Delta t = 0$, the induced voltage e would become infinitely large. This would violate Kirchhoff's first law, which states that at any instant the applied voltage in a circuit must equal the sum of the voltage drops around the circuit. Certainly, if $\Delta t = 0$, the voltage drop across the inductance would be greater than any applied voltage could be. And by extension, it may be seen that at any instant, no matter how fast the change of current or how great the value of inductance, the induced voltage cannot be greater than the applied voltage. On the other hand, if there were no change of current, that is, if Δt were equal to infinity, the circuit would be a d-c circuit, and e would be zero.

NOTES:

GROWTH AND DECAY OF CURRENT IN AN RL SERIES CIRCUIT

If a battery is connected across a pure inductance, the current builds up to its final value at a rate that is determined by the battery voltage and the internal resistance of the battery. The current buildup is gradual because of the counter emf generated by the self-inductance of the coil. When the current starts to flow, the magnetic lines of force move out, cut the turns of wire on the inductor, and build up a counter emf that opposes the emf of the battery. This opposition causes a delay in the time it takes the current to build up to steady value. When the battery is disconnected, the lines of force collapse, again cutting the turns of the inductor and building up an emf that tends to prolong the current flow.

A voltage divider containing resistance and inductance may be connected in a circuit by means of special switch, as shown in figure 10-7 (A). Such a series arrangement is called an RL series circuit.

If switch S1, is closed (as shown), a voltage, E, appears across the divider. A current attempts to flow, but the inductor opposes this current by building up a back emf that, at the initial instant, exactly equals the input voltage, E. Because no current can flow under this condition, there is no voltage across resistor R. Figure 10-7 (B) shows that all of the voltage is impressed across L and no voltage appears across R at the instant switch Sp is closed.

As current starts to flow, a voltage, e_r, appears across R, and e_L is reduced by the same amount. The fact that the voltage across L is reduced means that the growth current, i_g, is increasing and consequently e_r is increasing. Figure 10-7 (B) shows that e_L finally becomes zero when i_g stops increasing, while e_r builds up to the input voltage, E, as i_g reaches its maximum value. Under steady-state conditions, only the resistor limits the size of the current.

Electrical inductance is like mechanical inertia, and the growth of current in an inductive circuit can be likened to the acceleration of a boat on the surface of the water. The boat begins to move at the instant a constant force is applied to it. At this instant its rate of change of speed (acceleration) is greatest, and all the applied force is used to overcome the inertia of the boat. After a while the speed of the boat increases (its acceleration decreases)

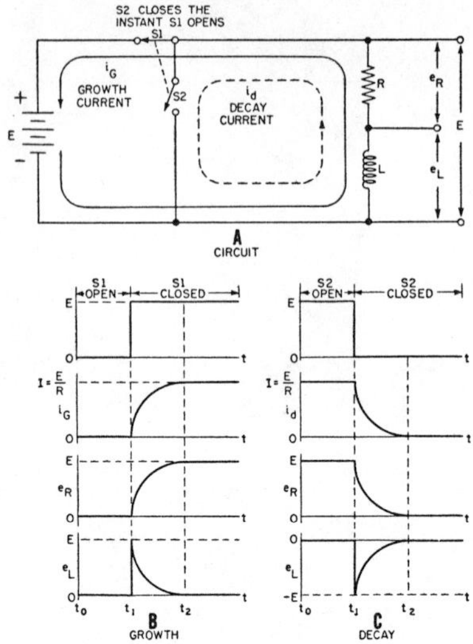

Figure 10-7.—Growth and decay of current in an RL series circuit.

and the applied force is used up in overcoming the friction of the water against the hull. As the speed levels off and the acceleration becomes zero, the applied force equals the opposing friction force at this speed and the inertia effect disappears.

When the battery switch in the R-1 circuit of figure 10-7 (A) is closed, the rate of current increase is maximum in the inductive circuit. At this instant all the battery voltage is used in overcoming the emf of self-induction which is a maximum because the rate of change of current is maximum. Thus the battery voltage is equal to the drop across the inductor and the voltage across the resistor is zero. As time goes on more of the battery voltage appears across the resistor and less across the inductor. The rate of change of current is less and the induced emf is less. As the steady-state condition of the current flow is approached the drop across the inductor approaches zero

NOTES:

and all of the battery voltage is used to overcome the resistance of the circuit.

Thus the voltages across the inductor and resistor change in magnitudes during the period of growth of current the same way the force applied to the boat divides itself between the inertia and friction effects. In both examples, the force is developed first across the inertia-inductive effect and finally across the friction-resistive effect.

If switch S2 is closed (source voltage E removed from the circuit), the flux that has been established around L collapses through the windings and induces a voltage, e_L, in L that has a polarity opposite to E and essentially equal to it in magnitude. The induced voltage, e_L, causes current i_d to flow through R in the same direction that it was flowing when S1 was closed. A voltage, e_r, that is initially equal to E, is developed across R. It rapidly falls to zero as the voltage, e_L, across L, due to the collapsing flux, falls to zero.

L/R Time Constant

The time required for the current through an inductor to increase to 63 percent (actually, 63.2 percent) of the maximum current or to decrease to 37 percent (actually, 36.7 percent) is known as the TIME CONSTANT of the circuit. An RL circuit and its charge and discharge graphs are shown in figure 10-8. The value of the time constant in seconds is equal to the inductance in henrys divided by the circuit resistance in ohms. One set of values is given in figure 10-8 (A). $\frac{L}{R}$ is the symbol used for this time constant.

Two useful relations used in calculating $\frac{L}{R}$ time constants are as follows:

$$\frac{L(\text{in henrys})}{R(\text{in ohms})} = t\ (\text{in seconds})$$

$$\frac{L(\text{in microhenrys})}{R(\text{in ohms})} = t\ (\text{in microseconds})$$

The time constant may also be defined as the time required for the current through the inductor to grow or decay to its final value IF it continued to grow or decay at its initial rate. As may be seen in figure 10-8 (B), the slope of the dotted tangent line, ox, indicates the initial rate of current growth with respect to time. At this rate, the current would reach its maximum value in $\frac{L}{R}$ seconds. Similarly, the slope of the dotted tangent line, YZ, indicates the initial rate of current decay with respect to time, and the decay would be completed in $\frac{L}{R}$ seconds.

The equation for the growth of current, i_L, through L is

$$i_L = \frac{E}{R}\left(1 - \frac{1}{2.718^{\frac{Rt}{L}}}\right)$$

where i_L is the instantaneous current through inductor L, E is the applied voltage (100 volts in this case), R is the resistance in ohms, t is the time in seconds, and L is the inductance in henrys. Figure 10-8 (B) shows a graph of this equation.

When $t = \frac{L}{R}$, the exponent $\frac{Rt}{L}$ in the preceding equation reduces to 1. Then $\frac{1}{2.718} = 0.368$. Therefore,

$$i_L = \frac{E}{R}(1-0.368) = 0.632\frac{E}{R}$$

In other words, when $t = \frac{L}{R}$, i_L is equal to 63.2 percent of the ratio $\frac{E}{R}$, which is the maximum current.

When the maximum current is 10 amperes (E = 100 and R = 10), the current through L grows to 6.32 amperes in $\frac{L}{R} = \frac{10}{10}$, or 1 second.

The equation for inductor voltage, e_L, on growth of current is

$$e_L = E\left(\frac{1}{2.718^{\frac{Rt}{L}}}\right)$$

The graph of this equation is also shown in figure 10-8 (B). When $t = \frac{L}{R}$, $e_L = 0.368E$; that is, $e_L = 0.368 \times 100 = 36.8$ volts.

MUTUAL INDUCTANCE

MUTUAL INDUCTANCE DEFINED

Whenever two coils are located so that the flux from one coil links with the turns of the other, a change of flux in one coil will cause

NOTES:

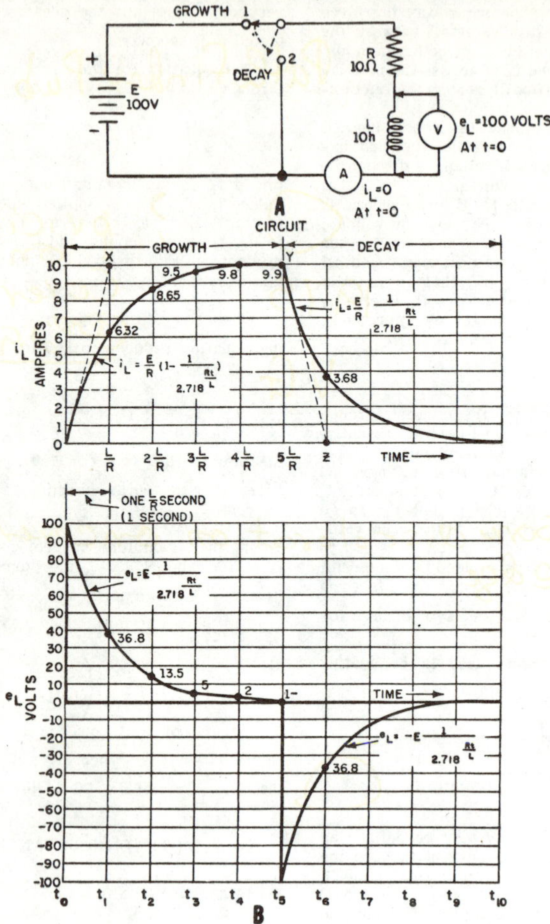

Figure 10-8.—L/R time constant.

NOTES:

an emf to be induced in other coil. The two coils have MUTUAL INDUCTANCE. The amount of mutual inductance depends on the relative position of the two coils. If the coils are separated a considerable distance, the amount of flux common to both coils is small and the mutual inductance is low. Conversely, if the coils are close together so that nearly all the flux of one coil links the turns of the other mutual inductance is high. The mutual inductance can be increased greatly by mounting the coils on a common iron core.

Two coils placed close together with their axes in the same plane are shown in figure 10-9. Coil A is connected to a battery through switch S, and coil B is connected to galvanometer G. When the switch is closed (fig. 10-9 (A)), the current that flows in coil A sets up a magnetic field that links coil B, causing an induced current and a momentary deflection of galvanometer G. When the current in coil A reaches a steady value, the galvanometer returns to zero. If the switch is opened (fig. 10-9 (B)), the galvanometer deflects momentarily in the opposite direction, indicating a momentary flow of current in the opposite direction in coil B. This flow of current in coil B, which is produced by the collapsing flux of coil A.

When current flows through coil A the flux expands in coil A producing a north pole nearest coil B. Some of the flux, that expands from left to right, cuts the turns of coil B. This flux induces an emf and current in coil B that opposes the growth of current and flux in coil A. Thus the current in B tries to establish a north pole nearest coil A (like poles repel).

When the switch is opened, the magnetic field produced by coil A collapses. The collapse of the flux cuts the turns of coil B in the opposite direction and produces a south pole nearest coil A (unlike poses attract). This polarity aids the magnetism of coil A, tending to prevent the collapse of its field.

FACTORS AFFECTING MUTUAL INDUCTANCE

The mutual inductance of two adjacent coils is dependent upon the—(1) physical dimensions of the two coils, (2) number of turns in each coil, (3) distance between the two coils, (4) relative positions of the axes of the two coils, and (5) the permeability of the cores.

If the two coils are positioned so that all the flux of one coil cuts all the turns of the

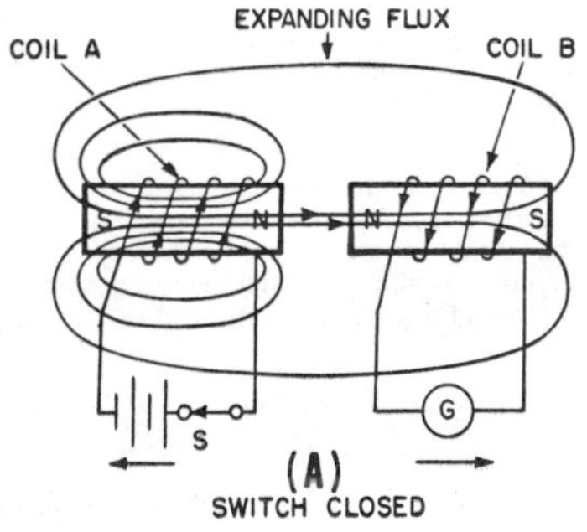

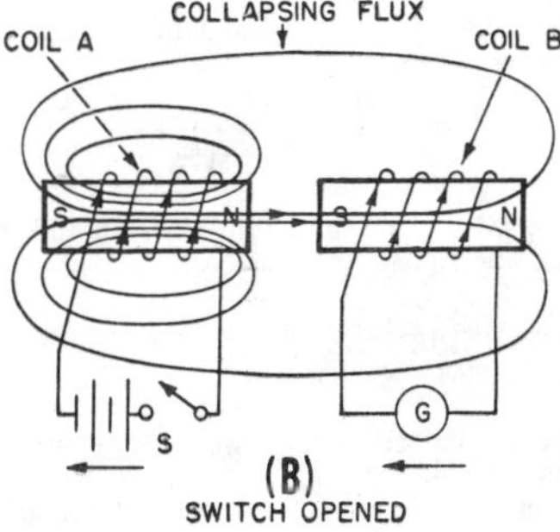

Figure 10-9.—Mutual inductance.

NOTES:

other, the mutual inductance can be expressed as follows:

$$M = \frac{0.4\pi\mu SN_1N_2}{10^8 l}$$

where

M equals mutual inductance of coils in henrys

N_1 and N_2 equal the number of turns in coils

1 and 2 respectively

S equals area of core in square cm

μ equals permeability of core

l equals length of core in cm

If the two coils are so positioned with respect to each other that all the flux of one coil cuts all the turns of the other, the coils have UNITY COEFFICIENT OF COUPLING. If all of the flux produced by one of the coils cuts only one-half the turns of the other coil, the coefficient of coupling is 0.5. Coefficient of coupling is designated by the letter, K. The coefficient of coupling is equal to the percentage of flux originating in one coil that cuts the other coil. It is never exactly equal to unity but it approaches this value in certain shell type transformers.

The mutual inductance between two coils, L_1 and L_2, may be expressed in terms of the inductance of each coil and the coefficient of coupling, K, as follows,

$$M = K\sqrt{L_1L_2}$$

where M is in the same units of inductance as L_1 and L_2.

SERIES INDUCTORS WITHOUT MAGNETIC COUPLING

When inductors are well shielded, or located far enough apart to make the effects of mutual inductance negligible, the inductance of the various inductors are added in the same manner that the resistances of resistors in series are added.

NOTES:

For example,

$$L_T = L_1 + L_2 + L_3 \ldots + L_N$$

where L_t is the total inductance; L_1, L_2, L_3 are the inductances of L_1, L_2, L_3; and L_N means that any number (N) of inductors may be used.

SERIES INDUCTORS WITH MAGNETIC COUPLING

When two inductors in series are so arranged that the field of one links the other, the combined inductance is determined as

$$L_T = L_1 + L_2 \pm 2M$$

where L_T is the total inductance, L_1 and L_2 are the self inductances of L_1 and L_2 respectively, and M is the mutual inductance between the two inductors. The plus sign is used with M when the magnetomotive forces of the two inductors are aiding each other. The minus sign is used with M when the mmf's of the two inductors oppose each other. The factor 2 accounts for the influence of L_1 on L_2 and L_2 on L_1.

If the coils are arranged so that one can be rotated relative to the other to cause a variation in the coefficient of coupling, the mutual inductance between them can be varied. The total inductance, L_T, may be varied by an amount equal to 4M.

PARALLEL INDUCTORS WITHOUT COUPLING

The total inductance, L_T, of inductors in parallel is calculated in the same manner that the total resistance of resistors in parallel is calculated, provided the coefficient of coupling between the coils is zero. For example,

$$\frac{1}{L_T} = \frac{1}{L_1} + \frac{1}{L_2} + \frac{1}{L_3} \cdots + \frac{1}{I_N}$$

where L_1, L_2, and L_3 are the respective inductances of inductors L_1, L_2, and L_3; and L_N means that any number (N) of inductors may be used.

EFFECTS OF INDUCTANCE IN AN ELECTRIC CIRCUIT

REACTION OF AN INDUCTOR TO FLUX CHANGE

Up to this point, the effects of inductance in an electric circuit have been considered as being controlled only by the particular inductor considered in each case. For instance, the time required for the growth and decay of current in the circuit shown in figure 10-7 is determined only by the inherent properties of the inductor L, namely, the ratio of its inductance to its resistance. When permitted to change at this inherent, or natural rate, the shaper of the L/R curve for any particular inductor will always remain the same. That is, it will always exhibit the same natural opposition to any change in current and flux, when left to its own tendencies. At this point, however, it must be pointed out that an inductor can be made to exhibit many different degrees of reaction. This is done by using external means to change the flux linking the inductor. These changes, such as variations in the applied frequency, usually do not occur at the same rate as the inductor's inherent rate. When the flux is changed RAPIDLY by some external means, the inductor's reaction is much greater than when the flux is changed gradually. That is, the inductor's self-induced emf is dependent on the RATE with respect to time at which the linking flux is changed.

The dependence of the induced voltage on the rate of change of flux with respect to time is shown by means of the simple apparatus in figure 10-10. When the switch is closed, current builds up to a maximum, and the lamp glows with its normal brilliance. If the iron core is inserted rapidly into the coil, the flux increases rapidly (because of the increased inductance of the coil), and the induced voltage opposes, according to Lenz's law, the source voltage. Therefore, less current flows through the lamp, and it dims momentarily. If the core is withdrawn rapidly from the coil, a portion of the flux that was established around the coil collapses. The resulting induced voltage opposes the decrease (again, according to Lenz's law) by aiding the source voltage, and the coil current increases. Consequently, the lamp burns brighter momentarily. The faster the iron core is moved the greater is the flux change per unit of time and the more noticeable is the effect on the lamp.

NOTES:

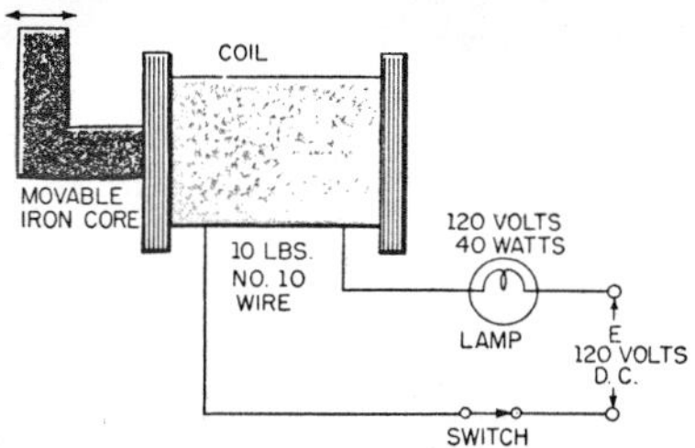

Figure 10-10.—Dependence of self-induced voltage on the rate of change of flux.

EFFECTS OF INDUCTANCE IN AN A-C CIRCUIT

When a circuit containing a coil is energized with direct current, the coil's effect in the circuit is evident only when the circuit is energized, or when it is deenergized. For instance, when the switch in figure 10-11 is placed in position ①, the inductance of coil L will cause a delay in the time required for the lamps to attain normal brilliance. After they have attained normal brilliance, the inductance has no effect on the circuit as long as the switch remains closed. When the switch is opened, an electric spark will jump across the opening switch contacts. The emf which produces the spark is caused by the collapsing magnetic field cutting the turns of the inductor.

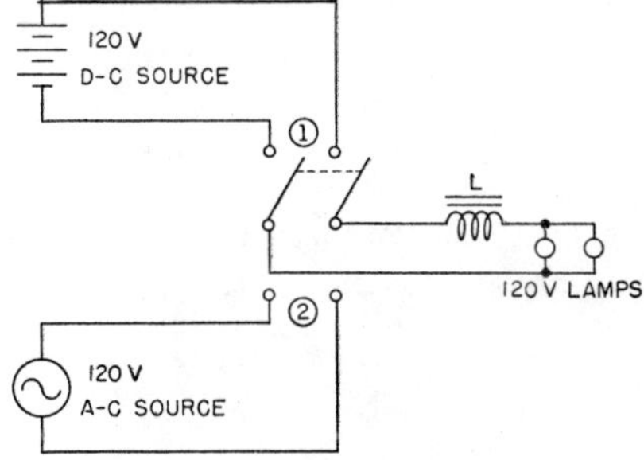

Figure 10-11.—Relative effects of inductance in d-c and a-c circuits.

When the inductive circuit is supplied with alternating current, however, the inductor's effect is continuous and much greater than when it was supplied with direct current. For equal applied voltages, the current through the circuit is less when a.c. is applied than when d.c. is applied, as may be demonstrated by the circuits of figure 10-11. The alternating current is accompanied by an alternating magnetic field around the coil, which cuts through the turns of the coil. This action induces a voltage in the coil that always opposes the changing current. When the switch is in position ① the lamps burn brightly on direct current but in position ②, although the effective value of the applied a-c voltage is equal to the d-c value, the lamps burn dimly because of the opposition developed across the inductance. Most of the applied voltage appears across L, with little remaining for the lamps.

RELATION BETWEEN INDUCED VOLTAGE AND CURRENT

As stated previously, any change in current, either a rise or a fall, in a coil causes a corresponding change of the magnetic flux around the coil (fig. 10-12 (A)). If the current is sinusoidal, the induced voltage will also have the form of a sine wave. Because the current changes at its maximum rate when it is going through its zero value at 0°, 180°, and 360° (fig. 10-12 (B)), the flux change is also greatest at those times. Consequently, the self-induced voltage in the coil is at its maximum value at these instants. According to Lenz's law, the induced voltage always opposes the change in current. Thus, when the current is rising in a positive direction at 0°, the induced emf is of opposite polarity to the impressed emf and opposes the rise in current. Later, when the current is falling toward its zero value at 180°, the induced voltage is of the same polarity as the current and tends to keep the current from falling. Thus the induced voltage can be seen to lag the current by 90°. The resistance of the coil is small and the principal **opposition** to the current flow through the coil is the induced voltage, E_{ind}. The applied voltage, E, is slightly larger than E_{ind} and diametrically opposed it, as indicated in the vector diagram (fig. 10-12 (C)).

The current lags the applied voltage in an inductive circuit by an angle of 90° and leads the induced voltage by 90°. The induced voltage

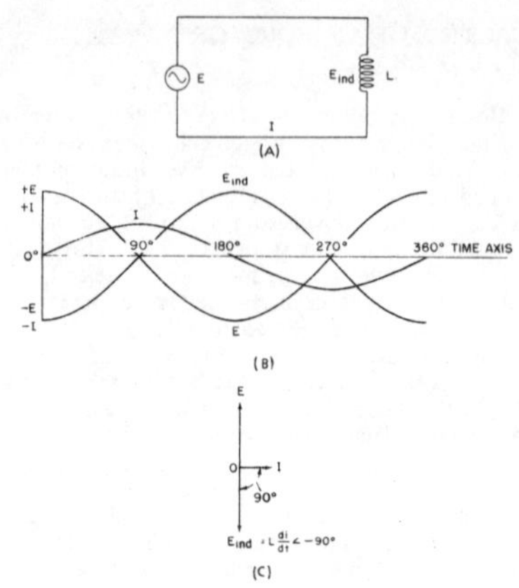

Figure 10-12.—Relation between induced voltage and current.

is always of opposite polarity to the applied voltage and is called a COUNTER EMF or a BACK EMF because it always opposes the change of current.

INDUCTIVE KICK

It has been shown that an inductor in which there is a changing current becomes a source of EMF, and that the direction of this EMF is such that it tends to oppose the change in current producing it. As a result of this action the current in the inductor does not rise to its full value the instant the switch is closed, but rises at a rate which depends upon the L/R ratio. Likewise when the switch is opened, the source removed, and a short placed across the circuit, the current does not instantaneously fall to zero; it slowly decays at a rate determined by the L/R ratio of the discharge circuit.

The circuits so far considered have been composed of ideal components. However,

NOTES:

components are not perfect, therefore no switch could be manufactured that would be capable of being moved from one position to another instantaneously. This leads to an explanation of the action of the induced voltage at the INSTANT a switch is opened. Because the magnitude of self-induced voltage can be extremely great even though the source voltage is very low, the development of high induced voltages when an inductive circuit is suddenly opened will now be studied.

Consider the circuit shown in figure 10-13 (A) consisting of a 6 volt battery, a switch, and a 30 henry coil. The resistor (R) represents the total circuit resistance, including the resistance of the coil. At the instant the switch is closed (time zero), 6 volts of CEMF will develop across the coil. This is shown in figure 10-13 (B) where $e_L = 6V$ at t_0.

Since at time zero the current, for all practical purposes has not started to flow, there is no voltage drop across the resistor. With an inductance of 30 henrys and a voltage of 6 volts, the initial rate of change (roc) will be:

$$roc = \frac{e}{L}$$

$$roc = \frac{6}{30}$$

$$roc = 0.2 \text{ amp/sec}$$

After the interval of time $(5\frac{L}{R})$ the current has reached its final steady value. For the circuit of figure 10-13 (A) this will be:

$$T = \frac{L}{R}$$

$$T = \frac{30}{1}$$

$$T = 30 \text{ seconds}$$

But: steady state value equals:

$$t = 5\frac{L}{R} = 5T$$

Where: t = elapsed time, in seconds

Therefore: steady state will be reached in

$$5 \times T = t$$

$$5 \times 30 = 150 \text{ sec}$$

NOTES:

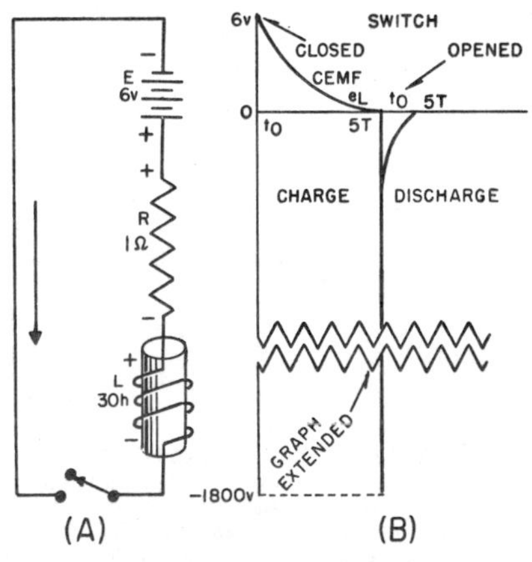

Figure 10-13.—Development of high induced voltage from low source.

The magnetic field around the coil is now fully established and steady. Figure 10-13 (B) shows no e_L (CEMF) at $5\frac{L}{R}$ or 5T), therefore, there must be no change in current at this time.

If the switch is now opened, one side of the battery is disconnected and there is no complete path through which the current can flow. As a result, the current supplied by the battery stops immediately. However, the inductance of the coil opposes the change. Since the current supplied by the battery is no longer available to support the magnetic field, it will start to collapse. As it collapses, the magnetic lines of force cuts the turns of the coil. Whereas expanding lines of force induced an EMF with the polarity on the coil as shown in Figure 10-13 (A), collapsing lines of force induce an EMF in the coil with the opposite polarity. The inductance is therefore acting as a source of voltage attempting to maintain the same current flow in the circuit just as if the battery was still connected.

The decay time will be the same as the growth time if the same L/R ratio is maintained during decay as existed during growth. For an understanding of the change in time constant, picture for a moment, the conditions existing in the circuit shown in figure 10-14 (A). The current flowing in this circuit is:

$$I = \frac{E}{R}$$

$$I = \frac{6}{1}$$

$$I = 6 \text{ amps}$$

The voltage drop across the resistor (E_R) is:

$$E_{R1} = IR$$

$$E_{R1} = 6 \times 1$$

$$E_{R1} = 6 \text{ volts}$$

Now if it were possible to maintain the battery at a constant 6 volts and briefly maintain the current at 6 amperes while INSTANTANEOUSLY removing the one ohm resistor (R_1) and inserting a large resistance (R_2), as in figure 10-14 (B), the resistive voltage drop would increase:

$$E_{R2} = IR$$

$$E_{R2} = 6 \times 2000$$

$$E_{R2} = 12{,}000 \text{ volts}$$

The above conditions are impossible to obtain in the resistive circuit discussed and are inserted here merely to show the possibility of obtaining a high voltage pulse when exchanging a high resistance for a low resistance and MOMENTARILY maintaining the same current. These conditions can be obtained in the circuit of figure 10-13. At the exact INSTANT the switch contacts represent the insertion of an extremely high resistance (open circuit) the action of the inductance is maintaining the current at very near the value built up with the small resistance in the circuit has been increased by a very large amount, the L/R time constant of the circuit for decay is not the original 30 seconds. Due to the insertion of the high resistance, the time constant has become a fraction of a second. Transposing the following equation and substituting values will give an approximation of the voltage developed across the coil by the collapsing magnetic field.

NOTES:

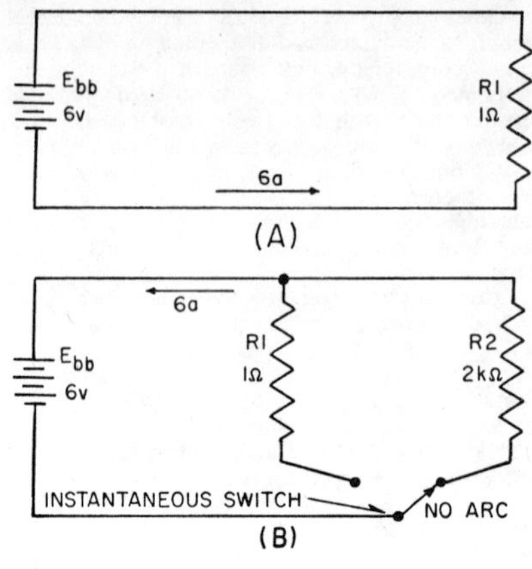

Figure 10-14.—Resistive demonstration circuit for high voltage pulse.

$$L = \frac{e}{\Delta i/\Delta t}$$

transposing:

$$e = L\frac{\Delta i}{\Delta t}$$

NOTE: Assuming for the purpose of illustration, that 0.00001 second (Δ t) after the switch is opened the current has decreased to 5.9994 amps, a change (Δi) of 0.0006 amp:

Substituting:

$$e = L\frac{\Delta i}{\Delta t}$$

$$e = 30 \times \frac{0.0006}{0.00001}$$

$$e = 30 \times 60$$

$$e = 1800 \text{ volts}$$

The energy contained in the collapsing magnetic field must be dissipated somewhere within the circuit. The voltage developed in the conductor is sufficient to create an arc across the switch contacts. The energy in the magnetic field is dissipated in the heat of this arc. The energy expended in the arc can seriously burn an individual, damage the switch contacts, or break down . the insulation of the coil. For these reasons care should be taken in the abrupt interruption of current in any incuctive circuit.

The development of a large voltage pulse from a low voltage source (INDUCTIVE KICK) is not always a disadvantage, but is commonly used in the spark coil circuits (ignition system) of most gasoline engines.

NOTES:

CHAPTER 11

CAPACITANCE

Every electrical circuit, no matter how complex, is composed of no more than three basic electrical properties; resistance, inductance, and capacitance. Therefore, a thorough understanding of each of these basic properties is a necessary step toward the understanding of electrical equipment. Since resistance and inductance have been covered, the last of the basic three, capacitance, will now be discussed.

Two conductors separated by a nonconductor exhibit the property called CAPACITANCE, because this combination can store an electric charge; whereas inductance was defined as a property of a circuit which opposes a change in CURRENT. Capacitance is a property of a circuit which opposes a change in VOLTAGE. Where inductance stored energy in an ELECTROMAGNETIC field, capacitance stores energy in an ELECTROSTATIC field.

REVIEW OF ELECTROSTATICS

In order to promote a clear understanding of capacitance, the reader should be thoroughly familiar with the theories and laws of electrostatics. For convenience, the main points of electrostatics will be briefly reviewed in this section.

When a charged body is brought into close proximity with another charged body, there is a force that causes the bodies to attract or repel one another. If the charged bodies possess the same sign of charge, a repelling force will exist between the two bodies. If they have unlike signs, there will be a force of attraction between them. The force of attraction or repulsion is caused by the electrostatic field that surrounds every charged body. If a material is charged positively, it has a deficiency of electrons. If it is charged negatively, it has an excess of electrons. The direction of the electrostatic field is represented by lines of force drawn perpendicular to the charged surface and shown originating from the positively charged material. Each line of force is drawn in the form of an arrow and is shown pointing from positive to negative.

The force between charges is described by Coulomb's law: "The force existing between two charged bodies is directly proportional to the product of the charges and inversely proportional to the square of the distance separating them."

If a test charge is inserted in an existing electrostatic field it will move toward one or the other of the charged areas which is causing the field to exist. The direction of movement will depend on whether the test charge is positive or negative. Previously it has been shown that a positive test charge placed in a field moves in the direction that the line of force points, from positive toward negative. In this case the test charge will be an electron and since the electron is negative it will move in a direction opposite to that of the positive charge. In other words, an electron in an electrostatic field will move AGAINST the arrow from negative toward positive. The above action is illustrated by figure 11-1.

If Coulomb's law is analyzed in connection with figure 11-1 it can be seen that the greater the distance between the electron and the positive charge the less the force of attraction. The importance of the distance between the charges creating the field will become apparent at a later time in this chapter.

One important characteristic of electrostatic lines of force is that they have the ability to pass through any known material.

THE CAPACITOR

CAPACITANCE is defined as the property of an electrical device or circuit that tends to

NOTES:

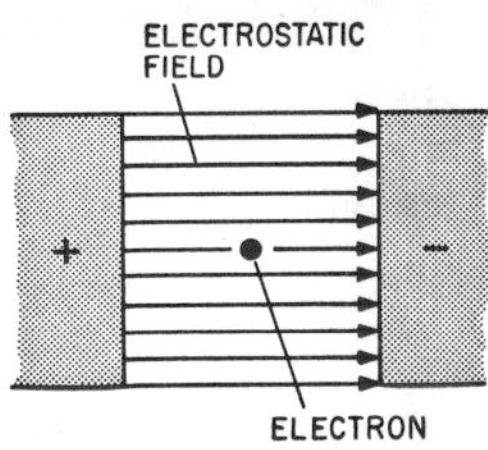

Figure 11-1.—Electron movement in an electrostatic field.

oppose a CHANGE in VOLTAGE. Capacitance is also a measure of the ability of two conducting surfaces, separated by some form of nonconductor, to store an electric charge. For the present time air will be used as the insulating material between the conducting surfaces.

The device used in electrical circuits to store a charge by virtue of an electrostatic field is called a CAPACITOR. (The larger the capacitor, the larger the charge that can be stored.)

The simplest type of capacitor consists of two metal plates separated by air. It has been established that a free electron inserted in an electrostatic field will move. The same is true, with qualifications, if the electron is in a bound state. The material between the two charged surfaces of figure 11-1 (air in this case) is composed of atoms containing bound orbital electrons. Since the electrons are bound they can not travel to the positively charged surface. Therefore, the resultant effect will be a distorting of the electron orbits. The bound electrons will be attracted toward the positive surface, and repelled from the negative surface. This effect is illustrated in figure 11-2. In figure 11-2 (A), there is no difference in charge placed across the plates; and the structure of the atom's orbits is undisturbed. If there is a difference in charge across the plates as shown in figure 11-2 (B), the orbits will be elongated in the direction of the positive charge.

An energy is required to distort the orbits, energy is transferred from the electrostatic field to the electrons of each atom between the charged plates. Since energy cannot be destroyed, the energy required to distort the orbits can be recovered when the electron orbits are permitted to return to their normal positions. This effect is analogous to the storage of energy in a stretched spring. A capacitor can thus "store" electrical energy.

An illustration of a simple capacitor and its schematic symbol is shown in figure 11-3. The conductors that form the capacitor are called PLATES. The material between the plates is called the DIELECTRIC. In figure 11-3 (B), the two vertical lines represent the connecting leads. The two horizontal lines represent the capacitor plates. Notice that the schematic symbol (B) and the simple capacitor diagram (A) are similar in appearance. In a practical capacitor, the parallel plates may be constructed in various configurations (circular, rectangular, etc.); but the cross-sectional area of the capacitor plates is tremendously large in comparison to the cross-sectional area of the connecting conductor. This means that there is an abundance of free electrons available in each plate of the capacitor. If the cross-sectional area and plate material of the capacitor plates are the same, the number of free electrons in each plate must be approximately the same. It should be noted that there is a possibility of the difference in charge becoming so large as to cause ionization of the insulating material to occur (cause bound electrons to be freed). This places a limit on the amount of charge that can be stored in the capacitor.

Figure 11-3 (A) depicts a capacitor in its simplest form. It consists of two metal plates separated by a thin layer of insulating material (dielectric). When connected to a voltage source (battery), the voltage forces electrons onto one plate, making it negative, and pulls them off the other, making it positive. Electrons cannot flow through the dielectric. Since it takes a definite quantity of electrons to "fill up," or charge, a capacitor, it is said to have a CAPACITY. This

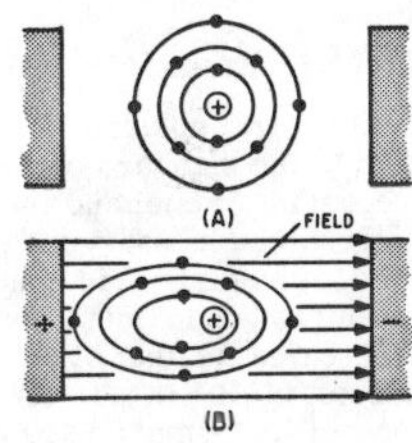

Figure 11-2.—Electron orbits with and without the presence of an electric field.

NOTES:

characteristic is referred to as CAPACITANCE.

DIELECTRIC MATERIALS

Various materials differ in their ability to support electric flux or to serve as dielectric material for capacitors. This phenomenon is somewhat similar to permeability in magnetic circuits. Dielectric materials, or insulators are rated in their ability to support electric flux in terms of a figure called the DIELECTRIC CONSTANT. The higher the value of dielectric constant (other factors being equal), the better is the dielectric material.

A vacuum is the standard dielectric for purposes of reference and is assigned the value of unity (or one). The dielectric constant of a dielectric material is also defined as the ratio of the capacitance of a capacitor having that particular material as the dielectric to the capacitance of the same capacitor having air as the dielectric. By way of comparison, the dielectric constant of pure water is 81; flint glass, 9.9; and paraffin paper, 3.5. The range of dielectric constants is much more restricted than is the range of permeabilities. Dielectric constants for some common materials are given in the following list:

Material	Constant
Vacuum	1.0000
Air	1.0006
Paraffin paper	3.5
Glass	5 - 10
Mica	3 - 6
Rubber	2.5 - 35
Wood	2.5 - 8
Glycerine (15 °C)	56
Petroleum	2
Pure water	81

Notice the dielectric constant for a vacuum. Since a vacuum is the standard of reference, it is assigned a constant of one; and the dielectric constants of all materials are compared to that of a vacuum. Since the dielectric constant of air has been determined experimentally to be approximately the same as that of a vacuum, the dielectric constant of AIR is also considered to be equal to one. The formula used to compute the value of capacitance using the physical factors just described is:

$$C \quad 0.2249 \left(\frac{KA}{d}\right) \qquad (11\text{-}1)$$

NOTES:

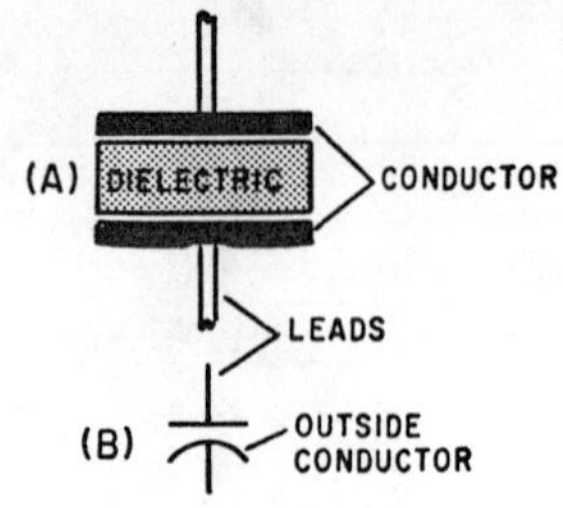

Figure 11-3.—Capacitor and schematic symbol.

Where C = capacitance, in picofarads (10^{-12})
A = area of one plate, in square inches
d = distance between plates, in inches
K = dielectric constant of insulating material.
0.2249 = a constant resulting from conversion from Metric to British units.

Example: Find the capacitance of a parallel plate capacitor with paraffin paper as the dielectric.

Given: $K = 3.5$

$d = 0.05$ inch

$A = 12$ square inches

Solution:

$$C = 0.2249 \left(\frac{KA}{d}\right) \qquad (11\text{-}1)$$

$$C = 0.2249 \left(\frac{3.5 \times 12}{0.05}\right)$$

$$C = 189 \text{ picofarads}$$

Using equation (11-1) it is easy to visualize the effects on capacitance of the physical factors involved. It can be seen that capacitance is a direct function of the dielectric constant and the area of the capacitor plates, and an inverse function of the distance between the plates.

UNIT OF CAPACITANCE

Capacitance is measured in a unit called the FARAD. This unit is a tribute to the memory of

Michael Faraday, a scientist who performed many early experiments with electrostatics and magnetism.

It was discovered that for a given value of capacitance, the ratio of charge deposited on one plate, to the voltage producing the movement of charge, is a constant value. This constant value is a measure of the amount of capacitance present. The symbol used to designate a capacitor is (C). The capacitance is equal to 1 farad when a voltage changing at the rate of 1 volt/per second causes a charging current of 1 amp to flow. This is expressed by the equation:

$$C = \frac{i}{\frac{\Delta e}{\Delta t}} \quad (11\text{-}2)$$

Where C = capacitance, in farads

i = instantaneous current in amp

$\frac{\Delta e}{\Delta t}$ = rate of change of voltage, in volts, with time, in seconds

Equation (11-2) may be clearer if expressed as follows:

$$\text{Capacitance equals 1 farad when} = \frac{\text{Charging current of 1 amp flows}}{\text{When voltage changes 1 volt in 1 second}}$$

The farad can also be defined in terms of charge and voltage. A capacitor has a capacitance of 1 farad if it will store 1 coulomb of charge when connected across a potential of 1 volt. This relationship can be expressed mathematically as:

$$C = \frac{Q}{E}$$

Where C = capacitance in farads

Q = charge in coulombs

E = applied potential in volts

Example: What is the capacitance of two metal plates separated by one centimeter of air, if .001 coulomb of charge is stored when a potential of 200 volts is applied to the capacitor?

Given: Q = 0.001 coulomb

E = 200 volts

$$\text{Solution: } C = \frac{Q}{E} \quad (11\text{-}3)$$

NOTES:

Converting to power of ten

$$C = \frac{10 \times 10^{-4}}{2 \times 10^{2}}$$

$$C = 5 \times 10^{-6}$$

$$C = 0.000005 \text{ farads}$$

Although this capacitance might appear rather small (five millionths of a farad), many electronic circuits require capacitors of much smaller value. Consequently the farad is a cumbersome unit, far too large for most applications. The MICROFARAD which is one millionth of a farad (1×10^{-6} farad) is a more convenient unit. The symbols used to designate microfarad are μf. In high frequency circuits even the microfarad becomes too large, and the unit MICROMICROFARAD (one millionth of a microfarad) is used. The symbols for micromicrofarads are $\mu\mu f$.

To avoid confusion and the use of double prefixes the name PICOFARAD (pf) is preferred in place of micromicrofarad. In powers of ten, 1 picofarad (or 1 micromicrofarad) is equal to 1×10^{-12} farad.

In using equation (11-3) one must not deduce the mistaken idea that capacitance is dependent upon charge and voltage. Capacitance is determined entirely by physical factors such as plate area, plate spacing, etc.

FACTORS AFFECTING THE VALUE OF CAPACITANCE

The capacitance of a capacitor depends on the three following factors:

1. The area of the plates.
2. The distance between the plates.
3. The dielectric constant of the material between the plates.

These three factors are related to the capacitance of a parallel-plate capacitor consisting of two plates by the formula

$$C = 0.2249\left(\frac{kA}{d}\right)$$

Where C is in picofarads, A is the area of one of the plates in square inches, d is the distance between the plates in inches, and k is the dielectric constant of the insulator separating the plates.

For example, the capacitance of a parallel-plate capacitor with an air dielectric and spacing of 0.0394 inch between the plates, each of which has an area of 15.5 square inches, is approximately

$$C = 0.225\left(\frac{1 \times 15.5}{0.0394}\right) = 88.5 \text{ picofarads}$$

From this formula it may seem that the capacitance increases when the plates are increased in area; it decreases if the spacing of the plates is increased; and it increases if the k-value is increased.

As previously mentioned, the dielectric constant, k, expresses the relative capacitance when materials other than air are used as the insulating material between the plates. For example, if mica is substituted for air as the dielectric, the capacitance increases from 3 to 6 times because the dielectric constant of mica is from 3 to 6 times greater than air (1).

If the capacitor is composed of more than two parallel plates, the capacitance is calculated by multiplying the preceding formula by N - 1, where N is the number of plates. The plates are interlaced as shown in figure 11-4, and the effect is that of increasing the capacitance of the two-plate capacitor by the factor N - 1. In the figure there are 11 plates, and the capacitance is 10 times that of a 2-plate capacitor of the same plate area, spacing, and dielectric material.

VOLTAGE RATING OF CAPACITORS

In selecting or substituting a capacitor for use in a particular circuit, consideration must be given to (1) the value of capacitance desired and (2) the amount of voltage to which the capacitor is to be subjected. If the voltage applied across the plates is too great, the dielectric will break down and arcing will occur between the plates.

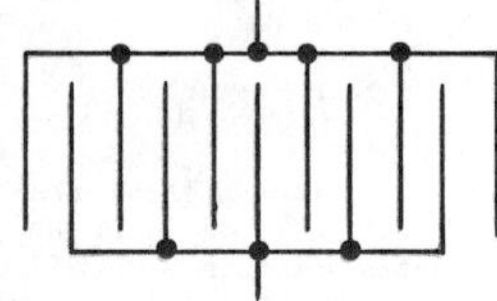

Figure 11-4.—Capacitor multiple-plate construction.

NOTES:

The capacitor is then short-circuited and the possible flow of direct current through it can cause damage to other parts of the equipment. Capacitors have a voltage rating that should not be exceeded.

The working voltage of the capacitor is the maximum voltage that can be steadily applied without danger of arc-over. The working voltage depends on (1) the type of material used as the dielectric, and (2) the thickness of the dielectric.

The voltage rating of the capacitor is a factor in determining the capacitance because capacitance decreases as the thickness of the dielectric increases. A high-voltage capacitor that has a thick dielectric must have a larger plate area in order to have the same capacitance as a similar low-voltage capacitor having a thin dielectric. The voltage rating also depends on frequency because the losses, and the resultant heating effect, increase as the frequency increases.

A capacitor that may be safely charged to 500 volts d.c. cannot be safely subjected to alternating or pulsating direct voltages whose effective values are 500 volts. An alternating voltage of 500 volts (r.m.s.) has a peak voltage of 707 volts, and a capacitor to which it is applied should have a working voltage of at least 750 volts. The capacitor should be selected so that its working voltage is at least 50 percent greater than the highest voltage to be applied to it. Effective (rms) voltage and the action of capacitors in a-c circuits are described in a later chapter.

CHARGING AND DISCHARGING A CAPACITOR

CHARGING

In order to better understand the action of a capacitor in conjunction with other components, the charge and discharge action of a purely capacitive circuit will be analyzed first. For ease of explanation the capacitor and voltage source used in figure 11-5 will be assumed to be perfect (no internal resistance, etc.) although this is impossible in practice.

In figure 11-5 (A) an uncharged capacitor is shown connected to a four-position switch. With the switch in position 1 the circuit is open and no voltage is applied to the capacitor. Initially each plate of the capacitor is a neutral body, and until a difference of potential is impressed across the capacitor no electrostatic field can exist between the plates.

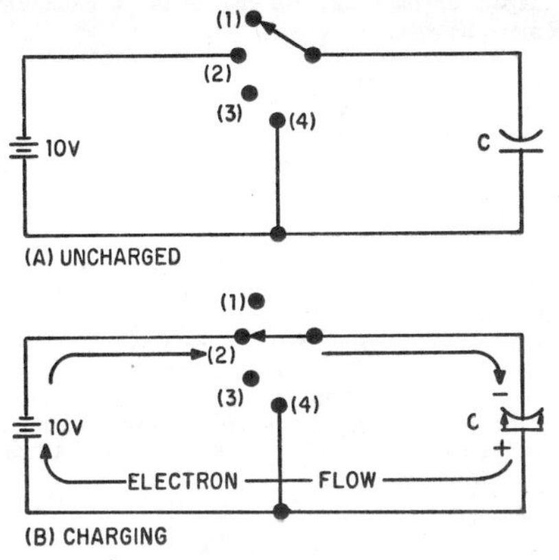

Figure 11-5.—Charging a capacitor.

To CHARGE the capacitor the switch must be thrown to position 2 which places the capacitor across the terminals of the battery. Under the given conditions the capacitor would reach full charge instantaneously, however, the charging action will be spread out over a period of time in the following discussion so that a step-by-step analysis can be made.

At the instant the switch is thrown to position 2 (fig. 11-5 (B)) a displacement of electrons will occur simultaneously in all parts of the circuit. This electron displacement is directed away from the negative terminal and toward the positive terminal of the source. An ammeter connected in series with the source will indicate a brief surge of current as the capacitor charges.

If it were possible to analyze the motion of individual electrons in this surge of charging current, the following action would be observed. (See fig. 11-6.)

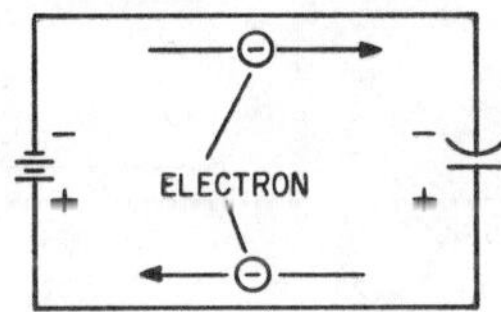

Figure 11-6.—Electron motion during charge.

At the instant the switch is closed, the positive terminal of the battery extracts an electron from the bottom conductor and the negative terminal of the battery forces an electron into the top conductor. At this same instant an electron is forced into the top plate of the capacitor and another is pulled from the bottom plate. Thus, in every part of the circuit a clockwise DISPLACEMENT of electrons occurs in the manner of a chain reaction.

As electrons accumulate on the top plate of the capacitor and others depart from the bottom plate, a difference of potential develops across the capacitor. Each electron forced onto the top plate makes that plate more negative, while each electron removed from the bottom causes the bottom plate to become more positive. Notice that the polarity of the voltage which builds up across the capacitor is such as to oppose the source voltage. The source forces current around the circuit of figure 11-6 in a clockwise direction. The emf developed across the capacitor, however, has a tendency to force the current in a counterclockwise direction, opposing the source. As the capacitor continues to charge, the voltage across the capacitor rises until it is equal in amount to the source voltage. Once the capacitor voltage equals the source voltage, the two voltages balance one another and current ceases to flow in the circuit.

In studying the charging process of a capacitor it must be emphasized that NO current flows THROUGH the capacitor. The material between the plates of the capacitor must be an insulator.

To an observer stationed at the source or along one of the circuit conductors, the action has all the appearances of a true flow of current even though the insulating material between the plates of the capacitor prevents having a complete path. The current which appears to flow in a capacitive circuit is called DISPLACEMENT CURRENT.

To provide a better understanding of charging action, a capacitor can be compared to the mechanical system in figure 11-7. Part A of the diagram shows a metal cylinder containing a flexible rubber membrane which blocks off the cylinder. The cylinder is then filled with round balls as shown. If an additional ball is now pushed into the left hand side of the tube, the membrane will stretch and a ball will be forced out of the right hand end of the tube. To an observer who could not see inside the tube the ball would have the appearance of traveling all the way through the tube. For each ball inserted into

NOTES:

the left hand side, one ball would leave the right hand side, although no balls actually pass all the way through the tube.

As more balls are forced into the tube it becomes increasingly difficult to force in additional balls, due to the tendency of the membrane to spring back to its original position.

If too many balls are forced into the tube, the membrane will rupture, and any number of balls can then be forced all the way through the tube.

A similar effect occurs in a capacitor when the voltage applied to the capacitor is too high. If an excessive amount of voltage is applied to a capacitor, the insulating material between the plates will break down and allow a current flow through the capacitor. In most cases this destroys the capacitor, necessitating its replacement.

When a capacitor is fully charged and the source voltage is equaled by the counter electromotive force (cem) across the capacitor, the electrostatic field between the plates of the capacitor will be maximum. Since the electrostatic field is maximum the energy stored in the dielectric will be maximum.

If the switch is now opened as shown in figure 11-8 (A), the electrons on the upper plate are isolated. Due to the intense repelling effect of these electrons, no electrons will return to the positive plate. Thus, with the switch in position 3, the capacitor will remain charged indefinitely. At this point it should be noted that the insulating dielectric material in a practical capacitor is not perfect and small leakage current will flow through the dielectric. This current will eventually dissipate the charge. A high quality capacitor may hold its charge for a month or more however.

To review briefly, when the capacitor is connected across a source, a surge of charging current will flow. This charging current develops a cemf across the capacitor which opposes the applied voltage. When the capacitor is fully charged the cemf will be equal to the applied voltage and charging current will cease. At full charge the electrostatic field between the plates is at maximum intensity and the energy stored in the dielectric is maximum. If the charged capacitor is disconnected from the source the charge will be retained for some period of time. The length of time the charge is retained depends on the amount of leakage current present. Since electrical energy is stored in the capacitor, a charged capacitor can act as a source.

DISCHARGING

To DISCHARGE a capacitor, the charges on the two plates must be neutralized. This is accomplished by providing a conducting path between the two plates (fig. 11-8 (B)). With the switch in position 4 the excess electrons on the negative plate can flow to the positive plate and neutralize its charge. When the capacitor is discharged the distorted orbits of the electrons in the dielectric return to their normal positions

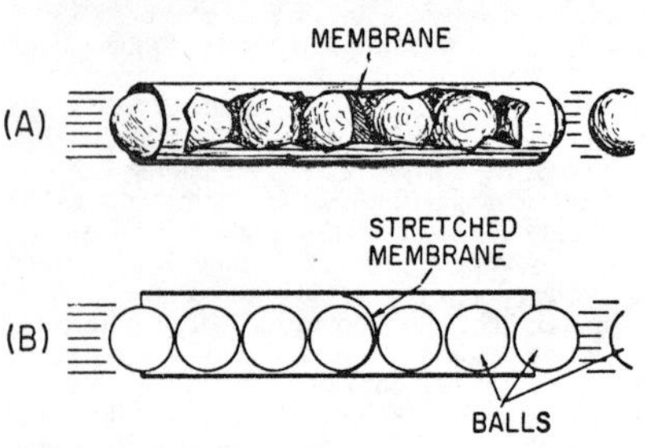

Figure 11-7.—Mechanical equivalent of a capacitor.

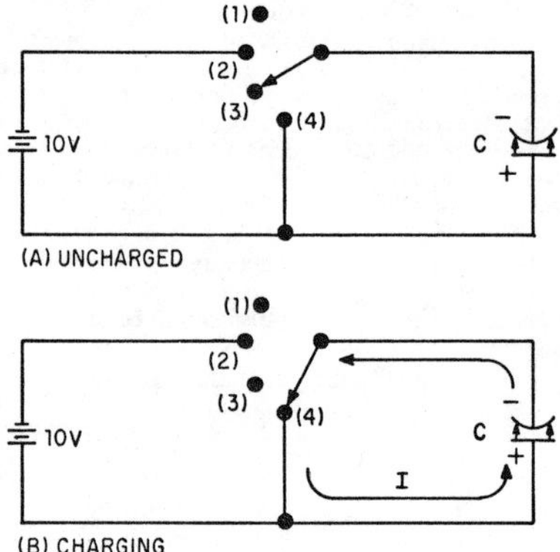

Figure 11-8.—Discharging a capacitor.

NOTES:

and the stored energy is returned to the circuit. It is important to note that a capacitor does not consume power. The energy the capacitor draws from the source is recovered when the capacitor is discharged.

CHARGE AND DISCHARGE OF AN RC SERIES CIRCUIT

Ohm's law states that the voltage across a resistance is equal to the current through it times the value of the resistance. This means that a voltage will be developed across a resistance ONLY WHEN CURRENT FLOWS through it.

A capacitor is capable of storing or holding a charge of electrons. When uncharged, both plates contain the same number of free electrons. When charged, one plate contains more free electrons than the other. The difference in the number of electrons is a measure of the charge on the capacitor. The accumulation of this charge builds up a voltage across the terminals of the capacitor, and the charge continues to increase until this voltage equals the applied voltage. The charge in a capacitor is related to the capacitance and voltage as follows:

$$Q = CE, \qquad (11\text{-}4)$$

in which Q is the charge in coulombs, C the capacitance in farads, and E the difference in potential in volts. Thus, the greater the voltage, the greater the charge on the capacitor. Unless a discharge path is provided, a capacitor keeps its charge indefinitely. Any practical capacitor, however, has some leakage through the dielectric so that the charge will gradually leak off.

A voltage divider containing resistance and capacitance may be connected in a circuit by means of a switch, as shown in figure 11-9 (A). Such a series arrangement is called an RC series circuit.

If S1 is closed, electrons flow counterclockwise around the circuit containing the battery, capacitor, and resistor. This flow of electrons ceases when C is charged to the battery voltage. At the instant current begins to flow, there is no voltage on the capacitor and the drop across R is equal to the battery voltage. The initial charging current, I, is therefore equal to $\frac{E_s}{R}$. Figure 11-9 (B) shows that at the instant the switch is closed, the entire input voltage, E_s, appears across R, and that the voltage across C is zero.

The current flowing in the circuit soon charges the capacitor. Because the voltage on

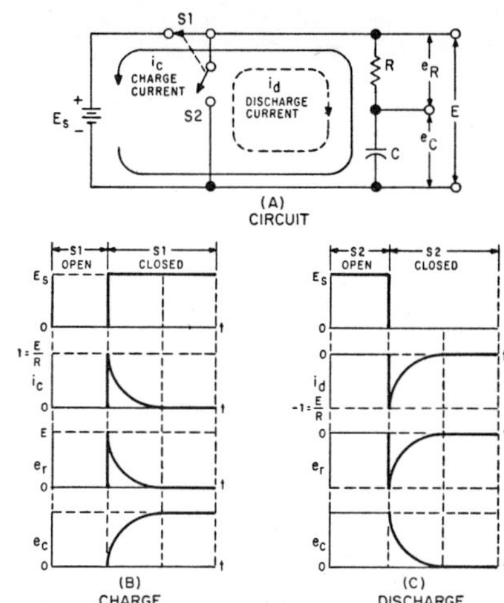

Figure 11-9.—Charge and discharge of an RC series circuit.

the capacitor is proportional to its charge, a voltage, e_c, will appear across the capacitor. This voltage opposes the battery voltage—that is, these two voltages buck each other. As a result, the voltage e_r across the resistor is E_s - e_c, and this is equal to the voltage drop (i_cR) across the resistor. Because E_s is fixed, i_c decreases as e_c increases.

The charging process continues until the capacitor is fully charged and the voltage across it is equal to the battery voltage. At this instant, the voltage across R is zero and no current flows through it. Figure 11-9 (B) shows the division of the battery voltage, E_s, between the resistance and capacitance at all times during the charging process.

If S2 is closed (S1 opened) in figure 11-9 (A), a discharge current, i_d, will discharge the capacitor. Because i_d is opposite in direction to i_c, the voltage across the resistor will have a polarity opposite to the polarity during the charging time. However, this voltage will have the same magnitude and will vary in the same manner. During discharge the voltage across the capacitor is equal and opposite to the drop across the resistor, as shown in figure 11-9 (C).

NOTES:

The voltage drops rapidly from its initial value and then approaches zero slowly, as indicated in the figure.

RC TIME CONSTANT

The time required to charge a capacitor to 63 percent (actually 63.2 percent) of maximum voltage or to discharge it to 37 percent (actually 36.8 percent) of its final voltage is known as the TIME CONSTANT of the circuit. An RC circuit with its charge and discharge graphs is shown in figure 11-10. The value of the time constant in seconds is equal to the product of the circuit resistance in ohms and its capacitance in farads, one set of values of which is given in figure 11-10 (A). RC is the symbol used for this time constant.

Some useful relations used in calculating RC time constants are as follows:

R (ohms) x C (farads) = t (seconds)
R (megohms) x C (microfarads) = t (seconds)
R (ohms) x C (microfarads) = t (microseconds)
R (megohms) x C (picofarads) = t (microseconds)

The time constant may also be defined as the time required to charge or discharge a capacitor completely IF it continues to charge or discharge at its initial rate. As may be seen in figure 11-10 (B), the slope of the dotted tangent line, OX, indicates the initial rate of charge. At this rate, the capacitor would be completely charged in RC seconds. Likewise, the slope of the dotted tangent line, YZ, indicates the initial rate of discharge with respect to time, and at this rate the capacitor would be completely discharged in RC seconds.

The equation for the rise in voltage, e_c, across the capacitor is

$$e_c = E\left(1 - \frac{1}{2.718^{\frac{t}{RC}}}\right)$$

where e_c, is the instantaneous voltage across the capacitor, E the applied voltage (100 volts in this case), t the time in seconds, R the resistance in ohms, C the capacitance in farads (0.000001 farad or 1μf in this case), and the number 2.718 the natural logarithm base. Figure 11-10 (B) shows a graph of this equation.

When t = RC, the exponent, $\frac{t}{RC}$, reduces to 1. Therefore,

$$e_c = E\left(1 - \frac{1}{2.718}\right) = 0.632E$$

In other words, when t = RC, e_c is equal to 63.2 percent of E, the maximum value. When the maximum is 100 volts, as shown, the voltage across the capacitor increases to 63.2 volts in RC seconds—that is, in 10 microseconds.

The equation for the charging current, i_c, is

$$i_c = \frac{E}{R}\left(\frac{1}{2.718^{\frac{t}{RC}}}\right)$$

The graph of this equation is shown in figure 11-10 (B). When t = RC, $i_c = 0.368 \times \frac{E}{R}$; that is, when t = 10 microseconds, $i_c = 0.368 \times \frac{100}{10} = 3.68$ ampere.

UNIVERSAL TIME CONSTANT CHART

Because the impressed voltage and the values of R and C or R and L usually will be known, a universal time constant chart (fig. 11-11) can be used. Curve A is a graph of the voltage across the resistor in series with the inductor on the growth of current. Curve B is a graph of capacitor voltage on discharge, capacitor current on charge, inductor current on decay, or the voltage across the resistor in series with the capacitor on charge. The graphs of resistor voltage and current and inductor voltage on discharge are not shown because negative values would be involved.

The time scale (horizontal scale) is graduated in terms of the RC or $\frac{L}{R}$ time constants so that the curves may be used for any value of R and C or L and R. The voltage and current scales (vertical scales) are graduated in terms of the fraction of the maximum voltage or current so that the curves may be used for any value of voltage or current. If the time constant and the initial or final voltage for the circuit in question are known, the voltages across the various parts of the circuit can be obtained from the curves for any time after the switch is closed, either on charge or discharge. The same reasoning is true of the current in the circuit.

NOTES:

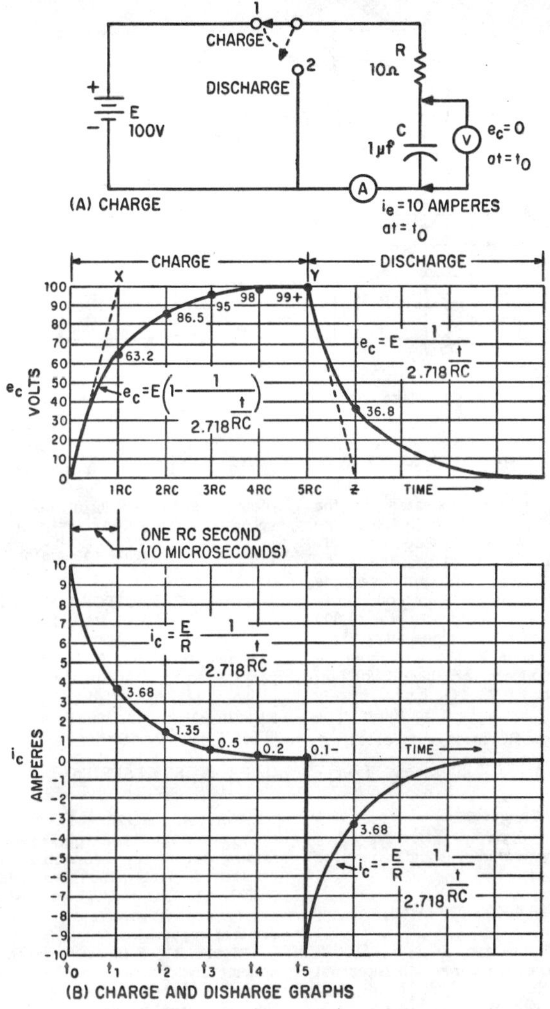

Figure 11-10.—RC time constant.

NOTES:

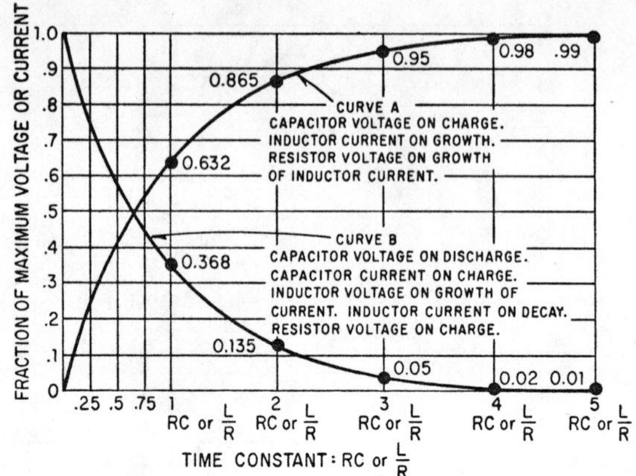

Figure 11-11.—Universal time constant chart for RC and RL circuits.

The following problem illustrates how the universal time constant chart may be used.

A circuit is to be designed in which a capacitor must charge to one-fifth (0.2) of the maximum charging voltage in 100 microseconds (0.0001 second). Because of other considerations, the resistor must have a value of 20,000 ohms. What size of capacitor is needed?

Curve A is first consulted to determine the RC time necessary to give 0.2 of the full voltage. The time is less than 0.25 RC, approximately 0.22 RC. If 0.22 RC must be equal to 100 microseconds, one complete RC must be equal to $\frac{100}{0.22}=$ 455 microseconds, or 0.000455 second. Therefore,

$$RC = 0.000455.$$

Substituting the known value of R and solving for C,

$$C=\frac{0.000455}{20,000}=0.000000023 \text{ farad},$$

or 0.023 microfarad.

The graphs shown in figure 11-10 are not entirely complete—that is, the charge or discharge (or the growth or decay) is not quite complete in 5 RC or 5 $\frac{L}{R}$ seconds. However, when the values reach 0.99 of the maximum (corresponding to 5 RC or 5 $\frac{L}{R}$) the graphs may be considered accurate enough for most purposes.

CAPACITORS IN SERIES AND PARALLEL

Capacitors may be connected in series or parallel to give resultant values, which may be either the sum of the individual values (in parallel) or a value less than that of the smallest capacitance (in series).

CAPACITORS IN SERIES

A circuit consisting of a number of capacitors in series is similar in some respects to one containing several resistors in series. In a series capacitive circuit the same DISPLACEMENT CURRENT flows through each part of the circuit and the applied voltage will divide across the individual capacitors.

Figure 11-12 shows a circuit containing a source and two series capacitors. When the switch is closed, current will flow in the direction indicated by the arrows on the diagram. Since there is only one path for current, the

NOTES:

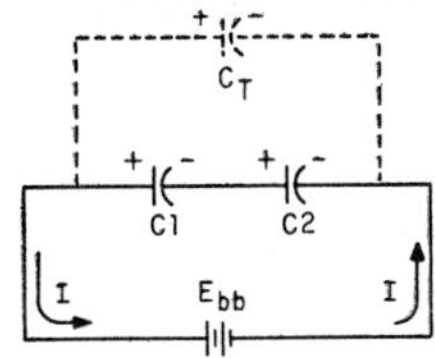

Figure 11-12.—Series capacitive circuit.

amount of charge current in motion is the same in all parts of the circuit. This current is of brief duration and will flow only until the total voltage across the capacitors is equal to the source voltage.

Since the charge (Q) is the same in all parts of the circuit:

$$Q_t = Q_1 = Q_2 = \ldots\ldots Q_n \quad (11\text{-}5)$$

also:

$$C = \frac{Q}{E} \quad (11\text{-}4)$$

transposing:

$$E = \frac{Q}{C} \quad (11\text{-}6)$$

Since the sum of the capacitor voltages must equal the source voltage (Kirchhoff's law):

$$E_t = E_1 + E_2 + \ldots E_n \quad (11\text{-}7)$$

Substituting equation (11-6) into (11-7)

$$\frac{Q_t}{C_t} = \frac{Q_1}{C_1} + \frac{Q_2}{C_2} + \cdots\cdots \frac{Q_n}{C_n} \quad (11\text{-}8)$$

Since by equation (11-5) all the charges are the same, dividing each term of (11-8) by Q_t yields:

$$\frac{1}{C_t} = \frac{1}{C_1} + \frac{1}{C_2} + \cdots\cdots \frac{1}{C_n} \quad (11\text{-}9)$$

Taking the reciprocal of both sides:

$$C_t = \frac{1}{\frac{1}{C_1} + \frac{1}{C_2} + \cdots\cdots \frac{1}{C_n}} \quad (11\text{-}10)$$

Where C_t, C_1, etc., are in farads.

Equation (11-10) is the general equation used to compute the total capacitance of capacitors connected in series. Notice the similarity between this equation and the one used to find equivalent resistance of parallel resistors. If the circuit contains only two capacitors the product over the sum formula can be used.

$$C_t = \frac{C_1 C_2}{C_1 + C_2} \quad (11\text{-}11)$$

Where: C_t, C_1, etc., are in farads.

As might be anticipated from the above equations, the total capacitance of series connected capacitors will always be smaller than the smallest of the individual capacitors.

Example. Determine the total capacitance of a series circuit containing three capacitors of 0.01 μf, 0.25 μf, and 50,000 pf respectively.

Given:

$C_1 = 0.01\ \mu f$
$C_2 = 0.25\ \mu f$
$C_3 = 50{,}000\ pf$

Solution:

$$C_t = \frac{1}{\frac{1}{C_1} + \frac{1}{C_2} + \frac{1}{C_3}} \quad (11\text{-}10)$$

$$C_t = \frac{1}{\frac{1}{.01\mu f} + \frac{1}{.25\mu f} + \frac{1}{50{,}000 pf}}$$

Converting to powers of ten.

$$C_t = \frac{1}{\frac{1}{1 \times 10^{-8}} + \frac{1}{25 \times 10^{-8}} + \frac{1}{5 \times 10^{-8}}}$$

$$C_t = \frac{1}{100 \times 10^6 + 4 \times 10^6 + 20 \times 10^6}$$

$$C_t = \frac{1}{124 \times 10^6}$$

$$C_t = 0.008 \mu f$$

The total capacitance of 0.008 μf is slight smaller than the smallest capacitor (0.01μ

CAPACITORS IN PARALLEL

When capacitors are connected in paral one plate of each capacitor is connected direc to one terminal of the source, while the otl plate of each capacitor is connected to the otl

NOTES:

terminal of the source. In figure 11-13, since all the negative plates of the capacitors are connected together, and all the positive plates are connected together, C_t appears as a capacitor with a plate area equal to the sum of all the individual plate areas. As previously mentioned, capacitance is a direct function of plate area. Connecting capacitors in parallel effectively increases plate area and thereby the capacitance.

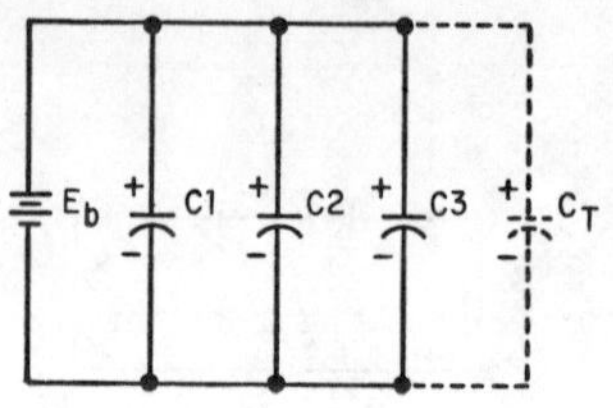

Figure 11-13.—Parallel capacitive circuit.

For capacitors connected in parallel the total charge is the sum of all the individual charges.

$$Q_t = Q_1 + Q_2 + Q_3 + \ldots . Q_n \qquad (11\text{-}12)$$

Transposing formula (11-4):

$$Q = CE \qquad (11\text{-}13)$$

Substitute: (11-13 into (11-12)

$$C_tE = C_1E + C_2E + C_3E \qquad (11\text{-}14)$$

Divide both sides by E:

$$C_t = C_1 + C_2 + C_3 + \ldots\ldots . C_n \qquad (11\text{-}15)$$

Where: all capacitances are in the same units.

Example: Determine the total capacitance in a parallel capacitive cirucit:

Given: $C_1 = 0.03$

$C_2 = 2\mu f$

$C_3 = 0.25\mu f$

Solution: $C_t = C_1 + C_2 + C_3$ (11-14)

$C_t = 0.03 + 2 + 0.25$

$C_t = 2.28\mu f$

SERIES PARALLEL CONFIGURATION

If capacitors are connected in a combination of series and parallel the total capacitance is found by applying equations (11-10) and 11-14) to the individual branches.

Example. Determine the total capacitance of the circuit in figure 11-14.

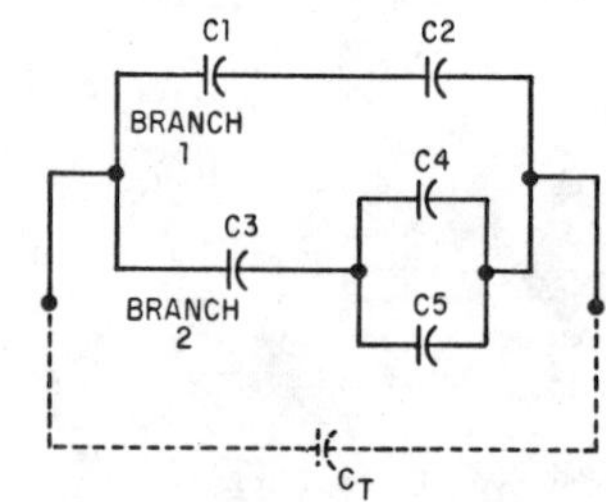

Figure 11-14.—Series-parallel capacitance configuration.

Given: $C_1 = 0.06\mu f$

$C_2 = 0.002\mu f$

$C_3 = 3\mu f$

$C_4 = 0.005\mu f$

$C_5 = 0.001\mu f$

Find: $C_t = ?$

Solution: Simplify the circuit into separate branches.

NOTES:

BRANCH 1 - consists of a series combination C_1 and C_2. To determine total capacitance of this branch only:

$$\text{Branch 1:}\quad C_{t1} = \frac{C_1 C_2}{C_1 + C_2} \qquad (11\text{-}11)$$

$$C_{t1} = \frac{0.6 \times 0.002}{0.06 + 0.002}$$

$$C_{t1} = \frac{0.00012}{0.062}$$

$$C_{t1} = 0.00193\mu f$$

BRANCH 2 - consists of a series-parallel combination. Solve for total capacitance of branch 2 only. The equivalent capacitance (C_{eq}) of the parallel combination of C_4 and C_5 must be determined first by the use of equation (11-14) for parallel configurations:

$$C_{eq} = C_4 + C_5$$

$$C_{eq} = 0.005 + 0.001$$

$$C_{eq} = 0.006\mu f$$

Branch 2 is now reduced to an equivalent series circuit consisting of C_3 and the equivalent capacitance of C_4 and C_5.

The circuit figure 11-14 has been simplified considerably and appears as the circuit in figure 11-15. To illustrate the use of the reciprocal method the total capacitance of branch 2 is determined using equation (11-10):

$$C_{t2} = \frac{1}{\frac{1}{C_3} + \frac{1}{C_{eq}}} \qquad (11\text{-}10)$$

$$C_{t2} = \frac{1}{\frac{1}{3} + \frac{1}{0.006}}$$

$$C_{t2} = \frac{1}{0.333 + 166.666}$$

$$C_{t2} = 0.00598\mu f$$

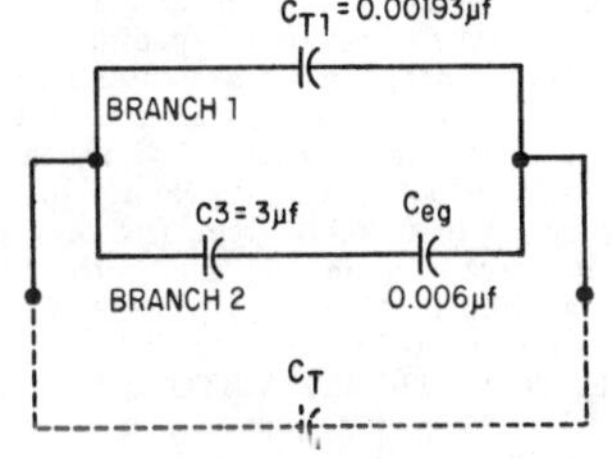

Figure 11-15.—Simplified series-parallel capacitance configuration.

NOTE: Equation (11-11) could also have been used to solve the previous problems.

The entire problem is now reduced to a simple parallel combination consisting of C_{t1} (capacity branch 1) and C_{t2} (capacity branch 2). Equation (11-15) is now used to determine the total capacitance of the original circuit.

$$C_t = C_{t1} + C_{t2}$$

$$C_t = 0.00193 + 0.00598$$

$$C_t = 0.00791\mu f$$

CAPACITOR LOSSES

Losses occurring in capacitors may be attributed to either dielectric leakage or dielectric hysteresis. Dielectric hysteresis may be defined as an effect in a dielectric material similar to the hysteresis found in magnetic materials. It is a result of the changes in orientation of electron orbits in the dielectric because of the rapid reversals of the polarity of the line voltage. The amount of loss due to hysteresis depends upon the type of dielectric used; vacuum shows the smallest loss.

Dielectric leakage occurs in a capacitor as the result of leakage of current through the dielectric. Normally, it is assumed that the dielectric will effectively prevent the flow of current through the capacitor. However, while the resistance of the dielectric is extremely high, a minute amount of current does flow. Ordinarily, this current is so small that for all practical purposes it is considered to be of no

NOTES:

consequence. However, if the leakage through the dielectric is abnormally high, there will be a rapid loss of charge and an overheating of the capacitor.

The power factor of a capacitor is determined by dielectric losses. If the losses are negligible and the capacitor returns the total charge to the circuit, it is considered to be a perfect capacitor with a power factor of zero. Therefore, the power factor of a capacitor is a measurement of its efficiency.

CAPACITOR TYPES

Capacitors may be divided into two major groups—fixed and variable.

FIXED CAPACITORS

Fixed capacitors are constructed in such manner that they possess a fixed value of capacitance which cannot be adjusted. They may be classified according to the type of material used as the dielectric, such as paper, oil, mica, and electrolyte.

A PAPER CAPACITOR is one that uses paper as its dielectric. It consists of flat thin strips of metal foil conductors, separated by the dielectric material. In this capacitor the dielectric used is waxed paper. Paper capacitors usually range in value from about 300 picofarads to about 4 microfarads. Normally, the voltage limit across the plates rarely exceeds 600 volts. Paper capacitors are sealed with wax to prevent the harmful effects of moisture and to prevent corrosion and leakage.

Many different kinds of outer covering are used for paper capacitors, the simplest being a tubular cardboard. Some types of paper capacitors are encased in a mold of very hard plastic; these types are very rugged and may be used over a much wider temperature range than the cardboard-case type. Figure 11-16 (A) shows the construction of a tubular paper capacitor; part (B) shows a completed cardboard encased capacitor.

A MICA CAPACITOR is made of metal foil plates that are separated by sheets of mica, which form the dielectric. The whole assembly is covered in molded plastic. Figure 11-17 (A) shows a cutaway view of a mica capacitor. By molding the capacitor parts into a plastic case, corrosion and damage to the plates and dielectric are prevented in addition to making the capacitor mechanically strong. Various

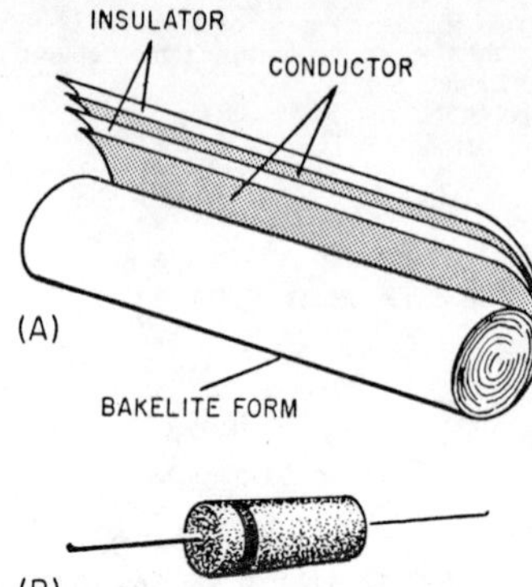

Figure 11-16.—Paper capacitor.

types of terminals are used to connect mica capacitors into circuits; these are also molded into the plastic case.

Mica is an excellent dielectric and will withstand higher voltages than paper without allowing arcing between the plates. Common values of mica capacitors range from approximately 50 picofarads to about 0.02 microfarad. Some typical mica capacitors are shown in figure 11-17 (B).

A CERAMIC CAPACITOR is so named because of the use of ceramic dielectrics. One type of ceramic capacitor uses a hollow ceramic cylinder as both the form on which to construct the capacitor and as the dielectric material. The plates consist of thin films of metal deposited on the ceramic cylinder.

A second type of ceramic capacitor is manufactured in the shape of a disk. After leads are attached to each side of the capacitor, the capacitor is completely covered with an insulating moisture-proof coating. Ceramic capacitors usually range in value between 1 picofarad and 0.01 microfarad and may be used with voltages as high as 30,000 volts. Typical capacitors are shown in figure 11-18.

ELECTROLYTIC CAPACITORS are used where a large amount of capacitance is required. As the name implies, electrolytic capacitors contain an electrolyte. This electrolyte can be in the form of either a liquid (wet electrolytic capacitor) or a paste (dry electrolytic capacitor).

NOTES:

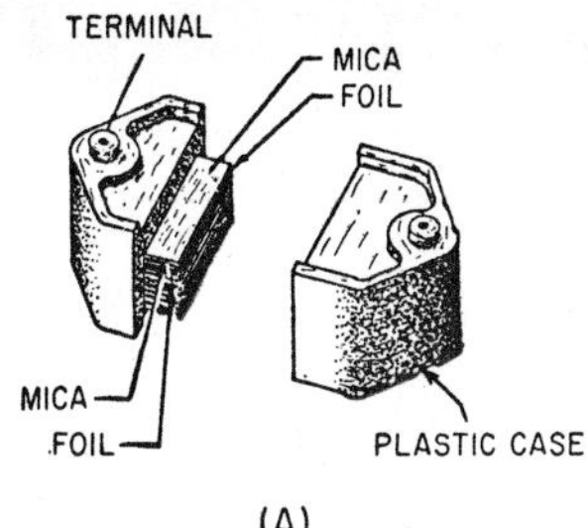

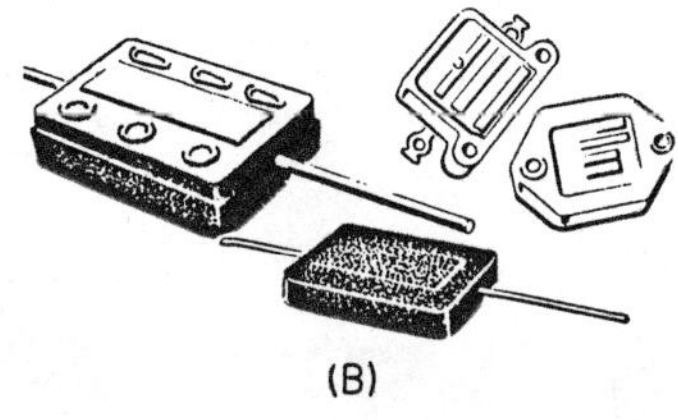

BE.214
Figure 11-17.—Typical mica capacitors.

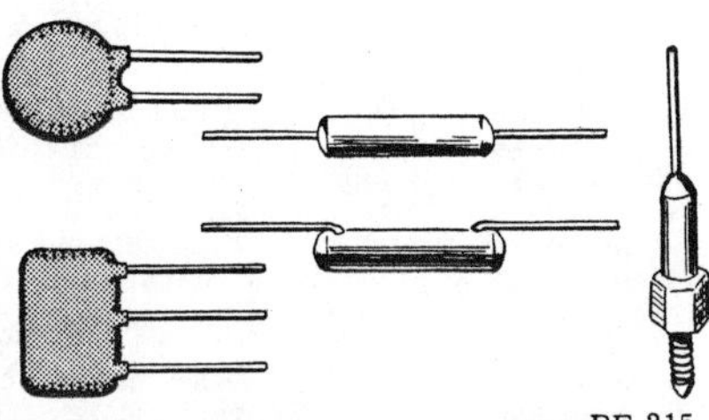

BE.215
Figure 11-18.—Ceramic capacitors.

Wet electrolytic capacitors are no longer in popular use due to the care needed to prevent spilling of the electrolyte.

Dry electrolytic capacitors consist essentially of two metal plates between which is placed the electrolyte. In most cases the capacitor is housed in a cylindrical aluminum container which acts as the negative terminal of the capacitor (fig. 11-19). The positive terminal (or terminals if the capacitor is of the multisection type) is in the form of a lug on the bottom end of the container. The size and voltage rating of the capacitor is generally printed on the side of the aluminum case.

An example of a multisection type of electrolytic capacitor is depicted in figure 11-19. The cylindrical aluminum container will normally enclose four electrolytic capacitors into one can. Each section of the capacitor is electrically independent of the other sections and one section may be defective while the other sections are still good. The can is the common negative connection with separate terminals for the positive connections identified by an embossed mark as shown in figure 11-19. The common identifying marks on electrolytic capacitors are the half moon, triangle, square, and no identifying mark. By looking at the bottom of the container and the identifying sheet pasted to the side of the container, the technician can easily identify each section.

Internally, the electrolytic capacitor is constructed similarly to the paper capacitor. The positive plate consists of aluminum foil covered with an extremely thin film of oxide which is formed by an electrochemical process. This thin oxide film acts as the dielectric of the capacitor. Next to, and in contact with the oxide, is placed

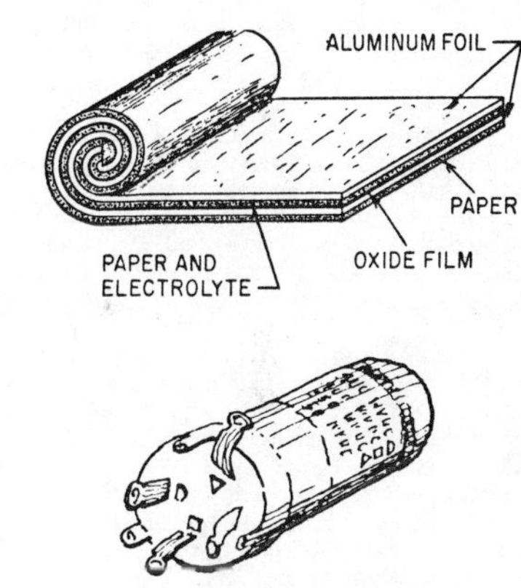

BE.216
Figure 11-19.—Construction of an electrolytic capacitor.

NOTES:

a strip of paper or gauze which has been impregnated with a paste-like electrolyte. The electrolyte acts as the negative plate of the capacitor. A second strip of aluminum foil is then placed against the electrolyte to provide electrical contact to the negative electrode (electrolyte). When the three layers are in place they are rolled up into a cylinder as shown in figure 11-19.

Electrolytic capacitors have two primary disadvantages in that they are POLARIZED, and they have a LOW LEAKAGE RESISTANCE. Should the positive plate be accidentally connected to the negative terminal of the source, the thin oxide film dielectric will dissolve and the capacitor will become a conductor (i.e., it will short). The polarity of the terminals is normally marked on the case of the capacitor. Since electrolytic capacitors are polarity sensitive, their use is ordinarily restricted to d-c circuits or circuits where a small a-c voltage is superimposed on a d-c voltage. Special electrolytic capacitors are available for certain a-c applications, such as motor starting capacitors. Dry electrolytic capacitors vary in size from about 4 microfarads to several thousand microfarads, and have a voltage limit of approximately 500 volts.

The type of dielectric used and its thickness govern the amount of voltage that can safely be applied to a capacitor. If the voltage applied to a capacitor is high enough to cause the atoms of the dielectric material to become ionized, an arc over will take place between the plates. If the capacitor is not self-healing, its effectiveness will be impaired. The maximum safe voltage of a capacitor is called its WORKING VOLTAGE and is indicated on the body of the capacitor. The working voltage of a capacitor is determined by the type and thickness of the dielectric. If the thickness of the dielectric is increased, the distance between the plates is also increased and the working voltage will be increased. Any change in the distance between the plates will cause a change in the capacitance of a capacitor. Because of the possibility of voltage surges (brief high amplitude pulses) a margin of safety should be allowed between the circuit voltage and the working voltage of a capacitor. The working voltage should always be higher than the maximum circuit voltage.

OIL CAPACITORS are often used in radio transmitters where high output power is desired. Oil-filled capacitors are nothing more than paper capacitors that are immersed in oil. The oil impregnated paper has a high dielectric constant which lends itself well to the production of capacitors that have a high value. Many capacitors will use oil with another dielectric material to prevent arcing between the plates. If an arc should occur between the plates of an oil-filled capacitor, the oil will tend to reseal the hole caused by the arc. These types of capacitors are often called SELF-HEALING capacitors.

VARIABLE CAPACITORS

Variable capacitors are constructed in such manner that their value of capacitance can be varied. A typical variable capacitor (adjustable capacitor) is the rotor/stator type. It consists of two sets of metal plates that are arranged so that the rotor plates move between the stator plates. Air is the dielectric. As the position of the rotor is changed, the capacitance value is likewise changed. This is the type capacitor used for tuning most radio receivers and it is shown in figure 11-20. Another type variable (trimmer) capacitor is shown in figure 11-21; it consists of two plates separated by a sheet of mica. A screw adjustment is used to change the distance between the plates, thereby changing the capacitance.

COLOR CODES FOR CAPACITORS

Although the value of a capacitor may be indicated by printing on the body of a capacitor, many capacitive values are indicated by the use of a color code. The colors used to represent the numerical value of a capacitor are the same as those used to identify resistance values.

The color coding system that is currently in popular use is the Radio Manufacturers' Association (RMA) system.

In this system a series of colored dots (sometimes bands) is used to denote the value of the capacitor. Mica capacitors are marked with either three dots or six dots. Both systems are similar, but the six dot system contains more information about the capacitor, such as working voltage, temperature coefficient, etc. Capacitors are manufactured in various sizes and shapes. Some are small tubular resistor like devices, and others are molded rectangular flat components.

An explanation of capacitor color codes is provided in appendix IV.

NOTES:

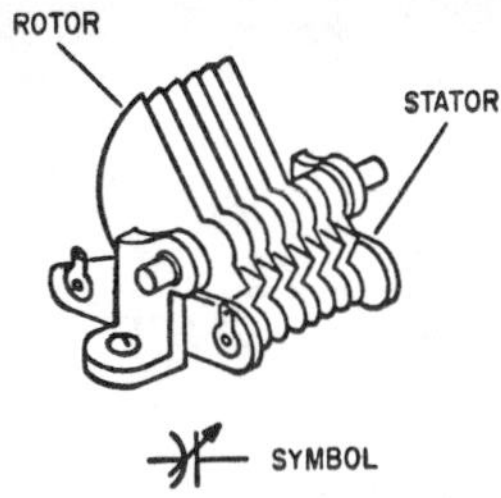

Figure 11-20.—Rotor-stator type variable capacitor.

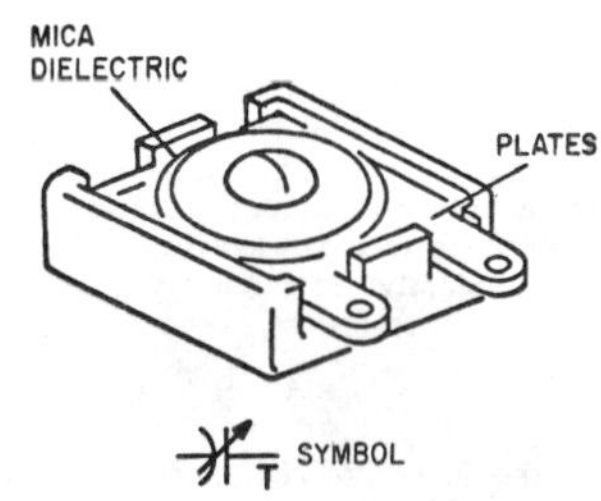

Figure 11-21.—Trimmer capacitor.

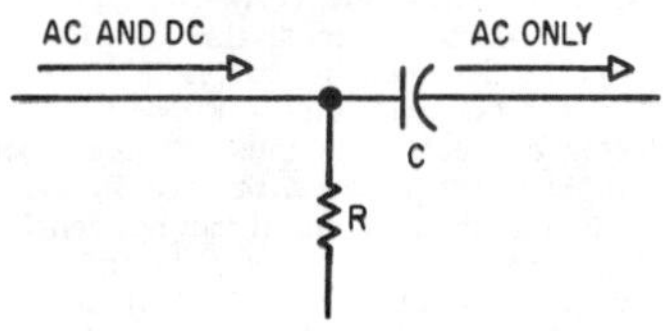

Figure 11-22.—Blocking capacitor.

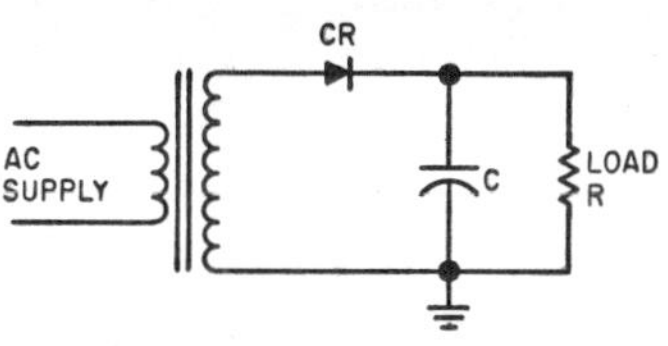

Figure 11-23.—Filter capacitor in a power supply circuit.

CAPACITOR APPLICATION

Capacitors are used in many ways in electrical and electronic equipment and circuits. A few of the more common applications are blocking direct current, filtering, and spark suppression.

The blocking capacitor in figure 11-22 is used in a circuit where both direct and alternating currents flow at the same time and it is necessary or desirable to pass the a.c. and block the d.c. This can be done by using a blocking capacitor as in figure 11-22. The capacitor C in the diagram allows alternating current to flow and blocks the flow of direct current.

The filter capacitor in figure 11-23 is used to maintain a steady d-c voltage by filtering out or removing undesired a-c or ripple voltages by capacitor action opposing any change in voltage. Filter capacitors are commonly used to filter power supply voltages.

Spark suppression is obtained by placing a capacitor across the contacts of relays and other movable points subject to electrical sparking when opening and closing. The capacitor minimizes the effects of the sparking and extends the life of the relay contacts or points. The buffer capacitor across the breaker points of an automobile ignition system is a fine example.

NOTES:

CHAPTER 12

INDUCTIVE AND CAPACITIVE REACTANCE

INDUCTIVE REACTANCE

Previously, the opposition that an inductance offers to a changing current was called self-induced voltage or CEMF and was measured in volts. However, opposition to current flow is normally measured in ohms, not in volts. Since a coil reacts to a current change by generating a CEMF, a coil is said to be reactive. The opposition of a coil is therefore called reactance (X) and is measured in ohms. Since more than one kind of reactance exists, the subscript L is added to denote INDUCTIVE REACTANCE. Thus, the opposition offered by a coil to alternating current is termed inductive reactance, designated X_L.

FACTORS AFFECTING INDUCTIVE REACTANCE

In the preceding chapter, it was shown that when an alternating voltage is applied to an inductor, the inductor reacts in such a way as to oppose alternating-current flow. This opposition is thus referred to in general as an inductive reactance. The amount of opposition exhibited by a coil depends on the magnitude of its self-induced voltage, E_{ind}. The effective value of this self-induced voltage, or counter emf in volts, is

$$E_{ind} = 2\pi fLI$$

where 2π is a constant, I is in effective amperes, L is in henrys, and f is in hertz. The induced voltage varies directly with the frequency, the inductance, and the current.

The amount of opposition to current flow offered by an inductor is referred to as its inductive reactance. This reactance is expressed as the ratio of the inductor's counter emf to the current through the inductor, or

$$X_L = \frac{E_{ind}}{I}$$

where X_L is the inductor's opposition to current flow measured in ohms, E_{ind} is the inductor's counter emf in volts, and I is the inductor current in amperes. Assuming the current and counter voltage have sine waveforms, the inductive reactance X_L, in ohms, is

$$X_L = 2\pi fL$$

Since the inductive reactance of a coil and its counter emf are closely related, anything which influences one must also influence the other. It has been shown that the counter emf of a coil depends on its inductance and the rate of flux change around the coil. Consequently, the inductive reactance must also be affected by the same influences. The rate of flux change per unit of time depends on the FREQUENCY of the inductor current. Flux must change more rapidly at high frequencies than at low frequencies. From the foregoing it can be seen that the inductive reactance of a coil depends primarily on (1) the coil's INDUCTANCE and (2) the FREQUENCY of the current flowing through the coil.

If frequency or inductance varies, inductive reactance must also vary. A coil's inductance does not vary appreciably after the coil is manufactured, unless it is designed as a variable inductor. Thus, frequency is generally the only variable factor affecting the inductive reactance of a coil. The coil's inductive reactance will vary directly with the applied frequency.

NOTES:

POWER IN AN INDUCTIVE CIRCUIT

The power in a d-c circuit is equal to the product of volts and amperes, but in an a-c circuit this is true only when the load is resistive, and has no reactance.

In a circuit possessing inductance only, the true power is zero (fig. 12-1). The current lags the applied voltage by 90°. The TRUE POWER is the average power actually consumed by the circuit, the average being taken over one complete cycle of alternating current. THE APPARENT POWER is the product of rms volts and rms amperes. Thus, in figure 12-1 (A), the apparent power is 100 x 10 = 1,000 volt-amperes. However, the power absorbed by the coil during the time the current is rising (fig. 12-1 (B)), is returned to the source during the time the current is falling, so that the average power is zero.

The product of instantaneous values of current and voltage yield the double frequency power curve P. (See fig. 12-1 (B).) The shaded areas under this curve above the X axis represent positive energy and the shaded areas below the X axis represent negative energy.

From 0° to 90° current is negative and falling; the magnetic field is collapsing and the energy in the field is being returned to the source (negative energy). The product of negative current and positive voltage is negative power.

From 90° to 180° the current is positive and rising; energy from the source is being stored in the rising magnetic field (positive energy). The product of positive current and positive voltage is equal to positive power.

From 180° to 270° current is positive and falling. Again energy is being returned to the source; the product of positive current and negative voltage is negative power.

From 270° to 360° current is negative and rising. Energy (positive) is being supplied by the source and stored in the magnetic field. The product of negative current and negative voltage is positive power.

Thus when current is rising, power is being supplied by the source and stored in the magnetic field; when current is falling, power is being returned to the source from the collapsing magnetic field. In the theoretically pure inductance shown in figure 12-1 the supplied power is equal to the returned power. Thus, average power used (true power) is zero.

The ratio of the true power to the apparent power in an a-c circuit is called the POWER FACTOR. It may be expressed as a percent or as a decimal. In the inductor of figure 12-1 (A), the power factor is $\frac{0}{1,000} = 0$. The power factor is also equal to the cos θ, where θ is the phase angle between the current and voltage. The phase angle between E and I in the inductor is

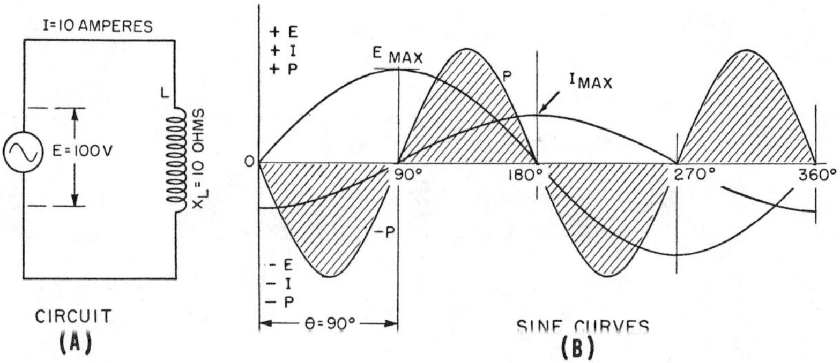

Figure 12-1.—Power in an inductive circuit.

NOTES:

$\theta = 90°$. Thus the power factor of an inductor of negligible losses is cos 90° = 0. The apparent power in a purely inductive circuit is called REACTIVE POWER and the unit of reactive power is called the VAR. This unit is derived from the first letters of the words volt-ampere-reactive.

RESISTANCE IN AN A-C CIRCUIT

In an a-c circuit containing only resistance, the current and voltage are always in phase ($\theta = 0°$). (See fig. 12-2.) The true power dissipated in heat in a resistor in an a-c circuit when sine waveforms of voltage and current are applied is equal to the product of the r.m.s. volts and the r.m.s. amperes. In the circuit of figure 12-2 (A) the power absorbed by the resistor is P = EI = 100 x 10 = 1,000 watts. The product of the instantaneous values of current and voltage (fig. 12-2 (B)) gives the power curve, P, the axis of which is displaced above the X axis by an amount that is proportional to 1,000 watts.

The true average power is

$$P = \frac{E_{max} \times I_{max}}{2} = E_{eff} \times I_{eff}$$

The power factor of a resistive circuit is cos 0° = 1, or 100 percent. The apparent power in the resistor is also equal to the true power. The reactive power in the resistive circuit is zero.

RESISTANCE AND INDUCTIVE REACTANCE IN SERIES

Because any practical inductor must be wound with wire that has resistance, it is not possible to obtain a coil without some resistance. The resistance associated with a coil may be considered as a separate resistor, R, in series with an inductor, L, which is purely inductive and contains only inductive reactance (fig. 12-3 (A)). The resistance has been exaggerated in this example to be of the same order of magnitude as the inductive reactance of the inductor in order to simplify the trigonometric solution.

If an alternating current, I_O, flows through the inductor a voltage drop occurs across both the resistor and the inductor. The voltage, E_r, across the resistor is in phase with the current, and the voltage drop, E_L, across the inductor leads the current by 90° (fig. 12-3 (B)). The voltage across the resistor is assumed to be 50 volts and the voltage across the inductor, 86.6

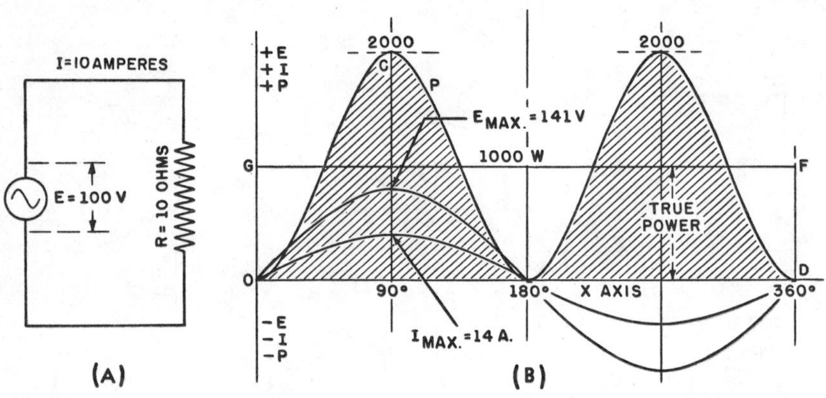

Figure 12-2.—Relation between E, I, and P in a resistive circuit.

NOTES:

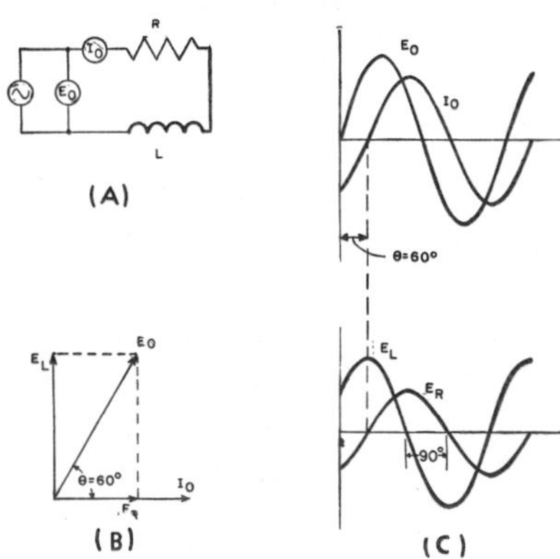

Figure 12-3.—Resistance and inductive reactance in series.

volts. These two voltages are 90° out of phase, as indicated in the figure. The applied voltage, E_O, is the hypotenuse of a right traingle, the sides of which are 50 and 86.6, respectively. Thus,

$$E_o = \sqrt{50^2 + 86.6^2} = 100 \text{ volts}$$

The sine curves of circuit current, I_O, applied voltage, E_O, the voltage across the coil, E_L, and the voltage across the resistor, E_r, are shown in figure 12-3 (C).

IMPEDANCE

Impedance is the total opposition to the flow of alternating current in a circuit that contains resistance and reactance. In the case of pure inductance, inductive reactance, X_L is the total opposition to the flow of current through it. In the case of pure resistance, R represents the total opposition. The combined opposition of R and X_L in series or in parallel to current flow is called IMPEDANCE. The symbol for impedance is Z.

NOTES:

The impedance of resistance in series with inductance is

$$Z = \sqrt{R^2 + X_L^2}$$

where Z, R, and X_L are the hypotenuse, base, and altitude respectively of a right triangle in which $\cos\theta = \frac{R}{Z}$, $\sin\theta = \frac{X_L}{Z}$, and $\tan\theta = \frac{X_L}{R}$. As mentioned before, cos θ is equal to the circuit power factor; sin θ is sometimes referred to as the reactive factor; and tan θ is referred to as the quality, or Q, of a circuit or a circuit component. The trigonometric functions including the sine, cosine, and tangent of angles between 0° and 90° are given in appendix IX at the end of this training manual.

POWER IN A SERIES CIRCUIT CONTAINING R and X_L

The true power in any circuit is the product of the applied voltage, the circuit current, and the cosine of the phase angle between them. Thus, in figure 12-4, the true power is

$$P = EI\cos\theta = 100 \times 7.07 \ (\cos 45° = 0.707)$$
$$= 500 \text{ watts}$$

The power curve is partly above the X axis (fig. 12-4 (B)) and partly below it. The axis of the power curve is displaced above the X axis an amount proportional to the true power, EI cos θ. The apparent power in this circuit is

$$100 \times 7.07 = 707 \text{ volt-amperes}$$

The power factor is

$$\cos\theta = \frac{\text{true power}}{\text{apparent power}} = \frac{500}{707}$$
$$= 0.707, \text{ or } 70.7 \text{ percent}$$

The reactive power in the LR circuit is the product of EI sin θ where sin θ is the reactive factor. Thus the reactive power is

$$100 \times 7.07 \ (\sin 45° = 0.707) = 500 \text{ VARS (lagging)}$$

SUMMARY OF E, Z, and P RELATIONS IN LR CIRCUIT

The relation between voltage, impedance, and power in series LR circuit with sine

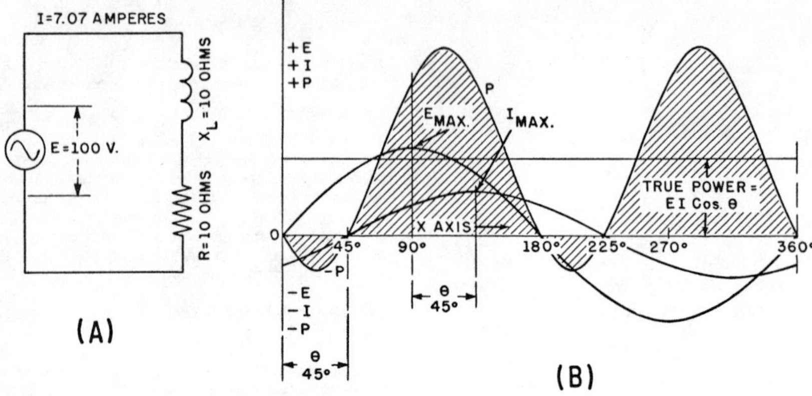

Figure 12-4.—Power in a circuit containing L and R in series.

waveforms applied is summarized in figure 12-5. In this example the circuit contains 12 ohms of resistance in series with 16 ohms of inductive reactance (fig. 12-5 (A)). The phase relations between the applied voltage; the voltage across the resistance, R; and the voltage across the inductance, L; are shown in figure 12-5 (B). The IR drop across R is 5 x 12 = 60 volts and forms the base of the right triangle. The altitude is the IX_L drop, or 5 x 16 = 80 volts, across L. The applied voltage represents the vector sum of the IR drop and the IX_L drop and is the hypotenuse of the right triangle. Its magnitude is 100 volts. All voltages are effective values.

The relation between resistance, inductive reactance, and impedance is shown by the vectors of figure 12-5 (C). The resistance of 12 ohms forms the base of the right triangle. The altitude of this triangle represents the inductive reactance of the coil and has a magnitude of 16 ohms. The combined impedance of the circuit is $Z = \sqrt{12^2 + 16^2} = 20$ ohms. In this triangle $\cos\theta = \frac{R}{Z}$. Thus $\cos\theta = \frac{12}{20} = 0.6$, from which $\theta = 53.1°$.

The relation between apparent power, true power, and reactive power is shown in figure 12-5 (D). The hypotenuse represents the apparent power and is equal to EI, or 100 x 5 = 500 volt-amperes. The base represents true power and is equal to EI cos θ, or 100 x 5 x 0.6 = 300

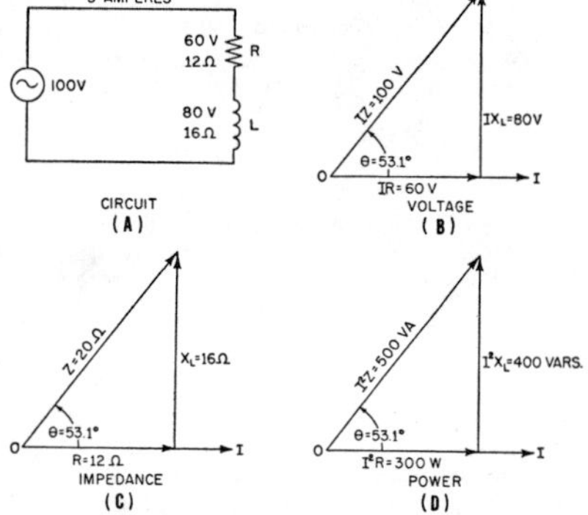

Figure 12-5.—Summary of relation between E, Z, and P in a series LR circuit.

NOTES:

watts, where 0.6 = cos 53.1°. The altitude represents reactive power and is equal to EI sin θ, or 100 x 5 x 0.8 = 400 VARS, where 0.8 = sin 53.1°.

In all three vector diagrams the right triangles are similar. The common factor of current, I, makes their corresponding sides protional. Thus the voltage triangle of figure 12-5 (B), is obtained by multiplying the corresponding sides of the impedance triangle of figure 12-5 (C), by I. The hypotenuse is equal to the applied voltage, or IZ = 5 x 20 = 100 volts. The base is equal to the voltage across R, or IR = 5 x 12 = 60 volts. The altitude is equal to the voltage across L, or IX_L = 5 x 16 = 80 volts.

Multiplying corresponding sides of the voltage triangle by I gives the power triangle of figure 10-5 (D). The hypotenuse is IZ times I, or $I^2Z = 5^2 \times 20$ = 500 volt-amperes of apparent power. The base is IR times I, or $I^2R = 5^2 \times 12$ = 300 watts of true power. The altitude is IX_L times I, or $I^2X_L = 5^2 \times 16$ = 400 VARS of reactive power.

In the three vector diagrams, the circuit power factor is equal to the following ratios: In the voltage diagram,

$$\cos\theta = \frac{IR}{IZ} = \frac{60}{100} = 0.6$$

In the impedance diagram,

$$\cos\theta = \frac{R}{Z} = \frac{12}{20} = 0.6$$

In the power diagram,

$$\cos\theta = \frac{I^2R}{I^2Z} = \frac{300}{500} = 0.6$$

In all three diagrams θ = 53.1° and the power factor is 60 percent.

CAPACITIVE REACTANCE

Capacitance was defined in chapter 11 as that quality of a circuit that enables energy to be stored in an electric field. A capacitor is a device that possesses the quality of capacitance. In simple form it has been shown to consist of two parallel metal plates separated by an insulator, called a DIELECTRIC. The electric field consists of parallel lines of electric force which terminate in a positive charge on one plate and a negative charge on the other. The positive and negative charges on the plates establish the electric field in the dielectric because the dielectric prevents these charges from neutralizing each other.

CURRENT FLOW IN A CAPACITIVE CIRCUIT

A capacitor that is initially uncharged tends to draw a large current when a d-c voltage is first applied. During the charging period, the capacitor voltage rises. After the capacitor has received sufficient charge the capacitor voltage equals the applied voltage and the current flow ceases. If a sine-wave a-c voltage is applied to a pure capacitance the current is a maximum when the voltage begins to rise from zero, and the current is zero when the voltage across the capacitor is a maximum (fig. 12-6 (A)). The current leads the applied voltage by 90°, as indicated in the vector diagram of figure 12-6 (B).

FACTORS THAT CONTROL CHARGING CURRENT

Because a capacitor of large capacitance can store more energy than one of small capacitance, a larger current must flow to charge a large capacitor than to charge a small one, assuming the same time interval in both cases. Also, because the current flow depends on the rate of charge and discharge, the higher the frequency, the greater is the current flow per unit time.

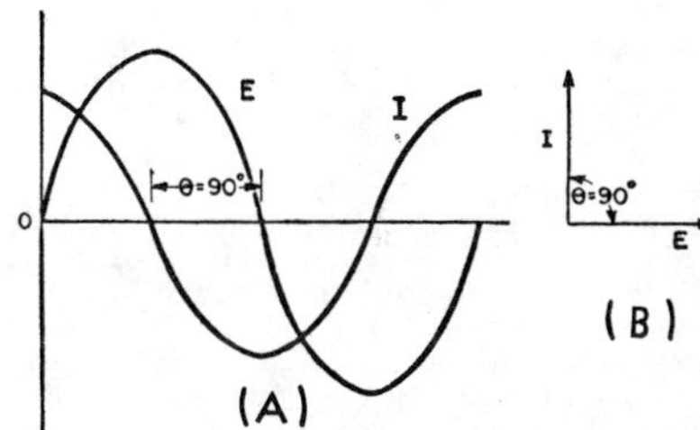

Figure 12-6.—Phase relation between E and I in a capacitive circuit.

NOTES:

The charging current in purely capacitive circuit varies directly with the capacitance, voltage, and frequency—

$$I = 2\pi fCE$$

where I is effective current in amperes, f is in hertz, C is the capacitance in farads, and E is in effective volts.

FORMULA FOR CAPACITIVE REACTANCE

The ratio of the effective voltage across the capacitor to the effective current is called the CAPACITIVE REACTANCE, X_c, and represents the opposition to current flow in a capacitive circuit of zero losses—

$$X_c = \frac{1}{2\pi fC}$$

When f is in hertz, and C is in farads, then X_c is in ohms.

Example: What is the capacitive reactance of a capacitor operating at a frequency of 60 hertz and having a capacitance of 133 microfarads, or 1.33×10^{-4} farads?

$$X_c = \frac{1}{2\pi fC} = \frac{1}{6.28 \times 69 \times 1.33 \times 10^{-4}} = 20 \text{ ohms}$$

POWER IN A CAPACITIVE CIRCUIT

With no voltage or charge, the electrons in the dielectric between the capacitor plates rotate around their respective nuclei in normally circular orbits. When the capacitor receives a charge the positive plate repels the positive nuclei and at the same time the electrons in the dielectric are strained toward the positive plate and repelled away from the negative plate. This distorts the orbits of the electrons in the direction of the positive charge. During the time the electron orbits are changing from normal to the strained position there is a movement of electrons in the direction of the positive charge. This movement constitutes the DISPLACEMENT CURRENT in the dielectric. When the polarity of the plates reverses, the electron strain is reversed. If a sine-wave voltage is applied across the capacitor plates the electrons will oscillate back and forth in a direction parallel to the electrostatic lines of force. DISPLACEMENT CURRENT is a result of the movement of bound electrons, whereas CONDUCTION CURRENT represents the movement of free electrons.

Figure 12-7 (A) shows a capacitive circuit and figure 12-7 (B) indicates the sine waveform of charging current, applied voltage, and instantaneous power. The effective voltage is 70.7 volts. The effective current is 7.07 amperes. Because the losses are neglected, the phase angle between current and voltage is assumed to be 90°. The true power is zero, as indicated by the expression

$$P = EI \cos \theta$$

$$= 70.7 \times 7.07 \ (\cos 90° = 0) = 0 \text{ watt}$$

Multiplying instantaneous values of current and voltage over one cycle, or 360°, gives the power curve, P. During the first quarter cycle (from 0° to 90°) the applied voltage rises from zero to a maximum and the capacitor is receiving a charge. The power curve is positive during this period and represents energy stored in the capacitor. From 90° to 180° the applied voltage is falling from maximum to zero and the capacitor is discharging. The corresponding power curve is negative and represents energy returned to the circuit during this interval. The third quarter cycle represents a period of charging the capacitor and the fourth quarter cycle represents a discharge period. Thus, the average power absorbed by the capacitor is zero. The action is like the elasticity of a spring. Storing a charge in the capacitor is like compressing the spring. Discharging the capacitor is like releasing the pressure on the spring, thus allowing it to return the energy that was stored within it on compression.

The apparent power in the capacitor is EI = 70.7 x 7.07 = 500 volt-amperes. The reactive power in the capacitor is

$$EI \sin \theta = 70.7 \times 7.07 \ (\sin 90° = 1)$$

$$= 500 \text{ VARS (leading)}$$

CAPACITIVE REACTANCE AND RESISTANCE IN SERIES

Losses that appears in capacitive circuits may be lumped in a resistor connected in series with the capacitor, as indicated in figure 12-8. In this example a 39.8-microfarad capacitor is connected in series with a 20-ohm resistor (fig 12-8 (A)). The applied voltage across the RC series circuit is 134 volts and the frequency is 100 hertz.

NOTES:

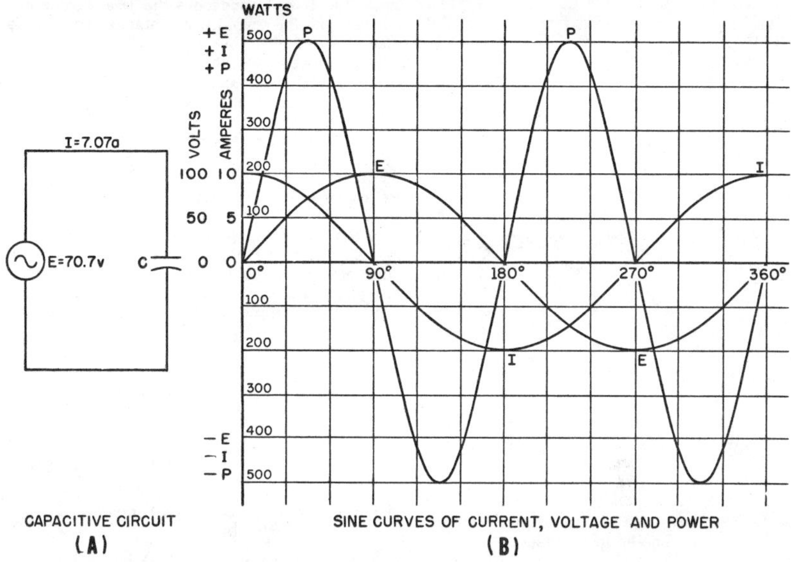

Figure 12-7.—Power in a capacitive circuit.

The capacitive reactance at this frequency is

$$X_C = \frac{1}{2\pi fC} = \frac{1}{6.28 \times 100 \times 39.8 \times 10^{-6}} = 40 \text{ ohms}$$

Voltage Relation

The voltage across R and C are 90° out of phase and equal to 60 volts and 120 volts respectively, as shown in the vector diagram of figure 12-8 (B). The voltage across C is represented as IX_C and is plotted vertically downward from the horizontal in order to indicate that the current leads the voltage across the capacitor by 90°. Angle between the voltage across the capacitor and the circuit current is represented as -90° because it is measured clockwise from the horizontal reference vector, OI. The total voltage is equal to the vector sum of IR and IX_C and is represented in the figure as the hypotenuse of a right triangle the base of which represents the voltage across R, having an effective value of 60 volts and the voltage drop across C, having an effective value of 120 volts. The total (applied) voltage is $\sqrt{60^2 + 120^2}$ = 134 volts.

Impedance

The total impedance, Z, of the series circuit is

$$Z = \frac{E}{I} = \frac{134}{3} = 44.7 \text{ ohms}$$

The impedance diagram is indicated in figure 12-8 (C). The base of the triangle represents the resistance, R, of the circuit and has a magnitude of 20 ohms. The capacitive reactance, X_C is represented as the altitude of the triangle and has a magnitude of 40 ohms. The total impedance may be represented as the hypotenuse of the triangle and has a magnitude of $\sqrt{20^2 + 40^2}$ = 44.7 ohms, which is also the ratio of the applied voltage to the circuit

NOTES:

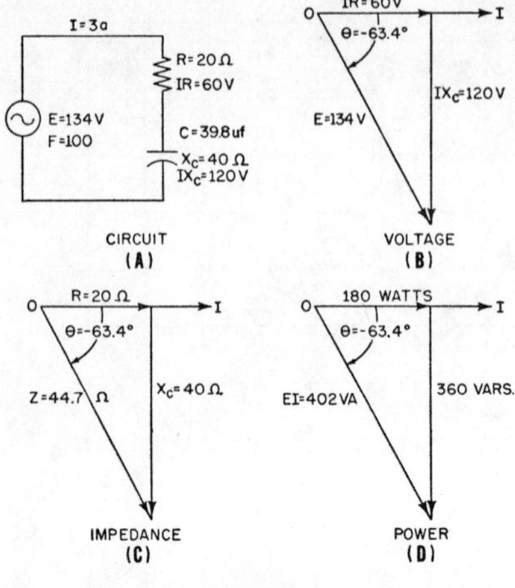

Figure 12-8.—Capacitive reactance and resistance connected in series.

current. The impedance diagram is a right triangle that is similar to the triangle representing the voltage relation in figure 12-8 (B).

Power

The power relations are indicated in the vector diagram of figure 12-8 (D). This diagram is also a right triangle similar to the other two. The base represents the true power absorbed by the circuit resistance and has a magnitude of $I^2R = 3^2 \times 20 = 180$ watts. The altitude represents the reactive volt-ampere (leading) and has a magnitude of $I^2X_c = 3^2 \times 40 = 360$ VARS. The total apparent power is

$$EI = 134 \times 3 = 402 \text{ volt-amperes}$$

Summary of E, Z, and P
Relations in RC Circuits

As mentioned previously, all three triangles are similar. The common factor is the circuit current; and the phase angle, θ, between E and I is equal to the same value in all three diagrams. Thus, the circuit power factor is equal to

$$\cos\theta = \frac{IR}{IZ} = \frac{60}{134} = 0.446 \text{ (fig. 12-8 (B))}$$

$$\cos\theta = \frac{R}{Z} = \frac{20}{44.7} = 0.446 \text{ (fig. 12-9 (C))}$$

$$\cos\theta = \frac{I^2R}{I^2Z} = \frac{180}{402} = 0.446 \text{ (fig. 12-8 (D))}$$

and angle $\theta = 63.4°$ in all three triangles. The true power is

$$EI\cos\theta = 134 \times 3 \times (\cos 63.4° = 0.446) = 180 \text{ watts}$$

The reactive power is

$$EI\sin\theta = 134 \times 3 \times (\sin 63.4° = 0.894) = 360 \text{ VARS}$$

The apparent power is

$$EI = 134 \times 3 = 402 \text{ volt-amperes}$$

NOTES:

CHAPTER 13

FUNDAMENTAL ALTERNATING-CURRENT CIRCUIT THEORY

In the preceding chapter, terms were clarified, a-c reactance was explained, and the effects of individual inductors and capacitors were described. To do this, only the simplest two-element RL and RC series circuits were employed. In this chapter, the subject coverage is expanded to include more complex reactive circuits.

RESISTANCE, INDUCTANCE, AND CAPACITANCE IN SERIES

RELATION OF VOLTAGES AND CURRENT IN AN RLC SERIES CIRCUIT

When resistive, inductive, and capacitive elements are connected in series, their individual characteristics are unchanged. That is, the current through and the voltage drop across the resistor are in phase while the voltage drops across the reactive components (assuming pure reactances) and the current through them are 90° out of phase. However, a new relation must be recognized with the introduction of the three-element circuit. This pertains to the effect on total line voltage and current when connecting reactive elements in series, whose individual characteristics are opposite in nature, such as inductance and capacitance. Such a circuit is shown in figure 13-1.

In the figure, note first that current is the common reference for all three element voltages, because there is only one current in a series circuit, and it is common to all elements. The common series current is represented by the dashed line in figure 13-1 (A). The voltage vector for each element, showing its individual relation to the common current, is drawn above each respective element. The total source voltage E is the vector sum of the individual voltages of IR, IX_L, and IX_C.

The three element voltages are arranged for summation in figure 13-1 (B). Since IX_L and IX_C are each 90° away from I, they are therefore 180° from each other. Vectors in direct opposition (180° out of phase) may be subtracted directly. The total reactive voltage E_X is the difference of IX_L and IX_C. Or, $E_X = IX_L - IX_C = 45 - 15 = 30$ volts. The final relationship of line voltage and current, as seen from the source, is shown in figure 13-1 (C). Had X_C been larger than X_L, the voltage would lag, rather than lead. When X_C and X_L are of equal values, line voltage and current will be in phase.

IMPEDANCE OF RLC SERIES CIRCUITS

The impedance of an RLC (three-element) series circuit is computed in exactly the same manner described for the two-element circuits in the preceding chapter. However, there is one additional operation to be performed. That is, the difference of X_L and X_C must be determined prior to computing total impedance. When employing the Pythagorean Theorem-based formula for determining series impedance, the net reactance of the circuit is represented by the quantity in parenthesis $(X_L - X_C)$. Applying this formula to the circuit in figure 13-1, the impedance is

$$Z = \sqrt{R^2 + (X_L - X_C)^2}$$

$$= \sqrt{40^2 + (45 - 15)^2}$$

$$= \sqrt{1,600 + 900}$$

$$= \sqrt{2,500}$$

$$Z = 50\Omega$$

Series impedance may also be determined by the use of vectorial layout, or triangulation. In an

NOTES:

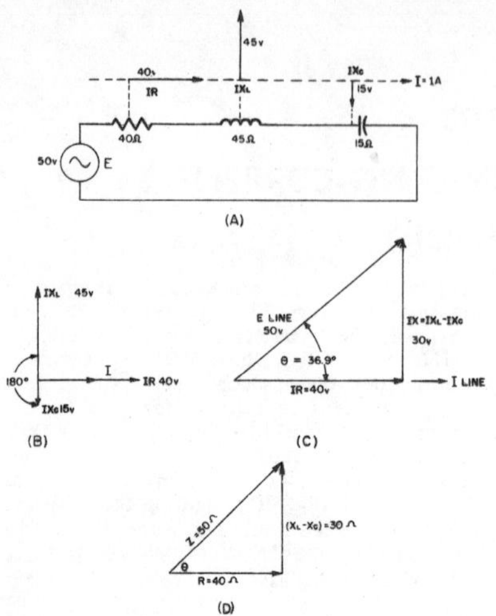

Figure 13-1.—Resistance, inductance, and capacitance connected in series.

impedance triangle (fig. 13-1 (D)) for a series circuit, the base always represents the series resistance, the altitude represents the NET reactance (X_L - X_C), and the hypotenuse represents total impedance.

It should be noted that as the difference of X_L and X_C becomes greater, total impedance also increases. Conversely, when X_L and X_C are equal, their effects cancel each other, and impedance is minimum, equal only to the series resistance. When X_L and X_C are equal, their individual voltages are 90° out of phase with current, but their collective effect is zero, because they are equal and opposite in nature. Therefore, when X_L and X_C are equal, line voltage and current are in phase. This condition is the same as if there were only resistance and no reactances in the circuit. A circuit in this condition is said to be at resonance.

POWER IN AN RLC SERIES CIRCUIT

Total true power in an RLC series circuit is the product of line voltage and current, times the cosine of the angle between them. When X_L and X_C are equal, total impedance is at a minimum, and thus current is maximum. When maximum current flows, the series resistor dissipates maximum power. When X_L and X_C are made unequal, total impedance increases, line current decreases, and moves out of phase with line voltage. Both the decrease in current and the creation of a phase difference cause a decrease in true power.

INDUCTANCE AND RESISTANCE IN PARALLEL

If the voltage applied across a resistor and inductor in parallel has a sine waveform, the currents in the branches will also have sine waveforms. In the parallel circuit of figure 13-2 (A), the applied voltage, E, has a magnitude of 100 volts (rms). Branch ① contains a 20-ohm resistor, and the current, I_1, is equal to $\frac{100}{20}$, or 5 amperes (rms). This current is in phase with E. Branch ② contains an inductance of 0.053 henry. The line frequency is 60 Hz and the inductive reactance is

$$X_L = 2\pi fL = 6.28 \times 60 \times 0.053 = 20 \text{ ohms}$$

The true power losses associated with the inductor in this example are considered negligible. Therefore, current I_2 is equal to $\frac{100}{20}$, or 5 amperes. This current lags E by an angle of 90°. (In an inductive circuit the current always lags the voltage by some angle.)

The sine waveforms of applied voltage and current are shown in figure 13-2 (B). I_1 is in phase with E. I_2 lags E by an angle of 90°. The total current, I_t, lags E by an angle of 45°.

CURRENT VECTORS

A polar-vector diagram representing the three currents and the applied voltage is shown in figure 13-2 (C). Vector OE represents the effective value of the applied voltage and is the horizontal reference vector for both branches because it is common to both branches. Vector I_1 is the effective current of 5 amperes in branch ①. This vector is in the same line as vector OE because the current and voltage in the resistive circuit are in phase. Vector I_2 is the effective current of 5 amperes in branch ② and lags vector OE by an angle of 90°. I_1 is called the energy component of the circuit current. It flows in the resistive branch in which

NOTES:

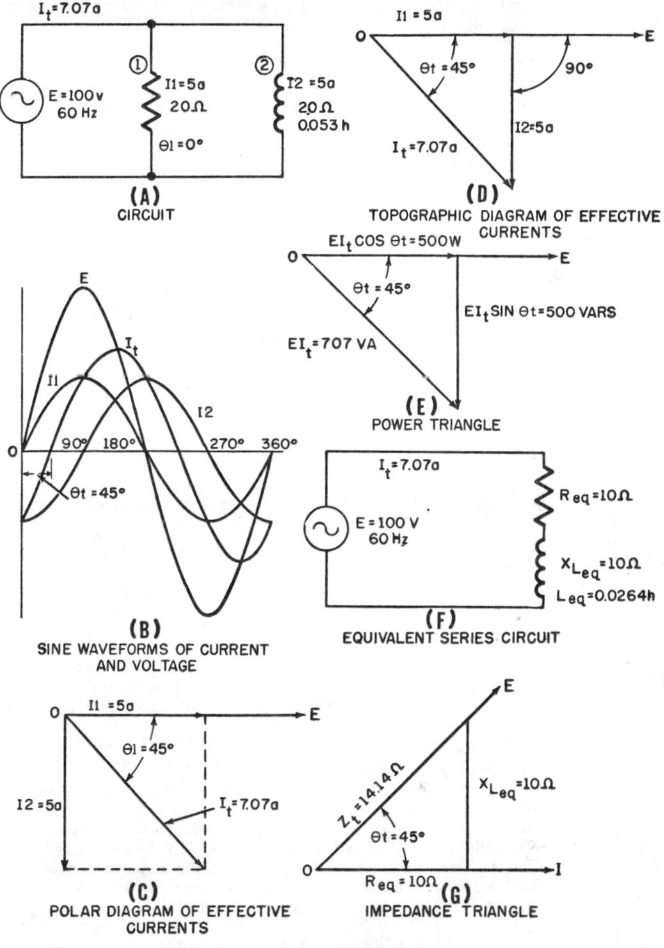

Figure 13-2.—Resistance and inductance, in parallel.

NOTES:

true power is absorbed from the source and is dissipated as heat. I_2 is called the nonenergy component of the circuit current. This current flows in the inductive branch where the true power is zero, and the reactive power is exchanged with the source twice during each cycle of applied voltage.

The total circuit current, I_t, is represented as the diagonal of the parallelogram, the sides of which are I_1 and I_2 (fig. 13-2 (C)). In this example, the sides are 5 amperes (rms) each and the diagonal is 7.07 amperes (rms).

A topographic-vector diagram representing the three currents and the applied voltage is shown in figure 13-2 (D). As in the polar diagram, OE is the reference vector. I_1, in phase with OE, is the base of the triangle. I_2, 90° out of phase with OE, is the altitude of the triangle, and is plotted downward to indicate lag. The resultant current, I_t, is the hypotenuse of the triangle. The hypotenuse is equal to the square root of the sum of the squares of the other two sides. Thus,

$$I_t = \sqrt{I_1^2 + I_2^2} = \sqrt{5^2 + 5^2} = 7.07 \text{ amperes}$$

the phase angle between I_t and E is the angle whose cosine is

$$\frac{I_1}{I_1} = \frac{5}{7.07} = 0.707$$

This angle is 45°.

POWER AND POWER FACTOR

The apparent power, true power, and reactive power in the parallel circuit are related as the hypotenuse, base, and altitude, respectively, of a right triangle in a similar manner to that described in the preceding chapter.

The relation between apparent power, true power, and reactive power is shown in figure 13-2 (E). The hypotenuse of the right triangle represents the apparent power and is equal to EI_t, or 100 x 7.07 = 707 volt-amperes. The base of the triangle represents the true power and is equal to $EI_t \cos \theta_t$, or 100 x 7.07 x 0.707 = 500 watts, where 0.707 = cos 45°. The altitude represents the reactive power and is equal to $EI_t \sin \theta_t$, or 100 x 7.07 x 0.707 = 500 vars, where 0.707 = sin 45°. The power triangle is similar to the current triangle and is related to it by the common factor of voltage (the voltage is the same across both branches of the parallel circuit).

Because branch ① (fig. 13-2 (A)) is purely resistive and branch ② is purely inductive, the true power of the circuit is absorbed in branch ① and the reactive power is developed in branch ②. In branch ① the true power is $EI_1 \cos \theta_1$, or 100 x 5 x 1 = 500 watts, where 1 = cos 0°. In branch ② the reactive power is $EI_2 \sin \theta_2$, or 100 x 5 x 1 = 500 vars, where 1 = sin 90°.

The true power in branch ① may also be calculated as $I_1^2R_1$, or 5^2 x 20 = 500 watts. The reactive power in branch ② may be calculated as $I_2^2X_{L2}$, or 5^2 x 20 = 500 vars. The total circuit power factor is $\cos \theta_t = \frac{\text{true power}}{\text{apparent power}} = \frac{500}{707} = 0.707$. The power factor of branch ① is $\cos \theta_1 = \cos 0° = 1$, and the power factor of branch ② is $\cos \theta_2 = \cos 90° = 0$.

EQUIVALENT SERIES CIRCUIT IMPEDANCE

The combined impedance of the parallel circuit is

$$Z_t = \frac{E}{I_t} = \frac{100}{7.07} = 14.14 \text{ ohms}$$

The combined impedance is also called the impedance of the equivalent series circuit. The equivalent series circuit (fig. 13-2 (F)) contains a resistor and an inductor in series that combine to give the same impedance as the total impedance, of the given parallel circuit. Thus, the current in the equivalent series circuit is equal to the total circuit current in the parallel circuit when rated voltage and frequency are applied to the circuits.

In figure 13-2 (G), the hypotenuse, Z_t, of the impedance triangle is 14.14 ohms and the phase angle, θ_t, between total current and line voltage, is 45°. The equivalent series resistance, R_{eq}, is the base of the impedance triangle an is equal to $Z_t \cos \theta$, or 14.14 x cos 45° = 10 ohms. The equivalent series reactance, X_{Leq}, is the altitude of the impedance triangle and is equal to $Z_t \sin \theta_t$, or 14.14 x sin 45 = 10 ohms.

The inductance of the equivalent series circuit is

$$L_{eq} = \frac{X_L}{2\pi f} = \frac{10}{6.28 \times 60} = 0.0264 \text{ henry}$$

NOTES:

Thus, in this example, the 20-ohm resistor in branch ① shunts the 0.053-henry inductor in branch ②, and the source "sees" an equivalent resistor of 10 ohms in series with an equivalent inductor of 0.0264 henry.

EQUIVALENT CIRCUIT OF A LOW-LOSS INDUCTOR

The losses in air-core inductor occur in the wire with which the inductor is wound. If the losses are small, the resistance of the wire will be small compared with the inductive reactance developed at the operating frequency. The coil resistance acts in series with the coil reactance.

An equivalent parallel circuit for a low-loss coil can be established by substituting an equivalent coil of zero losses, but having the same inductance as the given coil for one branch and a resistor for the other branch. The resistor is of such a magnitude that the same losses will occur in this branch as occur in the given coil when rated voltage and frequency are applied.

In the following example a low-loss circuit is established and then the equivalent parallel circuit is derived. A low-loss coil is indicated in figure 13-3 (A). It has an inductance of 1.59 henry and a d-c resistance of 10 ohms at an operating frequency of 100 Hz. The inductive reactance is

$$X_L = 2\pi fL = 6.28 \times 100 \times 1.59$$
$$= 1,000 \text{ ohms}$$

The equivalent series circuit (fig. 13-3 (B)), is shown with 10 ohms of resistance acting in series with 1,000 ohms of inductive reactance. The impedance triangle is shown in figure 13-3 (C).

A low-loss coil has a low power factor. Thus in the impedance triangle, $\cos\theta = \frac{R_{se}}{Z} = \frac{10}{1,000} = 0.01$, or 1 percent. As can be seen in this example, Z is approximately equal to X_L, and $\cos\theta = \frac{R_{se}}{X_L}$. Since $\tan\theta = \frac{X_L}{R_{se}}$, it follows that in this case $\cos\theta = \frac{1}{\tan\theta}$. Thus $\tan\theta = \frac{1,000}{10} = 100$, and the power factor $= \frac{10}{1,000}$, or 0.01, or 1 percent.

A useful relationship for finding angle θ in low-loss circuits is shown in figure 13-3 (D). This concept involves the relation between $\cos\theta$, expressed decimally, and the complementary angle, $90^\circ - \theta$, expressed in radians. The figure indicates a unit-radius circle in which any length of arc BC measures the corresponding angle $(90^\circ - \theta)$ which it subtends. The length of arc is equal to the angle in radians. Since 2π radians are equal to 360° it follows that 1 radian is approximately equal to $\frac{360^\circ}{2\pi}$, or 57.3°. Thus, from a knowledge of the angle in radians, the equivalent angle θ, in degrees, is $\theta = 57.3 \times$ angle in radians.

For angles greater than 84.3°, $\tan\theta$ is greater than 10 and $\cos\theta$ — which equals OA, in figure 13-3 (D) — is approximately equal to arc BC. Arc BC is a measure of the complementary angle $90 - \theta$ and is expressed in radians. Thus, $\cos\theta$ is approximately equal to the angle in radians by which the current fails to be exactly 90° out of phase with the voltage. In this example, the power factor is 0.01 and therefore the angle by which the current fails to be 90° out of phase with the voltage (complementary angle) is 0.01 radian. This angle, in degrees, becomes 57.3° x 0.01, or 0.573°. Therefore, θ is equal to 090° - 0.573°, or 89,427°, and cos 89.427° = 0.01.

The relationship described in the preceding example is expressed in general terms as follows: The complementary angle, $(90 - \theta)$ in radians, is equal to the power factor, $\cos\theta$, expressed decimally, where $\tan\theta$ is numerically equal to or greater than 10. From this relationship it is possible to estimate quickly the phase angle between current and voltage in low power-factor (low-loss) circuits.

The equivalent parallel circuit for the 1.59-henry coil discussed in this example is shown in figure 13-3 (E). The equivalent shunt resistor, R_{sh}, has a resistance of 100,000 ohms (to be derived later), and this resistor is connected in parallel with a 1.59-henry inductor having zero losses. With rated voltage and frequency applied, the input current to the parallel circuit has the same magnitude and phase with respect to the applied voltage as in the original coil.

The current in the resistive branch is $\frac{100}{100,000} = 0.001$ ampere, or 1 milliampere, and is the base of the current triangle (fig. 13-3 (F)).

NOTES:

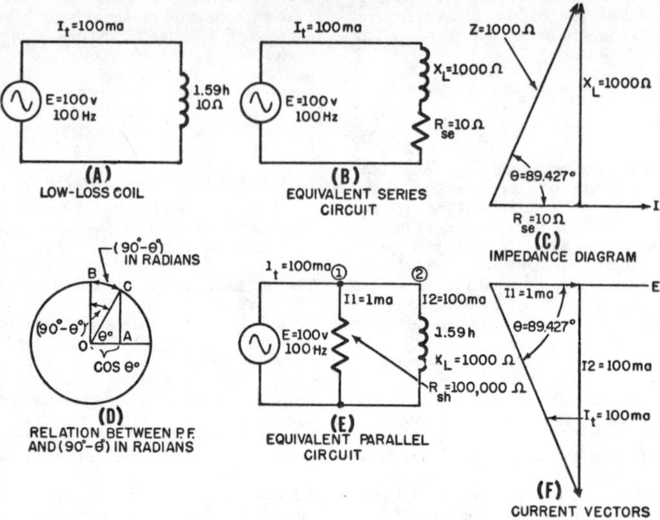

Figure 13-3.—Equivalent circuits of a low-loss coil.

The current in the inductive branch is $\frac{100}{1,000}$ = 0.1 ampere, or 100 ma, and is the altitude of the current triangle. The total current in the parallel circuit is equal to $\sqrt{1^2 + 100^2} = 100$ milliamperes (approx.), and is the hypotenuse of the current triangle. This current lags the line voltage by an angle of 90° - (57.3 x 0.01), or 89.427°.

The equivalent series-circuit impedance triangle (fig. 13-3 (C)) is similar to the current triangle (fig. 13-3 (F)). From the impedance triangle, $\cos\theta \approx \frac{R_{se}}{X_L}$; and from the current triangle, $\cos\theta \approx \frac{I_1}{I_2}$, where I_1 is the energy current and I_2 is the nonenergy current.

Therefore,

$$\frac{I_1}{I_2} = \frac{R_{se}}{X_L} \qquad (13\text{-}1)$$

For the resistive branch,

$$I_1 = \frac{E}{R_{sh}} \qquad (13\text{-}2)$$

and for the inductive branch,

$$I_2 = \frac{E}{X_L} \qquad (13\text{-}3)$$

Substituting equations 13-2 and 13-3 in equation 13-1,

$$\frac{\frac{E}{R_{sh}}}{\frac{E}{X_L}} = \frac{R_{se}}{X_L} \qquad (13\text{-}4)$$

Canceling E and transposing equation 13-4,

$$R_{sh} = \frac{X_L^2}{R_{se}} \qquad (13\text{-}5)$$

NOTES:

In the example of figure 13-3, R_{se} = 10 ohms, X_L = 1,000 ohms, and $R_{sh} = \frac{(1,000)^2}{10} = 100,000$ ohms.

The preceding relations are sufficiently accurate for low-loss inductive circuits in which θ is 84.3° or higher, and tan θ is 10 or higher.

COMBINING CURRENTS AT ACUTE ANGLES

Figure 13-4 (A) represents a 2-branch parallel circuit in which branch ① has a power factor of 0.50 and branch ② has a power factor of 0.866. The current in ① lags the applied voltage by an angle of 60° and the current in branch ② lags the applied voltage by an angle of 30°. The series resistance, R_1, of branch ① is 10 ohms, the series inductive reactance, X_{L1}, is 17.32 ohms, and the impedance, Z_1, is $\sqrt{10^2 + (17.32)^2}$ = 20 ohms. The series resistance, R_2, of branch ② is 17.32 ohms, the series inductive reactance, X_L2, is 10 ohms, and the impedance, Z_2, is $\sqrt{17.32)^2 + 10^2}$ = 20 ohms.

A topographic-vector diagram of the currents in the two branches and the total circuit current is shown in figure 13-4 (B). The current in branch ① is $I_1 = \frac{100}{20}$ = 5 amperes and is the hypotenuse of the right triangle of which the base (energy component) is 5 cos 60° = 2.5 amperes and the altitude (nonenergy component) is 5 sin 60° = 4.33 amperes. The current in branch ② is $I_2 = \frac{100}{20}$ = 5 amperes and is the hypotenuse of the right angle of which the base (energy component) is 5 cos 30° = 4.33 amperes and the altitude (nonenergy component) is 5 sin 30° = 2.5 amperes. The total circuit current is the vector sum of I_1 and I_2, and is the hypotenuse of the resultant right triangle, the base of which is the sum of the energy components of both branches, 2.5 + 4.33 = 6.83 amperes, and the altitude of which is the sum of the nonenergy components of both branches, 4.33 + 2.5 = 6.83 amperes. Thus,

$$I_t = \sqrt{(2.5 + 4.33)^2 + (4.33 + 2.5)^2}$$

$$= 9.66 \text{ amperes}$$

The circuit power factor is $\cos \theta t = \frac{2.5 + 4.33}{9.66}$ = 0.707, and θt = 45°.

NOTES:

The power relations for this circuit are shown in figure 13-4 (C). The apparent power in branch ① is EI_1 = 100 x 5 = 500 volt-amperes and is the hypotenuse of the right triangle, the base of which is $EI_1 \cos \theta_1$ = 100 x 5 cos 60° = 250 watts of true power, and the altitude of which is $EI_1 \sin \theta_1$ = 100 x 5 sin 60° = 433 VARS of reactive power.

The apparent power in branch ② is EI_2 = 100 x 5 = 500 volt-amperes and is the hypotenuse of the right triangle, the base of which is $EI_2 \cos \theta_2$ = 100 x 5 x cos 30° = 433 watts of true power and the altitude of which is $EI_2 \sin \theta_2$ = 100 x 5 x sin 30° = 250 VARS of reactive power.

The total apparent power of the combined parallel circuit is the hypotenuse of the resultant triangle, the base of which is equal to the sum of the true power components in both branches, and the altitude of which is equal to the sum of the reactive power components in both branches. Thus, the total apparent power = $\sqrt{(250 + 433)^2 + (433 + 250)^2}$ = 966 volts amperes.

The equivalent series circuit (fig. 13-4 (D)), representing the combined impedance of the given parallel circuit, has a total impedance, Z_t, that is determined by dividing the applied voltage, E, by the total circuit current, I_t.

$$Z_t = \frac{E}{I_t} = \frac{100}{9.66} = 10.35 \text{ ohms}$$

The hypotenuse of the impedance triangle (fig. 13-4 (E)) is 10.35 ohms. The impedance triangle is similar to the resultant-current right triangle (fig. 13-4 (B)), and angle θt, has the same magnitude in both triangles. This angle is equal to 45°, as determined in the current-triangle calculations.

The equivalent series resistance, R_{eq}, is the base of the impedance triangle and is equal to $Z_t \cos \theta t$, or 10.35 x cos 45 = 7.33 ohms.

The equivalent series reactance, X_{Leq}, is the altitude of the impedance triangle and is equal to $Z_t \sin \theta_t$ or 10.35 x sin 45° = 7.33 ohms.

The impedance triangle is also similar to the resultant power triangle for the combined parallel circuit (fig. 13-4 (C)). The common factor between the two triangles is the square of the total circuit current, I_t^2. The base of the resultant power triangle is $I_t^2 R_{eq}$ = $(9.66)^2$ x 7.33 = 683 watts of true power. This product may be checked by adding the true

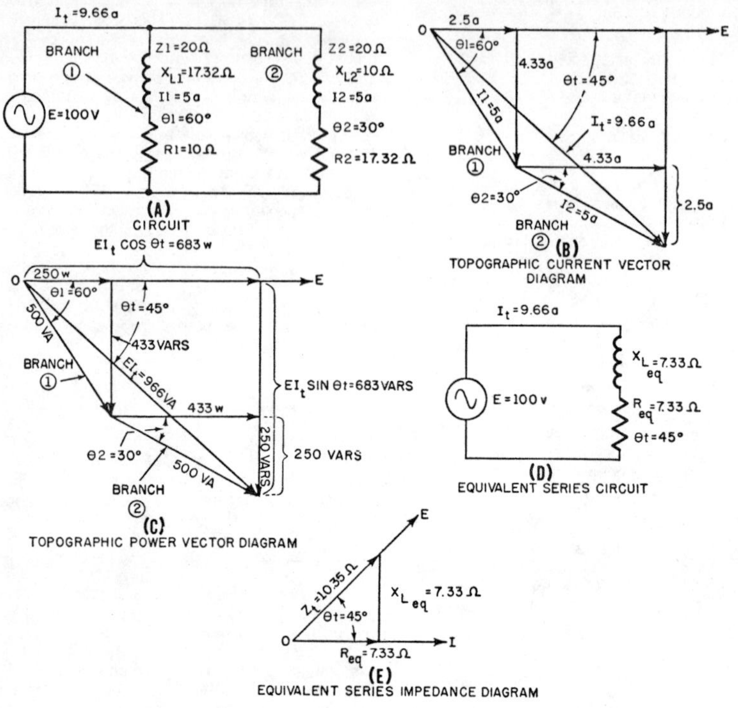

Figure 13-4.—Two-branch LR circuit with currents 30° out of phase.

power components in branches ① and ②. Thus, 250 + 433 = 683 watts of true power.

The altitude of the resultant power triangle is $I_t^2 X_{Leq} = (9.66)^2 \times 7.33 = 683$ VARS of reactive power. This produce may be checked by adding the reactive VAR components in branches ① and ②. Thus, 433 + 250 = 683 VARS of reactive power.

The hypotenuse of the resultant power triangle is $I_t^2 Z_t = (9.66)^2 \times 10.35 = 966$ volt-amperes of apparent power. This product may be checked by adding vectorially the apparent power in branch ① and the apparent power in branch ②. This produce is equal to the hypotenuse of the resultant power triangle (fig. 13-4 (C)). This produce cannot be checked accurately by adding arithmetically the apparent power in branch ① to the apparent power in branch ② because these quantities are not in phase with each other.

CAPACITANCE AND RESISTANCE IN PARALLEL

The action of a capacitor in an a-c series circuit was described in chapter 11 and is amplified further at this time as an introduction to

NOTES:

the a-c parallel RC circuit. In figure 13-5 (A), an a-c voltage of sine waveform is applied across a capacitor and a charging current of sine waveform flows around the circuit in a manner something like that of the flow of water in the hydraulic analogy shown in figure 13-5 (B). The pump at the left corresponds to the a-c source and the cylinder at the right corresponds to the capacitor. If the pump piston is driven by means of a crank turning at uniform speed, the resulting motion of the water will be sinusoidal. The motion of the water is transmitted through the flexible diaphragm in the cylinder and the resulting motion, first in one direction and then in the other, corresponds to the electron flow in the wires connecting the capacitor to the a-c source.

The mechanical stress in the diaphragm in the cylinder corresponds to the electric stress in the dielectric between the plates of the capacitor. Electron flow does not occur through the dielectric in the same way that water would flow through the flexible diaphragm if a hole were punctured in it. Instead, the electrons flow around the capacitor circuit on one alternation causing a negative charge to build up on one place, and a corresponding positive charge on the other, and on the next alternation causing a reversal of the polarity of the charges on the plates. Thus, the effective impedance which the capacitor offers to the flow of alternating current can be relatively low at the same time that the insulation resistance which the dielectric offers to the flow of direct current is extremely high.

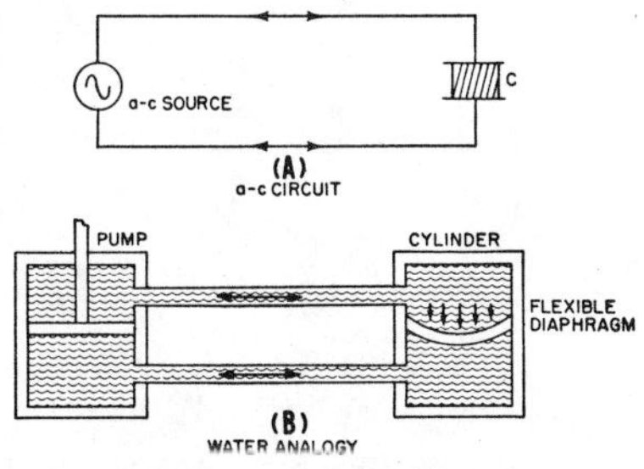

Figure 13-5.—Water analogy of a capacitor in an a-c circuit.

NOTES:

In the example of figure 13-6, a 2-branch circuit consists of a 100-ohm resistor and a 15.9-microfarad capacitor of negligible losses in parallel with a 100-volt a-c source. The frequency of the source voltage is 100 H_z and the voltage is assumed to have a sine waveform. The current in branch ① is $\frac{100}{100} = 1$ ampere (rms). The impedance of branch ② is composed of capacitive reactance; its resistance component is neglected. In branch ② $X_c = \frac{1}{2\pi fC} =$

$$\frac{1}{6.28 \times 100 \times 15.9 \times 10^{-6}} = 100 \text{ ohms.}$$

The current in branch ② is $\frac{100}{100} = 1$ ampere (rms). The sine waveform of the branch currents and the applied voltage, together with the resultant line current, are shown in figure 13-6 (B).

The current in the resistive branch is in phase with the applied voltage and has a peak value of $\frac{1}{0.707} = 1.41$ amperes. The current in the capacitor branch leads the applied voltage by an angle of 90° and has a peak value of $\frac{1}{0.707} = 1.41$ amperes. The resultant line current is the algebraic sum of the instantaneous values of the currents in the two branches and has a peak value of 2 amperes. The resultant line current, I_t, leads the applied voltage by an angle of 45°.

CURRENT VECTORS

A topographic-vector diagram of the effective values of these currents is shown in figure 13-6 (C). The base of the current triangle is 1 ampere and represents the current in branch ①. This current is in phase with the applied voltage and represents the energy component of the total line current.

The altitude of the current triangle is 1 ampere and represents the current in branch ②. This current leads the applied voltage by 90° and represents the nonenergy component of the total line current.

The hypotenuse of the triangle is 1.41 amperes and represents the total line current.

The reference vector, OE, for the current triangle is the line voltage, and in the RC parallel circuit, the altitude extends above the reference vector to indicate the sense of lead. This direction is opposite to that of the altitude of the current triangle for the RL parallel circuit in figure 13-2 (D). In both figures the vectors are assumed to rotate in a counterclockwise direction to indicate the sense of lead or lag of the currents with respect to the line voltage. In all single-phase circuits such as these, the current vectors are considered in their relative phase by angles that never exceed 90° with respect to their common reference voltage vector.

In figure 13-2 (D), the line current lags the line voltage by an angle of 45°, and in figure 13-6 (C), the line current leads the line voltage by an angle of 45°. If both inductance and capacitance exist in the same parallel circuit as separate branches, these branches will be 180° out of phase with each other, but the current in the inductive branch will never lag the line voltage by an angle in excess of 90°, and the current in the capacitive branch will never lead the line voltage by an angle in excess of 90°.

POWER AND POWER FACTOR

The power triangle for the parallel RC circuit is shown in figure 13-6 (D). The power in branch ① is

$$EI_1 \cos \theta_1 = 100 \times 1$$
$$\times (\cos 0° = 1) = 100 \text{ watts}$$

and forms the base of the triangle in line with the voltage vector, OE. the reactive power in branch ② is

$$EI_2 \sin \theta_2 = 100$$
$$\times 1 (\sin 90° = 1) = 100 \text{ VARS}$$

and is the altitude of the power triangle, perpendicular to OE. The reactive power in branch ① is

$$EI_1 \sin \theta_1 = 100$$
$$\times 1 (\sin 0° = 0) = 0 \text{ VAR}$$

and the power in branch ② is

$$EI_2 \cos \theta_2 = 100$$
$$\times 1 (\cos 90° = 0) = 0 \text{ watt}$$

The apparent power of the parallel RC circuit is

$$EI_t = 100 \times 1.41 = 141 \text{ volt-amperes}$$

and is the hypotenuse of the power triangle.

NOTES:

The power triangle (fig. 13-6 (D)) is similar to the current triangle (fig. 13-6 (C)), since θt has the same magnitude in both triangles. The total circuit power factor, as determined from the current triangle, is

$$\cos \theta_t = \frac{I_1}{I_t} = \frac{1}{1.41} = 0.707$$

The total circuit power factor, as determined from the power triangle, is

$$\cos \theta_t = \frac{\text{true power}}{\text{apparent power}} = \frac{100}{141} = 0.707$$

The power of the total circuit, as determined from the power triangle, is

$$EI_t \cos \theta_t = 100 \times 1.41$$
$$(\cos 45° = 0.707 = 100 \text{ watts}$$

This value is equal to the true power in branch ①.

The reactive power of the total circuit, as determined from the power triangle, is

$$EI_t \sin \theta_t = 100 \times 1.41$$
$$(\sin 45° = 0.707 = 100 \text{ VARS}$$

This value is equal to the reactive power in branch②.

EQUIVALENT SERIES IMPEDANCE

The total impedance of the parallel RC circuit is

$$Z_t = \frac{E}{I_t} = \frac{100}{1.41} = 70.7 \text{ ohms}$$

As mentioned previously, the total impedance is also the impedance of the equivalent series circuit (fig. 13-6 (E)). In figure 13-6 (F), the hypotenuse, Z_t, of the impedance triangle is 70.7 ohms and the phase angle, θ_t, between total current and line voltage is 45°. The equivalent series resistance, R_{eq}, is the base of the impedance triangle and has a magnitude of

$$R_{eq} = Z_t \cos \theta_t = 70.7$$
$$(\cos 45° = 0.707) = 50 \text{ ohms}$$

The equivalent series reactance, X_{Ceq}, is the altitude of the impedance triangle and has a magnitude of $Z_t \sin \theta_t = 70.7 (\sin 45° = 0.707) = 50$ ohms. The altitude is extended downward, in

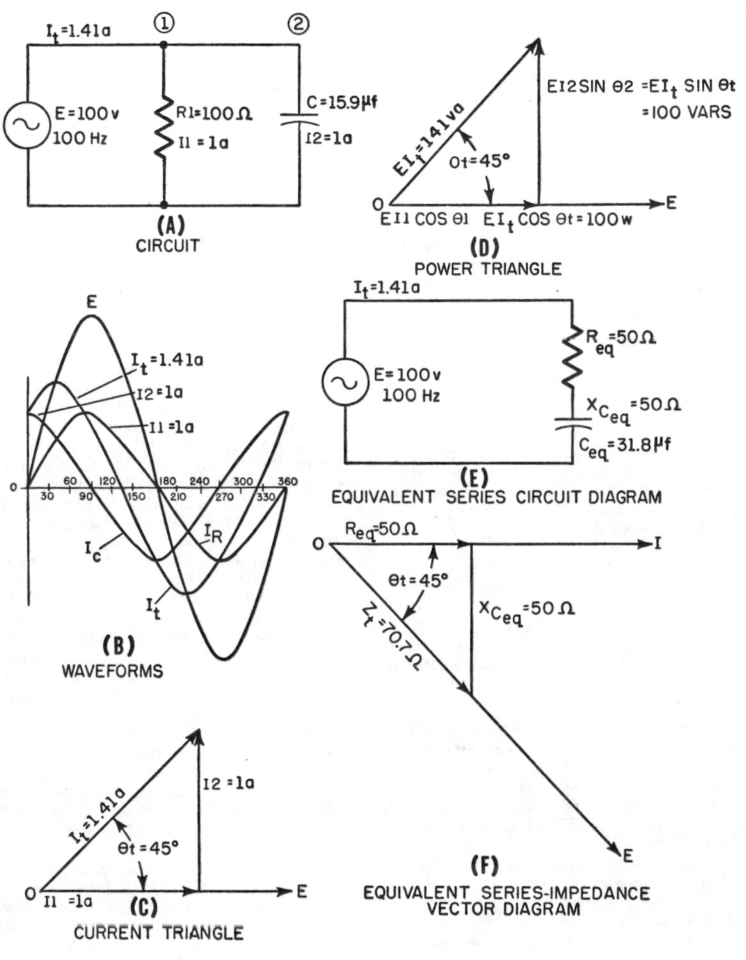

Figure 13-6.—Parallel RC circuit analysis.

NOTES:

contrast to the upward direction of the altitude of the current triangle in figure 13-6 (C), to maintain the sense of current lead for counterclockwise vector rotation. The line voltage is the horizontal reference vector, OE (fig. 13-6 (C)), and the line current is the horizontal reference vector, OI (fig. 13-6 (F)).

The capacitance in microfarads of the equivalent series impedance is

$$C_{eq} = \frac{10^6}{2\pi f X_{Ceq}} = \frac{10^6}{6.28 \times 100 \times 50}$$

$$= 31.8 \text{ microfarads.}$$

Thus, in this example, the 100-ohm resistor in branch ① shunts the 15.9-microfarad capacitor in branch ② and the source "sees" and equivalent resistor of 50 ohms in series with an equivalent capacitor of 31.8 microfarads (fig. 13-6 (E)).

EQUIVALENT CIRCUIT OF A LOW-LOSS CAPACITOR

The action of a capacitor in an a-c circuit was described as being like the flow of water in a cylinder having a flexible diaphragm. The diaphragm corresponds to the dielectric in a capacitor and the mechanical stresses in the diaphragm correspond to the electric stress in the dielectric.

Most of the heating produced in a capacitor is due to the loss in the dielectric. The movement of the electrons in the atoms of a solid dielectric is depicted in figure 13-7. The capacitor is connected to a source of a-c voltage of sine waveform and the conditions are indicated for three successive instants in the cycle of applied a-c voltage.

In figure 13-7 (A), the voltage across the capacitor plates is positive maximum and the capacitor is charged. The atoms in the dielectric are subjected to electric stress that causes their orbital electrons to move in paths that are charged from circular to elliptical patterns. The elliptical pattern forms as a result of the attractive force of the negatively charged plate on the positive nucleus and the simultaneous repulsion of the positively charged plate on the nucleus. At the same time the positive plate attracts the orbital electrons, the negative plate repels them.

In figure 13-7 (B), the charge on the plates is zero and the electric stress is removed from the dielectric. In this case the orbital electrons travel in circular paths about the nucleus with no external forces applied to them.

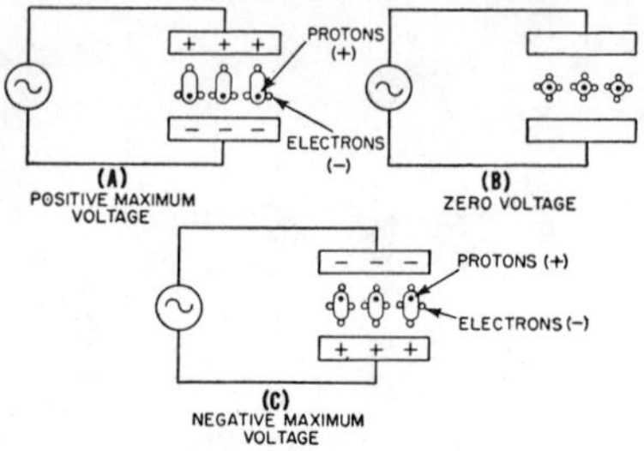

Figure 13-7.—Capacitor dielectric losses.

In figure 13-7 (C), the charges on the plates are reversed and the orbital electrons are again caused to move in elliptical paths. In this case the protons (entire nucleus) are moved toward the upper plate, the forces being opposite to those developed in figure 13-7 (A).

The rapid reversals of applied voltage accompanied by the change of the orbital electron paths from circular to elliptical and back to circular patterns cause heat to be developed in the dielectric. The change in pattern of the electron orbits constitutes a dielectric displacement current. This current is considered to be made up of two components, one leading the voltage by 90°, the other an energy component in phase with the applied voltage. Factors that determine the dielectric loss are the applied voltage, the dielectric constant, the capacitor power factor, and the frequency of the applied voltage. An air-dielectric capacitor has no appreciable loss and the power factor is 0. A mica capacitor, having a dielectric constant of 7 and a power factor of 0.0001, has relatively low loss and low dielectric heating at high voltages and high frequencies.

The product of the dielectric constant and the power factor is called the loss factor. The loss factor is low for good dielectrics that operate without much dielectric heating. The loss factor is the best indication of the ability of a material to withstand high voltages at high frequencies. The loss factor for the previously mentioned mica capacitor is 7 x

NOTES:

0.0001 = 0.0007. The loss factor for air dielectric is zero because the power factor is zero.

The equivalent circuits of a low power-factor capacitor are represented in figure 13-8. The circuits are derived by assuming a capacitor of the same capacitance as that of the original capacitor, but having no losses (zero power factor), to be connected (1) in series with a resistor that develops the same true power loss as in the original capacitor; and (2) a capacitor of the same capacitance as that of the original capacitor, but having no losses (zero power factor), to be connected in shunt with a resistor that develops the same true power loss as in the original capacitor. The capacitance of the capacitor in this example is 66.4 microfarads. The power factor of the capacitor is cos θ = 0.05. The effective voltage is 200 volts and the capacitor current is 5 amperes.

The equivalent series circuit is shown in figure 13-8 (B). At the operating frequency of 60 Hz the capacitive reactance of the capacitor is

$$X_C = \frac{1}{2\pi fC} = \frac{1}{6.28 \times 60 \times 66.4 \times 10^{-6}}$$

$$= 40 \text{ ohms approximately}$$

The impedance of the series circuit is

$$Z_t = \frac{E}{It} = \frac{200}{5} = 40 \text{ ohms.}$$

The equivalent series resistance is

$$R_{se} = Z_t \cos \theta_t = 40 \times 0.05 = 2 \text{ ohms}$$

(fig. 13-8 (B)). The base of the impedance triangle is 2 ohms of resistance (fig. 13-8 (C)), the altitude of the impedance triangle is 40 ohms of capacitive reactance, and the hypotenuse is approximately 40 ohms of impedance.

The power factor of this circuit is cos θ_t $= \frac{R_{se}}{Zt} = \frac{2}{40} = 0.05$, or 5 percent. In this example, as in the low-loss inductor of figure 13-3, Z_t is approximately equal to X_C, and cos θ_t = $\frac{R_{se}}{X_C}$. Since $\tan \theta_t = \frac{X_C}{R_{se}} = \frac{40}{2} = 20$, it follows that $\cos \theta_t = \frac{1}{\tan\theta}$, and the power factor of the capacitor is $\frac{2}{40}$, or 0.05.

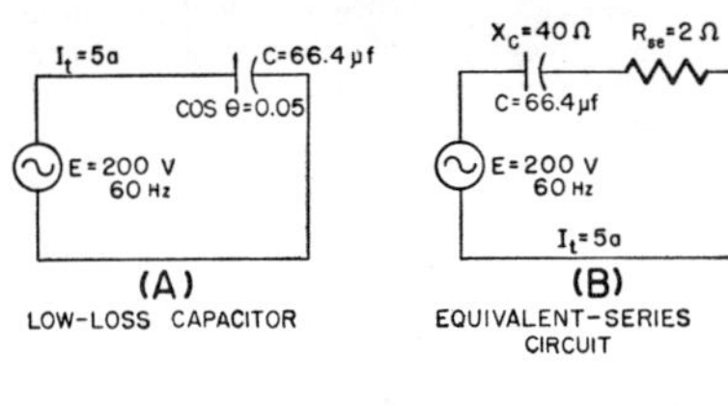

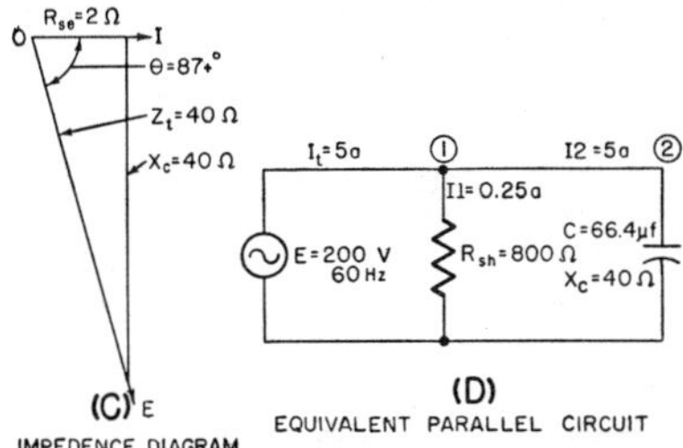

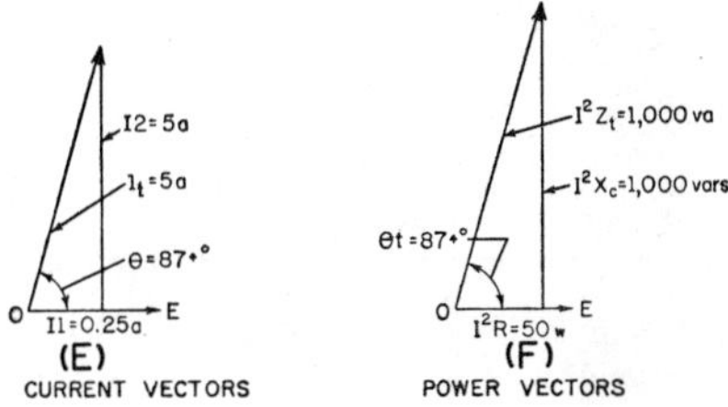

Figure 13-8.—Equivalent circuits of a low-loss capacitor.

The angle whose cosine is 0.05 may be closely approximated by the previously described relation between the power factor and the complementary angle in radians. In this example, the complementary angle by which the current fails to be 90° out of phase with the voltage is 0.05 radian, or 0.05 x 57.3 = 2.865°, and θ = 90° - 2.865° = 87.135°. This angle is indicated in all of the vector diagrams of figure 13-8 as 87+°.

The equivalent parallel circuit for the 66.4-microfarad capacitor discussed in this example

NOTES:

is shown in figure 13-8 (D). The equivalent series resistance and the equivalent shunt resistance are related in the equivalent capacitor circuits in the same way that R_{se} and R_{sh} are related in the low-loss inductor circuits of figure 13-3. This relation was stated in equation 13-5. Thus, in the capacitor circuits,

$$R_{sh} = \frac{(X_C)^2}{R_{se}} = \frac{(40)^2}{2} = 800 \text{ ohms}$$

The equivalent shunt resistance of 800 ohms is connected in parallel with a 66.4-microfarad capacitor of zero losses. With rated voltage and frequency applied, the input current to the parallel circuit will have the same magnitude and phase with respect to the applied voltage as in the original capacitor.

The energy current in the resistive branch is $I_1 = \frac{E}{R_{sh}} = \frac{200}{800} = 0.25$ ampere and is the base of the current triangle (figure 13-8 (E)). The nonenergy current in the capacitive branch is $I_2 = \frac{E}{X_C} = \frac{200}{40} = 5$ amperes and is the altitude of the current triangle. The total current in the parallel circuit is equal to

$$\sqrt{(0.35)^2 + (5)^2} = 5$$

amperes (approx.) and is the hypotenuse of the current triangle. This current leads the voltage by an angle of 87.135°.

The power relations are shown in figure 13-8 (F). The true power of the circuit may be found in a number of ways. Three methods are listed as follows:

1. $P = EI_t \cos\theta_t = 200 \times 5 \times 0.05 = 50$ watts (fig. 13-8 (A)).
2. $P = I_t^2 R_{se} = 5^2 \times 2 = 50$ watts (fig. 13-8 (B)).
3. $P = I_1^2 R_{sh} = (0.25)^2 \times 800 = 50$ watts (fig. 13-8 (D)).

The true power of 50 watts is the base of the power triangle (fig. 13-8 (F)).

The reactive power may be calculated in a number of ways. Three methods are indicated as follows:

1. $VARS = E_t I_t \sin\theta_t = 200 \times 5 \times (\sin 87+° = 1 \text{ approx.}) = 1{,}000$ VARS (fig. 13-8 (A)).
2. $VARS = I_t^2 X_C = 5^2 \times 40 = 1{,}000$ VARS (fig. 13-8 (B)).
3. $VARS = I_2^2 X_C = 5^2 \times 40 = 1{,}000$ VARS (fig. 13-8 (D)).

The reactive power of 1,000 VARS is the altitude of the power triangle (fig. 13-8 (F)).

The apparent power of the capacitor may be calculated as:

1. Apparent power $= EI_t = 200 \times 5 = 1{,}000$ VA (fig. 13-8 (A)).
2. Apparent power $= I_t^2 Z_t = 5^2 \times 40 = 1{,}000$ VA (fig. 13-8 (B)).

The apparent power of 1,000 volt-amperes is the hypotenuse of the power triange (fig. 13-8 (F)).

COMBINING CURRENTS AT ACUTE ANGLES

A 2-branch parallel circuit containing a 75-ohm resistor in branch ① and a series combination of a 79.6 microfarad capacitor and a 30-ohm resistor in branch ② is shown in figure 13-9 (A). The capacitor losses are assumed to be included with those of the 30-ohm resistor so that the power factor of that portion of branch ② represented by the capacitor is zero.

The capacitive reactance at the operating frequency of 50 Hz is

$$X_C = \frac{1}{2\pi fC} = \frac{1}{6.28 \times 50 \times 79.6 \times 10^{-6}}$$

$$= 40 \text{ ohms.}$$

The impedance, Z_2, of branch ② is the combined opposition of 30 ohms of resistance in series with 40 ohms of capacitive reactance and is the hypotenuse of the right triangle of which 30 ohms is the base and 40 ohms is the altitude (not shown in the figure). Thus,

$$Z_2 = \sqrt{30^2 + 40^2} = 50 \text{ ohms.}$$

The power factor of branch ② = is $\cos\theta_2 = \frac{R2}{Z2} = \frac{30}{50} = 0.60$ and the phase angle is 53.1°.

NOTES:

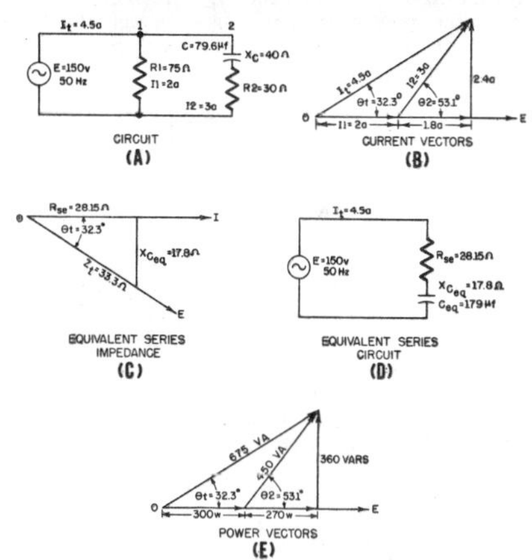

Figure 13-9.—Two-branch RC circuit with currents out of phase by an angle of 53.1°.

The current vectors for both branches are shown in figure 13-9 (B).

The current in branch ① is $I_1 = \frac{E}{Z_1} = \frac{150}{75}$ = 2 amperes, and is a portion of the base of the equivalent right triangle in line with the horizontal reference voltage, OE. This current is in phase with the applied voltage because there is no reactance present in branch ①

The current in branch ② is

$$I_2 = \frac{E}{Z_2} = \frac{150}{50} = 3 \text{ amperes}$$

and is the hypotenuse of the right triangle, the base of which is $I_2 \cos \theta_2$ = 3 (cos 53.1° = 0.6) = 1.8 amperes (energy current) and the altitude of which is $I_2 \sin \theta_2$ = 3 (sin 53.1° = 0.8) = 2.4 amperes (nonenergy current). The total current is the vector sum of I_1 and I_2 and is the hypotenuse of the equivalent right triangle, the base of which is the arithmetic sum of the current in branch ① and the energy component of current in branch ②, or 2 + 1.8 = 3.8 amperes. The altitude of the equivalent right triangle is equal to the nonenergy current in branch ②, or 2.4 amperes. The total current in the parallel circuit is

$$I_t = \sqrt{(3.8)^2 + (2.4)^2} = 4.5 \text{ amperes}$$

The phase angle, θ_t between the total circuit current and the applied voltage is the angle whose cosine is $\frac{3.8}{4.5}$ = 0.845. This angle is approximately 32.3°.

Thus, the total current leads the applied voltage by an angle of 32.3°.

The total impedance of the parallel circuit is

$$Z_t = \frac{E}{I_t} = \frac{150}{4.5} = 33.3 \text{ ohms}$$

and is the hypotenuse of the impedance triangle (fig. 13-9 (C)). The base of the triangle is

$$R_{se} = Z_t \cos \theta_t = 33.3$$

$$(\cos 32.3° = 0.845) = 28.15 \text{ ohms}$$

of resistance and is in line with the horizontal current reference vector for the equivalent series circuit (fig. 13-9 (D)). The altitude of the triangle is

$$X_{Ceq} = Z_t \sin \theta_t = 33.3$$

$$(\sin 32.3° = 0.534) = 17.8 \text{ ohms}$$

of capacitive reactance. The capacitance of the equivalent series circuit at the operating frequency of 50 H_z is

$$C_{eq} = \frac{10^6}{2\pi f X_C} = \frac{10^6}{6.28 \times 50 \times 17.8}$$

$$= 179 \text{ microfarads.}$$

The power relations are indicated in the triangles of figure 13-9 (E). The apparent power of the total circuit is EI_t = 150 x 4.5 = 675 volt-amperes and is the hypotenuse of the equivalent right triangle representing the total parallel circuit.

The true power in branch ① is

$$I_1^2 R_1 = 2^2 \times 75 = 300 \text{ watts}$$

The true power in branch ② is

$$I_2^2 R_2 = 3^2 \times 30 = 270 \text{ watts}$$

NOTES:

The total true power is

$$300 + 270 = 570 \text{ watts}$$

and is the base of the equivalent right triangle for the entire circuit. The reactive power in the total circuit—that is, the reactive power of branch ②—is

$$I_2^2 X_C = 3^2 \times 40 = 360 \text{ VARS}$$

and is the altitude of the equivalent triangle.

The apparent power in branch ① is equal to the true power because the power factor is unity. The apparent power in branch ② is $EI_2 = 150 \times 3 = 450$ volt-amperes and is the hypotenuse of the power triangle for this branch. The true power of 270 watts in branch ② is the base of this triangle.

The reactive power of 360 VARS in branch ② is the altitude of the triangle. The power factor for branch ② is $\cos\theta_2 = \frac{\text{true power}}{\text{apparent power}} = \frac{270}{450} = 0.60$, and the phase angle for branch ② is $\theta_2 = 53.1°$.

PARALLEL CIRCUITS CONTAINING L, R, AND C

Inductance in an a-c circuit causes the current to lag behind the applied voltage. Transformers and induction motors are essentially inductive in nature and the power factor, especially on light loads (as contrasted with full loads), is relatively low.

Most circuits, that supply electric power from the source to the consumer, transmit the power at relatively high voltage and low current in order to keep the I^2R loss in the transmission lines satisfactorily low. Transformers at the point of utilization reduce the voltage to the proper value to operate the equipment.

Capacitance in a-c circuits causes the current to lead the applied voltage and when placed in parallel with inductive components can produce a neutralizing effect so that lagging currents can be brought into phase with the applied voltage or may be made to lead the applied voltage, depending on the relative magnitude of the capacitance and inductance in parallel.

The true power of a circuit is $P = EI\cos\theta$; and for any given amount of power to be transmitted, the current, I, varies inversely with the power factor, $\cos\theta$. Thus, the addition of capacitance in parallel with inductance will, under the proper conditions, improve the power factor (make nearer unity power factor) of the circuit and make possible the transmission of electric power with reduced line loss and improved voltage regulation.

CURRENT VECTORS

In the example of figure 13-10, the parallel circuit contains three branches. Branch ① contains a 30-ohm resistor. Branch ② consists of an inductor of 0.0612 henry and a resistor of 6.6 ohms. Branch ③ contains a capacitance of 44.3 microfarads having negligible losses. The parallel circuit is shown in figure 13-10 (A). The current vectors are shown in figure 13-10 (B).

The current in branch ① is

$$I_{1en} = \frac{E}{R_1} = \frac{120}{30} = 4 \text{ amperes}$$

The corresponding 4-ampere vector in figure 13-10 (B), is drawn in the same horizontal reference line as the voltage vector because the phase angle between the current and voltage in this branch is zero.

The inductive reactance of branch ② at the operating frequency of 60 Hz is

$$X_L = 2\pi fL + 6.28 \times 60 \times 0.0612 = 23.1 \text{ ohms}$$

The impedance of branch ② is

$$Z2 = \sqrt{R_2^2 + X_L^2} = \sqrt{(6.6)^2 + (23.1)^2}$$
$$= 24 \text{ ohms}$$

The current in branch ②, at the operating voltage of 120 volts, is $I_2 = \frac{E}{Z_2} = \frac{120}{24} = 5$ amperes and is the hypotenuse of the current triangle for branch ②. The angle by which the current in branch ② lags the applied voltage is the angle whose cosine is

$$\frac{R_2}{Z_2} = \frac{6.6}{24} = 0.275$$

This angle is $\theta_2 = 74°$. The base of the current triangle for branch ② is

$$I_{2en} = I_2 \cos 74° = 5$$
$$(\cos 74° = 0.275) = 1.38 \text{ amperes}$$

NOTES:

The altitude of the current triangle for branch ② is

$$I_{2\,nonen} = I_2 \sin 74° = 5$$

$$(\sin 74° = 0.962) = 4.8 \text{ (approx) amperes}$$

and extends below the horizontal voltage reference vector to indicate current lag.

The capacitive reactance in branch ③ is

$$X_C = \frac{1}{2\pi fC} = \frac{1}{6.28 \times 60 \times 44.3 \times 10^{-6}}$$

$$= 60 \text{ ohms}$$

and with negligible losses, the impedance, Z_3, is also 60 ohms. The current in branch ③ is

$$I_{3\,nonen} = \frac{E}{Z_3} = \frac{120}{60} = 2 \text{ amperes}$$

The current in the capacitor branch leads the applied voltage by an angle of 90°. The nonenergy component of the current in the inductive branch lags the applied voltage by an angle of 90°. Therefore, these two currents are 180° out of phase with each other, and the capacity current vector, I_3, extends upward from the lower extremity of the vector representing the nonenergy component of current in branch ②.

The total current, I_t, in the parallel circuit is the vector sum of the currents in the three branches and is the hypotenuse of the equivalent current triangle, the base of which is the arithmetic sum of the energy components of the currents and the altitude of which is the algebraic sum of the nonenergy components of the currents. Thus, the total current is

$$I_t = \sqrt{(I_{1en} + I_{2en})^2 + (I_{2nonen} - I_{3nonen})^2}$$

$$= \sqrt{4 + (1.38)^2 + (4.8 - 2)^2}$$

$$= 6.06 \text{ amperes}$$

The phase angle, θ_t. between the total current and the applied voltage is the angle whose cosine is the ratio of the sum of the energy components of the currents in all the branches to the total current. Thus,

$$\cos \theta_t = \frac{4 + 1.38}{6.06} = 0.888$$

$$\theta_t = 27.5°$$

and the total circuit current lags the applied voltage by an angle of 27.5°.

EQUIVALENT SERIES IMPEDANCE

The total impedance of the parallel circuit (fig. 13-10 (A)) is equal to the total circuit voltage divided by the total circuit current, or

$$Z_t = \frac{E}{I_t} = \frac{120}{6.06} = 19.8 \text{ ohms}$$

and is the hypotenuse of the impedance triangle (fig. 13-10 (C)). This triangle represents the impedance relations existing in the equivalent series circuit (fig. 13-10 (D)). The phase angle, θ_t, of the equivalent series circuit has the same magnitude as the phase angle between the total circuit current and the applied voltage across the parallel circuit. The impedance of the equivalent series circuit has the same magnitude as the total impedance of the parallel circuit. The base of the triangle is

$$R_{eq} = Z_t \cos \theta_t = 19.8$$

$$(\cos 27.5° = 0.889) = 17.6 \text{ ohms}$$

and represents the resistive component of the equivalent series circuit. The altitude of the triangle is

$$X_{Leq} = Z_t \sin \theta_t = 19.8$$

$$(\sin 27.5° = 0.462) = 9.16 \text{ ohms}$$

and represents the inductive reactance component of the equivalent series circuit.

POWER AND POWER FACTOR

The true power of the parallel circuit (fig. 13-10 (A)) is the arithmetic sum of the power absorbed in each branch. The true power in branch ① is

$$P_1 = I_1^2R_1 = 4^2 \times 30 = 480 \text{ watts}$$

The true power in branch ② is

$$P_2 = I_2^2R_2 = 5^2 \times 6.6 = 165 \text{ watts}$$

The true power in branch ③ is negligible since the power factor of the capacitor is assumed to be zero. The total true power is

$$480 + 165 = 645 \text{ watts}$$

NOTES:

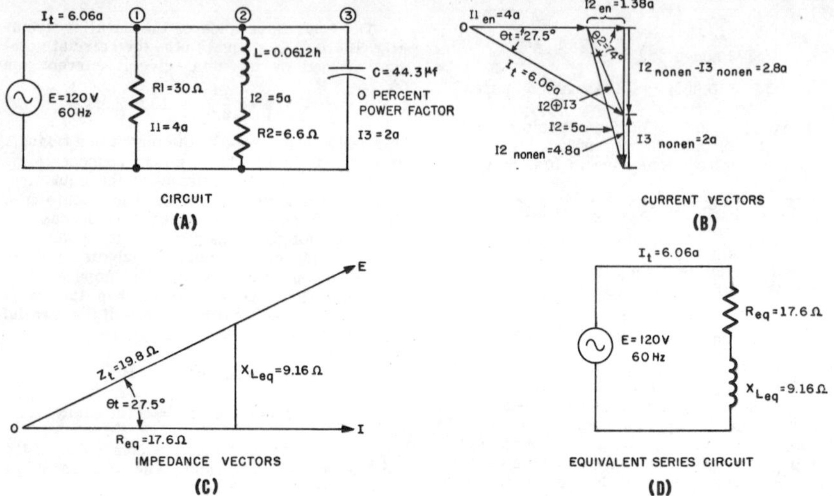

Figure 13-10.—Parallel circuit containing L, R, and C.

The total apparent power of the parallel circuit is the product of the total circuit current and the applied voltage. Thus, the apparent power is

$$EI_t = 120 \times 6.06 = 727.2 \text{ volt-amperes}$$

The parallel circuit power factor is the ratio of the total true power to the total apparent power. Thus, the circuit power factor is

$$\cos \theta_t = \frac{\text{true power}}{\text{apparent power}} = \frac{645}{727.2} = 0.888$$

POWER-FACTOR CORRECTION

Power-factor correction in parallel circuits is accomplished by placing a capacitor of the proper size in parallel with the circuit at the point where the power-factor correction is to be effected. The leading current of the capacitor branch supplies the lagging component of current in the inductive portion of the parallel circuit and reduces the line current accordingly. As mentioned before, this action improves the efficiency of transmission by reducing the line current and I^2R losses.

In the example of figure 13-11, the load is rated at 10 amperes, 1,000 volts, and a power factor of 50 percent lagging (the current lags the voltage). The true power absorbed by the load (fig. 13-11 (A)) is

$$P = EI \cos\theta = 1,000 \times 10 \times 0.5$$
$$= 5,000 \text{ watts.}$$

The load is supplied through a line having a resistance of 20 ohms and negligible reactance. The power loss in the line is

$$I^2R = 10^2 \times 20 = 2,000 \text{ watts,}$$

and the efficiency of transmission is

$$\frac{\text{output}}{\text{output + losses}} = \frac{5,000}{5,000 + 2,000}$$
$$= 71.4 \text{ percent.}$$

NOTES:

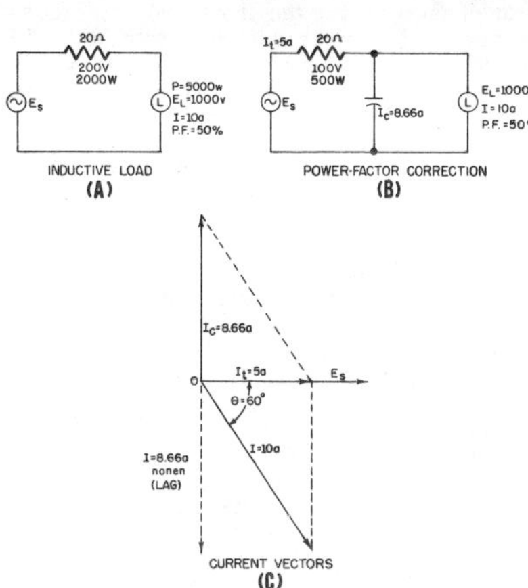

Figure 13-11.—Power-factor correction.

If a 1,000-volt capacitor of negligible losses supplying a leading current of 8.66 amperes is placed in parallel with the inductive load (fig. 13-11 (B)), the total line current will be reduced from 10 amperes to 5 amperes.

The current vectors are shown in figure 13-11 (C). The nonenergy component of the current in the inductive branch is $I_{nonen} = I \sin \theta = 10$ $(\sin 60^\circ = 0.866) = 8.66$ amperes (lagging) and is 180° out of phase with the capacitor current of 8.66 amperes (leading). These currents circulate between the capacitor and the inductive load and do not enter the line. The vector sum of the capacitor current and the total inductive load current is equal to the line current (I_t = 5 amperes), and in the vector diagram is represented as the diagonal of the parallelogram of which the 2 branch currents are the sides. The line current vector, I_t, is in the same horizontal reference as the load voltage vector, indicating that the line current is in phase with the voltage applied to the parallel combination of the inductive load and the capacitor.

The reduction in line current from 10 to 5 amperes reduces the line loss from 2,000 watts to $5^2 \times 20 = 500$ watts and increases the efficiency of transmission from 71.4 percent to $\frac{5{,}000}{5{,}000 + 500} = 91$ percent. Thus, the improvement in efficiency of operation of the line and load is demonstrated. The condition represented involves an interchange of energy between the inductive and capacitive branches known as parallel resonance and is described in the training manual, Basic Electronics, NavPers 10087-B, chapter 4, on the subject of resonance in LRC circuits.

VOLTAGE REDUCTION WITH RESISTANCE

The example of figure 13-12 illustrates the advantages and disadvantages of controlling the voltage on a load by means of resistance and inductance and also the effect of power-factor correction on the circuit efficiency.

In figure 13-12 (A), the voltage applied to the load is reduced from 100 volts to 50 volts through the action of the dropping resistor. The circuit

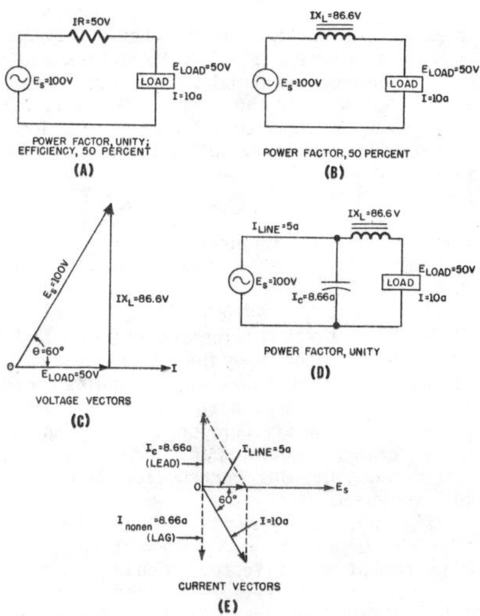

Figure 13-12.—Voltage reduction and power-factor correction.

NOTES:

current is 10 amperes, and the voltage across the dropping resistor is 50 volts. With this arrangement, the input power to the circuit is divided equally between the resistor and the load, and the circuit efficiency $\left(\frac{\text{output}}{\text{output} + \text{losses}} \times 100\right)$ is $\frac{500}{500+500} \times 100 = 50$ percent. This arrangement represents an inefficient method of voltage reduction.

VOLTAGE REDUCTION WITH INDUCTANCE

In figure 13-12 (B), the voltage applied to the load is reduced to 50 volts through the action of the series inductor. Neglecting the relatively small losses in the inductor, the circuit efficiency remains high compared to the efficiency of the previous circuit, but the series inductor lowers the power factor from unity (100 percent) to 50 percent. The circuit current is still 10 amperes, and the accompanying line loss between the source and the inductor unnecessarily high.

The voltage vectors for the LR circuit are shown in figure 13-12 (C). The load voltage of 50 volts is the base of the right triangle and is in phase with the load current. The voltage drop across the inductor is 86.6 volts and is the altitude of the voltage triangle. The source voltage, E_S = 100 volts, is the vector sum of the load voltage and the inductor voltage and is the hypotenuse of the voltage triangle. The circuit power factor is

$$\cos\theta = \frac{E_{load}}{E_{source}} = \frac{50}{100} = 0.5$$

and

$$\theta = 60^\circ$$

In figure 13-12 (D), the circuit power factor is improved to unity by the addition of a capacitor of negligible losses which supplies a leading current of 8.66 amperes. This current supplies the nonenergy component of the current in the branch containing the inductor and load, and reduces the line current from 10 amperes to 5 amperes.

The current vectors are shown in figure 13-12 (E). The capacitor current of 8.66 amperes is represented by the vector extending above the horizontal voltage reference vector to indicate a lead of 90°. The inductor branch (load) current extends below the horizontal at an angle of 60° to indicate lag. The vector sum of these currents is the diagonal of the parallelogram of which the branch currents are the sides, and is a horizontal vector of 5 amperes in phase with the source voltage. Thus, the power factor of the parallel circuit is unity and the line current is reduced from 10 amperes to 5 amperes.

ADVANTAGES OF INDUCTIVE ARRANGEMENT WITH POWER-FACTOR CORRECTION

In the example under consideration, voltage reduction with a series inductor and a shunt capacitor provides a means of supplying a 50-volt 10-ampere load from a 100-volt source with only the small losses associated with the reactive components. The line current is kept to the minimum value required to supply the 500-watt load and the circuit efficiency is high. Inductive control alone provides a means of reducing the load voltage by changing the phase of the applied voltage with respect to the load voltage. The addition of the capacitor reduces the current in the line without altering the load current, and thus reduces the line losses; at the same time the circuit power factor is increased to unity. In most circuits it is not economical to improve the power factor to unity, but to improve it, for example, from 50 percent lagging to 85 percent lagging. The reduction in line losses is most pronounced in this range and any further reduction in losses may not justify the added expense of the capacitance required to further improve the power factor in the range from 85 percent lagging to unity.

EFFECTIVE RESISTANCE-NONUSEFUL ENERGY LOSSES IN A-C CIRCUITS

ENERGY CONCEPT OF RESISTANCE

The energy stored in the magnetic field of a pure inductance, as the result of a rise in current through the coil, is returned to the circuit when the current decreases and the field collapses. Similarly, the energy stored in the electric field of a pure capacitance, as a result of the rise in voltage across the capacitor, is returned to the circuit when the voltage falls and the field collapses. Hence, in a pure inductance and a pure capacitance there is no loss or expenditure of energy.

When current flows through a conductor having appreciable resistance the flow is accompanied by the generation of heat. Work is done in moving the electrons through the conductor

NOTES:

resistance. The energy converted into heat is not returned to the circuit when the current falls, but is expended rather than stored. Thus, energy is stored periodically in inductance and capacitance but always expended in resistance.

Because resistance is the only circuit component capable of expending electrical energy, all energy expended in any circuit can be identified in electrical terms, one factor of which is effective resistance. The effective resistance, R_{ac}, of any circuit may be defined as the ratio of the true power absorbed by the circuit to the square of the effective current flowing in the circuit, or $R_{ac} = \frac{P}{I^2}$. When the power is expressed in watts and the current is expressed in amperes, the effective resistance will be in ohms. The d-c circuit resistance as measured by an ohmmeter or d-c bridge may be considerably lower than the effective a-c resistance as calculated from the readings on a wattmeter and an ammeter.

For example, assume that a motor draws 1 kilowatt from a 110-volt source. The input current is 10 amperes. The effective resistance between the motor terminals is

$$R_{ac} = \frac{P}{I^2} = \frac{1{,}000}{(10)^2} = 10 \text{ ohms}$$

The d-c resistance measured between the motor terminals, for example with an ohmmeter, is 0.5 ohm. Thus, in this example the effective a-c resistance is 20 times the d-c resistance. Most of the energy taken from the line is converted into mechanical energy and is not returnable to the electric circuit; hence, it is represented electrically as being expended in an effective a-c resistance of 10 ohms.

The source of power for the motor is unaware of the manner in which the motor expends the electrical energy. To the source, the motor appears as an impedance, $Z = \frac{E}{I} = \frac{110}{10} = 11$ ohms, having a resistive component of 10 ohms. The nature of the various energy conversions taking place inside the motor is important only when the motor itself is being analyzed. From this point of view a motor, electric light, loudspeaker, electron tube, or any other electrical device can be pictured as an equivalent electric circuit containing the fundamental components of inductance, capacitance, and resistance. The energy expended in the circuit is always interpreted in terms of the effective resistance component.

The energy expended in any electrical device may be divided into two parts: (1) That which is converted into useful form; and (2) that which is not useful. No machine has been built that is capable of perfect conversion—that is, one in which there are no losses. In the motor, for example, there are friction losses in the bearings and heat losses in the windings as a result of the current flow through the resistance.

The number of possible nonuseful losses in a-c circuits is much greater than in d-c circuits. These include: (1) ohmic-resistance loss, (2) skin-effect loss, (3) eddy-current loss, (4) dielectric loss, (5) magnetic-hysteresis loss, (6) corona loss, and (7) radiation loss.

EFFECTIVE RESISTANCE OF CONDUCTORS

The effective (a-c) resistance of electrical conductors is frequently higher than their d-c resistance especially when they are embedded in iron slots, as in the case of motor and generator armatures; and when they are being used in high-frequency circuits, as in radio transmitters and receivers.

Direct current is distributed uniformly throughout the cross-sectional area of a conductor. For example, if a conductor having a cross-sectional area of 1,000 circular mils is carrying one ampere of direct current then one-thousandth of an ampere (one milliampere) is flowing in each circular mil of cross-sectional area. However, when the current in the conductor varies in amplitude, this uniform distribution throughout the conductor cross section is no longer obtained. The accompanying magnetic field is strongest near the center of the conductor and weaker at the circumference. The varying field induces a voltage in the conductor that opposes the change in current. The voltage induced in that portion of the conductor near the center is greater than the voltage induced in the outer surface of the conductor. The total opposition to the current flow includes the effect of this induced emf and is greater near the center of the conductor than at the surface. Therefore, the current divides inversely with the opposition—more of the current flowing near the circumference, and less near the center of the conductor.

The overall result of this action is a decrease in the available area of cross section

NOTES:

to conduct the current and an increase in conductor resistance. The decrease in area and increase in resistance become pronounced at high frequencies, at high current densities, and at high magnetic flux densities. This action is called skin effect. It represents the tendency of a-c conductors to carry the circuit current on the surface, or skin, of the conductors rather than uniformly throughout their cross section. As a result of this tendency, many electrical conductors are made of hollow tubing in order to save the added weight and expense of the unused central portion of the solid conductor. The effective a-c resistance of an isolated circular conductor varies approximately as the product of the square root of the frequency and the length of the conductor, and inversely as the conductor diameter.

EFFECTIVE RESISTANCE OF INDUCTORS

When a conductor is wound in the form of a coil, the current is concentrated on the inner sides of the turns and into an area much smaller than would be the case in an isolated straight conductor. This action results in a large increase in effective resistance. The area in which the current is concentrated decreases as the frequency increases, hence, effective resistance will increase with frequency. When two or more conductors carrying alternating current are so placed that the magnetic field of one reacts with the field of the other, the resultant field around each conductor is no longer uniform. The change in current distribution in a conductor due to the action of an alternating current in a nearby conductor is called proximity effect.

The proximity effect decreases as the separation between conductors increases. Thus, to lower the effective resistance of radiofrequency inductance coils, it is common practice to space the turns a distance equal to the diameter of the conductor. This decreases the reaction between magnetic fields of adjacent turns and permits the current to distribute itself over a larger area in the cross section of each turn.

The inductance of a hollow-core coil operating at a frequency of 60 Hz is increased many fold when a laminated core of soft silicon steel is inserted in the coil. This increase is due to the high permeability of the transformer-iron laminations. Thus, the skin effect is also increased due to the increased field strength. In addition to the increased skin effect in the coil the effective a-c resistance is further increased because of the magnetic hysteresis loss in the iron. Thus, if the coil is connected to a constant-potential a-c source, the current in the coil will decrease when the iron core is inserted because of a small increase in effective resistance and a large increase in the coil reactance. If the laminated steel core is removed and a piece of steel shafting is inserted in the coil, the effective resistance is further increased due to the eddy-current losses and the larger hysteresis losses in the steel shaft. A wattmeter inserted in the coil circuit will indicate this increase in effective resistance by an increased deflection when the solid steel core is inserted in place of the laminated core.

Powdered iron cores are used in certain types of coils on frequencies as high as 100 megahertz in order to limit the effective resistance of the coil to a satisfactorily low value. The iron particles are separated from each other by an insulated coating and when compressed into cylindrical form and inserted in the coil the induced voltage in each iron particle is so small in relation to the resistance to the path for eddy currents that the accompanying heat loss is negligible. Eddy-current losses are reduced in generator and motor armature conductors of large size by laminating the conductors and insulating the adjacent laminations in a manner similar to that in which the iron of the armature core itself is laminated. Thus, the effective resistance of the armature conductors is reduced.

EFFECTIVE RESISTANCE OF CAPACITORS

The equivalent circuits of a low-loss capacitor were described earlier in this chapter and the factors affecting the equivalent series resistance noted. The effective a-c resistance of a capacitor is equal to its equivalent series resistance and represents the factor which when multiplied by the square of the effective capacitor charging current will equal the power expended in heat in the capacitor circuit.

As mentioned previously, most of the heating is produced in solid dielectrics, and only a negligible amount is produced in the capacitor plates themselves. The dielectric heating is produced by dielectric displacement currents described in

NOTES:

connection with figure 13-7. In most electrical circuits dielectric heating is a nonuseful loss. However, in one commercial application, dielectric heating has been put to good use—that of facilitating the gluing together of stacks of laminated plywood. The plywood laminations are stacked between the plates of a capacitor and moderately high-frequency voltage is applied across the plates. The resulting dielectric displacement currents heat the stack from the inside and the glue is quickly set—much more rapidly than in processes involving the external application of steam heat.

CORONA LOSS

Corona loss occurs as the result of the emission of electrons from the surface of electrical conductors at high potentials. It is dependent upon the curvature of the conductor surface, with most emission occurring from sharp points and the least emission occurring from surfaces having a large radius of curvature. Corona loss is frequently accompanied by a visual blue glow and an audible hissing sound as the electrons leak off the conductor surface into the atmosphere. Corona loss increases with voltage increase and decreases with increase in atmospheric pressure. This loss is held to a satisfactorily low value by (1) the use of large-diameter conductors, (2) not excessively high voltages, (3) smooth polished surfaces, (4) avoiding sharp points, bends, or turns, and (5) in some devices, for example, high-voltage capacitors, by the use of a compressed gas to retard the electron emission.

RADIATION LOSS

Radiation loss is not appreciable at power line frequencies, but in the field of communications this loss may become excessive. Power is radiated from transmitting antennas in the form of electric and magnetic fields, and its magnitude varies as the square of the antenna input current and as the so-called radiation resistance of the antenna. The transmission line that connects the transmitter and the antenna may, under certain circumstances, develop a radiation loss. This loss is discussed in chapter 12 of Basic Electronics, NavPers 10087-B, in connection with various types of transmission lines.

NOTES:

CHAPTER 14

CIRCUIT PROTECTIVE AND CONTROL DEVICES

Electricity, when properly controlled, is of vital importance to the operation of equipment. When it is not properly controlled, however, it can become dangerous and destructive. It can destroy components or complete units; it can injure personnel and even cause their death.

It is of the greatest importance, then, that all precautions necessary be taken to protect the electrical circuits and units and to keep this force under proper control at all times. In this chapter some of the devices that have been developed to protect and control electrical circuits are discussed.

PROTECTIVE DEVICES

When an electrical unit is built, the greatest care is taken to insure that each separate electrical circuit is fully insulated from all others so that the current in a circuit will follow its intended individual path. Once the unit is placed into service, however, there are many things that can happen to alter the original circuitry. Some of these changes can cause serious troubles if they are not detected and corrected in time.

Perhaps the most serious trouble we can find in a circuit is a direct short. Recall that this term is used to describe a situation in which some point in the circuit, where full system voltage is present, comes in direct contact with the ground or return side of the circuit. This establishes a path for current flow that contains no resistance other than that present in the wires carrying the current, and these wires have very little resistance.

According to Ohm's Law, if the resistance in a circuit is extremely small, the current will be extremely great. When a direct short occurs, then, there will be an extremely heavy current flowing through the wires. Suppose, for instance, that the two leads from a battery to a motor came in contact with each other. Not only would the motor stop running, because of the current going through the short, but the battery would become discharged quickly (perhaps ruined), and there would also be danger of fire.

The battery cables in our example would be very large wires, capable of carrying very heavy currents. Most wires used in electrical circuits are considerably smaller, and their current-carrying capacity is quite limited. The size of the wires used in any given circuit is determined by the amount of current the wires are expected to carry under normal operating conditions. Any current flow greatly in excess of normal, such as there would be in case of a direct short, would cause a rapid generation of heat.

If the excessive current flow caused by the short is left unchecked, the heat in the wire will continue to increase until something gives way. Perhaps a portion of the wire will melt and open the circuit so that nothing is damaged other than the wires involved. The probability exists, however, that much greater damage would result. The heat in the wires could char and burn their insulation and that of other wires bundled with them, which could cause more shorts. If a fuel or oil leak is near any of the hot wires, a disastrous fire might be started.

To protect electrical systems from damage and failure caused by excessive current, several kinds of protective devices are installed in the systems. Fuses, circuit breakers, and thermal protectors are used for this purpose.

DESCRIPTION AND PURPOSE

Circuit protective devices, as the name implies, all have a common purpose: to protect the units and the wires in the circuit. Some are designed primarily to protect the wiring. These open the circuit in such a way as to stop the current flow when the current becomes greater than the wires can safely carry. Other devices

NOTES:

are designed to protect a unit in the circuit by stopping current flow to it when the unit becomes excessively warm.

FUSES

The simplest protective device is a fuse. All fuses are rated according to the amount of current that is safely carried by the fuse element at a rated voltage. Usually, the current rating is in amperes, but some instrument fuses are rated in fractions of an ampere. When a fuse blows, it should be replaced with another of the same rated voltage and current capacity, including the same current-versus-time characteristic.

The most important fuse characteristic is its current-versus-time or "blowing" ability. Three time ranges for existence of overloads can be broadly defined as fast, medium, and delayed. FAST, may range from 5 microseconds through one-half second; MEDIUM, 1/2 to 5 seconds; DELAYED, 5 to 25 seconds.

Normally, when the circuit is overloaded, or a fault develops, the fuse element melts and opens the circuit that it is protecting. However, all fuse openings are not the result of current overload or circuit faults. Abnormal production of heat, aging of the fuse element, poor contact due to loose connections, oxides or corrosion forming within the fuse holder, and the heated condition of the surrounding atmosphere will alter the heating conditions and the time required for the element to melt.

Delayed-Action Fuses

Some equipment, such as an electric motor, requires more current during starting than for normal running. Thus, a fast-time or medium-time fuse rating that will give running protection might blow during the initial period when high starting current is required. Delayed-action fuses are used to handle these situations.

One type of delayed-action fuse has a heater element connected in parallel with the fuse element in order to get the delayed action. During normal operation, the heat developed in the fuse link is not great enough to melt the link. The melting, or opening, of the fuse link depends on the transfer of heat to the link from the heater. Therefore, more time is needed to melt the link than would be required if the link were directly heated.

Because the heater and fuse element are in parallel, the opening of the fuse element will cause the total circuit current to flow through the heater. The high current will cause the heater to burn out and completely open the circuit.

Another type of delayed-action fuse has the fuse element and heater connected in series. Current above that of the rated value for a short time will have no effect on the fuse or heater. However, prolonged overloads cause the heater section to become hot enough to melt the junction between the elements; this action opens the circuit.

Delayed-action fuses are sometimes called Time-Lag fuses; and three trade names "Slo Blo," "Fusestat," and "Fusetron" are in common use.

Plug Fuses

The plug fuse is constructed so that it can be screwed into a socket mounted on the control panel or distribution center. The fuse link is enclosed in an insulated housing of porcelain or glass. The construction is so arranged that the fuse link is visible through a window of mica or glass. Therefore, an open element may be located by visual examination. When found to be defective, the fuse is discarded and a new fuse installed in its place. The plug fuse is used primarily to protect low-voltage, low-current circuits. The operating ratings range from 0.5 to 30 amperes up to 150 volts.

Cartridge Fuses

In operation, the cartridge fuse is exactly the same as the plug fuse. In construction, the fuse link is enclosed in a tube of insulating material with metal ferrules at each end (for contact with the fuse holder). The dimensions of cartridge fuses vary with the current and voltage ratings.

Blown-Fuse Indicators

It is not always possible to detect a blown fuse by a visual examination. Hence, fuses are often equipped with a device that will provide a visual indication so that a blown-fuse condition can be readily detected (fig. 14-1). These devices consist of the spring-loaded and the neon-lamp types of blown-fuse indicators.

In the spring-loaded type (fig. 14-1), when the link opens, it releases a spring that is held under tension. This action exposes an indicator, which makes the visual location of the blown fuse possible.

NOTES:

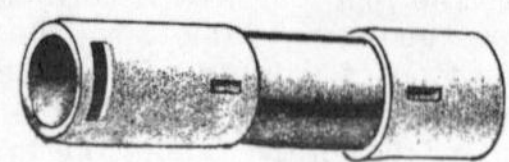

SPRING-LOADED INDICATOR

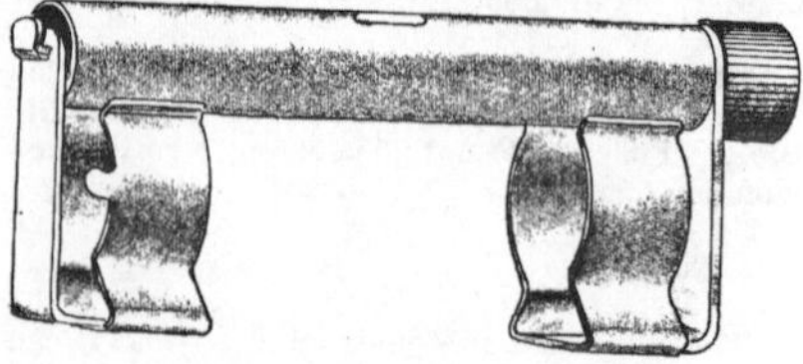

NEON-LAMP INDICATOR

Figure 14-1.—Blown-fuse indicators.

The neon lamp type (fig. 14-1) is designed to be mounted on the fuse. When the link opens, a neon lamp glows to show a blown fuse.

When no indicator is used, it is necessary to test the fuse continuity with a megger, ohmmeter, or voltmeter. Various methods of testing will be described later in this chapter.

Most fuse panels and switchboards are of the enclosed panel type. The term "dead-front" means that all fuses and bus connections are enclosed in a metal cabinet when the cover is closed. The use of this type of construction reduces the possibility of equipment damage and danger to personnel. Modern switchboards are of the "dead-front" type.

However, the complete enclosure of the equipment makes it less accessible for test purposes. Therefore, most fuses used on "dead-front" switchboards have indicators that show when a fuse is blown. The fuse holder consists of a molded phenolic base, plug, and cap with a built-in indicator lamp (blown-fuse indicator). The lamp is usually a small neon bulb, which normally is shunted by the fuse element. When the fuse opens, the shunt is removed, causing an increase in the voltage across the neon lamp. The lamp then glows, indicating the open fuse.

TROUBLESHOOTING FUSED CIRCUITS

An electrical system may consist of a comparatively small number of circuits or, in the larger systems, the installation may be equal to that of a fair sized city.

Regardless of the size of the installation, an electrical system consists of a source of power (generators or batteries) and a means of delivering this power from the source to the various loads (lights, motors, and other electrical equipments).

From the main power supply the total electrical load is divided into several feeder circuits and each feeder circuit is further divided into several branch circuits. Each final branch circuit is fused to safely carry only its own load while each feeder is safely fused to carry the total current of its several branches. This reduces the possibility of one circuit failure interrupting the power for the entire system. The feeder distribution boxes and the branch distribution boxes contain fuses to protect the various circuits.

The distribution wiring diagram showing the connections that might be used in a lighting system is illustrated in figure 14-2. An installation might have several feeder distribution boxes, each supplying six or more branch circuits through branch distribution boxes.

Fuses F_1, F_2, and F_3 (fig. 14-2) protect the main feeder supply from heavy surges such as short circuits or overloads on the feeder cable. Fuses A-A1 and B-B1 protect branch No. 1. If trouble develops and work is to be done on branch No. 1, switch S_1 may be opened to isolate this branch. Branches 2 and 3 are protected and isolated in the same manner by their respective fuses and switches.

Branch Circuit Tests

Usually, receptacles for portable equipment and fans are on branch circuits separate from lighting branch circuits. Test procedures are the same for any branch circuit. Therefore, a description will be given of the steps necessary to (1) locate the defective circuit and (2) follow through on that circuit and find the trouble.

Assume that, for some reason, several of the lights are not working in a certain section. Because several lights are out, it will be reasonable to assume that the voltage supply has been interrupted on one of the branch circuits.

To verify this assumption, first locate the distribution box feeding the circuit that is inoperative. Then make sure that the inoperative circuit is not being supplied with voltage. Unless the circuits are identified in the distribution box,

NOTES:

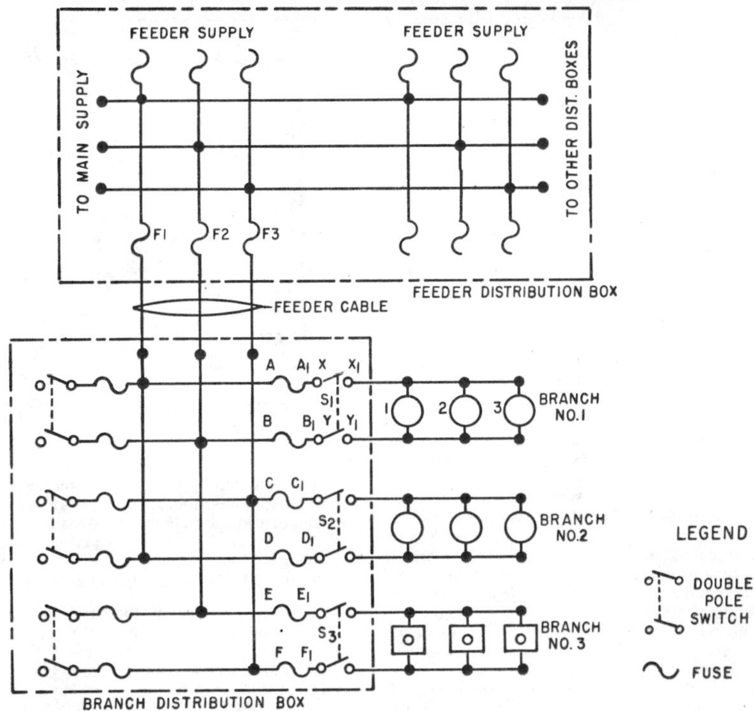

Figure 14-2.—Three-phase distribution wiring diagram.

the voltage at the various circuit terminations will have to be measured. For the following procedures, use the circuits shown in figure 14-2 as an example circuit.

To pin down the trouble, connect the voltage tester to the load side of each pair of fuses in the branch distribution box. No voltage between these terminals indicates a blown fuse or a failure in the supply to the distribution box. To find the defective fuse, make certain S_1 is closed, then connect the voltage tester across A-A1, and next across B-B1 (fig. 14-2). The full-phase voltage will appear across an open fuse, provided circuit continuity exists across the branch circuit. However, if there is an open circuit at some other point in the branch circuit, this test is not conclusive. If the load side of a pair of fuses does not have the full-phase voltage across its terminals, place the tester leads on the supply side of the fuses. The full-phase voltage should be present. If the full-phase voltage is not present on the supply side of the fuses, the trouble is in the supply circuit from the feeder distribution box.

Assume that you are testing at terminals A-B (fig. 14-2) and that normal voltage is

NOTES:

present. Move the test lead from A to A1. Normal voltage between A1 and B indicates that fuse A-A1 is in good condition. To test fuse B-B1, place the tester leads on A and B, and then move the lead from B to B1. No voltage between these terminals indicates that fuse B-B1 is open. Full-phase voltage between A and B1 indicates that the fuse is good.

This method of locating blown fuses is preferred to the method in which the voltage tester leads are connected across the suspected fuse terminals, because the latter may give a false indication if there is an open circuit at any point between either fuse and the load in the branch circuit.

CIRCUIT BREAKERS

A circuit breaker is designed to break the circuit and stop the current flow when the current exceeds a predetermined value. It is commonly used in place of a fuse and may sometimes eliminate the need for a switch. A circuit breaker differs from a fuse in that is "trips" to break the circuit and it may be reset, while a fuse melts and must be replaced.

Several types of circuit breakers are commonly used. One is a magnetic type. When excessive current flows in the circuit, it makes an electromagnet strong enough to move a small armature which trips the breaker. Another type is the thermal overload switch or breaker. This consists of a bimetallic strip which, when it becomes overheated from excessive current, bends away from a catch on the switch lever and permits the switch to trip open.

Some circuit breakers must be reset by hand, while others reset themselves automatically. When the circuit breaker is reset, if the overload condition still exists, the circuit breaker will trip again to prevent damage to the circuit.

One common type of circuit breaker now being used is depicted in figure 14-3. This breaker is designed for front or rear connections as required and may be mounted so as to be removable from the front without removing the circuit breaker cover. The voltage ratings of this breaker are 500 volts a.c., 60 Hz, or 250 volts d.c. with a maximum current capacity of 250 amperes. Trip units (fig. 14-4) for this breaker are available with current ratings of 125, 150, 175, 225, and 250 amperes.

The trip unit houses the electrical tripping mechanisms, the thermal element for tripping the circuit breaker on overload conditions, and

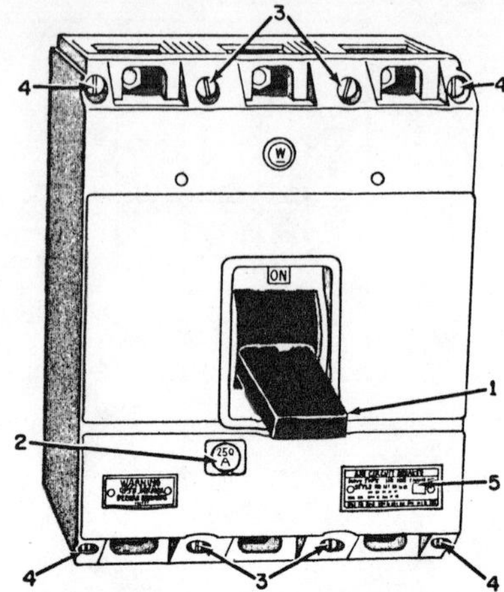

1. Operating handle shown in latched position.
2. Ampere rating maker.
3. Mounting screws.
4. Cover screws.
5. Breaker nameplate.

Figure 14-3.—Circuit breaker, front view.

the instantaneous trip for tripping on short circuit conditions. The automatic trip devices of this circuit breaker are "trip free" of the operating handle; this means the circuit breaker cannot be held closed by the operating handle if an overload exists. When the circuit breaker has tripped due to overload or short circuit, the handle rests in a center position. To reclose after automatic tripping, the handle must be moved to the extreme OFF position which resets the latch in the trip unit; then the handle must be physically moved to the ON position. Metal locking devices are available that can be attached to the handles of circuit breakers to prevent accidental operation.

THERMAL PROTECTORS

A thermal protector, or switch, is a device used to protect a motor. It is designed to open the circuit automatically whenever the temperature of the motor becomes excessively high. It has two positions, open and closed. The most

NOTES:

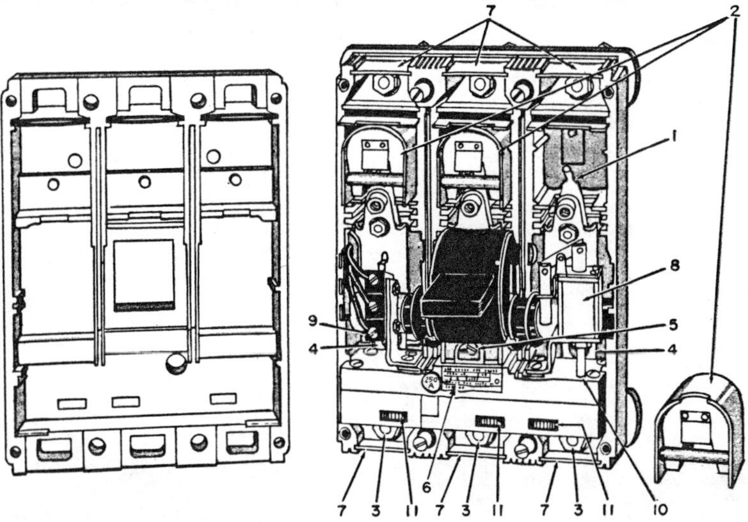

1. Stationary contact.
2. Arc suppressors.
3. Terminal stud nuts and washers.
4. Trip unit line terminal screw-outer poles.
5. Trip unit line terminal screw-center pole.
6. Trip unit nameplate.
7. Terminal barriers.
8. Shunt trip.
9. Auxiliary switch.
10. Hole for shunt trip undervoltage release plunger.
11. Instantaneous trip adjusting wheels.

Figure 14-4.—Circuit breaker, cover and arc suppressor removed.

common use for a thermal switch is to keep a motor from overheating. If some malfunction in the motor causes it to overheat, the thermal switch will break the circuit intermittently. If the trouble is a locked rotor, the intermittent opening and closing of the circuit may release the rotor and allow the motor to resume normal operation.

The thermal switch contains a bimetallic disk, or strip, which bends and breaks the circuit when it is heated. This happens because one of the metals expands more than the other when they are subjected to the same temperature. When the strip or disk cools, the metals contract and the strip returns to its original position and closes the circuit.

OVERLOAD DEVICE

A prolonged overloaded electrical system can be damaged beyond repair by the resulting heat and flame. Therefore, it is expedient to use a device which can detect an overload before damage occurs and either warns the operator of

NOTES:

the hazardous conditon or automatically turns off the power. Relays have been designed which are capable of accomplishing these protective functions. A few of these relays that are commonly used as overload devices are discussed in greater detail later in this manual.

CONTROL DEVICES

Control devices are those electrical accessories which govern (in some predetermined way) the power delivered to any electrical load.

In its simplest form the control applies voltage to, or removes it from a single load. In more complex control systems, the initial switch may set into action other control devices that govern motor speeds, servomechanisms, temperatures, and numerous other equipments. In fact, all electrical systems and equipment are controlled in some manner by one or more controls. A controller is a device or group of devices which serves to govern, in some predetermined manner, the device to which it is connected.

In large electrical systems, it is necessary to have a variety of controls for operation of the equipment. These controls range from simple pushbottons to heavy duty contactors that are designed to control the operation of large motors. The pushbutton is manually operated while a contactor is electrically operated.

SWITCHES

A switch may be described as a device used in an electrical circuit for making, breaking, or changing connections under conditions for which the switch is rated. Switches are rated in amperes and volts; the rating refers to the maximum voltage and current of the circuit in which the switch is to be used. Because it is placed in series, all the circuit current will pass through the switch. Because it opens the circuit, the applied voltage will appear across the switch in the open circuit position. Switch contacts should be opened and closed quickly to minimize arcing; therefore, switches normally utilize a snap action.

Many types and classifications of switches have been developed. A common designation is by the number of poles, throws, and positions they have. The number of poles indicates the number of terminals at which current can enter the switch. The throw of a switch signifies the number of circuits each blade or contactor can complete through the switch. The number of positions indicates the number of places at which the operating device (toggle, plunger, etc.) will come to rest. Figure 14-5 presents the schematic diagrams of some often used switches.

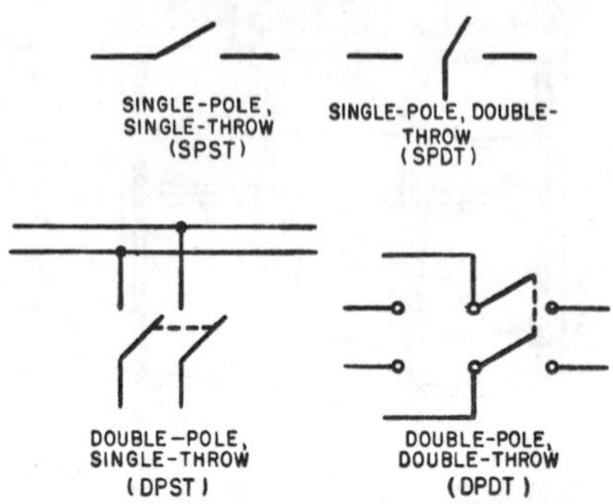

Figure 14-5.—Schematic diagrams of commonly used switches.

An example of the switch position designation is a toggle switch which comes to rest at either of two positions, opening the circuit in one position and completing it in another. This is called a two-position switch. A toggle switch which is spring loaded to the OFF position and must be held in the ON position to complete the circuit is called a momentary contact two-position switch. If the toggle switch will come to rest at any of three positions, it is called a three-position switch.

Another means of classyfying switches is the method of actuation; that is, toggle, pushbutton, sensitive, and rotary types. Further classification can be accomplished by a description of switch action such as on-off, momentary on-off, on-momentary off, etc. Momentary contact switches hold a circuit closed or open only as long as the operator deflects the actuating control.

Manually Operated Switches

One of the most common types of switches is the toggle. Toggle switches have their moving parts enclosed. A double-pole, double-throw, ON-OFF-ON toggle switch is shown in

NOTES:

figure 14-6. These switches have many uses and are used especially for applying power to various circuits. They are often provided with a luminous tip on the lever so as to be visible in the dark.

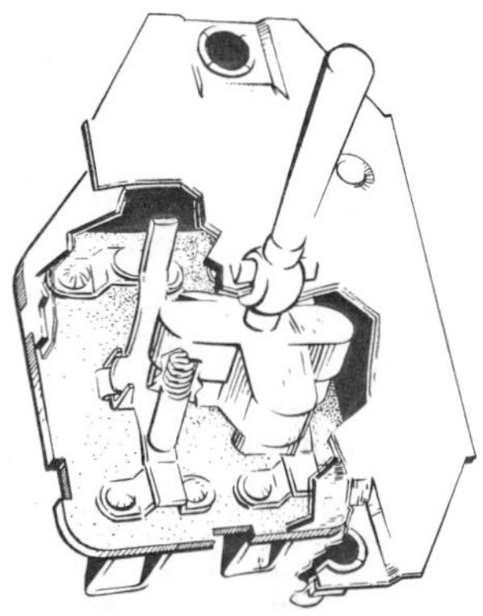

Figure 14-6.—Toggle switch.

Pushbutton switches have one or more stationary contacts and one or more movable contacts. The movable contacts are attached to the pushbutton by an insulator. This switch is usually spring loaded and is of the momentary contact type. These switches have many uses, such as indicator light checks and circuit reset.

A rotary selector switch may perform the functions of a number of switches. As the knob of a rotary selector switch is rotated, it opens one circuit and closes another. This can be seen from an examination of figure 14-7. Some rotary switches have several layers of wafers. By adding wafers, the switch can be made to operate as a large number of switches. Ignition switches and voltmeter selector switches are typical examples of this type.

Mechanically Operated Switches

Mechanically operated switches are used in many applications. They are widely used because of their small size, light weight, and excellent dependability. The term Micro switch, although frequently used in referring to all switches of this type, is a trade name for the switches made by the Micro Switch Division of the Minneapolis Honeywell Regulator Company.

These switches will open or close a circuit with a very small movement of the tripping device (1/16 inch or less). They are usually of the pushbutton variety and depend upon one or more springs for their snap action. For example, the heart of the Micro switch is a beryllium copper spring, heat-treated for long life and unfailing action. The simplicity of the one-piece spring contributes to the long life and dependability of this switch. The basic Micro switch is shown in figure 14-8.

The versatility of the snap-action switch is shown in figure 14-9, which shows how the basic switch may be used with different type enclosures and actuators. The particular type switching unit used depends upon the function it is to perform and environmental conditions. Figure 14-9 (A) shows a lightweight aluminum enclosure which may contain one or more plastic enclosed basic switches. The plastic enclosure provides electrical insulation between the housing and the energized electrical parts, support for the terminals, and a dusttight box around the electrical contacts.

Figure 14-9 (B) shows one of the various types of switch actuators that may be used with basic switches when complete enclosures are not needed. They provide protection and mounting means. They relay the operating motion from a cam or slide to the basic switch in a way that assures long life and dependability. Figure 14-9 (C) shows a toggle type switch which uses one or more of the basic subminiature switches. These subminiature switches are smaller than the conventional Micro switch and are finding wide use in various switching arrangements.

Pressure-operated switches usually have Bourdon tubes, syphons, or diaphrams against which the fluids or air operate to actuate the switch. Some uses of pressure switches are in connection with fuel, oil, and hydraulic pressure signals and electric heaters.

Thermal switches usually incorporate a bimetallic sheet that bends or snaps at a desired temperature to actuate the switch. They are used extensively as circuit breakers and also find application in controlling the igniter circuit on heater and operating signal lights at critical temperatures.

NOTES:

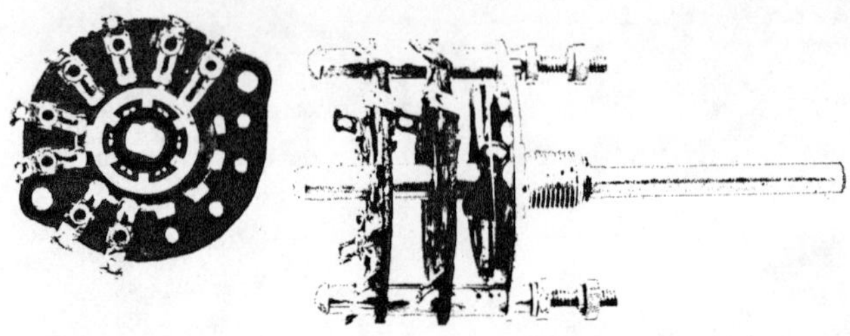

Figure 14-7.—Rotary selector switch.

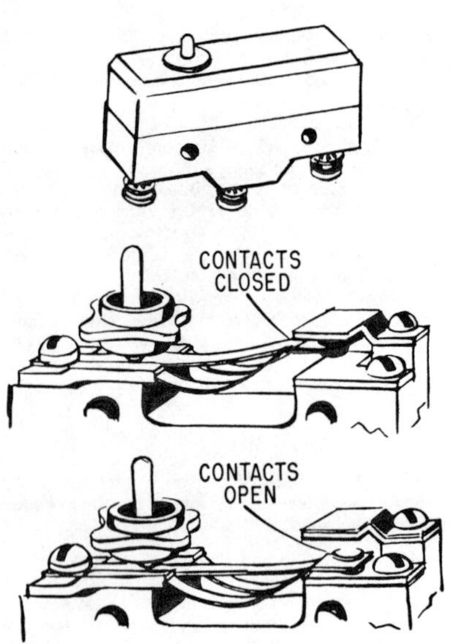

Figure 14-8.—Micro switch.

Maintenance of Switches

While the switch itself is relatively simple to check, it sometimes offers difficulty in maintenance because of its location in inaccessible places. After a visual inspection of the connections and the switch, a continuity test will indicate any malfunctions. When the switch mechanism is found to be defective, it normally is not repairable and therefore should be replaced.

When enclosed switches are used, failure to seal properly around cable openings may cause difficulty. Altitude changes permit "breathing" of moist air into enclosures with improperly sealed cable openings, and the moisture in the air may condense within the switch enclosure. The condensation can short across the switch terminals and can corrode the switch actuators in a manner that may make them inoperative. This difficulty can be corrected by careful sealing of openings or by using hermetically sealed switches. Hermetically sealed switches will also prevent dust and dirt from reaching the contacts and thereby reduce the possibility of high resistance and open circuits.

Some switches are damaged during installation, particularly those with plastic housings. Proper care in installing or replacing plastic enclosed switches will eliminate this.

Some switch assemblies are equipped with adjustments which enable them to operate at a preset time or pressure. Caution should be exercised in making these adjustments; if they are not accurate, damage can result.

NOTES:

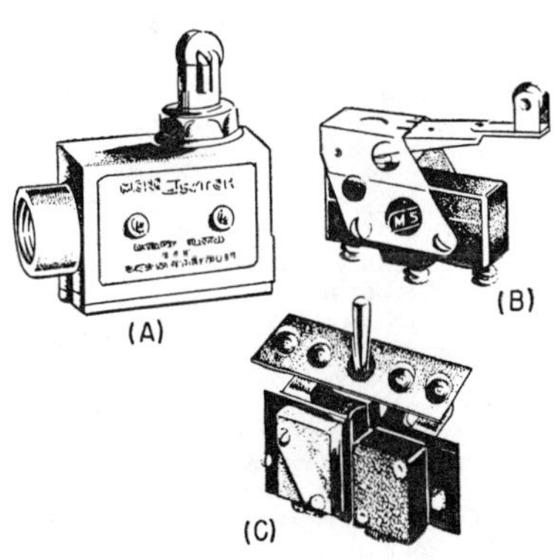

Figure 14-9.—Snap-action switching units.

RELAYS

Relays are electrically operated switches that are classified according to their use as control relays, power relays, or sensing relays. The power relays are the workhorses of a large electrical system. As such, they control the heavy power circuits.

The function of a control relay is to take a relatively small amount of electrical power and use it either to signal or to control a large amount of power. Where multipole relays are used, several circuits may be controlled simultaneously.

The use of relays saves space and weight by permitting the use of small switches at remote control stations. These switches permit the operator to control large amounts of current at other locations and the heavy power cables need to be run only to the point of use. Only lightweight control wires are connected to the control switches. Safety is also an important factor in using relays, since high power circuits can be switched remotely without danger to the operator.

Control relays, as their name implies, are frequently used in the control of other relays, although the small control relays find many other uses. With these, small currents can control the larger currents necessary to operate electrical devices. They find frequent use in automatic relaying circuits, where a small electric signal sets off a chain reaction of successively acting relays performing various functions. Control relays can also be used in so-called "lockout" action to prevent certain functions from occurring at the improper time. Various electrical operations in the equipment which must not occur simultaneously can be "interlocked" by control relays. Another important function of control relays in equipment is for "sensing." Control relays are used for sensing undervoltage and overvoltage, reversal of current, and excessive currents.

Another possible classification of relays is open, semisealed, and sealed. Semisealed relays have protextive covers and are gasketed against entrance of salt, dust, and foreign material into the contact or mechanism area. These relays are still considered satisfactory for certain applications in current equipment. Open relays are seldom used outside of black boxes.

For other applications in today's complex equipment, however, it is necessary to go beyond the protection offered by the open type and the semisealed relays. When such relays are used, quick changes in altitude, humidity, or temperature can cause condensation of water vapor within the unit. Subsequent low temperature will then freeze the moisture on the contactor with a resultant inability to carry electric current.

Hermetically sealed relays were developed to answer the demands of complex and delicate equipment. A true hermetic seal is generally considered one that is metal to metal or glass to metal. Plastic or plastic rubber type gasketed seals are not generally considered true hermetic seals. However, both semisealed and hermetically sealed relays are used. There are applications where a gasket type sealed relay may be adequate, but the true hermetically sealed type is generally considered to be more permanent. Besides being independent of environmental changes, the hermetically sealed relay also has the advantage of being protected from improper adjustments.

NOTES:

In general, the basic components of a relay are as follows: the coil or solenoid, the iron core, the fixed and movable contacts, and the mounting (and if sealed, the can). A manual switch, limit switch, or other small control device starts and stops the flow of current to the magnet coil. The flow of electric current through the coil creates a strong magnetic field around and within the coil. This magnetic field moves a clapper or plunger which completes the magnetic circuit. Figure 14-10 (A) shows a basic single coil clapper type relay. The dashed lines indicate the magnetic lines of flux.

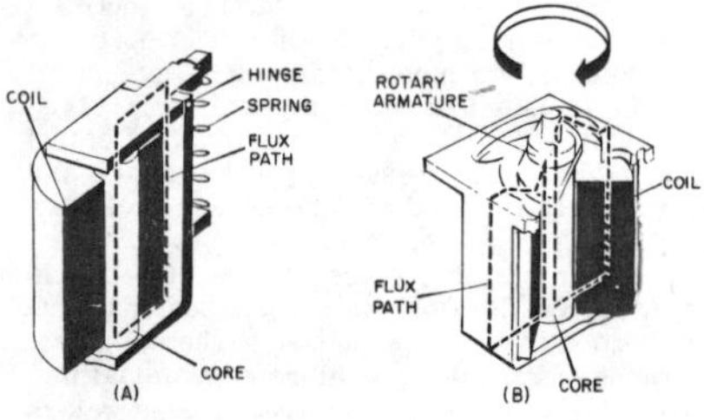

Figure 14-10.—Basic types of relays.

The second basic type of relay is the rotary. (See fig. 14-10 (B).) Although this type of construction is not as common as the clapper type, the rotary type has greater vibration and shock resistance than the others. The disadvantage is that they are somewhat sluggish and require higher operating power for many purposes. The rotary relay operates on the principle of an electric motor, but through only a small arc. The problem of hanging contacts on such a mechanism is a difficult one, and therefore the use of these devices is limited to applications where high shock warrants the larger size and weight. When used with standard wafer switch assemblies, this type of relay provides a means for assembling a switching device of any degree of complexity.

Occasionally relays operating from an a-c supply are encountered. These a-c relays depend upon the same fundamental principles as the d-c relay; that is, magnetic fields. When a.c. is applied to an electromagnet, the current will pass through zero twice every cycle. Since the pull on the armature is proportional to the current through the electromagnet, the armature tends to open every time the current nears zero, causing chatter. To remedy this, shading coils (sometimes called shaded poles) are used.

A shading coil consists of copper band or stamping which is short-circuited and embedded around part of the electromagnet pole face. By being placed around part of the pole face, it acts as a shorted transformer secondary. The current in the main coil lags the applied voltage by approximately 90°, and the flux is in phase with the current. The voltage of the shading coil is induced voltage and lags the current in the main coil by 90°. Since the shading coil acts like a shorted secondary (resistive), the current in the shading coil is in phase with the induced voltage. Therefore, the magnetic field of the shading coil lags the magnetic field of the main coil by 90°. This means that flux will exist in the electromagnet even when the main coil current becomes zero. Thus, chattering is prevented.

The arrangements of relay contacts are found in many different forms. Usually the number and sequence of switching operations to be performed dictates the contact arrangement.

It is often desirable to introduce time delays by use of relays. One method is to use a thermal relay for a time delay. Due to its simple mechanism, it can be made very small and hermetically sealed, making it ideal for use in aircraft. Because the thermal relay is activated by heat, it can be used on either a.c. or d.c.

Maintenance of Relays

The relay is one of the most dependable electromechanical devices in use, but like any other mechanical or electrical device, relays occasionally wear out or become inoperative for one reason or another. Should relay inspection determine that a relay has exceeded its safe life, the relay should be removed immediately and replaced with another of the same type. Care should be exercised in obtaining the same type replacement because relays are rated in voltage, amperage, type of service, number of contacts, continuous or intermittent duty, and similar characteristics.

For spotting potential relay trouble during preventive maintenance, the following guides are suggested: check for charred or burned insulation on the relay and for darkened or charred terminal leads coming from the relay. Both of these indicate overheating. If there is even a slight indication that the relay has overheated

NOTES:

it should be replaced with a new relay of the same type. An occasional cause of relay trouble is not the fault of the relay at all, but is due to overheating caused by the power terminal connectors not being tight enough. This should always be checked during preventive maintenance.

It is recommended that covers not be removed from semisealed relays in the field. Removal of a cover in the field, although it might give useful information to a trained eye, may result in entry of dust or other foreign material which may cause contact discontinuity.

NOTES:

CHAPTER 15

PRACTICAL ELECTRICAL WIRING

FUNDAMENTALS AND PROCEDURES OF ELECTRICAL WIRING

FUNDAMENTALS OF ELECTRICITY

Throughout this chapter, emphasis is placed on the constructional aspects of electric wiring. The term "phase" is used when referring to the angular displacement between two or more like quantities, either altenating electromotive force (EMF) or alternating currents. It is likewise used in distinguishing the different types of alternating current generators. For example, a machine designed to generate a single EMF wave is called a single-phase alternator, and one designed to generate two or more EMF waves is called a polyphase alternator.

Power generators will produce single or three-phase voltages that may be used for electrical power systems at generated voltages or through transformer systems.

1. Single-phase generators are normally used only for small lighting and single-phase motor loads. If the generated voltage is 120 volts then a two-wire system is used. One of the conductors is grounded and the other is ungrounded or hot. The generated single-phase voltage may be 240 volts. This voltage is normally used for larger single-phase motors. In order to provide power for lighting loads, the 240-volt phase is center-tapped to provide a three-wire single-phase system. The center tap is the grounded neutral conductor. The voltage from this grounded conductor to either of the two ungrounded (hot) conductors is 120 volts. This is one-half of the total phase value. The voltage between the two ungrounded conductors is 240 volts. This system provides power for both lighting and single-phase 240-volt motors.

2. The most common electrical system is the three-phase system. The generated EMF's are 120 degree apart in phase. Often three-phase systems may be carried by three or four wires. If connected in a delta (Δ), the common phase voltage is 240 volts. Some systems generate 480 or 600 volts. If the delta has a grounded center tap neutral, then a voltage equal to one-half the phase voltage is available. If the phases are wye (y) connected then the phase voltage is equal to V 3 (1.73) times the phase-to-neutral voltage.

Single-phase three-wire and three-phase four-wire systems provide voltages for both lighting and power loads. If the load between each of the three phases or between the two ungrounded conductors and their grounded center tapped neutral are equal, a balanced circuit exists. When this occurs there is no current flowing in the neutral conductor. Because of this, two ungrounded conductors and one grounded neutral may be used to feed two circuits. Thus, three conductors may be used where otherwise four are normally required.

The preceding discussion leads logically into one-, two-, and three-phase electric light circuits. Electric lamps for indoor lighting in the United States are generally operated at 110 to 120 volts from constant-potential circuits. Two- and three-wire distribution systems, either direct current or single-phase alternating current, are widely used for lighting installations.

These systems of distribution are capable of handling both lamp and motor loads connected in parallel between the constant-potential lines. The three-wire system provides twice the potential difference between the outside wires than it does between either of the outside wires and the central or neutral wire. This system makes it possible to operate the larger motors at 240 volts while the lamps and smaller motors operate at 120 volts. When the load is unbalanced, a current in the neutral wire will correspond to the difference in current taken by the two sides. A balance of load is sought in laying out the wiring for lighting installations.

NOTES:

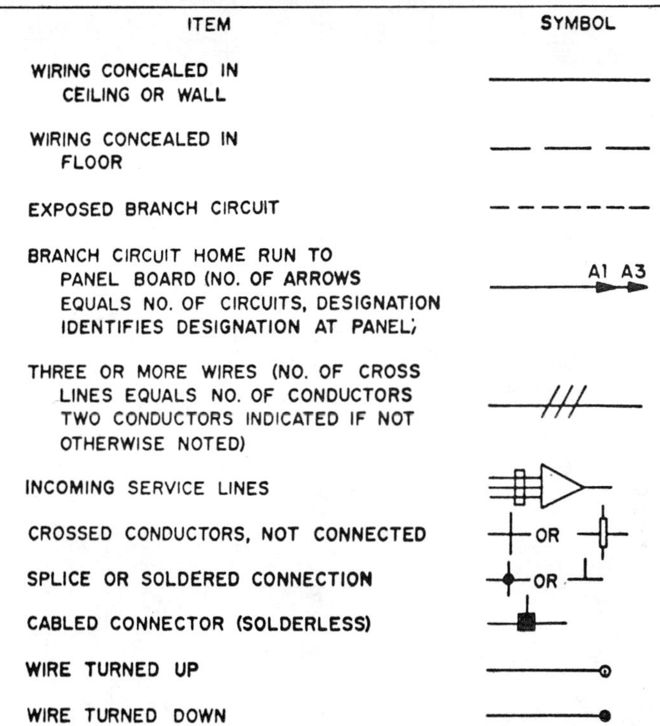

Figure 15-1.—Standard electrical symbols (Sheet 1).

DRAWING SYMBOLS AND BLUEPRINT READING

The electrician must be able to interpret simple blueprints since his construction orders will ordinarily be in that form. He needs the ability to make simple engineering sketches to describe work for which he receives only verbal orders.

Symbols. The more common symbols and line conventions used in wiring plans are shown in figure 15-1. These symbols enable precise location of any electrical equipment in a building from the study of a drawing.

Schematic Wiring Diagrams. Electrical plans show what items are to be installed, their approximate location, and the circuits to which they are to be connected. A typical electrical plan for a post exchange is shown in figure 15-2. The plan shows that the incoming service consists of three No. 8 wires and that two circuit-breaker panels are to be installed. Starting at the upper left, the plan shows that nine ceiling lighting outlets and two duplex wall outlets are to be installed in the bulk storage area. The arrow designated "B2" indicates that these outlets are to be connected to circuit 2 of circuit-breaker panel B. Note that three wires are indicated from this point to the double home-run arrows designated "B1, B2." These are, the hot wire from the bulk storage area to circuit 2 of panel B, the hot wire from the administration area to

NOTES:

ITEM	SYMBOL	ILLUSTRATION
LIGHTING OUTLETS*-	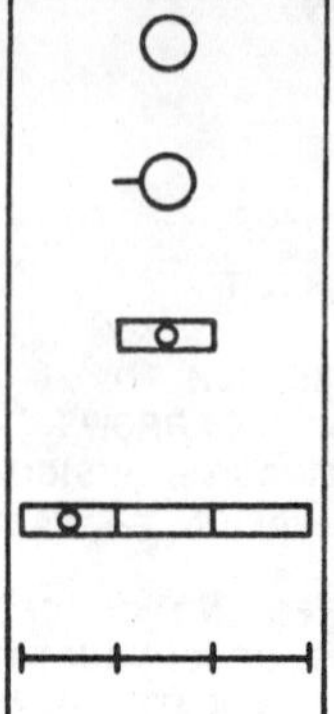	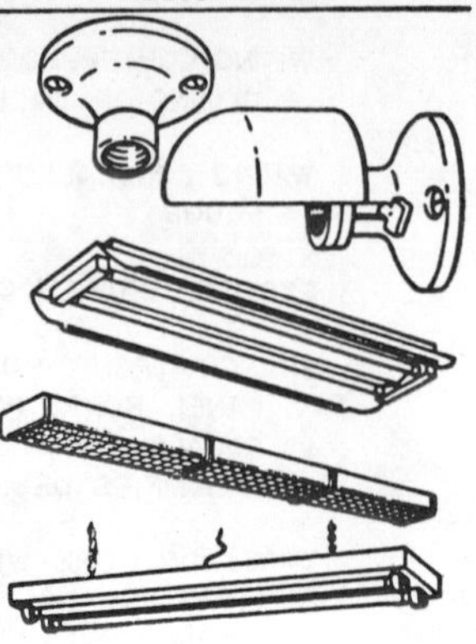
CEILING		
WALL		
FLUORESCENT FIXTURE		
CONTINUOUS ROW FLUORESCENT FIXTURE		
BARE LAMP FLUORESCENT STRIP		

* LETTERS ADDED TO SYMBOLS INDICATE SPECIAL TYPE OR USAGE

J- JUNCTION BOX R- RECESSED
L- LOW VOLTAGE X- EXIT LIGHT

ITEM	SYMBOL	ILLUSTRATION
RECEPTACLE OUTLETS**-	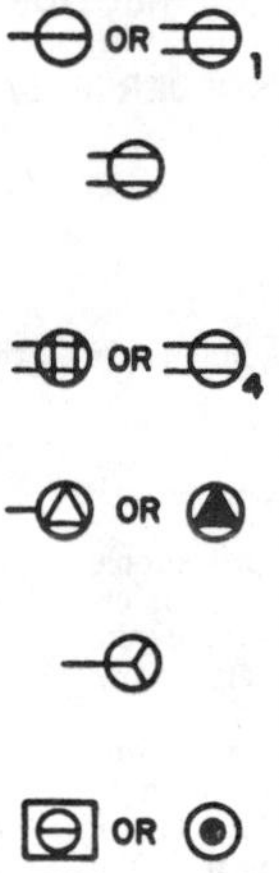	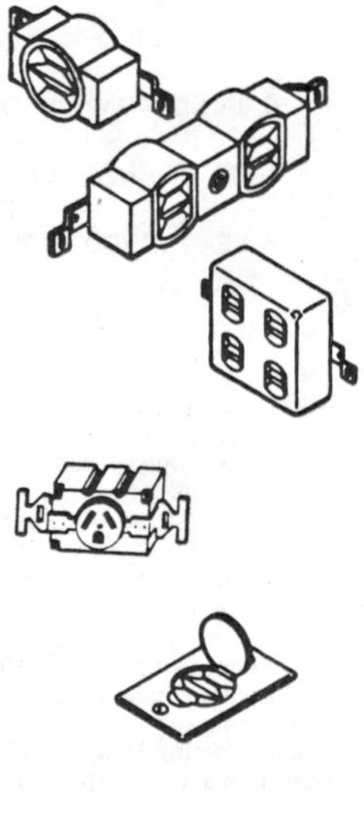
SINGLE OUTLET	OR 1	
DUPLEX OUTLET		
QUADRUPLEX OUTLET	OR 4	
SPECIAL PURPOSE OUTLET	OR	
20-AMP, 250-VOLT OUTLET		
SINGLE FLOOR OUTLET (BOX AROUND ANY OF ABOVE INDICATES FLOOR OUTLET OF SAME TYPE)	OR	

** LETTER G NEXT TO SYMBOL INDICATES GROUNDING TYPE

Figure 15-1.—(Continued, Sheet 2).

NOTES:

ITEM	SYMBOL	ILLUSTRATION
SWITCHES -		
SINGLE POLE SWITCH	S	
DOUBLE POLE SWITCH	S_2	
THREE WAY SWITCH	S_3	
SWITCH AND PILOT LAMP	S_P	
CEILING PULL SWITCH	Ⓢ	
PANEL BOARDS AND RELATED EQUIPMENT		
PANEL BOARD AND CABINET		
SWITCHBOARD, CONTROL STATION OR SUBSTATION		
SERVICE SWITCH OR CIRCUIT BREAKER	OR OR ⊗	
EXTERNALLY OPERATED DISCONNECT SWITCH		
MOTOR CONTROLLER	OR MC	
MISCELLANEOUS -		
TELEPHONE		
THERMOSTAT	Ⓣ	
MOTOR	Ⓜ	

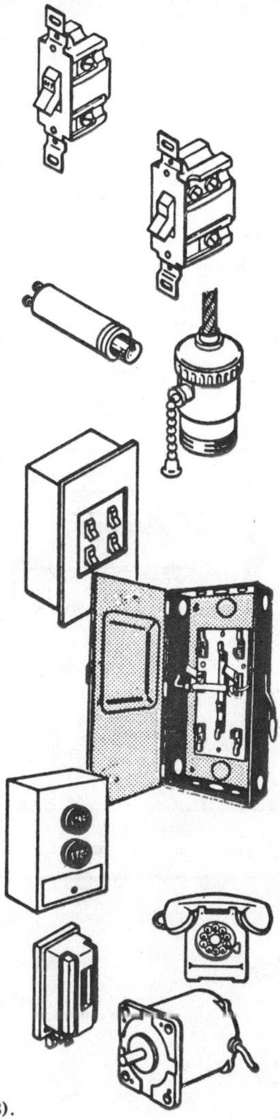

Figure 15-1.—(Continued, Sheet 3).

NOTES:

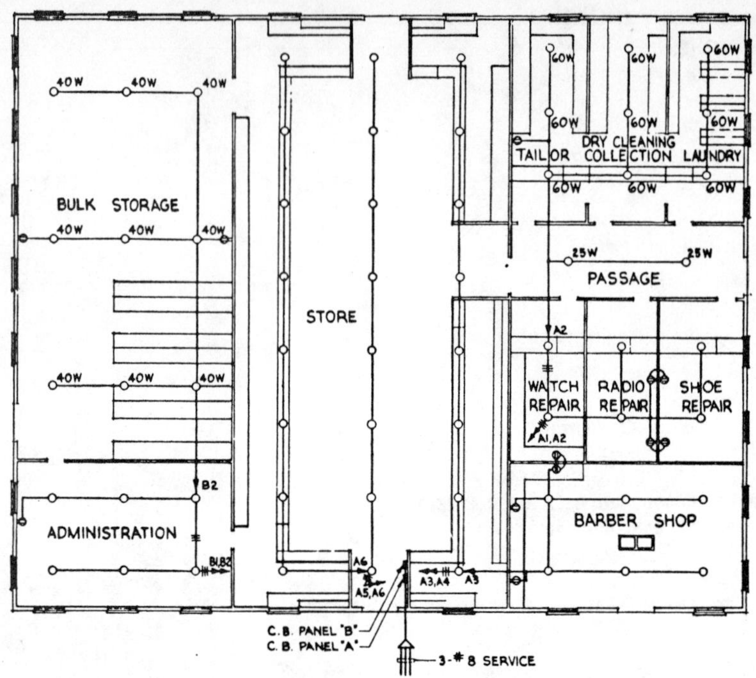

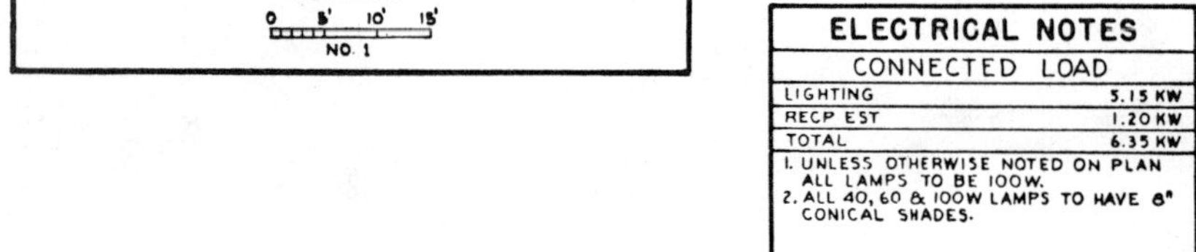

ELECTRICAL NOTES

CONNECTED LOAD	
LIGHTING	5.15 KW
RECP EST	1.20 KW
TOTAL	6.35 KW

1. UNLESS OTHERWISE NOTED ON PLAN ALL LAMPS TO BE 100W.
2. ALL 40, 60 & 100W LAMPS TO HAVE 8" CONICAL SHADES.

Figure 15-2.—Typical wiring diagram.

NOTES:

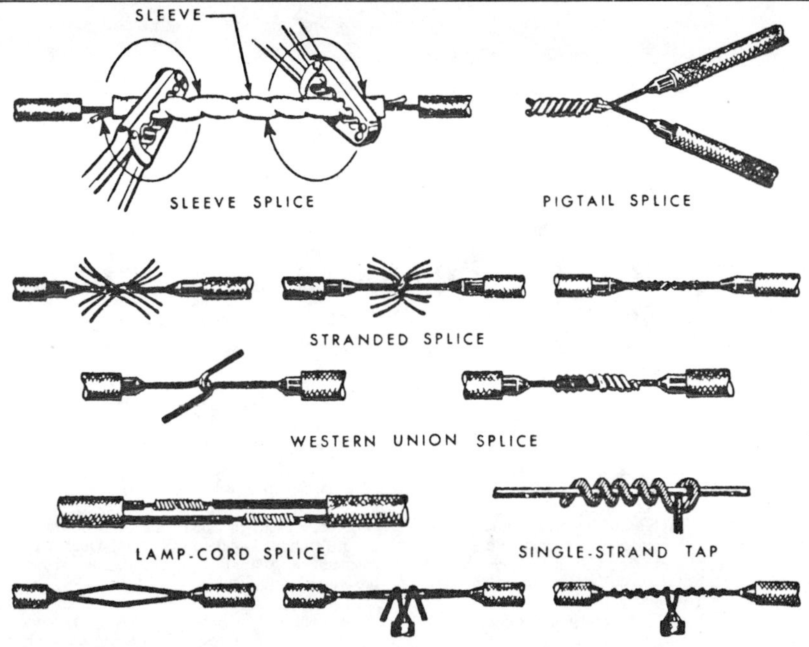

Figure 15-3.—Typical wire splices and taps.

circuit 1 of panel B, and a common neutral. The two hot conductors must be connected to different phases at the panel. This allows a cancellation of current in the neutral when both circuits are fully loaded. From the double arrowhead, these wires are run to the circuit breaker panel without additional connections.

Schematic Wiring Diagrams. This is the form of wiring plans used most frequently for construction drawings (figure 15-2). Single lines indicate the location of wires connecting the fixtures and equipment. Two conductors are indicated in a schematic diagram by a single line. If there are more than two wires together, short parallel lines through the line symbols indicate the number of wires represented by the line. Connecting wires are indicated by placing a dot at the point of intersection. No dot is used where wires cross without connecting. The electrician may encounter drawings in which the lines indicating the wiring have been omitted. In this type of drawing only the fixture and equipment symbols are shown; the location of the actual wiring is to be determined by the electrician. No actual dimensions or dimension lines are shown in electrical drawings. Location dimensions and spacing requirements are given in the form of notes or follow the standard installation principles shown in figure 15-1.

Drawing Notes. A list of drawing notes is ordinarily provided on a schematic wiring diagram to specify special wiring requirements and indicate building conditions which alter standard installation methods.

COLOR CODING

The National Electrical Code requires that a grounded or neutral conductor be identified by an

NOTES:

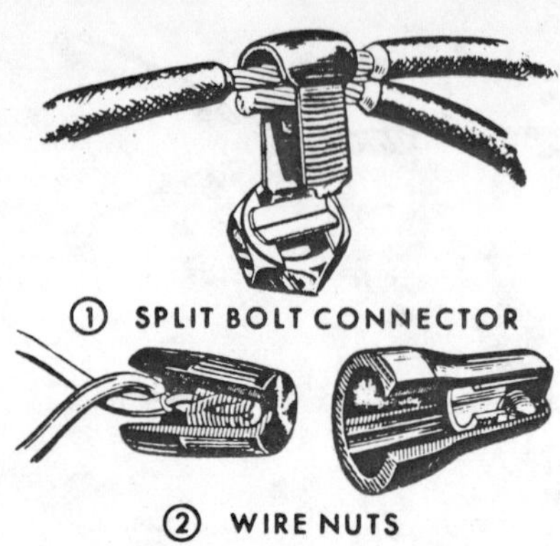

Figure 15-4.—Solderless connectors.

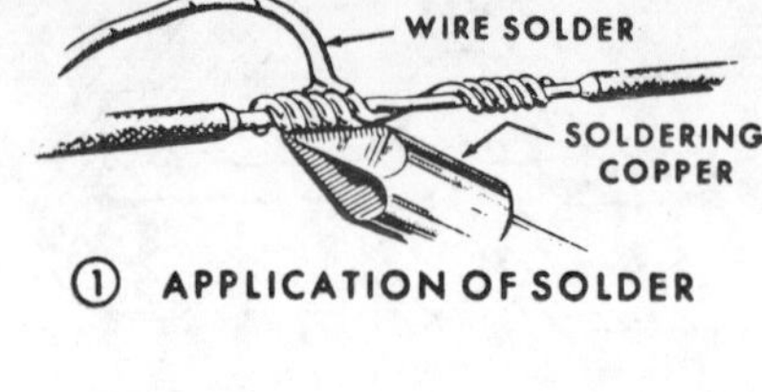

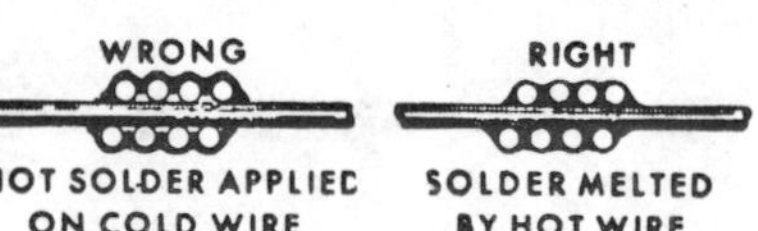

② RIGHT AND WRONG SOLDER JOINT

Figure 15-5.—Soldering and solder joints.

outer color of white or natural gray for Number 6 wire or smaller. For larger conductors the outer identification of white or natural gray may be used, or they should be identified by white markings at the terminals. The ungrounded conductors of a circuit should be identified with insulation colored black, red, and blue, used in that order, in two-, three-, or four-wire circuits, respectively. All circuit conductors of the same color shall be connected to the same ungrounded (hot) feeder conductor throughout the installation. A grounding conductor, used solely for grounding purposes, should be bare or have a green covering.

SPLICES

A spliced wire must be as good a conductor as a continuous conductor. Figure 15-3 shows many of the variations of splicing used to obtain an electrically secure joint. Though splices are permitted wherever accessible in wiring systems, they should be avoided whenever possible. The best wiring practice is to run continuous wires from the service box to the outlets. UNDER NO CONDITIONS SHOULD SPLICES BE PULLED THROUGH CONDUIT. SPLICES SHOULD BE PLACED IN APPROPRIATE ELECTRICAL BOXES.

SOLDERLESS CONNECTORS

Figure 15-4 illustrates connectors used in place of splices because of their ease of installation. Since heavy wires are difficult to splice and solder properly, split-bolt connectors (1, figure 15-4) are commonly used for wire joining. Solderless connectors, popularly called wire nuts, which are used for connecting small-gage and fixture wires, are illustrated in 2, figure 15-4. One design shown consists of a funnel-shaped metalspring insert molded into a plastic shell, into which the wires to be joined are screwed. The other type shown has a removable insert which contains a setscrew to clamp the wires. The plastic shell is screwed onto the insert to cover the joint.

SOLDERING

When a solderless connector is not used, the splice must be soldered before it is considered to be as good as the original conductor. The primary requirements for obtaining a good solder joint are a clean soldering iron, a clean joint, and a nonacid flux. These requirements can be satisfied by using pure rosin on the joint or by using a rosin core solder.

To insure a good solder joint, the electric heated or copper soldering iron should be applied to the joint (1, figure 15-5) until the joint melts the solder by its own heat. Two, figure 15-5 shows the difference between a good and bad solder joint. The bad joint has a weak crystalline structure.

Figure 15-6 illustrates dip soldering. This method of soldering is frequently used by ex-

NOTES:

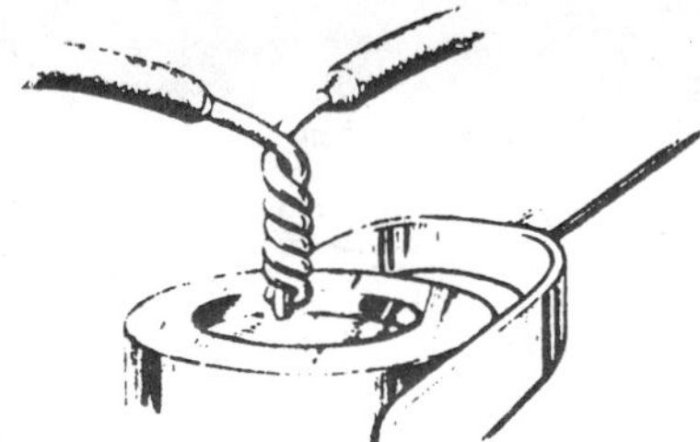

Figure 15-6.—Dip soldering.

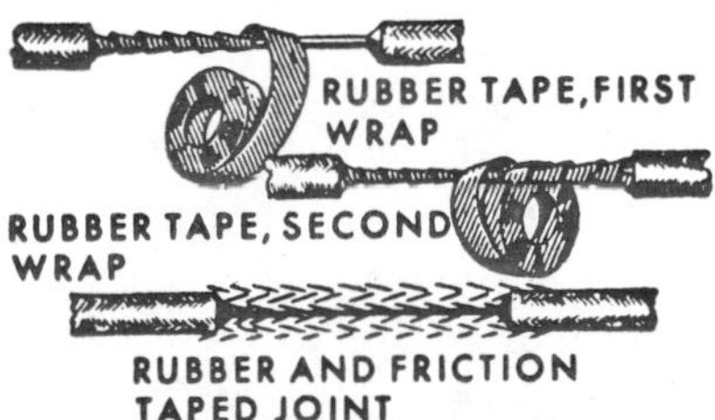

Figure 15-7.—Rubber and friction-tape insulating.

perienced electricians because of its convenience and relative speed for soldering pigtail splices.

TAPING JOINTS

Every soldered joint must be covered with a coating of rubber, or varnished cambric, and friction tape to replace the wire insulation of the conductor. In taping a spliced solder joint (figure 15-7), the rubber or cambric tape is started on the tapered end of the wire insulation and advanced toward the other end, with each succeeding wrap, by overlapping the windings. This procedure is repeated from one end of the splice to the other until the original thickness has been restored. The joint is then covered with several layers of friction tape.

Though the method above for taping joints is still considered to be standard, the plastic electrical tape, which serves as an insulation and a protective covering, should be used whenever available. This tape materially reduces the time required to tape a joint and reduces the space needed by the joint because a satisfactory protective and insulation covering can be achieved with three-layer taping.

INSULATION AND MAKING WIRE CONNECTIONS

When attaching a wire to a switch or an electrical device or when splicing it to another wire, the wire insulation must be removed to bare the copper conductor. One, figure 15-8 shows the right and wrong way to remove insulation. When the wire-stripping tool is applied at right angles to the wire, there is danger that the wire may be nicked and thus weakened. Therefore extreme caution must be used to make sure the wire is not nicked. To avoid nicks, the cut is made at an angle to the conductor. After the protective insulation is removed, the conductor is scraped or sanded thoroughly to remove all traces of insulation and oxide on the wire.

Two and three, figure 15-8 show the correct method of attaching the trimmed wire to terminals. The wire loop is always inserted under the terminal screw, as shown, so that tightening the screw tends to close the loop. The loop is made so that the wire insulation ends close to the terminal.

JOB SEQUENCE

Scope. The installation of interior wiring is generally divided into two major divisions called roughing-in and finishing. Roughing-in is the installation of the outlet boxes, cable, wire, and conduit. Finishing is the installation of the switches, receptacles, covers, fixtures, and the completion of the service. The interval between these two work periods is used by other trades for plastering, enclosing walls, finishing floors, and trimming.

Roughing-In:

1. The first step in the roughing-in phase of a wiring job is the mounting of outlet boxes. The mounting can be expedited if the locations of all boxes are first marked on the studs and joists of the building.

2. All the boxes are mounted on the building members on their own or by special brackets. For concealed installation, all boxes must be installed with the forward edge or plaster ring of the boxes flush with the finished walls.

3. The circuiting and installation of wire for open wiring, cable, or conduit should be the next step. This involves the drilling and cutting-out of the building members to allow for the passage of the conductor or its protective covering. The production-line method of drilling the holes for all runs (as the installations between boxes are called) at one time, and then installing all of the wire,

NOTES:

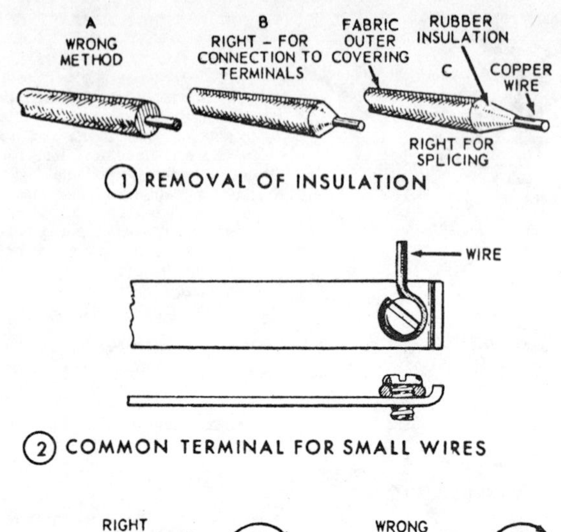

Figure 15-8.—Removing insulation and attaching wire to terminals.

cable, or conduit, will expedite the job.

4. The final roughing-in step in the installation of conduit systems is the pulling-in of wires between boxes. This can also be included as the first step in the finishing phase and requires care in the handling of the wires to prevent the marring of finished wall or floor surfaces.

Finishing:

1. The splicing of joints in the outlet and junction boxes and the connection of the bonding circuit is the initial step in the completion phase of a wiring job.

2. Upon completion of the first finishing step, the proper leads to the terminals of switches, ceiling and wall outlets, and fixtures are then installed.

3. The devices and their cover plates are then attached to the boxes. The fixtures are generally supported by the use of special mounting brackets called fixture studs or hickeys.

4. The service-entrance cable and fusing or circuit breaker panels are then connected and the circuits fused.

5. The final step in the wiring of any building requires the testing of all outlets by the insertion of a test prod or test lamp, the operation of all switches in the building, and the loading of all circuits to insure proper circuiting has been installed.

NOTES:

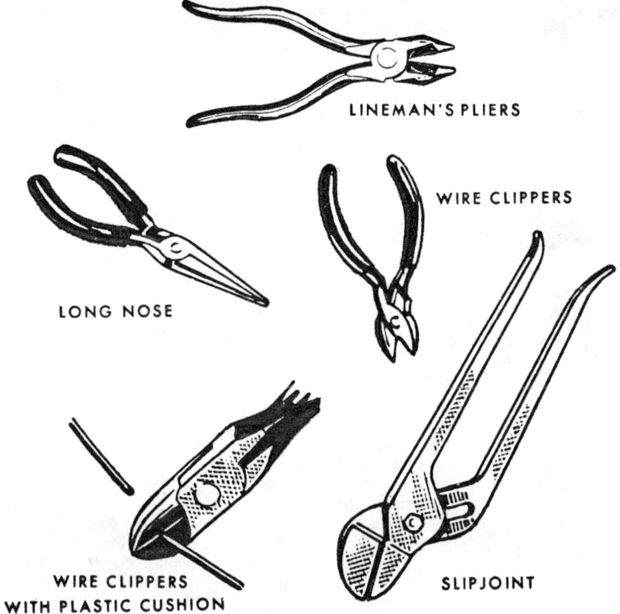

Figure 15-9.—Pliers

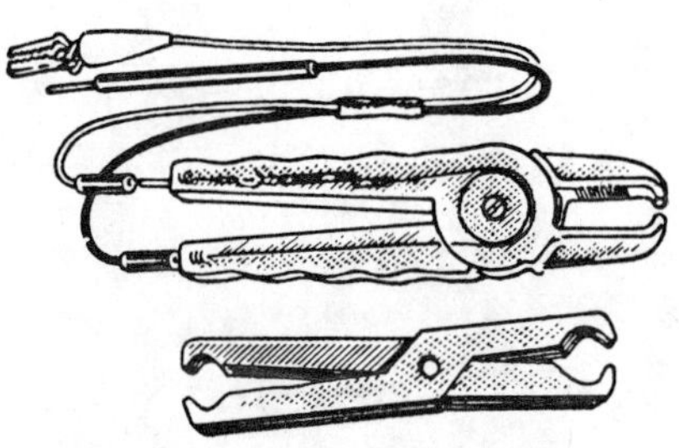

Figure 15-10.—Fuse pullers.

ELECTRICIAN'S TOOLS AND EQUIPMENT

PURPOSE

The electrical apparatus and materials that an electrician is required to install and maintain are different from other building materials. Their installation and maintenance require the use of special handtools. This section describes and illustrates the tools normally used by an electrician in interior wiring.

PLIERS

Pliers are furnished with either uninsulated or insulated handles. Although the insulated handle pliers are always used when working on or near "hot" wires, they must not be considered sufficient protection alone and other precautions must be

NOTES:

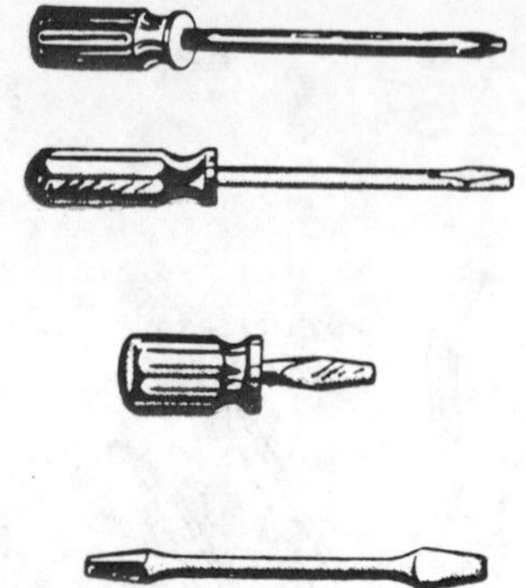

Figure 15-11.—Screwdrivers.

taken. Long-nose pliers are used for close work in panels or boxes. Wire clippers are used to cut wire to size. One type of wire clippers shown in figure 15-9 has a plastic cushion in the cutting head which grips the clipped wire end and prevents the clipped piece from flying about and injuring anyone. The slip joint pliers are used to tighten locknuts or small nuts on devices.

FUSE PULLER

The fuse puller shown in figure 15-10 is designed to eliminate the danger of pulling and replacing

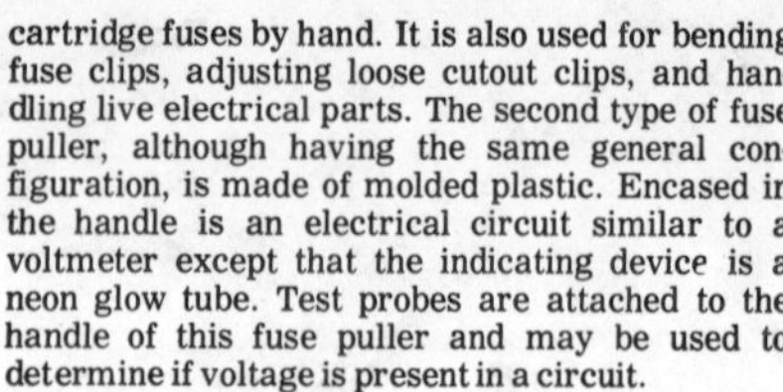

cartridge fuses by hand. It is also used for bending fuse clips, adjusting loose cutout clips, and handling live electrical parts. The second type of fuse puller, although having the same general configuration, is made of molded plastic. Encased in the handle is an electrical circuit similar to a voltmeter except that the indicating device is a neon glow tube. Test probes are attached to the handle of this fuse puller and may be used to determine if voltage is present in a circuit.

SCREWDRIVERS

Screwdrivers (figure 15-11) are made in many sizes and tip shapes. Those used by electricians should have insulated handles. Generally the electrician uses screwdrivers in attaching electrical devices to boxes and attaching wires to terminals. One variation of the screwdriver is the screwdriver bit which is held in a brace and used for heavy-duty work. For safe and efficient application, screwdriver tips should be kept square and properly tapered and should be selected to match the screw slot.

WRENCHES

Figure 15-12 shows four types of wrenches used by electricians. Adjustable open-end wrenches, commonly called crescent wrenches, open end, closed end, and socket wrenches are used on hexagonal and square fittings such as machine bolts, hexagon nuts, or conduit unions. Pipe wrenches are used for pipe and conduit work and should not be used where crescent, open end, closed

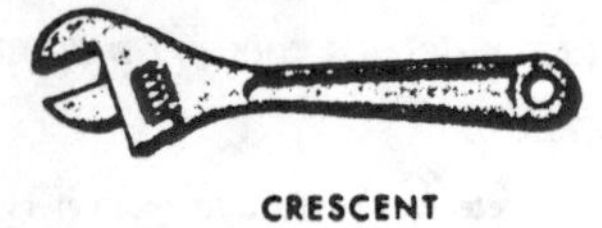

Figure 15-12.—Wrenches.

NOTES:

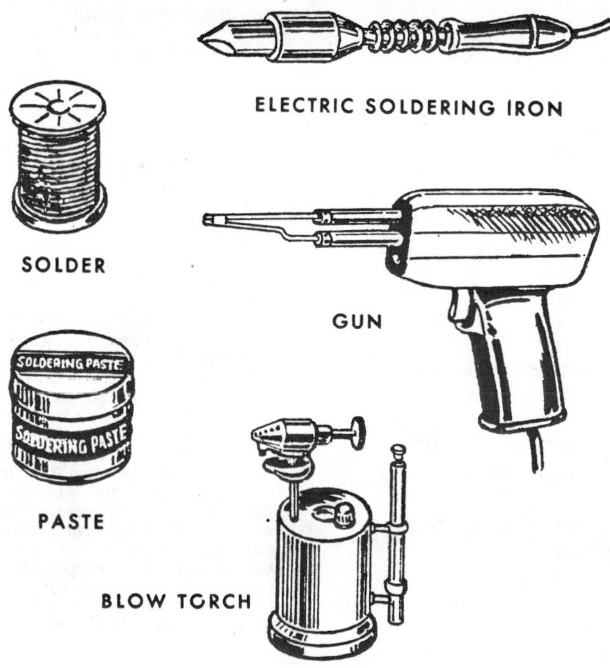

Figure 15-13.—Soldering equipment.

end, or socket wrenches can be used. Their construction will not permit the application of heavy pressure on square or hexagonal material, and the continued misuse of the tool in this manner will deform the teeth on the jaw faces and mar the surfaces of the material being worked.

SOLDERING EQUIPMENT

A standard soldering kit (figure 15-13) used by electricians consists of nonelectric or electric soldering irons or both, a blowtorch (for heating a nonelectric soldering iron and pipe or wire joints), a spool of solid tin-lead wire solder or flux core solder, and soldering paste. An alcohol or propane torch may also be used in place of the blowtorch. Acid core solder should never be used in electrical wiring.

DRILLING EQUIPMENT

Drilling equipment (figure 15-14) consists of a brace, a joist-drilling fixture, an extension bit to allow for drilling into and through deep cavities, an adjustable bit, and a standard wood bit. These are required in electrical work to drill holes in building structures for the passage of conduit or wire in new or modified construction. Similar equipment is required for drilling holes in sheet-metal cabinets and boxes. In this case high speed drills should be used. Carbide drills are used for tile or concrete work. Electric power drills aid in this phase of an electrician's work.

NOTES:

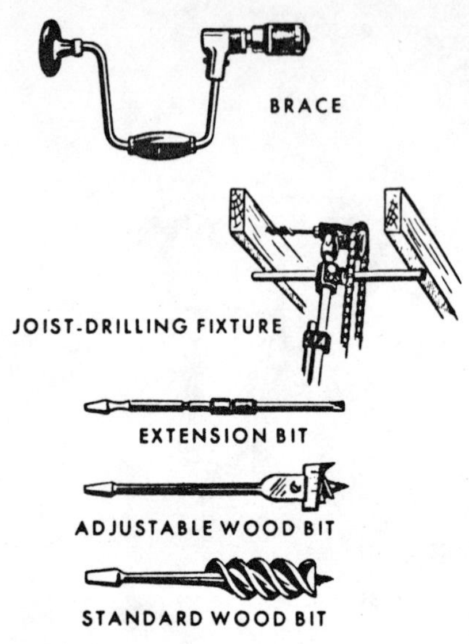

Figure 15-14.—Drilling equipment.

WOODWORKING TOOLS

The crosscut and keyhole saws and wood chisels shown in figure 15-15 are used by electricians to remove wooden structural members obstructing a wire or conduit run and to notch studs and joists to take conduit, cable, or box-mounting brackets. They are also used in the construction of wood-panel mounting brackets. The keyhole saw may again be used to cut openings in walls of existing buildings where boxes are to be added.

METALWORKING TOOLS

The cold chisels and center punches shown in figure 15-16, besides several other types of metalworking tools employed by the electrical trade, are used when working on steel panels. The knockout punch is used either in making or enlarging a hole in a steel cabinet or outlet box. The hacksaw is usually used by an electrician to cut conduit, cable, or wire too large for wire cutters. A light steady stroke of about 40 to 50 times a minute is best. A new blade should always be inserted with the teeth pointing away from the handle. The tension wingnut is tightened until the blade is rigid. Care must be taken because insufficient tension will cause the blade to twist and jam whereas too much tension will cause the blade to break. Blades have 14, 18, 24 and 32 teeth per inch. The best blade for general use is one having 18 teeth per inch. A blade with 32 teeth per inch is best for cutting thin material. The mill file shown in the figure is used in

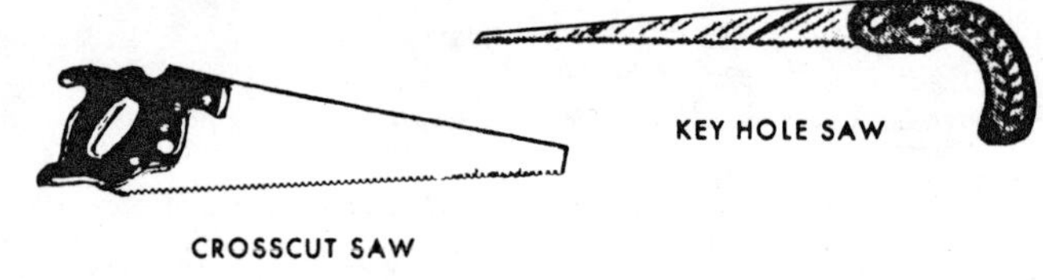

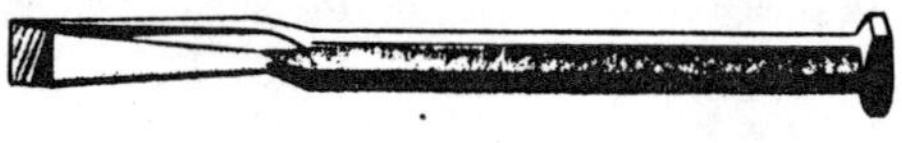

Figure 15-15.—Woodworking tools.

NOTES:

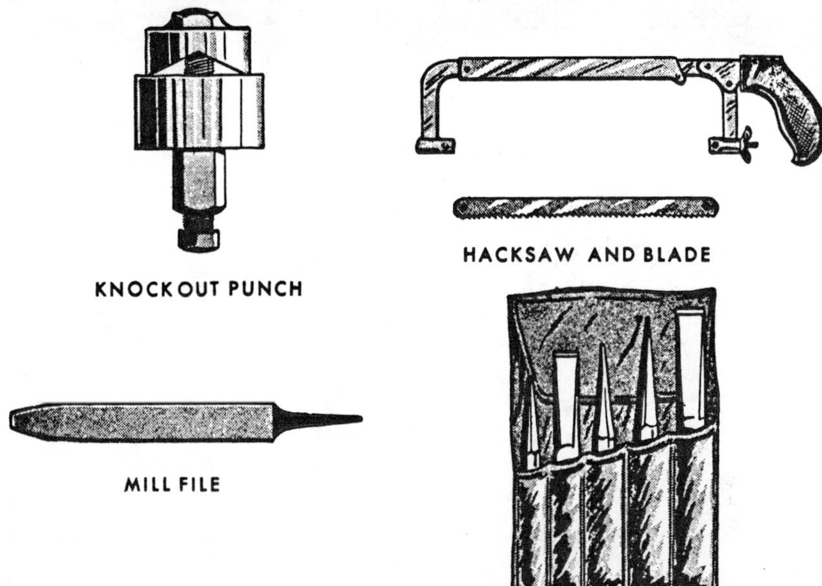

Figure 15-16.—Metalworking tools.

filing the sharp ends of cutoffs as a precaution against short circuits.

MASONRY-WORKING TOOLS

An electrician should have several sizes of masonry drills in his tool kit. These normally are carbide-tipped and are used to drill holes in brick or concrete walls either for anchoring apparatus with expansion screws or for the passage of conduit or cable. Figure 15-17 shows the carbide-tipped bit used with a power drill and a hand-operated masonry drill.

CONDUIT THREADERS AND DIES

Rigid conduit is normally threaded for installation. Figure 15-18 illustrates one type of conduit threader and dies used in cutting pipe threads on conduit. The tapered pipe reamer is used to ream the inside edge of the conduit as a precaution against wire damage. The conduit cutter is used when cutting thin-wall conduit and has a tapered blade attachment for reaming the conduit ends.

KNIVES AND OTHER INSULATION-STRIPPING TOOLS

Wire and cable insulation is stripped or removed with the tools shown in figure 15-19. The knives and patented wire strippers are used to bare the wire of insulation before making connections. The scissors shown are used to cut insulation and tape. A multipurpose tool designed to cut and skin wires, attach terminals, gage wire, and cut small

NOTES:

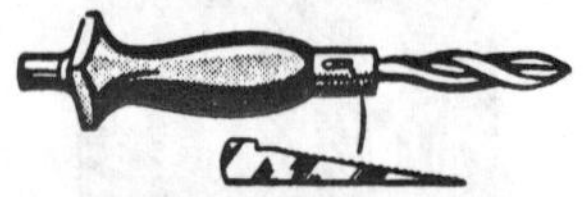

Figure 15-17.—Masonry tools.

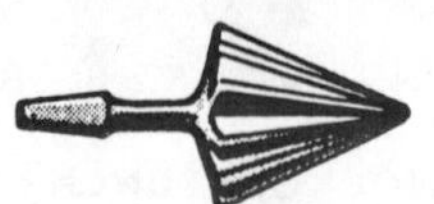

Figure 15-18.—Conduit threader, reamers and cutter.

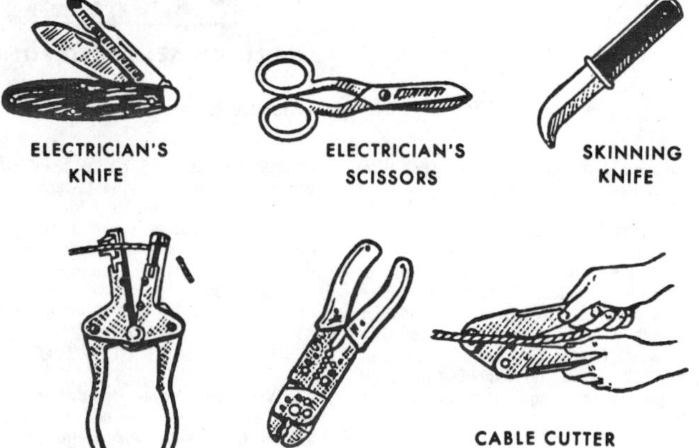

Figure 15-19.—Insulation-stripping tools.

NOTES:

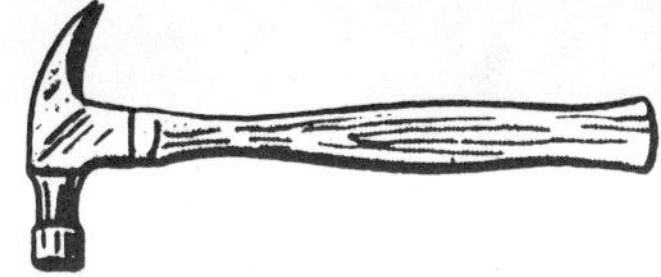

Figure 15-20.—Hammers.

Figure 15-21.—Wire grip and splicing clamp.

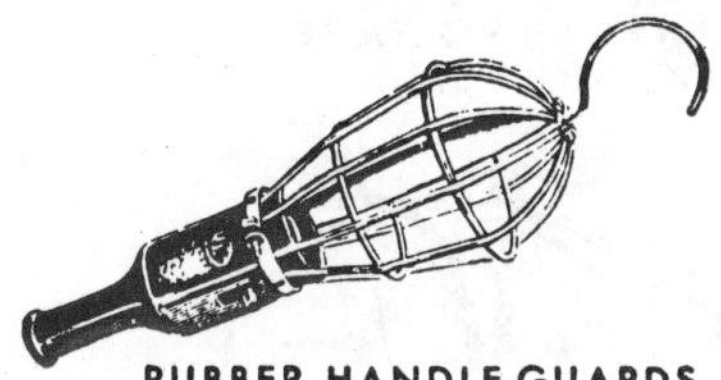

Figure 15-22.—Extension light (without bulb).

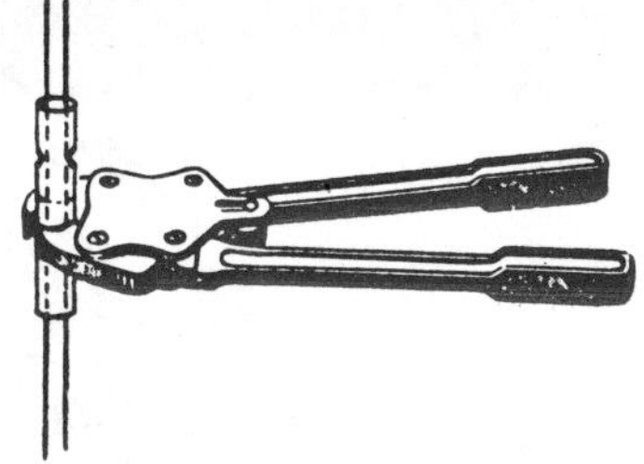

Figure 15-23.—Thin-wall conduit impinger.

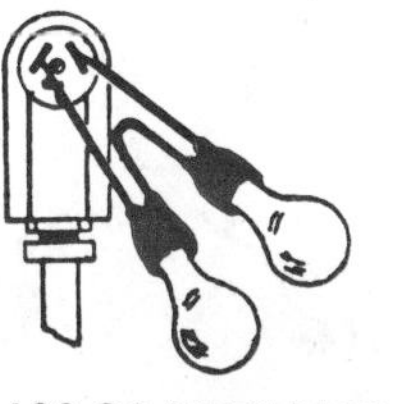

Figure 15-24.—Test lamps.

bolts may also be used. The armored-cable cutter may be used instead of a hacksaw in removing the armor from the electrical conductors at box entry or when cutting the cable to length.

HAMMERS

Hammers are used either in combination with other tools such as chisels or in nailing equipment to building supports. Figure 15-20 shows a carpenter's clawhammer and a machinist's ball peen hammer, both of which can be advantageously used by electricians in their work.

TAPE

Various types of tapes are used to replace insulation and wire coverings. Friction tape is a cotton tape impregnated with an insulating adhesive compound. It provides weather resistance and limited mechanical protection to a splice already insulated. Rubber or varnished cambric tape may be used as an insulator when replacing wire covering. Plastic electrical tape is made of a plastic material with adhesive on one face. It has replaced friction and rubber tape in the field for 120- and 208-volt circuits, and as it serves a dual

NOTES:

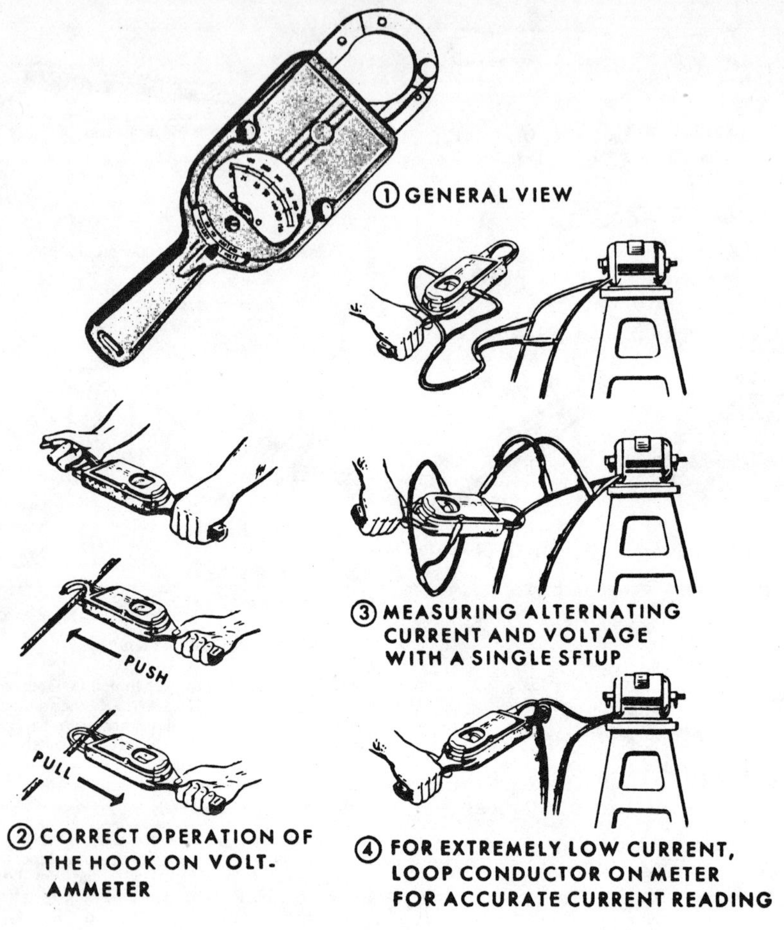

Figure 15-25.—Hook-on volt-ammeter.

NOTES:

purpose in taping joints, it is preferred over the former methods.

FISH WIRE AND DROP CHAIN

Fish Wire. Fish wires are used primarily to pull wires through conduits. Many pulls are quite difficult and require a fish-wire "grip" or "pull" to obtain adequate force on the wire in pulling. The fish wire is made of tempered spring steel about ¼-inch wide and is available in lengths to suit requirements. It is stiff enough to preclude bending under normal operation but can be pushed or pulled easily around the bends or conduit elbows.

Drop Chain. When pulling wires and cables in. existing buildings, the electrician will normally employ a fish wire or drop chain between studs. A drop chain consists of small chain links attached to a lead or iron weight. It is used only to feed through wall openings in a vertical plane.

RULER AND MEASURING TAPE

As an aid in cutting conduit to exact size as well as in determining the approximate material quantities required for each job, the electrician should be equipped with a folding rule and a steel tape.

WIRE CLAMPS AND GRIPS

To pull wire through conduit and to pull open-wire installations tight, the wire grip shown in figure 15-21 is an invaluable aid. As seen in the figure, the wire grip has been designed so that the harder the pull on the wire, the tighter the wire wil be gripped. Also shown in the figure is the splicing clamp used to twist the wire pairs into a uniform and tight joint when making splices.

EXTENSION CORD AND LIGHT

The extension light shown in figure 15-22 normally is supplied with a long extension cord and is used by the electrician when normal building lighting has not been installed or is not functioning.

THIN-WALL CONDUIT IMPINGER

When the electrician uses indenter type couplings and connectors with thin-wall conduit, an indenter tool (a thin-wall conduit impinger shown in figure 15-23) must be used to attach these fittings permanently to the conduit. This tool has points which, when pressed against the fittings, form indentations in the fitting and are pressed into the wall of the tubing to hold it on the conduit. The use of these slip-on fittings and the impinger materially reduces the installation time required in electrical installations and thus reduces the cost of thin-wall conduit installations considerably.

WIRE CODE MARKERS

Tapes with identifying numbers or nomenclature are avilable for the purpose of permanently identifying wires and equipment. These are particularly valuable to identify wires in complicated wiring circuits, in fuse circuit breaker panels, or in junction boxes.

METERS AND TEST LAMPS

Test Lamps. An indicating voltmeter or test lamp is useful when determining the system voltage or locating the ground lead, and for testing circuit continuity through the power source. They both have a light which glows in the presence of voltage. Figure 15-24 shows a test lamp used as a voltage indicator.

Hook-On Volt-Ammeter. A modern method of measuring current flow in a circuit uses the hook-on volt-ammeter (figure 15-25) which does not need to be hooked into the circuit. Two, figure 15-25 shows its ease of operation. To make a measurement, the hook-on section is opened by hand and the meter is placed against the conductor. A slight push on the handle snaps the section shut; a pull springs the hook on the C-shaped current transformer open and releases the conductor. Applications of this meter are shown in 3, figure 15-25 where voltage is being measured using the meter leads. Current is measured using the hook-on section. With three coils around the meter (4, figure 15-25) the current reading will be three times the actual current flowing through the wire. To obtain the true current, therefore, this reading is divided by 3. The hook-on volt-ammeter can be used only on alternating current circuits and can measure current only in a single conductor.

Wattmeter. The basic unit of measurement for electric power is the watt. In the power ratings of electric devices used by domestic consumers of electricity, the term watts signifies that, when energized at the normal line voltage, the apparatus will use electricity at the specified rate. In alternating-current circuits, power is the product of three quantities: the potential (volt), the current (amperes), and the power factor (percent). Power is measured by a wattmeter (figure 15-26). This

NOTES:

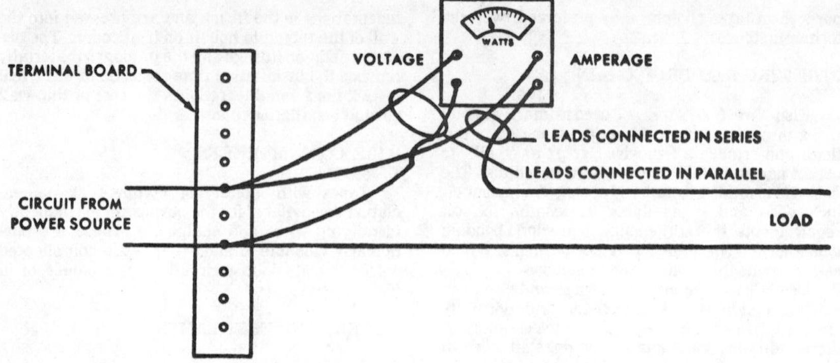

Figure 15-26.—Wattmeter connection.

instrument is connected (figure 15-26) so that the current in the measured circuit flows through the stationary field coils in the wattmeter and the voltage across the measured circuit is impressed upo the wattmeter-armature circuit, which includes movable coils and a fixed resistor. The power factor is automatically included in the measurement because the torque developed in the wattmeter is always proportional to the product of the instantaneous values of current and voltage. Consequently, the instrument gives a true indication of the power, or rate at which energy is being utilized.

WIRING MATERIALS

ELECTRICAL CONDUCTORS

Single Conductors. Electrical conductors that provide the paths for the flow of electric current generally consist of copper or aluminum wire or cable over which an insulating material is formed. The insulating material insures that the path of current flow is through the conductor rather than through extraneous paths, such as conduits, water pipes, and so on. The wires or conductors are initially classified by type of insulation applied and wire gage. The various types of insulation are in turn subdivided according to their maximum operating temperatures and nature of use. Figure 15-27 illustrates the more common single conductors used in interior wiring systems.

Wire Sizes. The wire sizes are denoted by the use of the American Wire Gage (AWG) standards. The largest gage size is No. 0000. Wires larger than this are classified in size by their circular mil cross-sectional area. One circular mil is the area of a circle with a diameter of 1 1,000 of an inch. Thus, if a wire has a diameter of 0.10 inch or 100 mil, the cross sectional area is 100 x 100 or 10,000 circular mils. The most common wire sizes used in interior wiring are 14, 12, and 10 and they are usually of solid construction. Some characteristics of the numbering system are—

1. As the numbers become larger, the size of the wire decreases.
2. The sizes normally used have even numbers, e.g., 14, 12 and 10.
3. Numbers 8 and 6 wires, which are furnished either solid or stranded, are normally used for heavy-power circuits or as service-entrance leads to buildings. Wire sizes larger than these are used for extremely heavy loads and for poleline distributions.

Multiconductor Cables. There are many types of installations of electrical wiring where the use of individual conductors spaced and supported side by side becomes an inefficient as well as hazardous practice. For these installations, multiconductor cables have been designed and manufactured. Multiconductor cables consist of the individual conductors as outlined above, arranged in groups of two or more. An additional insulating or protective shield is formed or wound around the group of conductors. The individual conductors are color coded for proper identification. Figure 15-28

NOTES:

illustrates some of the types of multiconductors. The description and use of each type are given in 1. through 5. below.

1. Armored cable, commonly referred to as BX, can be supplied either in two- or three-wire types and with or without a lead sheath. The wires in BX, matched with a bare equipment ground wire, are initially twisted together. This grouping, totaling three or four wires with the ground, is then wrapped in coated paper and a formed self-locking steel armor. The cable without a lead sheath is widely used for interior wiring under dry conditions. The lead sheath is required for installation in wet locations and through masonry or concrete building partitions where added protection for the copper conductor wires is required.

2. Nonmetallic sheathed cable consists of two or three rubber- or thermoplastic-insulated wires, each covered with a jute type of filler material which acts as a protective insulation against mishandling. This in turn is covered with an impregnated cotton braid. The cable is light in weight, simple to install, and comparatively low

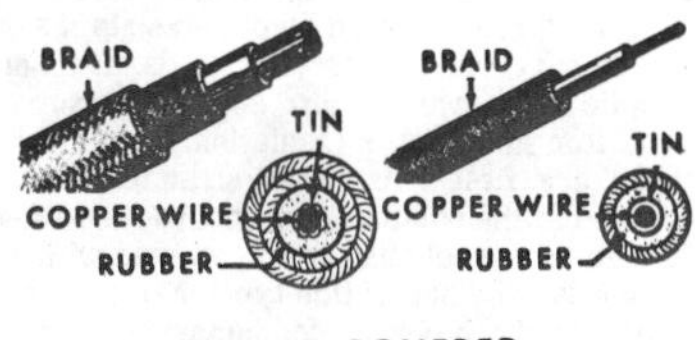

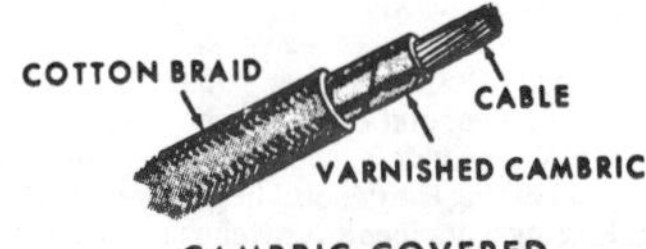

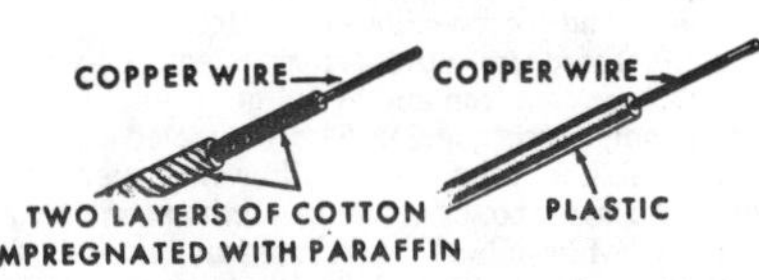

Figure 15-27.—Single conductors.

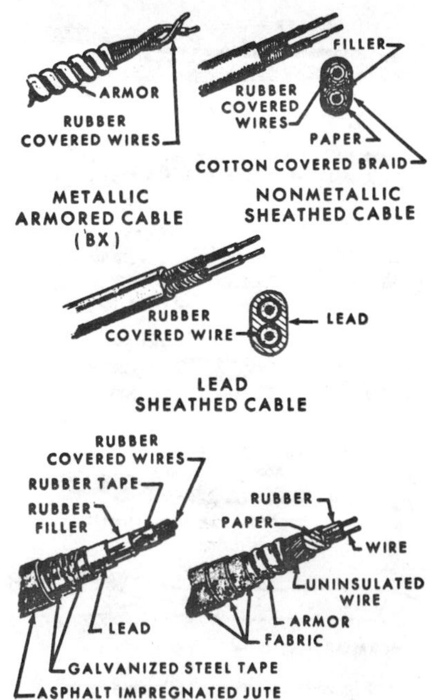

Figure 15-28.—Multiconductor cables.

priced. It is used quite extensively in interior wiring, but is not approved for use in wet locations. A dual-purpose plastic sheathed cable with solid copper conductors can be used underground outdoors or indoors. It needs no conduit, and its flat shape and gray or ivory color makes it ideal for surface wiring. It resists moisture, acid, and corrosion and can be run through masonry or between studding.

3. Lead-covered cable consists of two or more rubber-covered conductors surrounded by a lead sheathing which has been extruded around it to permit its installation in wet and underground locations. Lead-covered cable can also be im-

NOTES:

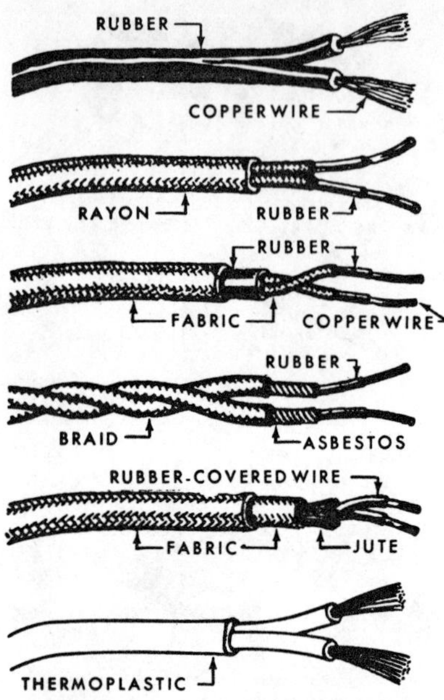

Figure 15-29.—Types of flexible cords.

mersed in water or installed in areas where the presence of liquid or gaseous vapors would attack the insulation on other types.

4. Parkway cable provides its own protection from mechanical injury and therefore can be used for underground services by burying it in the ground without any protecting conduit. It normally consists of rubber-insulated conductors enclosed in a lead sheath and covered with a double spiral of galvanized steel tape which acts as a mechanical protection for the lead. On top of the tape, a heavy braid of jute saturated with a waterproofing compound is applied for additional weather protection.

5. Service-entrance cable normally has three wires with two insulated and braided conductors laid parallel and wound with a bare conductor. Protection against damage for this assembly is obtained by encasing the wires in heavy tape or armor, which serves as an inner cushion, and covering the whole assembly with braid. Though the cable normally serves as a power carrier from the exterior service drop to the service equipment of a building, it may also be used in interior-wiring circuits to supply power to electric ranges and water heaters at voltages not exceeding 150 volts to ground provided the outer covering is armor. It may also be used as a feeder to other buildings on the same premises under the same conditions, if the bare conductor is used as an equipment grounding conductor from a main distribution center located near the main service switch.

Cords. Many items using electrical power are either of the pendant, portable, or vibration type. In these cases the use of cords as shown in figure 15-29 is authorized for delivery of power. These can be grouped and designated as either lamp, heater, or heavy-duty power cords. Lamp cords are supplied in many forms. The most common types are the single-paired rubber-insulated and twisted-paired cords. The twisted-paired cords consist of two cotton-wound conductors which have been covered with rubber and rewound with cotton braid. Heater cords are similar to this latter type except that the first winding is replaced by heat-resistant asbestos. Heavy-duty or hard-service cords are normally supplied with two or more conductors surrounded by cotton and rubber insulation. In manufacture, these are first twisted or stranded. The voids created in the twisting process are then filled with jute and the whole assembly covered with rubber. All cords, whether of this type or of the heater or lamp variety, have the conductors color coded for ease of identification.

ELECTRICAL BOXES

Design. Outlet boxes bind together the elements of a conduit or armored cable system in a continuous grounded system. They provide a means of holding the conduit in position, a space for mounting such devices as switches and receptacles, protection for the device, and space for making splices and connections. Outlet boxes are manufactured in either sheet steel, porcelain, bakelite, or cast iron and are either round, square, octagonal, or rectangular. The fabricated steel box is available in a number of different designs. For example, some boxes are of the sectional or "gang" variety, while others have integral brackets for mounting on studs and joists. Moreover, some boxes have been designed to receive special cover plates so that switches, receptacles, or lighting fixtures may be more easily installed. Other designs facilitate installation in plastered surfaces.

NOTES:

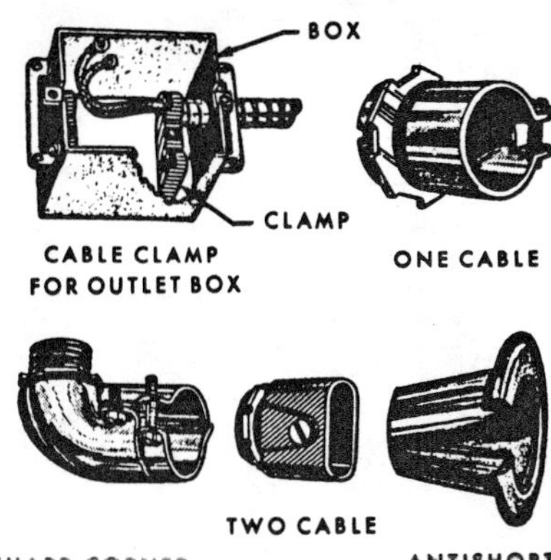

Figure 15-30.—Armored-cable fittings.

Regardless of the design or material, they all should have sufficient interior volume to allow for the splicing of conductors or the making of connections. For this reason the allowable minimum depth of outlet boxes is limited to 1½ inches in all cases except where building-supporting members would have to be cut. In this case the minimum depth can be reduced to ½ inch.

Selection. The selection of boxes in an electrical system should be made in accordance with the maximum allowable conductor capacity for each type of box. In these tables a conductor running through the box is counted along with each conductor terminating in the box. For example, one conductor running through a box and two terminating in the box would equal three conductors in the box. Consequently, any of the boxes listed would be satisfactory. The tables apply for boxes that do not contain receptacles, switches, or similar devices. Each of these items mounted in a box will reduce by one the maximum number of conductors allowable as shown in the tables.

Outlet Boxes for Rigid and Thin-Wall Circuit and Armored Cable. Steel or cast iron outlet boxes are generally used with rigid and thin-wall conduit or armored cable. The steel boxes are either zinc- or enamel-coated, the zinc coating being preferred when installing conduit in wet locations. All steel boxes have "knockouts." These knockouts are indentations in the side, top, or back of an outlet box, sized to fit the standard diameters of conduit fittings or cable connectors. They usually can be removed with a small cold chisel or punch to facilitate entry into the box of the conduit or cable. Boxes designed specifically for armored-cable use also have integral screw clamps located in the space immediately inside the knockouts and thus eliminate the need for cable connectors. This reduces the cost and labor of installation. Box covers are normally required when it is necessary to reduce the box openings, provide mounting lugs for electrical devices, or to cover the box when it is to be used as a junction. Figure 15-30 illustrates several types of cable connectors and also a cable clamp for use in clamping armored cable in an outlet box. The antishort bushing shown in the figure is inserted between the wires and the armor to protect the wire from the sharp edges of the cut armor when it is cut with a hacksaw or cable cutter.

Outlet Boxes for Nonmetallic Sheathed Cable and Open Wiring.

1. Steel. Steel boxes are also used for nonmetallic cable and open wiring. However, the methods of box entry are different from those for conduit and armored-cable wiring because the electrical conductor wires are not protected by a hard surface. The connectors and interior box clamps used in nonmetallic and open wiring are formed to provide a smooth surface for securing the cable rather than being the sharp-edged type of closure normally used.

2. Nonmetallic. Nonmetallic outlet boxes made of either porcelain or bakelite may also be used with open or nonmetallic sheathed wiring. Cable or wire entry is generally made by removing the knockouts of preformed weakened blanks in the boxes.

3. Special. In open wiring, conductors should normally be installed in a loom from the last support to the outlet box. Although all of the boxes described in 1. and 2. above are permissible for open wiring, a special loom box is available which has its back corners "sliced off" and allows for loom and wire entry at this sliced-off position.

Attachment Devices for Outlet Boxes. Outlet boxes which do not have brackets are supported by wooden cleats or bar hangers as illustrated in figure 15-31.

1. Wooden cleats. Wooden cleats are first cut to size and nailed between two wooden members. The boxes are then either nailed or screwed to these cleats through holes provided in their back plates.

2. Strap hangers. If the outlet box is to be mounted between studs, mounting straps are

NOTES:

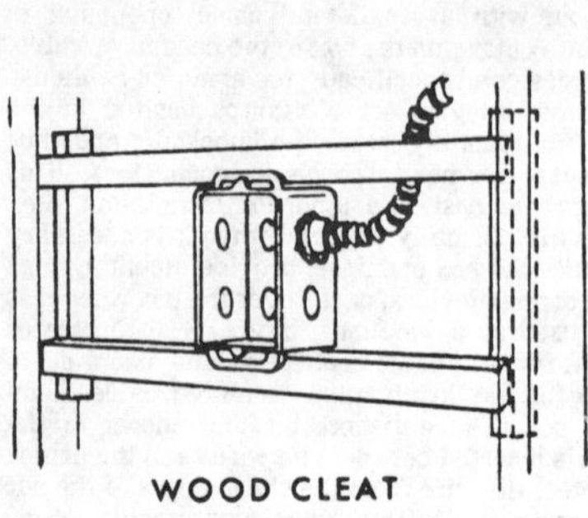

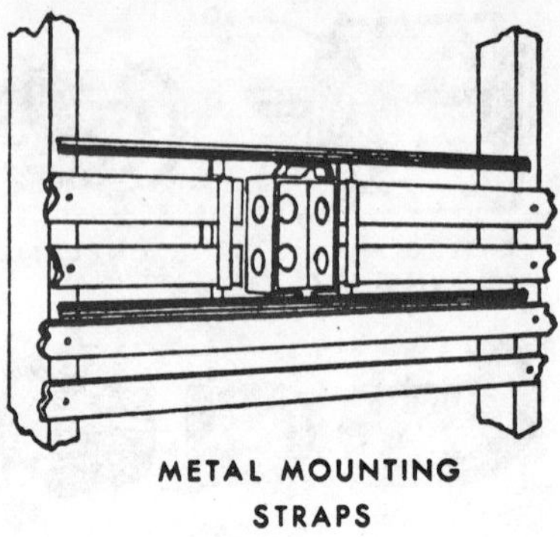

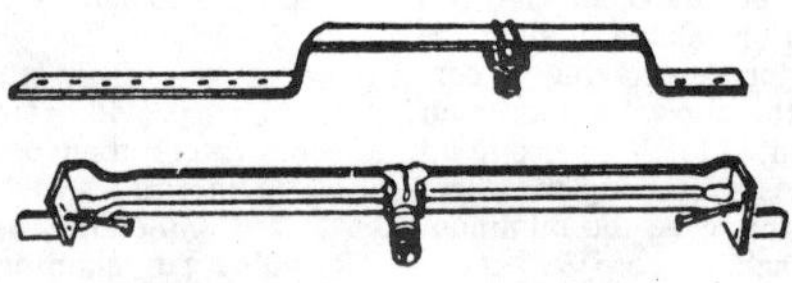

Figure 15-31.—Typical box mountings.

necessary. The ready-made straps are handy and accommodate not only a single box, but a 2, 3, 4, or 5 gang box.

3. Bar hangers. Bar hangers are prefabricated to span the normal 16-inch and 24-inch joist and stud spacings and are obtainable for surface or recessed box installation. They are nailed to the joist or stud exposed faces. The supports for recessed boxes normally are called offset bar hangers.

4. Patented supports. When boxes have to be installed in walls that are already plastered, several patented supports can be used for mounting. These obviate the need for installing the boxes on wooden members and thus eliminate extensive chipping and replastering.

KNOBS, TUBES, CLEATS, LOOM, AND SPECIAL CONNECTORS

Open wiring requires the use of special insulating supports and tubing to insure a safe installation. These supports, called knobs and cleats, are smooth-surfaced and made of porcelain. Knobs and cleats support the wires which are run singly or in pairs on the surface of the joists or studs in the buildings. Tubing or tubes, as they are called, protect the wires from abrasion when passing through wooden members. Insulation of loom of the "slip-on" type is used to cover the wires on box entry and at wire-crossover points. The term "loom" is applied to a continuous flexible tube

NOTES:

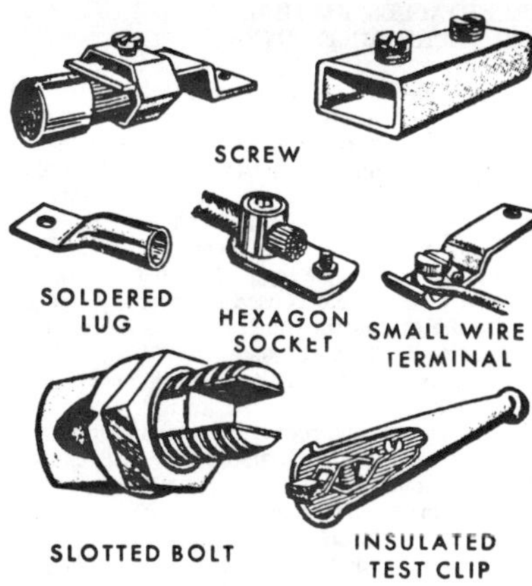

Figure 15-32.—Cable and wire connectors.

woven of cambric material impregnated with varnish. At points where the type of wiring may change and where boxes are not specifically required, special open wiring to cable or conduit wiring connectors should be used. These connectors are threaded on one side to facilitate connection to a conduit and have holes on the other side to accommodate wire splices but are designed only to carry the wire to the next junction box.

CABLE AND WIRE CONNECTORS

Code requirements state that "Conductors shall be spliced or joined with splicing devices approved for the use or by brazing, welding, or soldering with a fusible metal or alloy. Soldered splices shall first be so spliced as to be mechanically and electrically secure without solder and then soldered." Soldering or splicing devices are used as added protection because of the ease of wiring and the high quality of connection of these devices. Assurance of high quality is the responsibility of the electrician who selects the proper size of connector relative to the number and size of wires. Figure 15-32 shows some of the many types of cable and wire connectors in common use.

STRAPS AND STAPLES

Policy. All conduits and cables must be attached to the structural members of a building in a manner that will preclude sagging. The cables must be supported at least every 4½ feet for either a vertical or horizontal run and must have a support in the form of a strap or staple within 12 inches of every outlet box. Conduit-support spacings vary with the size and rigidity of the conduit.

Cable Staples. A very simple and effective method of supporting BX cables on wooden members is by the use of cable staples as shown in 1, figure 15-33.

Insulating Staples. Bell or signal wires are normally installed in pairs in signal systems. The operating voltage and energy potential is so low in these installations (12 to 24 volts) that protective

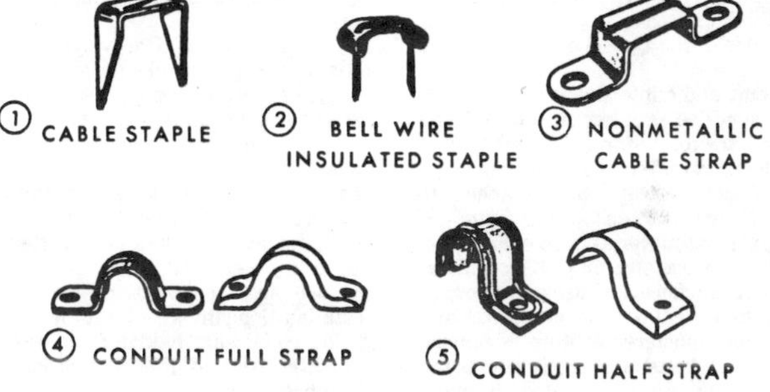

Figure 15-33.—Straps and staples.

NOTES:

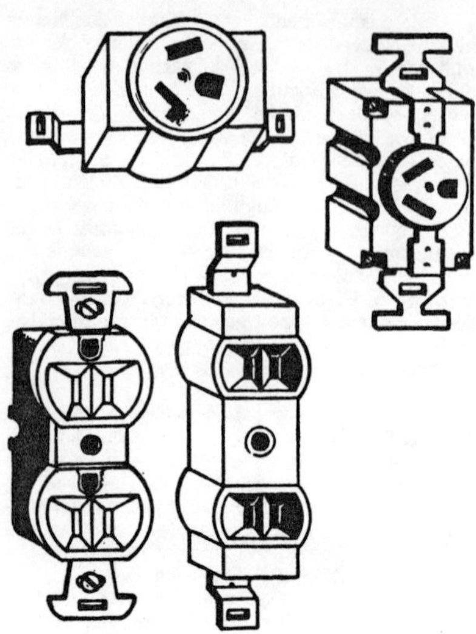

Figure 15-34—Types of wall receptacles.

coverings such as conduit or loom are not required. To avoid any possibility of shorting in the circuit, they are normally supported on wood joists or studs by insulated staples of the type shown in 1, figure 15-33.

Straps. Conduit and cable straps (3, 4, and 5, figure 15-33) are supplied as either one-hole or two-hole supports and are formed to fit the contour of the specific material for which they are designed. The conduit· and cable straps are attached to building materials by "anchors" designed to suit each type of supporting material. For example, a strap is attached to a wood stud or joist by nails or screws. Expanding anchors are used for box or strap attachment to cement or brick structures and also to plaster or plaster-substitute surfaces. Toggle and "molly" bolts are used where the surface wall is thin and has a concealed air space which will allow for the release of the toggle or expanding sleeve.

NOTES:

RECEPTACLES, FIXTURES, AND RECEPTACLE COVERS

Applicability. Portable appliances and devices are readily connected to an electrtical supply circuit by means of an outlet called a receptacle. For interior wiring these outlets are installed either as single or duplex receptacles. Receptacles previously installed, and their replacements in the same box, may be two-wire receptacles. All others must be the three-wire type. The third wire on the three-wire receptacle is used to provide a ground lead to the equipment which receives power from the receptacle. This guards against dangers from current leakage due to faulty insulation or exposed wiring and helps prevent accidental shock. The receptacles are constructed to receive plug prongs either by a straight push action or by a twist-and-turn push action. Fixtures are similar to receptacles but are used to connect the electrical supply circuit directly to lamps inserted in their sockets.

Knob-and-Tube Wiring. Receptacles with their entire enclosures made of some insulating material, such as bakelite, may be used without metal outlet boxes for exposed, open wiring or nonmetallic sheathed cable.

Conduit and Cable. The receptacles (figure 15-34) commonly used with conduit and cable installations are constructed with yokes to facilitate their installation in outlet boxes. In this case they are attached to the boxes by metal screws through the yokes, threaded into the box. Wire connections are made at the receptacle terminals by screws which are an integral part of the outlet. Receptacle covers made of either brass, steel, or nonmetallic materials are then attached to box and receptacle installations to afford complete closure at the outlets.

Surface Metal Raceways. These provide a quick inexpensive electrical wiring installation method since they are installed on the wall surface instead of inside the wall (figure 15-35).

1. Surface metal raceway is basically of two types: one-piece construction or two-piece construction. When working with the one-piece construction type, the metal raceway is installed like conduit, then the wires are "pulled" to make the necessary electrical connections. If working with the two-piece construction type, the base piece is installed along the wiring run. Wiring is next laid in the base piece and held in place with clamps. After the wires are laid, the capping is snapped on and the job is complete.

2. A multioulet system, with grounding inserts if desired, has outlets spaced every few inches so

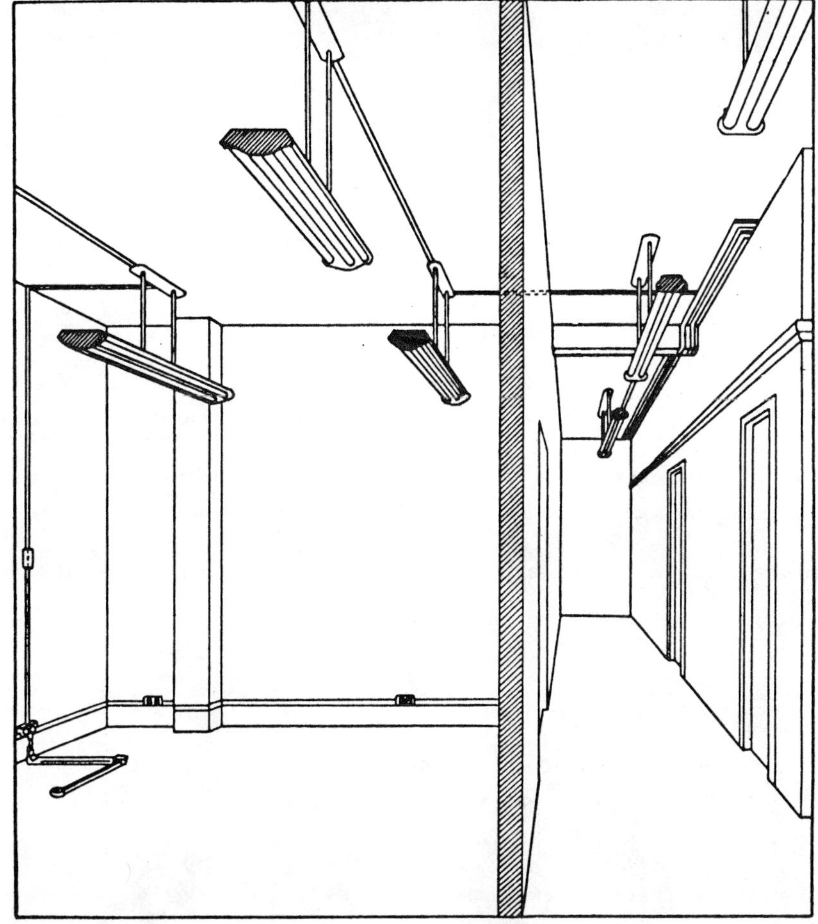

Figure 15-35.—Surface metal raceways.

that several tools or pieces of equipment can be used simultaneously. An overfloor metal raceway system handles telephone and signal or power and light wiring where the circuits must be brought to locations in the middle of the floor area. These systems are all designed so that they can be installed independently of other wiring systems, or may be economically connected to existing

NOTES:

Figure 15-36.—Attachment plugs.

Figure 15-37.—The Underwriters knot.

systems.

PLUGS AND CORD CONNECTORS

Plugs. Portable appliances and devices that are to be connected to receptacles have their electrical cords equipped with plus (figure 15-36) that have prongs which mate with the slots in the outlet receptacles. A three-prong plug can fit into a two-prong receptacle by using an adapter. If the electrical conductors connected to the outlet have a ground system, the lug on the lead wire of the adapter is connected to the center screw holding the receptacle cover to the box. Many of these plugs are permanently molded to the attached cords. There are other types of cord-grips that hold the cord firmly to the plug. Twist-lock plugs have patented prongs that catch and are firmly held to a mating receptacle when the plugs are inserted into the receptacle slots and twisted. Where the plugs do not

Figure 15-38.—Switches and covers.

NOTES:

have cord-grips, the cords should be tied with an Underwriters knot (figure 15-37) at plug entry to eliminate tension on the terminal connections when the cord is connected and disconnected from the outlet receptacle. Figure 15-37 shows the steps to be used in tying this type of knot.

Cord Connectors. There are some operating conditions where a cord must be connected to a portable receptacle. This type of receptacle, called a cord connector body or a female plug, is attached to the cord in a manner similar to the attachment of the male plug outlined above.

SWITCHES AND COVERS

Definition. A switch is a device used to connect and disconnect an electrical circuit from the source of power. Switches may be either one-pole or two-pole for ordinary lighting or receptacle circuits. If they are of the one-pole type, they must be connected to break the hot or ungrounded conductor of the circuit. If of the two-pole type, the hot and ground connection can be connected to either pole on the line side of the switch. Switches are also available that can be operated in combinations of two, three or more in one circuit. These are called three-way and four-way switches.

Open and Nonmetallic Sheathed Wiring. Switches used for exposed open wiring and nonmetallic sheathed cable wiring are usually of the tumbler type with the switch and cover in one piece. Other less common ones are the rotary-snap and pushbutton types. These switches are generally nonmetallic in composition (1, 2, and 3, figure 15-38).

Conduit and Cable Installations. The tumbler switch and cover plates (4 and 5, figure 15-38) normally used for outlet-box installation are mounted in a manner similar to that for box type receptacles and covers and are in two pieces. Foreign installations may still use pushbutton switches as shown in 6 and 7, figure 15-38.

Entrance Installations. at every power-line entry to a building a switch and fuse combination or circuit breaker switch of a type similar to that shown in figure 15-39 must be installed at the service entrance. This switch must be rated to disconnect the building load while in use at the system voltage. Entrance or service switches, as they are commonly called, consist of one "knife" switch blade for every hot wire of the power supplied. The switch is generally enclosed and sealed in a sheet-steel cabinet. When connecting or disconnecting the building circuit, the blades are operated simultaneously through an exterior

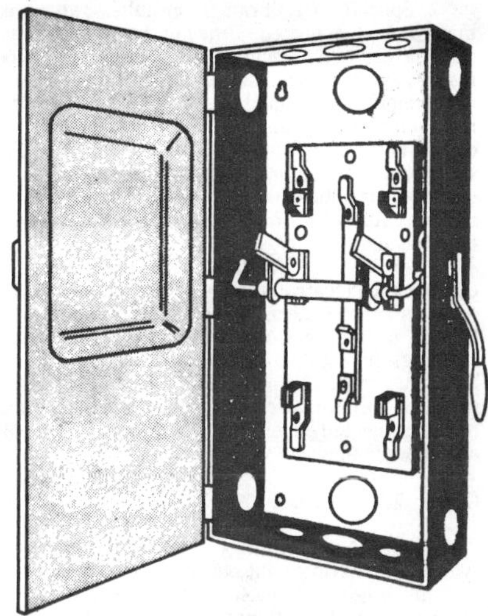

Figure 15-39.—Service switch box.

handle by the rotation of a common shaft holding the blades. The neutral or grounded conductor is not switched but is connected at at a neutral terminal within the box. Many entrance switches are equipped with integral fuse blocks or circuit breakers which protect the building load. The circuit breaker type of entrance switch is preferred particularly in field installations, because of its ease of resetting after the overload condition in the circuit has been cleared.

FUSES AND FUSE BOXES

Fuses. The device for automatically opening a circuit when the current rises beyond the safety limit is technically called a cutout, but more commonly is called a fuse. All circuits and electrical apparatus must be protected from short circuits or dangerous overcurrent conditions through correctly rated fuses.

1. Standard. The cartridge type fuse is used for current rating above 30 amperes in interior wiring systems. The ordinary plug or screw type fuse is satisfactory for incandescent lighting or heating appliance circuits.

NOTES:

2. Special. On branch circuits, wherever motors are connected, time-lag fuses should be used instead of the standard plug or cartridge type fuse. These fuses have self-compensating elements which maintain and hold the circuit in line during a momentary heavy ampere drain, yet cut out the circuit under normal short-circuit conditions. The heavy ampere demand normally occurs in motor circuits when the motor is started. Examples of such circuits are the ones used to power oil burners or air conditioners.

Fuse Boxes. As a general rule the fusing of circuits is concentrated at centrally located fusing or distribution panels. These panels are normally located at the service-entrance switch in small buildings or installed in several power centers in large buildings. The number of service centers or fuse boxes in the latter case would be determined by the connected power load. Fuses and a fuse box are shown in figure 15-40.

CIRCUIT BREAKER PANELS

Circuit breakers are devices resembling switches that trip or cut out the circuit in case of overamperage. They perform the same function as fuses and can be obtained with time-lag opening features similar to the special fuses outlined in paragraph 15-42. Based on their operation, they may be classified as a thermal, magnetic, or combination thermal-magnetic reaction type. A thermal type circuit breaker has a bimetallic element integrally built within the breaker that responds only to fluctuations in temperature within the circuit. The element is made by bonding together two strips of dissimilar metal, each of which has a different coefficient of expansion. When a current is flowing in the circuit, the heat created by the resistance of the bimetallic element will expand each metal at a different rate causing the strip to bend. The element acts as a latch in the circuit as the breaker mechanism is adjusted so that the element bends just far enough under a specified current to trip the breaker and open the circuit. A magnetic circuit breaker responds to changes in the magnitude of current flow. In operation an increased current flow will create enough magnetic force to "pull up" an armature, opening the circuit. The magnetic circuit breaker is usually used in motor circuits for closer adjustment to motor rating while the circuit conductors are protected, as usual, by another circuit breaker. The thermal-magnetic breaker, as the name implies, combines the features of the thermal and magnetic types. Practically all of the molded case circuit breakers used in lighting panelboards are of this type. The thermal element protects against overcurrents in the lower range and the magnetic element protects against the higher range usually occurring from short circuits. During the last decade, circuit breakers have been used to a greater extent than fuses because they can be

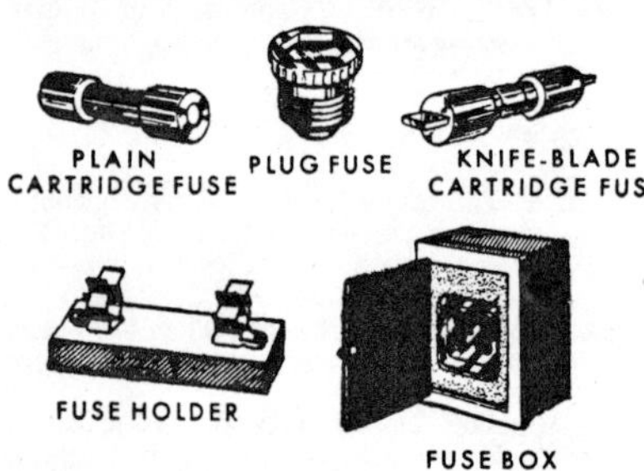

Figure 15-40.—Typical fuses and fuse box.

Figure 15-41.—Typical circuit breaker box.

NOTES:

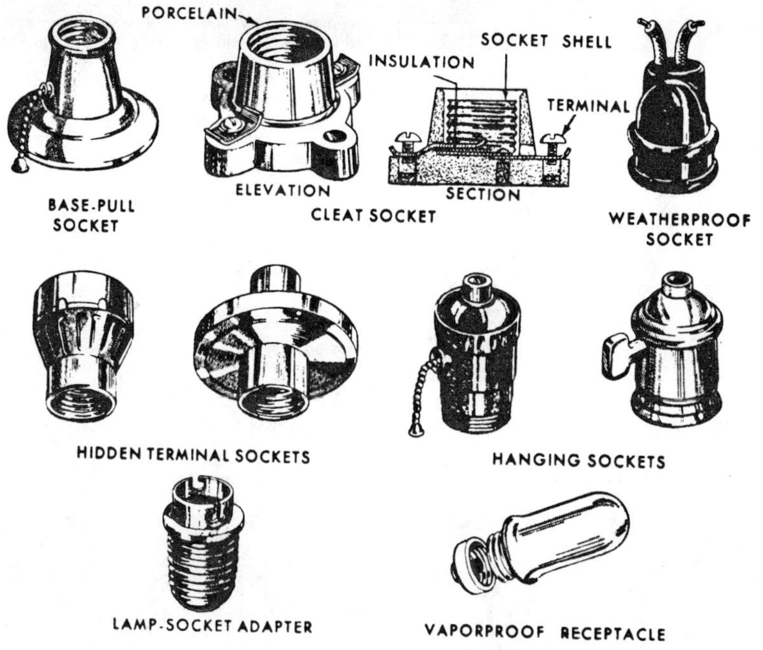

Figure 15-42.—Lampholders and sockets.

manually reset after tripping, whereas fuses require replacement. Fuses may easily be replaced with higher capacity ones that do not protect the circuit. This is difficult to do with circuit breakers. In addition they combine the functions of fuse and switch, and when tripped by overloads or short circuits, all of the ungrounded conductors of a circuit are opened simultaneously. Each branch circuit must have a fuse or circuit breaker protecting each ungrounded conductor. Some installations may or may not have a main breaker that disconnects everything. As a guide during installation, if it does not require more than six movements of the hand to open all the branch circuit breakers, a main breaker or switch is not required ahead of the branch-circuit breaker. However, if more than six movements of the hand are required, a separate disconnecting main circuit breaker is required ahead of the branch-circuit breaker. Each 120-volt circuit requires a single-pole (one-pole) braker which has its own handle. Each 208-volt circuit requires a double-pole (two-pole) breaker to protect both ungrounded conductors. You can, however, place two single-pole breakers side by side, and tie the two handles together mechanically to give double-pole protection. Both handles can then be moved by a single movement of the hand. A two-pole breaker may have one handle or two handles which are mechanically tied together, but either one requires only one movement of the hand to break the circuit. Figure 15-41 illustrates a typical circuit breaker panel.

LAMPHOLDERS AND SOCKETS

Lamp sockets as shown in figure 15-42 are generally screw-base units placed in circuits as holders for incandescent lamps. A special type of

NOTES:

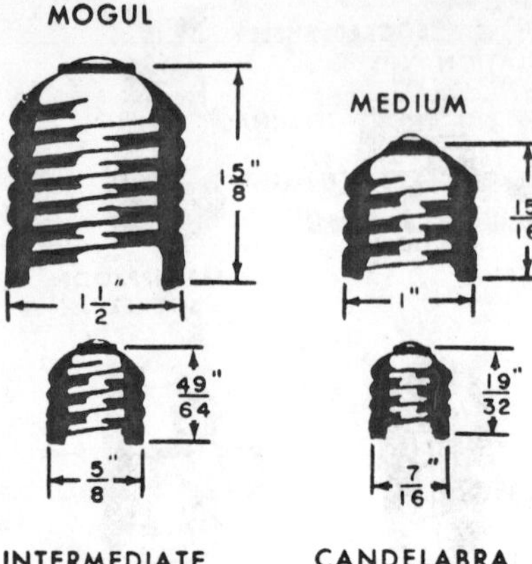

Figure 15-43.—General lamp-socket sizes.

lampholder has contacts, rather than a screw base, which engage and hold the prongs of fluorescent lamps when they are rotated in the holder. The sockets can generally be attached to a hanging cord or mounted directly on a wall or ceiling in open wiring installations by using screws or nails in the mounting holes provided in the nonconducting material which is molded or formed around the lamp socket. The two mounting holes in a porcelain lamp socket are spaced so the sockets may also be attached to outlet box "ears" or a plaster ring with machine screws. The screw threads molded or rolled in the ends of the lampholder sockets also facilitate their ready integration in other types of lighting fixtures such as table lamps, floor lamps, or hanging fixtures which have reflectors or decorative shades. In an emergency, a socket may also be used as a receptacle. The socket is converted to a receptable by screwing in a female plug. One type of ceiling lampholder has a grounded outlet located on the side. Lamp sockets are produced in many different sizes and shapes. A few of the most common sizes are shown in figure 15-43.

SIGNAL EQUIPMENT

Figure 15-44 illustrates the most common components in interior wiring signal systems. Their normal operating voltages are 6, 12, 18, or 24 volts, ac or dc. As a general rule they are connected by open-wiring methods and are used as interoffice or building-to-building signal systems.

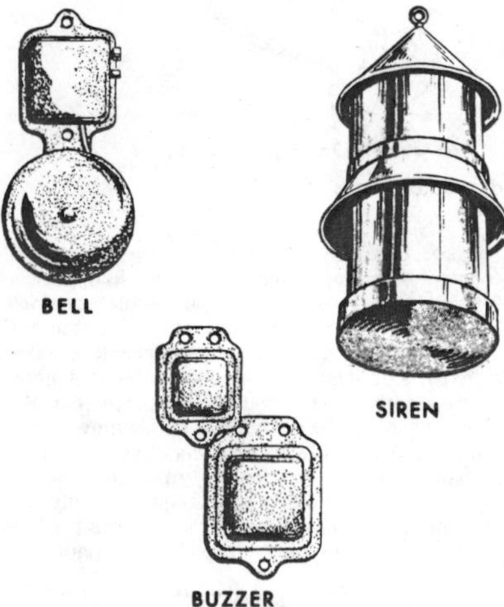

Figure 15-44.—Types of signal equipment.

Figure 15-45.—Types of reflectors.

NOTES:

REFLECTORS AND SHADES

Figure 15-45 shows several types of reflectors and shades which are used to focus the lighting effect of bulbs. Of these, some are used to flood an area with high intensity light and are called floodlights. Others, called spotlights, concentrate the useful light on a small area. Both floodlights and spotlights can come in two- or three-light clusters with swivel holders. They can be mounted on walls or posts or on spikes pushed into the ground. One and two, figure 15-45 illustrate reflectors that deliver normal building light of average intensity in a pattern similar to the floodlight shown in 3, figure 15-45.

INCANDESCENT LAMPS

The most common light source for general use is the incandescent lamp. Though it is the least efficient type of light, its use is preferred over the fluorescent type because of its low initial cost, ease of maintenance, convenience, and flexibility. Its flexibility and convenience is readily seen by the wide selection of wattage ratings that can be inserted in one type socket. Further, since its emitted candlepower is directly proportional to the voltage, a lower voltage application will dim the light. A high rated voltage application from a power source will increase its intensity. Although an incandescent light is economical, it is also inefficient because a large amount of the energy supplied to it is converted to heat rather than light. Moreover, it does not give a true light because the tungsten filament emits a great deal more red and yellow light than does the light of the sun. Incandescent lamps are shown in figure 15-46. Incandescent lights are normally built to last 1,000 hours when operating at their rated voltage.

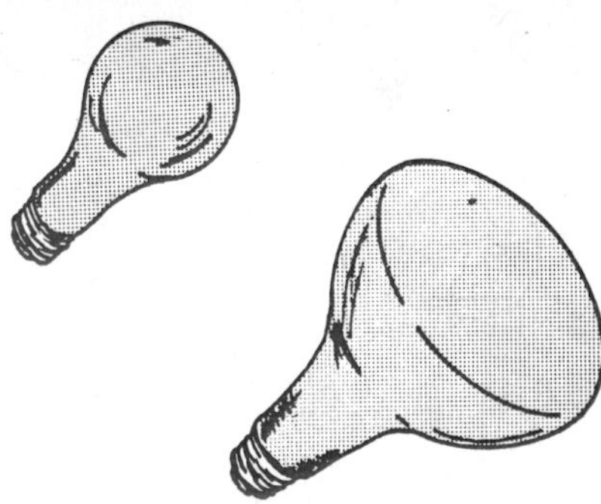

Figure 15-46.—Incandescent lamps.

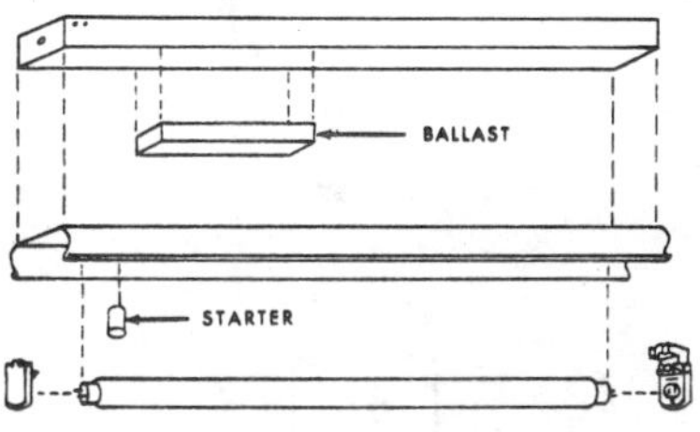

Figure 15-47.—Fluorescent light accessories.

FLUORESCENT LAMPS

Fluorescent lamps (figure 15-47) are either of the conventional "hot cathode" or "cold cathode" type. The "hot cathode" lamp has a coiled wire type of electrode, which when heated gives off electrons. These electrons collide with mercury atoms, provided by mercury vapor in the tubes, which produces ultraviolet radiation. Fluorescent powder coatings on the inner walls of the tubes absorb this radiation and transform the energy into visible light. The "cold cathode" lamp operates in a similar manner except that its electrode consists of a cylindrical tube. It receives its name because the heat is generated over a larger area and, therefore, the cathode does not reach as high a temperature as in the "hot cathode" tube. The "cold cathode" is less efficient but has a longer life than the "hot cathode" unit. It is used most frequently on flashing circuits. Because of the higher light output per watt input, more illumination and less heat is obtained per watt from fluorescent lamps than from incandescent ones. Light diffusion is also better, and surface brightness is lower. The life of fluorescent lamps is also longer compared to filament types. However, the fluorescent lamp, because of its design, cannot control its beam of light as well as the incandescent type and has a tendency to produce stroboscopic effects which are counteracted by phasing arrangements. Moreover,

NOTES:

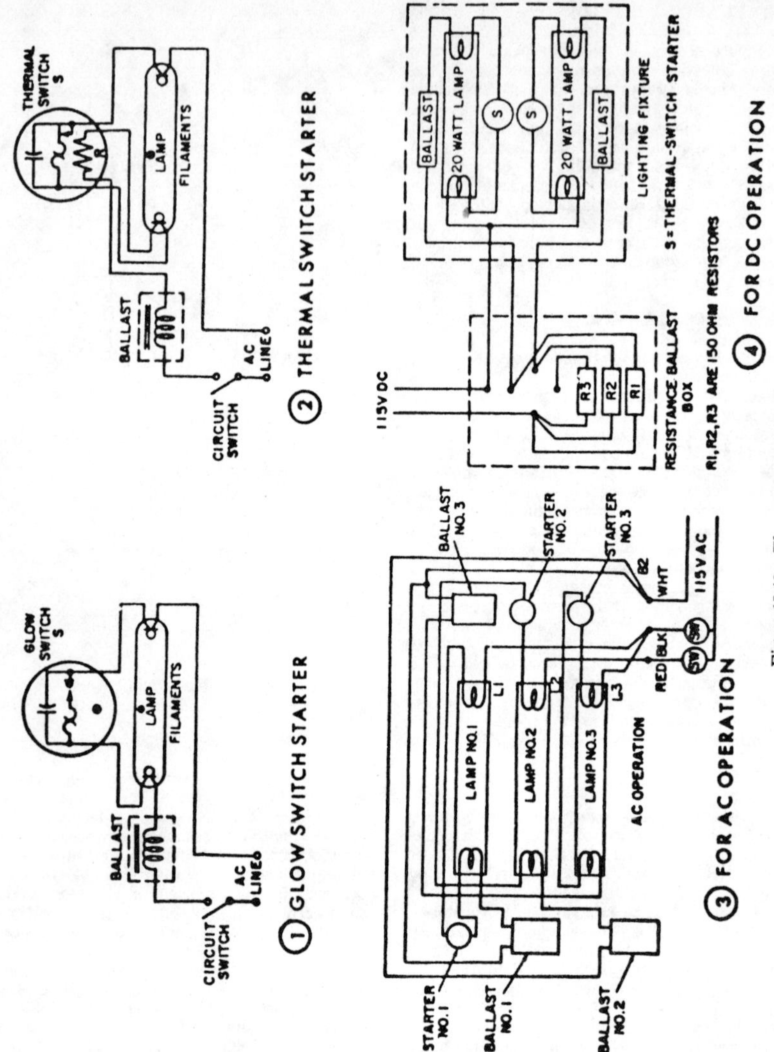

Figure 15-48.—Fluorescent schematic.

NOTES:

when voltage fluctuations are severe, the lamps may go out prematurely or start slowly. Finally, the higher initial cost in fluorescent lighting, which requires auxiliary equipment such as starters, ballasts, special lampholders, and fixtures (figure 15-47), is also a disadvantage compared with other types of illumination.

Construction. The fluorescent lamp is an electric discharge lamp that consists of an elongated tubular bulb with an oxide-coated filament sealed in each end to comprise two electrodes (figure 15-48). The bulb contains a drop of mercury and a small amount of argon gas. The inside surface of the bulb is coated with a fluorescent phosphor. The lamp produces invisible, short wave (ultraviolet) radiation by the discharge through the mercury vapor in the bulb. The phosphor absorbs the invisible radiant energy and reradiates it over a band of wavelengths that are sensitive to the eye.

1. Detail illumination is required where the intensity of general illumination is not sufficient, and in engineering spaces for examination of gages. The fixtures for detail illumination commonly use single fluorescent lamps. One and two, figure 15-48, illustrates the wiring arrangement for these single units, and three, figure 15-48, shows a multiple unit.

2. Because of greater cost and shorter life of 8-watt fluorescent lamps, as compared to 15-watt and 20-watt lamps, fixtures with 8-watt lamps are used only for detail illumination and general illumination within locations where space is restricted.

3. Although the fluorescent lamp is basically an ac lamp, it can be operated on dc with the proper auxiliary equipment. The current is controlled by an external resistance in series with the lamp (4, figure 15-48). Since there is no voltage peak, starting is more difficult and thermal switch starters are required. The lamp tends to deteriorate at one end due to the uniform direction of the current. This may be partially overcome by reversing the lamp position or the direction of current periodically.

4. Because of the power lost in the resistance ballast box in the dc system, the overall lumens per watt efficiency of the dc system is about 60 percent of the ac system. Also, lamps operated on dc may provide as little as 80 percent of rated life.

5. The fluorescent lamp, like all discharge light sources, requires special auxiliary control equipment for starting and stabilizing the lamp. This equipment consists of an iron-core choke coil, or ballast, and an automatic starting switch connected in series with the lamp filaments. The starter (starting switch) can be either a glow switch or a thermal switch. A resistor must be connected in series with the ballast in dc circuits because the ballast alone does not offer sufficient resistance to maintain the arc current steady.

6. Each lamp must be provided with an individual ballast and starting switch, but the auxiliaries for two lamps are usually enclosed in a single container. The auxiliaries for fluorescent lighting fixtures are mounted inside the fixture above the reflector. The starting switches (starters) project through the reflector so that they can be replaced readily. The circuit diagram for the fixture appears on the ballast container.

Operation. A fluorescent lamp equipped with a glow-switch starter is illustrated in 1, figure 15-48. The glow-switch starter is essentially a glow lamp containing neon or argon gas and two metallic electrodes. One electrode has a fixed contact, and the other electrode is U-shaped, bimetal strip having a movable contact. These contacts are normally open.

1. When the circuit switch is closed, there is practically no voltage drop across the ballast, and the voltage across the starter, S, is sufficient to produce a glow around the bimetallic strip in the glow lamp. The heat from the glow causes the bimetal strip to distort and touch the fixed electrode. This action shorts out the glow discharge and the bimetal strip starts to cool as the starting circuit of the fluorescent lamp is completed. The starting current flows through the lamp filament in each end of the fluorescent tube, causing the mercury to vaporize. Current does not flow across the lamp between the electrodes at this time because the path is short circuited by the starter and because the gas in the bulb is nonconducting when the electrodes are cold. The preheating of the fluorescent tube continues until the bimetal strip in the starter cools sufficiently to open the starting circuit.

2. When the starting circuit opens, the decrease of current in the ballast produces an induced voltage across the lamp electrodes. The magnitude of this voltage is sufficient to ionize the mercury vapor and start the lamp. The resulting glow discharge (arc) through the fluorescent lamp produces a large amount of ultraviolet radiation that impinges on the phosphor, causing it to fluoresce and emit a relatively bright light. During normal operation the voltage across the fluorescent lamp is not sufficient to produce a glow in the starter. Hence, the contacts remain open and the starter consumes no energy.

NOTES:

3. A fluorescent lamp equipped with a thermal-switch starter is illustrated in 1, figure 15-48. The thermal-switch starter consists of two normally closed metallic contacts and a series resistance contained in a cylindrical enclosure. One contact is fixed, and the movable contact is mounted on a bimetal strip.

4. When the circuit switch is closed, the starting circuit of the fluorescent lamp is completed (through the series resistance, R) to allow the preheating current to flow through the electrodes. The current through the series resistance produces heat that causes the bimetal strip to bend and open the starting circuit. The accompanying induced voltage produced by the ballast starts the lamp. The normal operating current holds the thermal switch open.

5. The majority of thermal-switch starters use some energy during normal operation of the lamp. However, this switch insures more positive starting by providing an adequate preheating period and a higher induced starting voltage.

Characteristics. The failure of a hot-cathode fluorescent lamp usually results from loss of electron-emissive material from the electrodes. This loss proceeds gradually throughout the life of the lamp and is accelerated by frequent starting. The rated average life of the lamp is based on normal burning periods of 3 to 4 hours. Blackening of the ends of the bulb progresses gradually throughout the life of the lamp.

1. The efficiency of the energy conversion of a fluorescent lamp is very sensitive to changes in temperature of the bulb. The maximum efficiency occurs in the range of 100° F. to 120° F., which is the operating temperature that corresponds to an ambient room temperature range of 65° to 85° F. The efficiency decreases slowly as the temperature is increased above normal, but also decreases very rapidly as the temperature is decreased below normal. Hence, the fluorescent lamp is not satisfactory for locations in which it will be subjected to wide variations in temperature. The reduction in efficiency with low ambient room temperature can be minimized by operating the fluorscent lamp in a tubular glass enclosure so that the lamp will operate at more nearly the desired temperature.

2. Fluorescent lamps are relatively efficient compared with incandescent lamps. For example, a 40-watt fluorescent lamp produces approximately 2800 lumens, or 70 lumens per watt. A 40-watt fluorescent lamp produces six times as much light per watt as does the comparable incandescent lamp.

3. Fluorescent lamps should be operated at voltage within 8 percent of their rated voltage. If the lamps are operated at lower voltages, uncertain starting may result, and if operated at higher voltages, the ballast may overheat. Operation of the lamps at either lower or higher voltages results in decreased lamp life. The characteristic curves for hot-cathode fluorescent lamps show the effect of variations from rated voltage on the condition of lamp operation. Also, the performance of fluorescent lamps depends to a great extent on the characteristics of the ballast, which determines the power delivered to the lamp for a given line voltage.

4. When lamps are operated on ac circuits, the light output executes cyclic pulsations as the current passes through zero. This reduction in light output produces a flicker that is more noticeable in fluorescent lamps than in incandescent lamps at frequencies of 50 and 60 cycles and may cause unpleasant stroboscopic effects when moving objects are viewed. The cyclic flicker can be minimized by combining two or three lamps in a fixture and operating the lamps on different phases of a three-phase system. Where only single-phase circuits are available, leading current may be supplied to one lamp and lagging current to another through a lead-lag ballast circuit so that the light pulsations compensate each other.

5. The fluorescent lamp is inherently a high power-factor device, but the ballast required to stabilize the arc is a low power-factor device. The voltage drop across the ballast is usually equal to the drop across the arc, and the resulting power factor for a single-lamp circuit with ballast is about 50 percent. The low power factor can be corrected in a single-lamp ballast circuit by a capacitor shunted across the line. This correction is accomplished in a two-lamp circuit by means of a "tulamp" auxiliary that connects a capacitor in series with one of the lamps to displace the lamp currents, and, at the same time, to remove the unpleasant stroboscopic effects when moving objects come into view.

Glow Lamps. Glow lamps are electric discharge light sources, which are used as indicator or pilot lights for various instruments and on control panels. These lamps have relatively low light output, and thus are used to indicate when circuits are energized or to indicate the operation of electrical equipment installed in remote locations.

1. The glow lamp consists of two closely spaced metallic electrodes sealed in a glass bulb that contains an inert gas. The color of the light emitted

NOTES:

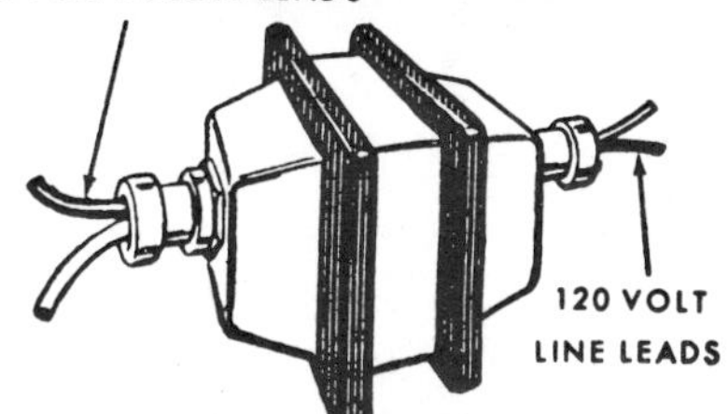

Figure 15-49.—Transformer.

by the lamp depends on the gas. Neon gas produces a blue light. The lamp must be operated in series with a current-limiting device to stabilize the discharge. This current-limiting device consists of a high resistance that is usually contained in the lamp base.

2. The glow lamp produces light only when the voltage exceeds a certain striking voltage. As the voltage is decreased somewhat below this value, the glow suddenly vanishes. When the lamp is operated on alternating current, light is produced only during a portion of each half cycle, and both electrodes are alternately surrounded with a glow. When the lamp is operated on direct current, light is produced continuously, and only the negative electrode is surrounded with a glow. This characteristic makes it possible to use the glow lamp as an indicator of alternating current and direct current. It has the advantages of small size, ruggedness, long life, and negligible current consumption, and can be operated on standard lighting circuits.

TRANSFORMERS

The transformer is a device for changing alternating current voltages into either high voltages for efficeint powerline transmission or low voltages for consumption in lamps, electrical devices, and machines. Transformers vary in size according to their power handling rating. Their selection is determined by input and output voltage and load current requirements. For example, the transformer used to furnish power for a doorbell reduces 115-volt alternating current to about 6 to 10 volts. This is accomplished by two primary wire leads which are permanently connected to the 115-volt circuit and two secondary screw terminals from the low voltage side of the transformer. Figure 15-49 shows a common type of signal system transformer. It is used to lower the building voltage of 120 volts or 240 volts ac to the 6, 12, 18, or 24 volts ac. The wires shown are input and output leads. In figure 15-49 the input leads are smaller than the output leads because the current in the output circuit is greater than in the input circuit.

NOTES:

CHAPTER 16

CABLE WIRING

ARMORED CABLE WIRING

ADVANTAGES AND USES

Armored cable wiring, commonly called BX, is permissible by Code for all interior installations, except where it is exposed to saturation by liquid or is in contact with acid fumes. In wet areas a lead-covered cable is required. From an economic viewpoint the labor costs for armored cable installations compare favorably with the open or knob-and-tube wiring installation. The material requirements for armored cable wiring are greater, and thus overall cost is generally higher. This increased cost is often warranted because an armored conductor has greater mechanical damage protection, and thus eliminates the need for porcelain insulators and loom which are required in open wiring.

MATERIALS

Cable. As outlined in chapter 15, armored cable construction is made up in two or three rubber- or thermoplastic-covered wire combinations encased in flexible steel armor. It is obtained from the manufacturer as Type AC without a lead sheath, and Type ACL with a lead sheath under the armor. Type AC cables have a copper or aluminum bonding strip. One of the conductors of armored cable is always white. Because of this color coding, the code allows a white wire in a switch installation (with armored cable and also nonmetallic sheathed cable) to be used as a hot wire and thereby allows its connection to a black wire. This is shown in the wiring of lampholder No. 3 and switch No. 4, figure 16-1.

Three-Wire Armored Cable.

1. Service. Three-wire armored cable is used to carry power from the service-entrance switch to the fuse panels or to local load centers if the system in a building is three-wire 120--240 or 120--208 volts. The minimum size conductor recommended for this use is No. 10 AWG. In this type of service the neutral wire can be of the same gage as the two hot wires because its maximum current will be no greater than the maximum for either of the other two wires. A three-wire cable connection is shown at the load center, No. 1, in figure 16-1.

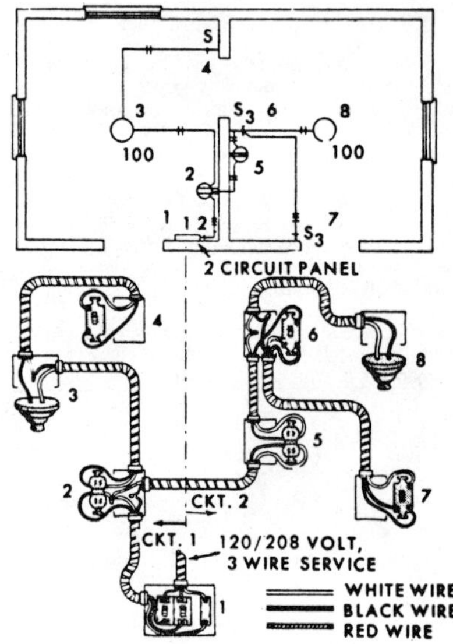

Figure 16-1.—Typical armored cable connections.

NOTES:

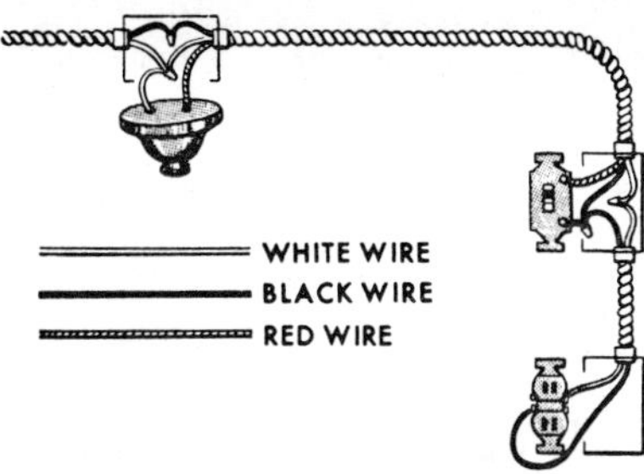

Figure 16-2.—Three-wire cable, two-circuit use.

2. Two-Circuit. Sometimes in laying out and circuiting armored cable installations, several circuits feed out in the same general direction from the protection (fuse) panel. When the power distribution system uses three-wire 110 to 220 volts, and the circuits are not fed from a common hot-line wire, it is advantageous, both from voltage-drop and economic standpoints, to install three-wire cable as a two-circuit carrier as far as possible. An example of this use of three-wire cable is shown in figure 16-1, between the circuit breaker panel, No. 1, and the receptacle box, No. 2. Similarly, many installations are made wherein a switch and an outlet receptacle are connected from wires originating in an overhead light to be switch controlled. Rather than use two two-wire cables for circuiting these devices, a three-wire installation from the light to the switch (figure 16-2) is made.

Supports. Armored cable may be fastened to wooden building members with a one- or two-hole type mounting strap formed to fit the contour and size of the cable, or by staples made specifically for armored cable used. The cable is normally supported at the box entry by integral BX clamps built into the boxes or by BX connectors.

Boxes and Devices. Boxes and devices recommended for use with armored cable wiring are detailed in chapter 15. For quick installation the electrical boxes with the integral cable clamps and attached mounting brackets are used.

INSTALLATION

Cable Support. Whenever possible an armored cable installation should be run through holes centrally drilled in the building structural members, and the holes should be at least 1/8 inch oversize to facilitate easy "pull through" of the BX. The flush type mounting of BX accomplished by notching the joists and studs should be avoided whenever possible. This type installation exposes the BX to possible short circuits, by locating the cable in a position where it could be accidentally pierced by nails, and materially weakens the structural member. When armored cable is run between joists and studs, it should be supported by staples or straps at least every 4½ feet along the length of the cable run. These supports must also be installed within 12 inches of each box entry, unless the support interferes with installations that require extreme flexibility. This requirement also assures the continuance of a satisfactory box connection by relieving the strain on the splices and connection within the outlet box. Cable runs installed across the bottom of ceiling joists and studding faces at least 7 feet above the floor must be supported on each joist or stud. If preferred they may also be installed on running boards similar to those used in open wiring installations.

Damage Protection. When armored cable is installed on the top of floor joists or studding in accessible locations (attics and temporary buildings) at a distance less than 7 feet from the floor, guard strips at least as high as the cable must be installed.

Armored Cable Bending. When installing armored cable, care must be used to avoid bending or shaping the cable in a manner that damages the protective armor. This type of installation damage may occur in drilled holes for BX, in corner runs, or when locating boxes on studs and joists. To prevent this, the radius of the inner edge of any bend must not be less than five times the cable diameter. Figure 16-3 illustrates the acceptable armored cable bend at box entry.

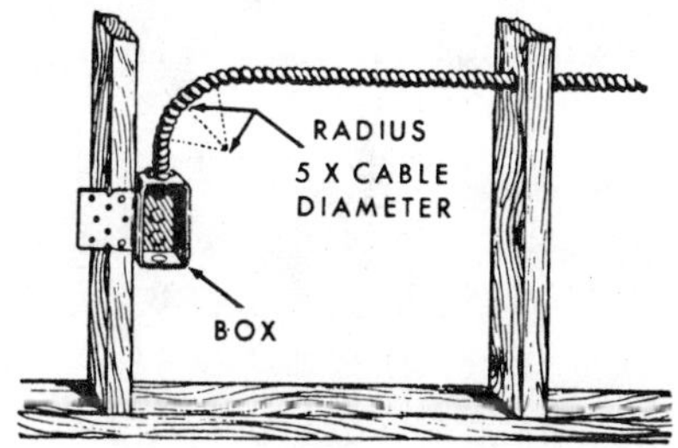

Figure 16-3.—Armored cable bend.

NOTES:

Box Connection.

1. Procedure. armored cable may be spliced or connected to devices only in standard junction or outlet boxes. All the cable used, therefore, must be cut long enough to run from box to box. To prevent cutting the cable too short, the BX should first be threaded through the mounting holes drilled in the joists or studs and attached to one box. The slack is taken out of the cable by using just enough force to maintain the proper bends. Keeping this tension, the cable is cut from the roll and connected to the box. The procedure to be followed in preparing and attaching the cable to a box is shown in figure 16-4.

2. Cable cutting (1, figure 16-4). Though armored cable can be cut with a BX cutter specifically designed for the job, the majority of electricians generally use a hacksaw. In making an outlet connection, the cable should first be cut completely through about 8 inches longer on each end than required for the run. In removing the armored cable from the wire, the armor should be cut approximately 8 inches from the cable end so that ample wire will be inserted in the box for connecting to the outlet device. These lead lengths may be increased when the wire run terminates in a fuse or circuit breaker panel box and a longer cable is required. The cutting of cable armor is a simple operation, but care must be taken to avoid damaging the wire insulation or the metal bonding strip when making the cut. With the hacksaw in one hand and the cable end held firmly in the other, the cut should be made with the blade of the hacksaw placed at right angles to the lay of the armor strip. The hacksaw and cable should form two legs of a 60-degree triangle. When the blade has cut almost through the armor strip, the cable end should be bent back and forth several times until it breaks. The loose armor can then be stripped from the wire leads by a twisting and pulling action. If one end of the cable has already been attached to a box, the cable should be pulled tight enough to assure a steady sawing surface. When cut from coil, the armor is held firmly by stepping on the coil cable end and pulling it tight. Rough or sharp ends of the cut are then smoothed with a file.

3. Unwrapping paper (2, figure 16-4). The fiber paper that is twisted around the conductors before the metallic armor is attached must be removed to allow free wire movement. Normally two or three turns of the paper are removed from under the armor by tearing the loose paper away from the wire at the armor end by a jerking action. The free space between the armor and the wires facilitates mounting the antishort bushing.

4. Attaching the antishort bushing (3, figure 16-4)). When the ends of the cut armor are filed, only the outer burred edges are removed. The inner edges are always sharp and jagged at the cut end, and if not covered, would tend to puncture the wire insulation and cause short circuits and grounds. To prevent this, a tough fiber bushing, commonly called an antishort, must be inserted between the armor and the wire to protect the wire against damage.

5. Attaching cable to box (4, figure 16-4).

A. When a BX connector is used in attaching the armored cable to a box, the cable is first inserted in the connector and the holding screw or screws are tightened against the armor, securely connecting the cable and connectors. The BX connector is then inserted through a box knockout

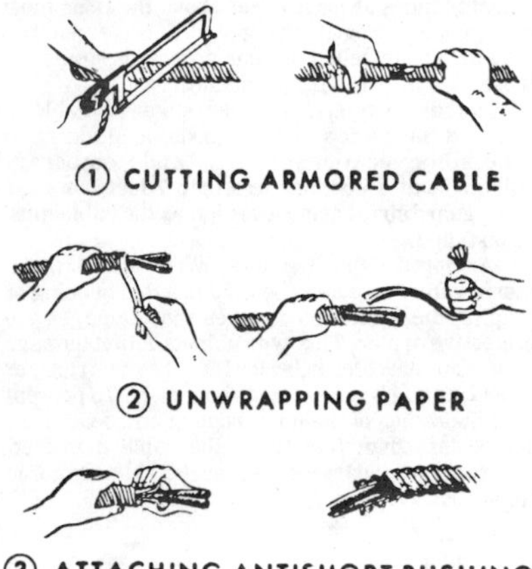

Figure 16-4.—Procedure for preparing and attaching cable.

NOTES:

opening and is secured to the box by a locknut threaded on the connectors from inside the box.

B. When the cable is used with a box having integral cable clamps, the knockout at point of entry must first be pried out. Next the clamp-holder screw is loosened and the cable is inserted through the knockout opening and the leads threaded through the clamp (figure 16-5). The armor is then forced snugly against the clamp end and the clamp screw retightened, forcing the clamp into the ridges of the cables.

ADDITIONS TO EXISTING WIRING

Circuiting. Additions to existing armored cable layouts require analysis to determine whether additional circuit capacity is needed to handle the new load. These considerations are the same as those required for other types of installations as outlined in chapter 17.

Cable Connection. Armored cable additions must always originate and terminate in electrical boxes. The junction box used for the addition should be located close enough to the desired outlets so that the voltage drop to the new device is within allowable limits. The box from which the additional outlet or outlets are to originate must also have both a neutral and hot wire of the same circuit for the new load connection. This means that the conductors from an added outlet can be connected only to the conductors of an existing cable in an outlet box (white to white and black to black) if the existing conductors can be traced to the fuse or circuit breaker without interruption. Figure 16-6 illustrates two methods of connecting to existing conductors, one in a switch box and the other in a receptacle box.

Installation of Armored Cable Additions.

1. Exposed. The installation of exposed armored cable additions to existing wiring must be patterned according to rules outlined for original installations. If an armored cable installation is to be made into another type of wiring system, the changeover must be made in a junction box specifically installed for the purpose or in an existing outlet box, provided its conductor capacity will allow the entry of additional wires.

2. Concealed. Armored cable is preferred over all other types of wiring when additional outlets are required on completed buildings. The armor provides damage protection and, together with the bonding strip, adequate continuous ground to the

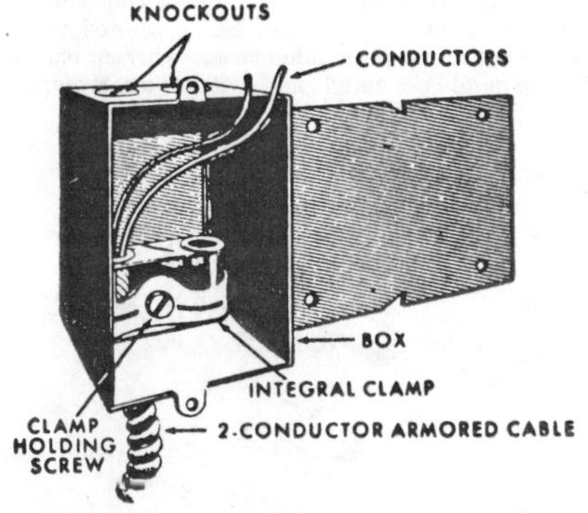

Figure 16-5.—Cable connection to box with integral clamps.

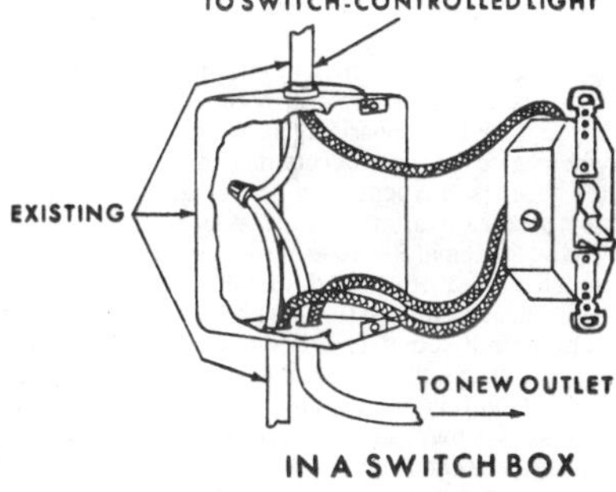

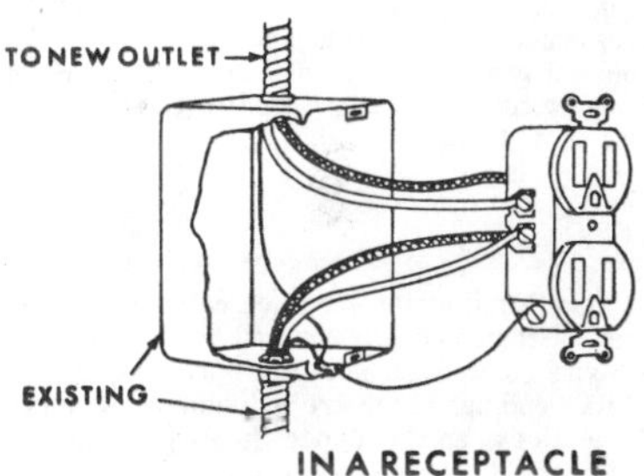

Figure 16-6.—Additions to existing wiring.

NOTES:

metal outlet boxes. BX is also flexible enough to allow feeding it through small openings from attic or basement areas to boxes mounted on walls and ceilings. The cable is usually pulled into the concealed box with a fish wire or drop chain. The fish wire is used when the cable is to be fed from below the box location, whereas the drop chain is used when the installation is to be made from above. In these cases the junction box in which the power tap is to be made should be in a clear, readily accessible area since the fishing and catching of a fish wire and drop chain become a tedious and time-consuming operation if the junction box is concealed. If it is difficult because of building construction to feed into the power tap box, the finished wall may have to be removed to allow entry. This will necessitate a replastering job after the addition has been installed.

NONMETALLIC SHEATHED CABLE WIRING

ADVANTAGES AND USES

The conventional nonmetallic sheathed cable (NM) is approved for use in concealed or exposed dry indoor locations, and is recommended for use where a good system ground is not available. Since the cable is inexpensive, light weight, and requires no special installation tools, it is suited for many wiring systems. Because of its construction, it is not approved for imbedded installation in masonry, concrete, fill, or plaster. It should NOT be installed in potentially dangerous areas where wire damage may occur, such as commercial garages, theaters, storage-battery rooms, and hoistways. It is not used in humid or wet areas such as ice plants or cold storage warehouses.

A newer nonmetallic sheathed cable (NMC) is the dual-purpose plastic sheathed cable with solid copper conductors. It needs no conduit, and its flat shape and gray or ivory color make it ideal for surface wiring. It resists moisture, acid, and corrosion and can be run through masonry or between studding.

MATERIALS

Cable. Nonmetallic sheathed cable consists of rubber- or thermoplastic-covered wires in two- or three-wire combinations with a bare copper wire used for bonding. These are individually wrapped with a thick spiral paper tape for damage protection and covered with a woven fabric braid that has been saturated with a moisture-resistant and flame-retardant compound. The entire assembly is then coated with wax. The local codes in some areas also require the addition of a bare uninsulated conductor in the nonmetallic sheathed cable. This bare wire provides the same type of equipment ground or bonding at the outlet boxes as the armor in armored cable installation. The bare wire is attached to the outlet box by clamping it either under the connector locknut at box entry or under one of the screws of the cable clamp (figure 16-7).

Supports. Nonmetallic sheathed cable is generally mounted on wooden building members with one- or two-hole mounting straps formed to fit the contour of the cable. BX staples are not approved for this type of installation because of the danger of possible cable damage.

Boxes and Devices. The boxes and devices used in nonmetallic sheathed cable wiring are similar to those used with conduit. They are made of metal or nonmetallic materials such as porcelain or bakelite. They can be obtained with built-in clamps or knockout holes for the cable connectors. The knockouts in porcelain boxes however are designed for cable entry only. It is recommended that metal boxes with integral clamps be used whenever possible to assure a safe and efficient installation. Insulated switches, outlets, and lampholder devices may be used without boxes in exposed nonmetallic sheathed cable wiring. The cable-entry holes to these devices must clamp the cable securely, and the device must fully enclose the section of the cable from which the outer sheathing has been removed. No splicing can be done in these devices. Consequently, since all wires must be connected to terminals, use of these devices is limited to installation in rural or other areas wherein only a small number of outlets and switches are required.

Figure 16-7.—Three-wire nonmetallic sheathed ground.

NOTES:

INSTALLATION

Cable Support. Nonmetallic sheathed cable installation should be supported in a manner similar to that outlined for armored cable. As shown in figure 16-8, the cable can be installed either on running boards, in holes drilled in the center of the joists, or on the sides of joists and studs. When running boards or the sides of joists and studs are used, straps should support the cable at distances not greater than 4½ feet, and a cable strap should be attached within 12 inches of a box. When the cable run is to be made at an angle in an overhead installation and is supported on the edge of the joists, at least two No. 6 gage or three No. 8 gage wires must be used in the wire assembly. If smaller size wires are used, they must be installed through holes bored in the joists or mounted on running boards.

Damage Protection. If the cable is installed across the top of a floor or floor joists, it must be protected by guard strips at least as high as the cable. When the wire installation is made in a location not normally used, such as an attic or crawl space under a building, damage-protection devices such as guard strips are required only within 6 feet of the entrance. Concealed nonmetallic cable installations should not be installed near baseboards, door and window casings, or other possible locations of trim or equipment because of the possibility of damage from building nails. If thermal insulation is to be installed where nonmetallic sheathed cable is in place, only noncorrosive, noncombustible, nonconductive insulation should be used. During the installation of the insulation, care must be used to prevent adding additional strain on the cable, its supports, or its terminal connections. This is especially necessary if the NM installation includes porcelain outlet boxes. (NM is the code designation for nonmetallic sheathed cable.)

Nonmetallic Sheathed Cable Bending. To prevent accidental damage to the sheathing on nonmetallic sheathed cable, the minimum allowable radius of bends is five times the cable diameter. Though this bend limit is similar to the armored cable requirement, nonmetallic sheathed cable can be bent in a smaller arc. This is true because the cable diameters are smaller for the same wire gage combinations.

Box Connection. Cable runs must be continuous from outlet to outlet because wire splices are only permitted inside a box. NM cable is prepared for box connection in the same manner as outlined for armored cable. In removing the protective

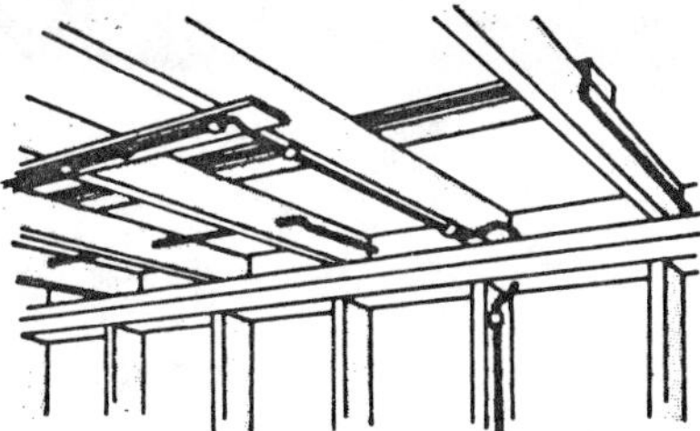

Figure 16-8.—Nonmetalic sheathed cable installation.

sheathing from the conductors for connection, an electrician's knife rather than a hacksaw is used. In removing the covering, a slit should be cut in the sheathing parallel to the wires without touching the individual wire insulation. A cut approximately 8 inches long for cable entry to ordinary boxes is satisfactory but can be increased to suit entry to panels. The knife is then used to remove the slitted sheathing. The moisture-preventive paper should also be removed from the wires. One, figure 16-9 illustrates the slitting of a cable end and figure 16-9(2) shows a special tool called a cable stripper which can be used instead of a knife to remove the sheathing from NM cable, lead-covered cable, and portable cords. In operation the stripper is inserted over the cable, squeezed together, and then pulled off the conductor. This action rips off the outer braid quickly and efficiently. The use of a stripper instead of a knife for outer braid removal is recommended since it cannot damage the wire insulation.

ADDITIONS TO EXISTING WIRING

Circuiting. The factors pertinent for additions to existing wiring systems outlined in paragraph 3-16 are the same as those which should be considered for NM wiring.

Cable Connections. Connection additions for nonmetallic sheathed cable are the same as the connection additions for armored cable outlined above.

Installation of Nonmetallic Sheathed Cable Additions.

1. Exposed. The installation of exposed NM cable additions to existing wiring must conform to the same requirements outlined for original installations. If a wiring system other than non-

NOTES:

metallic sheathed cable is to be extended with nonmetallic sheathed cable the systems must be coupled with a junction box. Existing boxes with available spare conductor capacity can be used.

2. Concealed. Additions to concealed nonmetallic sheathed cable are similar in method and procedure to those outlined above with the exception that insulated switches, outlets, and lampholders may be installed without boxes on the wall surfaces. In these installations the cable is fished through the wall and fed to the device at the point of entry.

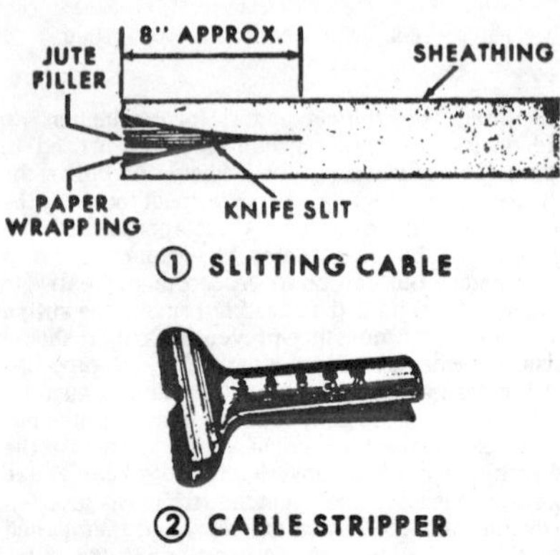

Figure 16-9.—Removal of sheathing.

RIGID CONDUIT INSTALLATION

USES AND ADVANTAGES

Either black enameled or galvanized rigid metal conduit is approved for use under all conditions and locations except that metal conduit and fittings protected from corrosion only by enamel may be used only indoors and in areas not subject to severe corrosion. Though it is generally the most expensive type of wiring installation, its inherent strength permits installation without running boards and damage protection. Its conductor capacity facilitates carrying more conductors in one run than in any other system, and its rigidity permits installation with fewer supports than the other types of wiring systems. Moreover, the sizes of conduit used in the system's installation generally provide for the possible addition of several more conductors in the conduit when additional circuits and outlets are required in the run.

MATERIALS

Scope. Though the materials used in rigid conduit wiring have been outlined in detail in chapter 15, we will review the advantages of these standard materials as well as their limitations.

Rigid Conduit. Rigid conduit (1, figure 16-10) has the same size designations as water pipe. Under the Code limitations no conduit smaller than ½ inch may be used except in finished buildings where extensions are to be made under plaster. In these installations 5/16-inch conduit or tubing is permitted. The size of conduit is determined by the inside diameter. For example, ½-inch conduit has an inside diameter of approximately ½ inch (0.622). Standard conduit sizes used in interior wiring are ½, ¾, 1, 1¼, 1½, 2, and 2½ inches. Larger sizes are available up to 6 inches for special use in certain commercial and factory installations. Though conduit is made in dimensions similar to water pipe it differs from water pipe in a number of ways. It is softer than water pipe and thus can be bent fairly easily. In addition the inner surface is smooth to prevent damage to wires being pulled through it. Moreover, the finish is rust-resistant. Black enamel conduit is used for dry and indoor installations, and exterior galvanized conduit is used in outside installations to provide moisture protection for the conductors. For wiring installations in corrosive atmospheres aluminum, copper alloy, or plastic-jacketed conduit is available.

Conductors. Rubber-covered insulated-type R or RH wire is used with conduit in the majority of interior wiring installations, though the thermoplastic insulation types T or TW are gaining favor because of their superior insulating characteristics. Underground or wet installations require the insertion of lead-covered cables in rigid galvanized conduit for permanent protection.

Supports. The conduit straps illustrated and described in chapter 15 are preferred for use in mounting conduit in interior wiring systems. According to Code requirements, the conduit should be supported on the proper spacings.

Fittings. There are two types of fittings. These are the standard ordinary size outlet box and the small junction or pull boxes called condulets. The standard outlet box fittings are classified as type F and are used normally in exposed installations to house receptacles or switches where high quality of installation is desired. The junction or pull box fittings (2, figure 16-11) are used to either provide

NOTES:

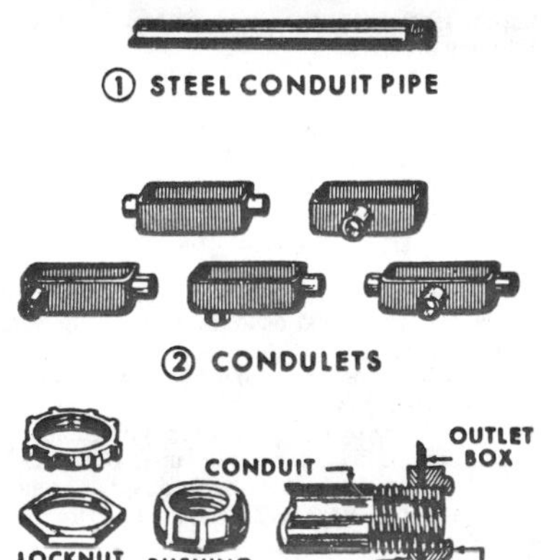

Figure 16-10.—Rigid conduit and fittings.

either intermediate points in long conduit runs for pull-through of wire or junctions for several concealed installations where they will not be accessible. They are classified by the manufacturers as follows:

1. Service entrance, type SE.
2. Elbow or turn fittings, type L.
3. Through fittings, type C.
4. Through fittings with 90-degree takeoff, type T.

Boxes and Connectors. Steel or cast iron outlet boxes are used in rigid conduit installations. Boxes normally used are supplied with knockouts which are removable for conduit insertion. Bushings and locknuts are provided for attachment of the conduit to the boxes as shown in 3, figure 16-11. Boxes to be used in wet or hazardous locations must have threaded hubs into which the conduit is screwed.

Devices. The devices used in conduit installations are all box-mounted units and are covered in chapter 15.

Conduit Accessories.

1. Threaded couplings. A threaded coupling (1, figure 16-12) is furnished with each length of rigid conduit.

2. Threadless couplings. Rigid conduit may be installed using threadless couplings (2, figure 16-12) provided the couplings are installed tightly.

3. Elbows. Standard conduit elbows (3, figure 16-12) are manufactured for use where 90-degree bends are required.

4. Conduit unions. To permit the opening of a conduit at any point without sawing or breaking the conduit run, conduit unions (1, figure 16-12) are installed. By the use of unions, conduit may be started from two outlets and joined together at any convenient place in the run.

INSTALLATION

Bends. Bends of rigid conduit must be made without collapsing the conduit wall or reducing the internal diameter of the conduit at the bend.

1. Most bends are made on the job by the electrician as an integral part of the installation procedure. These are called field bends. The radius of the curve of the inner edge of any field bend must be at least six times the internal diameter of the conduit for rubber-, braid- or thermoplastic-covered conductors, and not less than 10 times the internal diameter of the conduit for lead-covered conductors. The maximum number of quarter bends for a conduit run between two openings is four. Moreover, a 10-foot length of conduit should have no more than three quarter bends. Factory-made bends may be used instead of bending conduit on the job. However, they are not generally used since they increase the wiring cost. This is because more conduit cutting and threading is required and additional couplings must be used.

2. Conduit up to and including ¾ inch is usually bent with a hand conduit bender called a hickey as shown in 1, figure 16-13. This can be slipped over the conduit. Conduit bending forms are also available as built-in units of pipe-vise stands. If either of these tools is not available, bends can be made

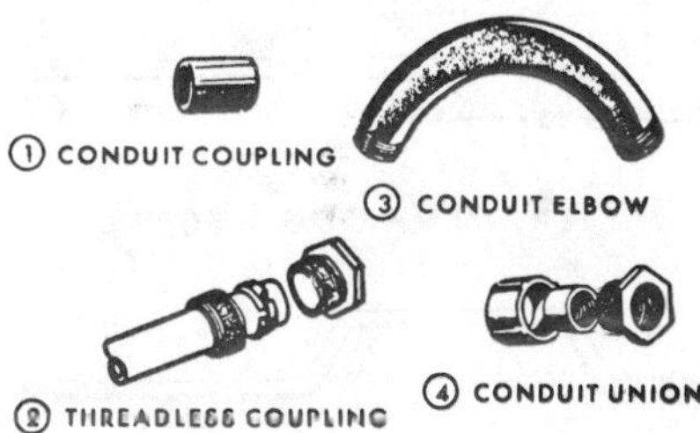

Figure 16-11.—Rigid conduit accessories.

NOTES:

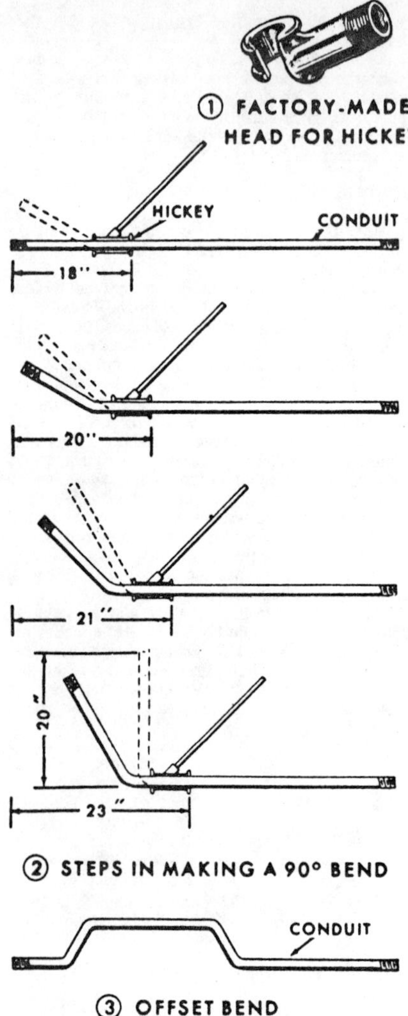

Figure 16-12.—Bending rigid conduit.

NOTES:

using the lever advantage between two fixed posts or building members.

3. The procedure in (A) through (E) below, illustrated in 2, figure 16-13, is recommended as one method of making a right-angle bend in a length of ½-inch conduit. If a 90-degree bend is to be made in a length of conduit at a distance of 20 inches from one end, the electrician must:

(A) Mark off 20 inches from the end of the conduit.

(B) Place the conduit hickey 2 inches in front of the 20-inch mark and bend the conduit about 25 degrees.

(C) Move the bender to the 20-inch mark and bring the bend up to 45 degrees.

(D) Move the bender about 1 inch behind the 20-inch mark and bring the conduit up to 70 degrees.

(E) Move the hickey back about 2 inches behind the 20-inch mark and bring the bend up to 90 degrees.

4. Miscellaneous conduit bends (offset bends, 3, figure 16-13) can be made more accurately if the contour of the bend is drawn with chalk on the floor and the bend in the pipe is matched with the chalk diagram as the bend is formed. Conduit in excess of 1 inch is usually bent by a hydraulic bender.

Cutting Conduit. Conduit can be cut with either a hacksaw or standard pipe cutter. When a hand hacksaw is used, the conduit should be held in a vise, and care should be taken that the cut is at right angles to the length of the pipe. If a large number of conduits are to be cut, a power hacksaw is recommended. Though pipe cutters may be used and are considered standard equipment, a hacksaw is recommended for electrical conduit cutting since considerable time is required to remove the burr left in the inside of a pipe by a pipe cutter. Cutting oil should always be used when cutting pipe.

Reaming Conduit. Irrespective of the cutting method used, a sharp edge always remains inside the conduit after cutting. Consequently, to avoid conductor damage, this edge must be removed before the conduit is installed. Pipe reamers or files as illustrated in chapter 15 are generally used for the reaming operation.

Cutting Threads. Since the outside and inside diameters of rigid conduit are the same as those of gas, water, or steam pipes, the standard thread forms, and consequently similar threading tools and dies, are used. Normally the smaller sizes of pipe are threaded with dies that cut a thread for every turn of the die. For larger sizes (1½ inch and over) electricians generally use a ratchet type cutter. Motor-driven pipe-threading machines are also available when large installations are made and when considerable conduit must be threaded. Good practice requires an electrician to examine, before installation, each piece of threaded conduit for—

1. Foreign matter inside the pipe. This should be removed to prevent conductor damage.

2. Thread condition. Mishandling, extraneous paint, or dirt may require the conduit to be rethreaded before installation. Cutting oil should always be used when threading conduit.

Conduit Installation. Conduit should be run as straight and direct as possible. When a number of conduits are to be installed parallel and adajacent to each other in exposed multiple-conduit runs, they should be erected simultaneously instead of installing one line before starting the others. Conduit installed on building surfaces can be supported by either pipe straps or pipe hangers. On wooden surfaces, nails or wood screws can be used to secure the straps. On brick or concrete surfaces, holes must be drilled first with a star or carbide drill and expansion anchors installed. The strap is then secured to the surface with machine screws. On tile or other hollow material the straps are secured with toggle bolts. If the installation is made on a metal surface, holes for the straps or hangers can be drilled and tapped into the metal, and the supports secured by machine screws to the metal surface. An adequate number of supports should be provided. The conduit run, as the conduit between boxes is called, must be cut to proper length, threaded, reamed, and then bent to suit the building contours. The conduit-run ends are then attached to the boxes. Figure 16-14 illustrates a typical rigid conduit exposed installation. In a concealed installation the building members may be notched sufficiently to allow placing the conduit behind the wall surface. Care must be taken to avoid undue weakening of the structure.

Box Connection.

1. When the boxes are of threaded-hub construction, the conduit ends are screwed into the box hubs and the conduit runs are connected at some midpoint by a coupling.

2. If the boxes are of knockout type construction, they should be loosely located in the required position on studs and joists. A bowed locknut having teeth on one side should then be screwed onto the threads at the run ends of the conduit with the teeth of the locknut adjacent to the box. The conduit ends should next be inserted into the knockout openings. Bushings, which have smooth surfaces on their inside diameter to insure damage-free conductor installation, should then be screwed tightly onto the conduit ends in the boxes.

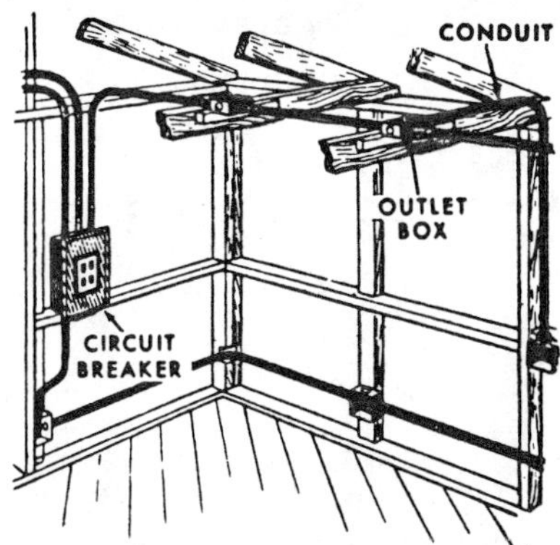

Figure 16-13.—Typical installation of conduit wiring.

Finally the locknuts should be tightened against the boxes so that the teeth will dig into the metal sides of the box. This operation can be accomplished by driving a drive punch against one of the locknut lugs and forcing the locknut to move on the threaded conduit against the box. Figure 16-15 shows a standard box connections for conduit using locknuts and bushings. After this connection and all other box connections have been made, the box can be fastened securely to the building.

Wire Pulling. Upon installation of the boxes and conduit runs the conductor wires should be pulled into the conduit. For short runs with few wires conductors can be paired and pushed through the conduit run from box to box. When the conduit run has several bends and more than two conductors, a fish wire must be used in pulling wire. For normal runs the fish wire (or tape) is pushed through the conduit run from one end to the other. Occasionally, on long conduit runs, separate fish wires are used from either end of the conduit as shown in 1, figure 16-16. After the conductor ends are bared of insulation, they are wrapped around the fish wire (2, figure 16-16) and taped (3, figure 16-16) for pulling through the conduit. Taping of the fish wire and conductor junction is required to preclude damaging the conduit interior and existing conductors in the conduit. In taping, the joint is also compacted and strengthened, thus insuring easier pulling. For efficient and safe operation, wire pulling is generally a two-man operation. One

NOTES:

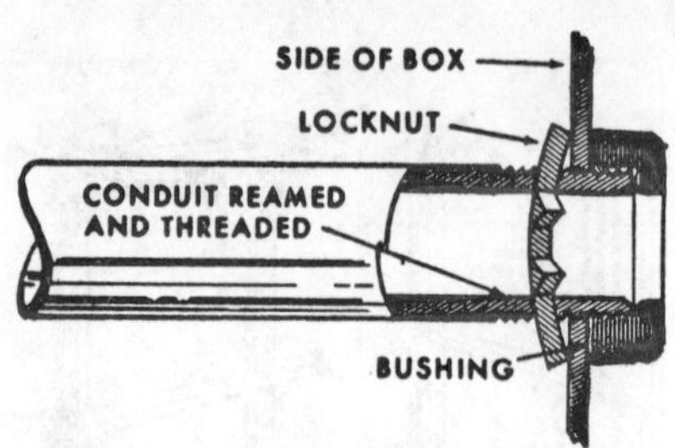

Figure 16-14.—Conduit and box connection.

electrician is required to pull the conductors through the conduit while the other feeds the conductors into the conduit. In this operation, care must be used in feeding and pulling the wires so they maintain their same relative position in the conduit throughout the run length, thus avoiding insulation injury. For ease of operation a wire lubricant such as powdered soapstone may be rubbed on the conductors or blown into the conduit. In intricate runs, wire pulling may be performed in sections between boxes. This procedure requires a large amount of additional splicing to be made in the boxes and requires that more time be taken in wiring. The preferred practice in wire pulling is to pull the conductors from the source through to the last box in the conductor run. Loops which extend about 8 inches from the box openings are made for each conductor which is to be tapped or connected to a device in the box. Conductors which are not to be tapped are pulled directly through the box to their connection.

Splices. Wire splices in conduit installations are not under tension and a simple pigtail splice, carefully made to obtain a good electrical joint, can be used. No wire splices which will be concealed in the conduit runs are to be made. This requirement is necessary because splices would reduce the pulling area in a conduit and would easily be a source of electrical failure.

CIRCUITING

Availability of different sizes of conduit along with their varying conductor capacities makes the wiring installation for conduit somewhat different from that of the open or cable types. For example, where cable installation requires several runs in a particular location, a conduit installation would use a single conduit with multiple conductors. Consequently, conduit layouts and runs should be planned to use the minimum amount of conduit possible and also keep the conductor runs to each outlet short enough to maintain a low voltage drop. Figure 16-17 shows a typical wiring layout in conduit.

Conductor Connection. No exceptions to the standard color coding of wires as outlined in the other systems are permitted in conduit wiring. All load utilization devices (fixtures, receptacles) operating at line to neutral voltage in a grounded neutral system must be connected to both a white and a black (or substitute color) wire. The white wire is always the grounded neutral wire. Black wires are the hot leads which are fused and connected to the switch when controlling power to a lampholder or outlet. Red-, blue-, and orange-colored insulation wire can be used as substitutes for black wire when wire combinations are combined in a conduit or circuit. The white wire must never be connected to a black or substitute color wire. The white wire must not be fused or switched except in a multipole device opening all conductors of the circuit simultaneously. A green-colored insulated conductor denotes a wire used to provide an auxiliary equipment ground. As an expedient measure the ends of the wire insulation may be painted to obtain proper color coding when the

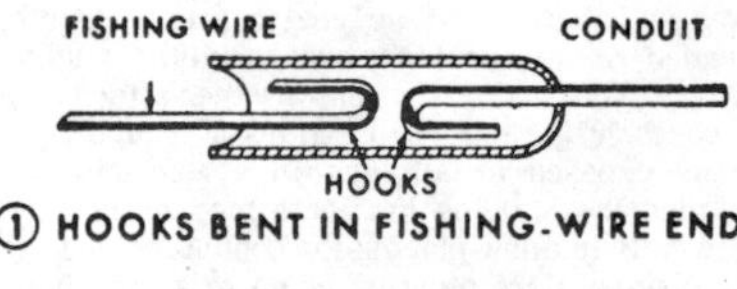

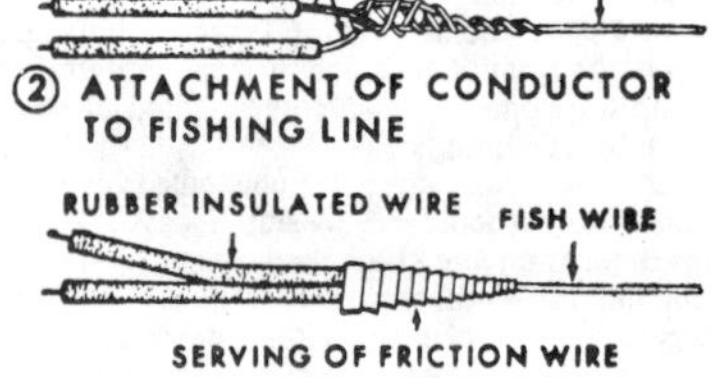

Figure 16-15.—Fish-wire pulling.

NOTES:

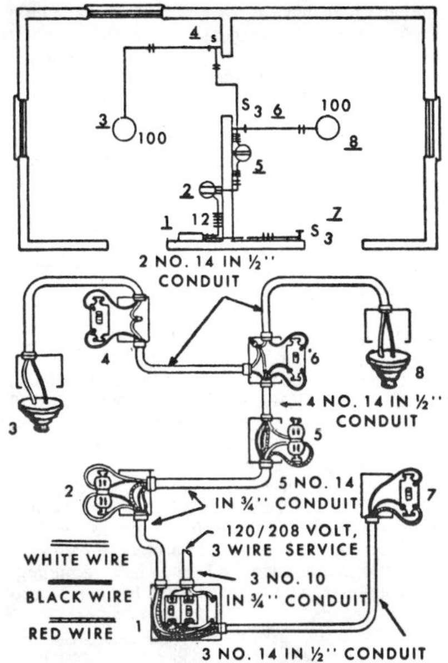

Figure 16-16.—Typical wiring layout in conduit.

colored insulation is not available. They may also be identified by the use of wire code markers (chapter 15).

Conduit Capacity. Cable wiring is normally limited to two or three standard combinations of wire sizes. Conduit, however, has the capacity to accommodate many more conductors than two or three in one run. For example, six No. 14 gage wires would require the installation of a ¾-inch conduit run. In many installations it is necessary to use more than one size wire in a conduit run. In such cases the conductors cannot have a combined or cross-sectional area equal to more than the allowable percent of cross-sectional area of the conduit. For each size conduit, the percent of conduit cross-sectional area in square inches available for conductor use must be determined. For example, if three No. 10 gage type R and four No. 8 gage type R conductors are to be inserted in a conduit, their combined cross-sectional area is 3 x 0.0460 + 4 x 0.0760 or 0.4420 square inches. The proper size conduit for this installation is 1¼ inches. This can be found by looking in the proper table furnished by the manufacturer for the total area under the headings "not lead covered" and "4 cond. and over." It is seen that 0.4420 lies between 0.34 square inch (for 1-inch conduit) and; 60 square inch (for 1¼-inch conduit). Consequently, as the 1-inch conduit is too small, the 1¼-inch size is selected.

Circuit Wiring. A fundamental law of electricity generation can be restated for wiring purposes as follows: When a conductor carrying current changes position or the current reverses direction in the conductor, it induces a current in an iron or steel conduit carrying the conductor. Consequently, if this conductor were isolated in an iron or steel conduit, the conduit would be heated by the induced current. This would result in considerable power loss. In an alternating current system, both wires of a circuit are encased in a single conduit thereby causing the induced current of each to balance and cancel each other. To eliminate any possibility of induced heating of iron or steel conduit, both the wires of a circuit must travel in the same conduit. If the conductors in the circuit (figure 16-18) are run separately in this conduit, induced current (indicated by arrows) will flow through the conduit.

ADDITIONS TO EXISTING WIRING

Increase of Circuit Amperage. A standard conduit installation has enough flexibility to accommodate a normal increase in circuit load even if an increase in circuit amperage is required. For example, a ½-inch conduit in a standard conduit-wiring installation generally carries two No. 14 gage conductors, which have a 15-ampere capacity. It is also evident that the ½-inch conduit can also accommodate two No. 12 gage conductors having a 20-ampere capacity. Consequently, if the load in an existing circuit must be increased, the No. 14 gage wire can be replaced by two No. 12 gage wires. Hence, when all the wires in the circuit are replaced, the amperate for the fuse or circuit breaker in the central fuse panel for the circuit can be safely increased from 15 to 20 amperes to accommodate the additional load.

Addition of New Circuit. When adding a new load to an existing building with conduit wiring and the circuit analysis indicates the need for a new circuit, the existing conduit in many cases can be used to carry the new circuit most of its distance. Figure 16-19 illustrates this principle. The new circuit is installed by pulling in an additional wire

NOTES:

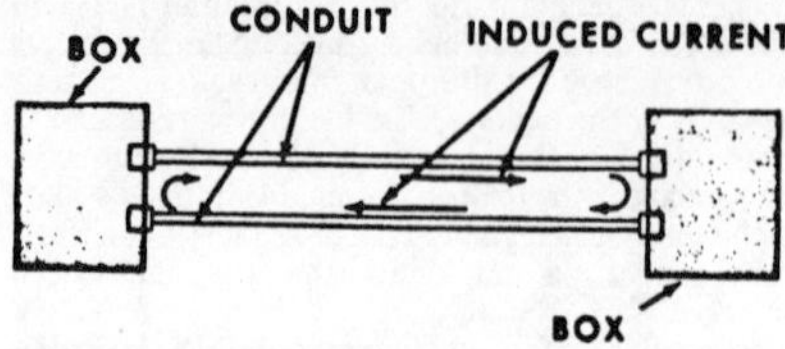

Figure 16-17.—Circulating current in conduit, showing induced-current flow.

(red) from the circuit breaker panel to the existing outlet, and then adding the required outlet box beyond this location. The new load is connected to the additional circuit. The proper tables furnished by the manufacturer or supplier should be used to determine whether the existing conduit could accommodate an additional wire. The installation of the additional outlet box and conduit should conform to the rules and practices as outlined above. In this type of installation a common grounded neutral wire is used.

THIN-WALL CONDUIT WIRING

USES AND ADVANTAGES

Thin-wall conduit is a metallic tubing which can be used for either exposed or concealed electrical installations. Its use should be confined to dry interior locations. This is necessary because it has a very thin plating which does not protect it from rusting when exposed to the elemnts or humid conditions. It is less expensive than rigid conduit and much easier to install. The process of bending requires less effort, and the ends are not required to be threaded. In comparison with the other systems of wiring, it ranks behind rigid conduit, but ahead of the other types of wiring when considering the quality and durability of the installation. For this reason, and because of the decreased cost in materials and labor, it is most generally specified for homebuilding construction. It is installed in the same manner as rigid conduit except that pressure-type couplings and connectors are used instead of threaded units.

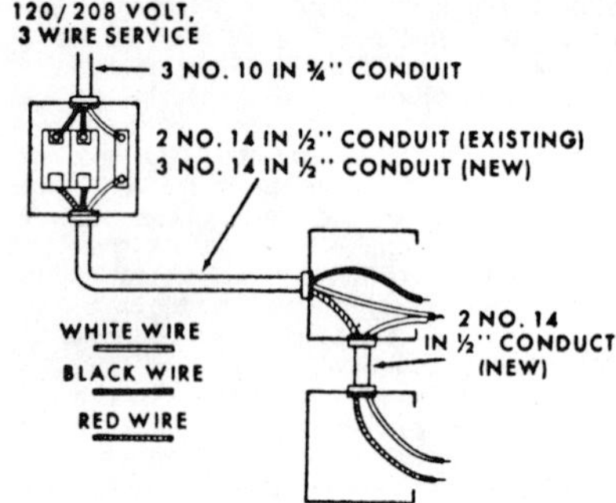

Figure 16-18.—Circuit addition in existing conduit.

MATERIALS

Thin-wall Conduit and Fittings. Electrical metallic tubing (EMT) commonly called "thin-wall conduit" is more easily installed than rigid conduit. This conduit, as its name implies, has a thinner wall than rigid conduit but has the same interior diameter and cross-sectional area. EMT is available in sizes from ⅜ to 2 inches. The ⅜-inch size is used only for underplaster extensions. The inside surface is enameled to protect the wire insulation and minimize the friction in wire pulling. All couplings and connections to boxes are

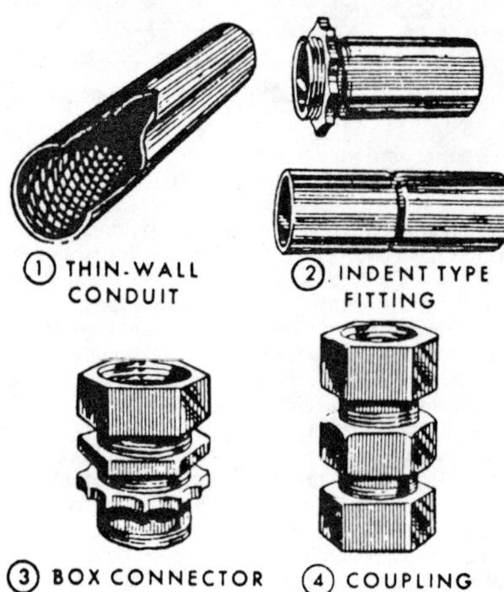

Figure 16-19.—Thin-wall conduit and fittings.

NOTES:

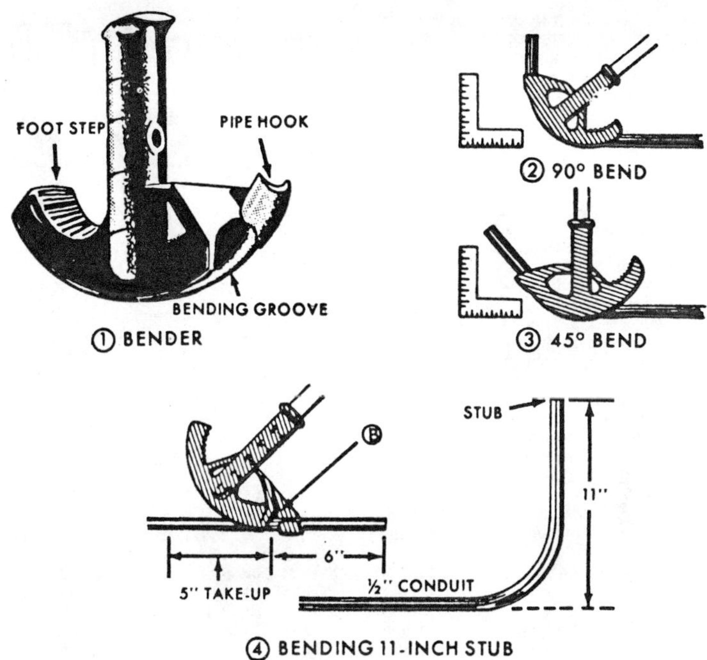

Figure 16-20.—Bending thin-wall conduit.

threadless and are of either the clamp or compression type. Figure 16-20 illustrates thin-wall conduit (1) and the fittings commonly used. Some fittings are similar to sleeves (2) and are secured to the conduit by an impinger tool which pinches circular indentations in the fitting. This holds it firmly against the conduit. Others have threaded bushings which, when tightened, force a tapered sleeve firmly against the tubing. Three and four, figure 16-20, shows a box connector and coupling fitting of this type. An adapter for joining rigid conduit to thin-wall conduit is also shown.

BENDING

Policy. Extreme care must be used when bending metallic tubing to avoid kinking the pipe or reducing its inside area. The radius of the curve of the inner edge of any field bend must not be less than six times the internal diameter of the tubing when braid-covered conductors are used, and not less than 10 times the interior diameter of the tubing when lead-covered conductors are used. This is the same rule that applies to rigid conduit.

Construction. The thin-wall conduit bender (1, figure 16-21) has a cast steel head which is attached to a steel-pipe handle approximately 4 feet long. It is used in the field to form thin-wall conduit into standard and offset bends. Benders are made for each size of conduit and must be used only on those sizes for which they are designed. Each size bends the conduit to the recommended safe radius. The projection on the head of the bender, sometimes called a "foot step", is used to steady the bender in

NOTES:

operation, and reduces the pressure required on the handle for bending. The numbers cast on the bender shaft are inch measurements and are used to check the depths of offset bends.

Operation. In operating the bender, first place the conduit on a level surface and hook the end of the proper size tube bender under the conduit's stub end. Then, with the bending groove over the conduit and using a steady and continuous force, while firmly holding the conduit and bender with the body, push down on the handle and step on the foot step, bending the conduit to the desired angle. To make a 45-degree bend in this manner (3, figure 16-21), move the bending tool until the handle is vertical. For accurately bending conduit stubs, the bender must be placed at a predetermined distance from the end of the conduit. This distance is equal to the required stub dimension minus an amount commonly called a takeup height. This takeup height is based on a constant allowance determined by the bending radii for various size conduit. The take height is 5 inches for ½-inch conduit, 6 inches for ¾-inch conduit and 8 inches for 1-inch conduit. In the bending of an 11-inch stub in a ½-inch conduit, for example (4, figure 16-21), the takeup height of 5 inches is first subtracted from the 11-inch dimension of the stub. The mark "B" on the bender is then set at the resultant value of 6 inches and the bends made.

INSTALLATION

Thin-wall conduit may be cut either with a hacksaw or with a special thin-wall cutter. As with rigid conduit, the sharp edge in thin-wall tubing should also be reamed after cutting to prevent premature wire damage. Exposed thin-wall conduit is supported in a similar manner and with the same type of supports as used with rigid conduit. Since there is no positive link between the couplings, box connectors, and thin-wall conduit, care must be taken during the installation to make sure each conduit joint is electrically and mechanically secure. All conduit ends must be inserted into the fittings until they touch the inner limiting edges. The fittings should then be tightened firmly, securely gripping the conduit walls. Care is also necessary to prevent the loosening of the conduit from the fittings which could cause a loose connection, short circuit, or electrical fire at the point of wire and conduit contact. A mechanically loose

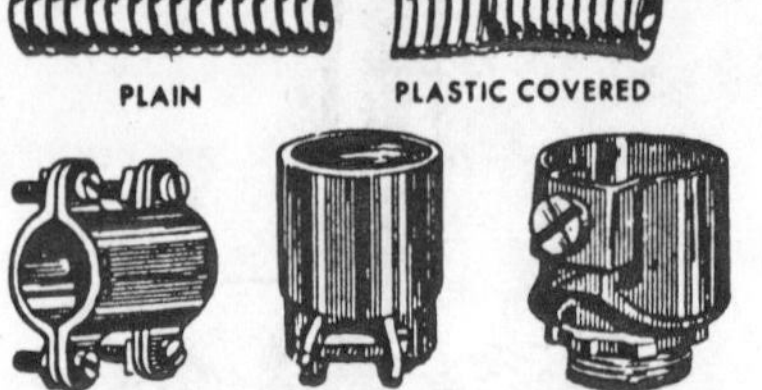

Figure 16-21.—Greenfield flexible conduit and fittings.

conduit joint will not maintain the ground continuity required in an electrical wiring installation. This could also create an operating hazard.

FLEXIBLE CONDUITS

MATERIALS

Flexible metal conduit is generally called "Greenfield." It resembles armored cable in appearance but it is more adaptable than cable since various sizes and numbers of wires can be pulled into it after it is installed. Plastic-covered Greenfield may be used where the internal conductors are exposed to oil, gasoline, or other materials which have a deteriorative effect on the wire insulation. This metal conduit has a thermoplastic outer-sheath covering similar to that used on type T wire. Figure 16-22 shows Greenfield conduit and the various fittings available.

INSTALLATION

Flexible conduit installation is similar to that covered above for thin-wall conduit except that Greenfield must be supported more frequently, as an armored cable. Its prohibitive cost limits its use to connections between rigid wiring systems and movable or vibrating equipment such as motors or fans. it may also be installed where the construction requires a conduit bend which is either difficult or impossible to make.

NOTES:

CHAPTER 17

DESIGN AND LAYOUT OF INTERIOR WIRING

INTRODUCTION

The different wiring systems in common use are cable and conduit systems. Chapter 16 covers the installation details for these systems. Many installation methods and procedures used in the wiring processes are common to both systems, and these are described in this chapter. In most wiring installations the type of wiring to be installed will be specified on the blueprints. If not so specified, the installation method must be determined. In general the type of wiring used should be similar to that installed in adjacent or nearby buildings.

TYPE OF DISTRIBUTION

The electrical power load in any building cannot be properly circuited until the type and voltage of the central power-distribution system is known. The voltage and the number of wires from the powerlines to the buildings are normally shown or specified on the blueprints. However, the electrician should check the voltage and type of distribution at the power-service entrance to every building in which wiring is to be done. This is especially necessary when he is altering or adding circuits. The voltage checks are usually made with an indicating voltmeter at the service-entrance switches or at the distribution load centers. The type of distribution is determined by visual check of the number of wires entering the building.

If only two wires enter the building, the service is either direct current or single-phase alternating current. The voltage is determined by an indicating voltmeter.

When three wires enter a building from an alternating current distribution system, the service can either be single-phase, three-phase, or two ungrounded conductors and a neutral of a three-phase system (V-phase).

1. If the service is single-phase, two of the conductors are hot and the third is ground. A voltmeter reading between the two hot conductors will be twice as great as the reading between either hot conductors and the neutral or ground conductor.

2. If the service is three-phase, the voltage between any two of the conductors is the same. Normally one of the conductors is grounded to establish a ground reference voltage for the system.

3. Two ungrounded conductors and a neutral or a V-phase system is the most common service for theater of operations construction. The distribution system is described below. The voltage between the two hot conductors will be the V 3 or 1.732 times greater than the voltage between either hot conductor and the neutral or ground.

Four-wire distribution denotes three-phase and neutral service. When tested, voltages between the neutral conductor and each of the three hot conductors should be all the same. The voltage readings between any two of these three wires are similar and should equal the neutral to hot wire voltage multiplied by 1.732. Common operating voltages for this type service are 120 and 208 volts or 277 and 480 volts.

LOAD PER OUTLET

The first step in planning the circuit for any wiring installation is to determine connected load per outlet. It is best to use volt-amperes as the method of determining electrical needs. This eliminates power factor considerations. The power needed for each outlet or load per outlet is used to find the number of circuits. It is also used to find the power needed for the whole building. The load per outlet can be obtained in several different ways:

A. The most accurate method of determining load per outlet is to obtain the stated value from the blueprints or specifications.

1. Commonly, the lighting outlets shown on the blueprints are listed in the specifications along with

NOTES:

their wattage ratings.

If the lights used are of the incandescent type, this figure represents the total wattage of the lamp.

When fluorescent type lights are specified, the wattage drain (also called load per outlet) should be increased approximately 20 percent to provide for the ballast load. For example, when the figure is rated as a two-lamp, 40-watt unit, the actual wattage drain is 80 watts, plus approximately 16 watts for both lamp ballasts, or a total load of 96 watts.

2. If the specifications are not available, the blueprints in many cases designate the type of equipment to be connected to specific outlets. Though the equipment ultimately used in the outlet may come from a different manufacturer, equipment standards provide the electrician with assurance that the outlets will use approximately the same wattage. If the equipment is available, the nameplate will list the wattage used or ampere drain. If not, a manufacturer or distributor should be able to supply the information to obtain the average wattage consumption of electrical appliances.

B. To provide adequate wiring for systems where the blueprints or specifications do not list any special or appliance loads, the following general rules will apply:

1. For heavy duty outlets or mogul size lampholders, the load per outlet should be figured at 5 amperes each.

2. For all other outlets, both ceiling and wall, the wattage drain (load per outlet) should be computed at 1.5 amperes per outlet.

C. The total outlet load may also be determined on a watts-per-square-foot basis. In this load-determination method the floor area of the building to be wired is computed from the outside dimensions of the building. This square footage area is then multiplied by the standard watts-per-square-foot requirement based on the type of building to be wired.

CIRCUITING THE LOAD

If all the power load in a building were connected to a single pair of wires and protected by a single fuse, the entire establishment would be without power in case of a breakdown, a short circuit, or a fuse blowout. In addition the wires would have to be large enough to handle the entire load, and, therefore, too large in some cases to make connections to individual devices. Consequently, the outlets in a building are divided into small groups known as branch circuits. These circuits normally are rated in amperes. Normally the total load per circuit should not exceed 80 percent of the circuit rating.

The method of circuiting the building load varies with the size of the building and the power load.

1. In a small building with little load, the circuit breakers or fuses are installed at the power-service entrance and the individual circuits are run from this location.

2. For buildings of medium size with numerous wiring circuits, the fuse box should be located at the center of the building load so that all the branch runs are short, minimizing the voltage drop in the lines.

3. When buildings are large or have the loads concentrated at several remote locations, the ideal circuiting would locate fuse boxes at each individual load center. It is assumed that the branch circuits would be radially installed at each of these centers to minimize the voltage drops in the runs.

The number of circuits required for adequate wiring can be determined by adding the connected load in watts and dividing the total by the wattage permitted on the size of branch circuit selected. This method should not include special heavy loads, such as air conditioners, requiring separate circuits. The total wattage is obtained from the sum of the loads of each individual outlet determined by one of the three methods outlined in paragraph 3—3. For example, if 20-ampere, 120-volt circuits are to be used, 80 percent of this rating or 16 amperes per circuit are allowed. The maximum wattage permitted on each circuit equals 16 times 120 or 1920 watts. If the total connected load is assumed to be 18,000 watts, 18,000 divided by 1920 shows that 9.375 circuits are required. Since we can only have whole circuits, ten 20-ampere circuits should be used to

Figure 17-1.—Motor switches.

NOTES:

carry the connected load. The number of circuits determined by this method should be the basic minimum. For long-range planning in permanent installations the best practice requires the addition of several circuits to the minimum required, or the installation of the next larger modular-size fusing panel to allow for future wiring additions. If additional circuits over the minimum required are used, the number of outlets per circuit can be reduced, therefore the electrical installation is more efficient. This is true because the voltage drop in the system is reduced, allowing the apparatus to operate more efficiently.

Motors that are used on portable appliances are normally disconnected from the power source either by removal of the appliance plug from its receptacle or by the operation of an attached built-in switch. Some large-horsepower motors, however, require a permanent power installation with special controls. Motor switches, some of which are shown in figure 17-1, are rated in horsepower capacity. In a single motor installation a separate circuit must be run from the fuse or circuit breaker panel to the motor, and individual fuses or circuit breakers installed. For multiple motor installations the National Electrical Code requires that "Two or more motors may be connected to the same branch circuit, protected at not more than 20 amperes at 125 volts or less or 15 amperes at 600 volts or less, if each does not exceed 1 horsepower in rating and each does not have a full load rating in excess of 6 amperes. Two or more motors of any rating, each with individual overcurrent protection (provided integral with the motor start switches or an individual unit), may be connected on one branch circuit provided each motor controller and motor-running overcurrent device be approved for group installation and the branch circuit fusing rating be equal to the rating required for the largest motor plus an amount equal to the sum of the full load ratings of the other motors."

BALANCING THE POWER LOAD ON A CIRCUIT

The ideal wiring system is planned so that each wiring circuit will have the same ampere drain at all times. Since this can never be achieved, the circuiting is planned to divide the connected load as evenly as possible. Thus, each individual circuit uses approximately the average power con-

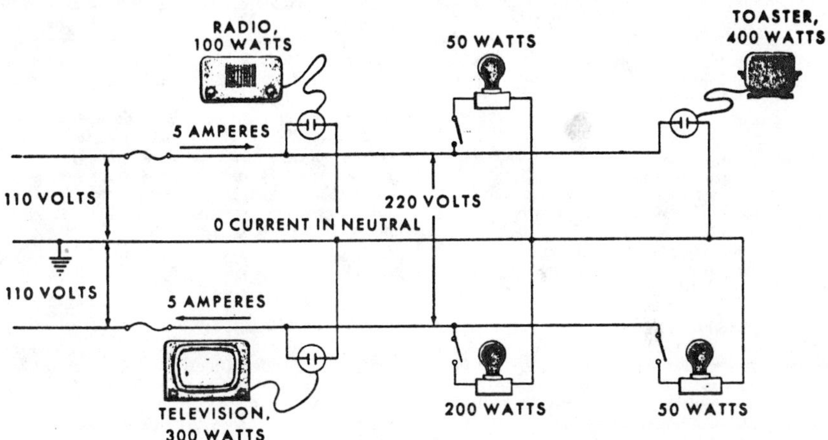

Figure 17-2.—Diagram showing circuit balancing.

NOTES:

sumption for the total system. This will make for minimum service interruption. Figure 17-2 demonstrates the advantage of a balanced circuit when a three-wire single-phase, 110-220 volt distribution system is used.

The current used is known as alternating current or ac because the current in each wire changes or alternates continually from positive to negative to positive to negative and so on. The change from positive to negative and back again to positive is known as a "cycle." Usually this takes place 60 times every second, and such current is then known as 60-cycle current. Sixty times every second each wire is positive, and 60 times every second it is negative, and 120 times every second there is no voltage at all on the wire. The voltage is never constant but is always gradually changing from zero to maximum with the average being about 120 volts. The current in the neutral conductor of a balanced three-wire single-phase system will remain at zero as a result of the applied alternating current.

LOAD PER BUILDING

Maximum Demand. In some building installations the total possible power load may be connected at the same time. In this case the demand on the power supply, which must be kept available for these buildings, is equal to the connected load. In the majority of building installations, the maximum load which the system is required to service is much less than the connected load. This power load, which is set at some arbitrary figure below the possible total connected load, is called the "maximum demand" of the building.

Demand Factor. The ratio of maximum demand to total connected load in a building expressed as a percentage is termed "demand factor." The determination of building loads can be obtained by the use of standard demand factors. For example, if the connected load in a warehouse is 22,500 watts, the actual building load can be obtained as follows: 100 percent of the first 12,500 watts equals 12,500, 50 percent of the remaining 10,000 watts equals 5,000; therefore, the total building load is 12,500 plus 5,000 watts or 17,500 watts. Other factors can be discovered by consulting the proper information source.

BALANCING THE POWER LOAD OF A BUILDING

The standard voltage distribution system from a generating station to individual building installation is the three- or four-wire, three-phase type. Distribution transformers on the powerline poles change the voltage to 120 or 240, and are designed to deliver three-wire single-phase service. The transformers are then connected across the distribution phase leads in a balanced arrangement. Consequently, for maximum transformer efficiency, the building loads assumed for power distribution should also be balanced, as previously illustrated in figure 17-2.

WIRE SIZE

Wire sizes No. 14 and larger are classified according to their maximum allowable current-carrying capacity based on their physical behavior when subjected to the stress and temperatures of operating conditions. Fourteen-gage wire is the smallest wire size permitted in interior wiring systems.

The determination of the conductor size to be used in feeder and branch circuits is dependent on the maximum allowable current-carrying capacity and the voltage drop coincident with each wire size. The size of the conductors for branch circuits (that portion of the wiring system extending beyond the last overcurrent device protecting the circuit) should be such that the voltage drop will not exceed 3 percent to the farthest outlet for power, heating, or lighting loads. The maximum voltage drop for feeders is also 3 percent, provided that the total voltage drop for both feeder and branch circuit does not exceed 5 percent.

The minimum gage for service-wire installation is No. 8. The service-wire sizes are increased because they must not only meet the voltage-drop requirement but also be inherently strong enough to support their own weight, plus any additional loading caused by climatic conditions (ice, branches, and so on).

GROUNDING AND BONDING

Requirements.

1. All electrical systems must have the neutral conductor grounded if the voltage between the hot lead and the ground is less than 150 volts.

2. It is recommended that all systems have a grounded neutral where the voltage to ground does not exceed 300 volts.

3. Interior circuits operating at less than 50 volts need not be grounded, provided the transformer supplying the circuit is connected to a grounded system.

NOTES:

Types of Grounding.

1. A system ground is the ground applied to a neutral conductor. It reduces the possibility of fire and shock by reducing the voltage of one of the wires of a system to 0 volts potential above ground.

2. An equipment ground is an additional ground that should be attached to all appliances and machinery. An equipment ground is advantageous because the appliances and machinery can be maintained at zero voltage, and if a short circuit does occur in a hot load, the fuse protection opens the circuit and prevents serious injury to operating personnel.

Methods of Grounding. A system ground is provided by placing a No. 6 copper (or No. 4 aluminum) wire between the neutral wire, service box, bonding wire, and a grounding electrode at the building service entrance. The grounding electrode may be a buried water pipe, the metal frame of a building, a local underground system, or a fabricated device. If more than one electrode is used, they must be placed a minimum of 6 feet apart.

Water pipe. An underground water piping system will always be used as the grounding electrode if such a piping system is available. If the piping system is less than 10 feet deep, then supplemental electrodes will be used. Interior metallic cold water piping systems will always be bonded to the grounding electrode or electrodes.

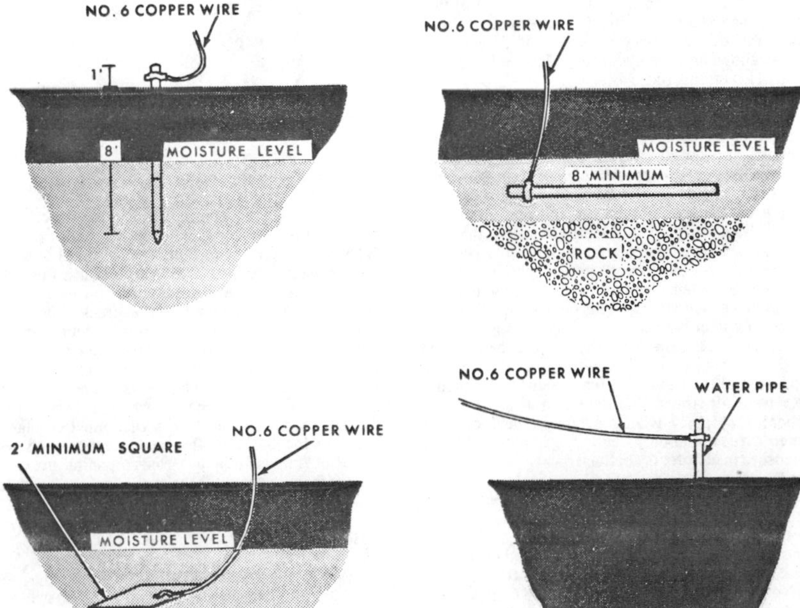

Figure 17-3.—Typical grounding fixtures.

NOTES:

2. Plate electrodes. Each plate electrode will have at least 2 square feet of surface exposed to the soil. Iron or steel electrodes shall be at least ¼-inch thick and nonferrous metal shall be at least 0.06-inch thick. Plates should be buried below the permanent moisture level if possible.

3. Pipe electrodes. Clean metallic pipe or conduit at least ¾-inch trade size may be used. Each pipe must be driven to a depth of at least 8 feet. If this cannot be done, the electrodes may be buried in a horizontal trench. In this case the electrode must be at least 8 feet long.

4. Rod electrodes. Rod electrodes of steel or iron must be at least ⅜-inch in diamater. Rods of nonferrous material must be at least ½-inch in diameter. The standard ground rod for the military is a ⅝-inch steel rod with three sections of 3 feet each. Installation of rod electrodes is the same as stated above for pipe electrodes. Figure 17-3 shows typical grounding fixtures.

Ground Resistance.

1. Electrodes should have a resistance to ground of 25 ohms or less. Underground piping systems and metal frames of buildings normally have resistances to ground of less than 25 ohms. Resistances to ground of fabricated electrodes will vary greatly depending on the soil and method of installation. Burying an electrode below the permanent moisture level of the soil normally reduces the resistance to ground to acceptable values. Grounding systems should be tested by using a "Megger" ground tester. This device may be found in post engineer type organizations.

2. The voltage between the neutral or grounded conductor of a grounded system and ground (water pipes, metal frame of building, ground rod), should be zero at all times. The detection of any voltage indicates a faulty wiring system. Short circuits which are commonly called grounds will be discussed in chapter 8, section II.

Bonding.

1. Bonding is a method of providing a continuous and separate electrical circuit between all metallic circuit elements (conduit, boxes, etc.) and the service entrance ground. The ground or neutral conductor is attached to the bonding circuit *only* at the service entrance. At this point the neutral wire and a bonding wire from the service box are attached to the grounding electrode. All other metallic circuit elements are connected to this grounded point through physical connections of the metallic conduit, BX, or flexible metal conduit. Nonmetallic conduit or cable must have a conductor in it to provide this bonding circuit. This conductor would be attached to the metal boxes or fixtures where the nonmetallic cable or conduit terminates.

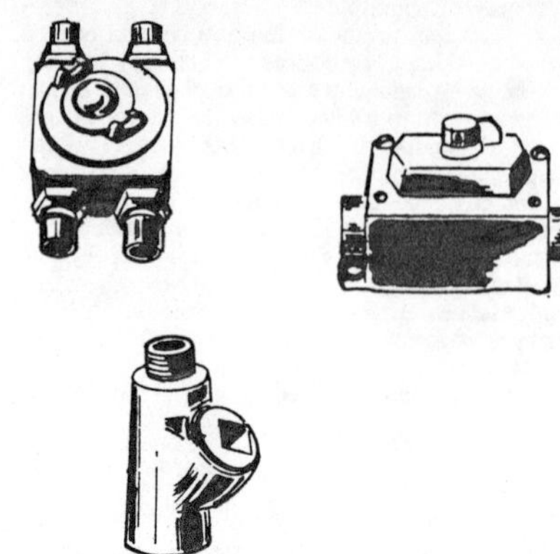

Figure 17-4.—Explosionproof fittings.

2. To insure a continuous bonding circuit, all metal to metal connections must be made up tight. When the electrical continuity of a metal to metal connection is doubtful, a wire jumper must be used between the two metal pieces. As an example, the equipment ground wire of a receptacle (the green wire) is connected to the bonding circuit through the two screws fastening the receptacle to the box. If the box is recessed, then in order to obtain a flush receptacle, the connecting screws are not tightened but are used as adjusting screws to properly locate the receptacle. In this case a wire must be placed between the box and the equipment ground terminal of the receptacle. Proper bonding prevents shocks from metal surfaces of an electrical system.

WIRING FOR HAZARDOUS LOCATIONS

Special materials and procedures must be used when installing electrical systems in areas where a spark could cause a fire or explosion. Typical areas of this type are hospital operating rooms and acetylene storage or production facilities. Hazardous locations are divided into three classes. Each class is further subdivided into two divisions. In division one of each class the hazardous material is present in free air so that the atmosphere is dangerous. In division two of each class the

NOTES:

hazardous material is in containers and dangerous mixtures in air occur only through accidents. The following is a description of the classes:

Class I. Locations in which flammable gasses or vapors are or may be present in the air in quantities sufficient to produce explosive or ignitable mixtures.

Class II. Locations where combustible dust is present.

Class III. Locations where easily ignitable fibers or flyings are present but are not suspended in air in sufficient quantities to produce ignitable mixtures.

INSTALLATION IN HAZARDOUS LOCATIONS

Detailed information on installation material and procedures must be obtained from theater commands and standard plans or from the National Electrical Code. Figure 17-4 shows typical explosion-proof fittings. The information below gives general guidance as to the type of installation required for each class and division of hazardous locations.

Class I. In division one location, all wiring must be in threaded rigid metal conduit with explosionproof fittings or mineral-insulated metal-sheathed (type MI) cable. Division two locations may have flexible metal fittings, flexible metal conduit, or flexible cord approved for hard usage. All equipment such as generators, controllers, motors, fuses, and circuit breakers must be enclosed in explosionproof housings.

Class II. In division one, wiring must be in threaded rigid conduit or type MI cable, with flexible metal conduit and threaded fittings where necessary. Equipment must be in dustproof cabinets with motors and generators in totally enclosed fan-cooled housings. In division two locations electrical metallic tubing may also be used.

Class III. In division one the same requirements exist as for class I division one. In division two open wiring may be permitted. Motors and generators must be totally enclosed.

INSTALLATION OF SIGNAL EQUIPMENT

Signal equipment may occasionally be supplied for 120-volt operation, in which case it must be installed in the same manner as outlets and sockets operating on this voltage. Most bells and buzzers are rated to operate on 6, 12, 18 or 24 volts ac or dc. They can be installed with minimum consideration for circuit insulation since there is no danger of shock to personnel or fire due to short circuits. The

NOTES:

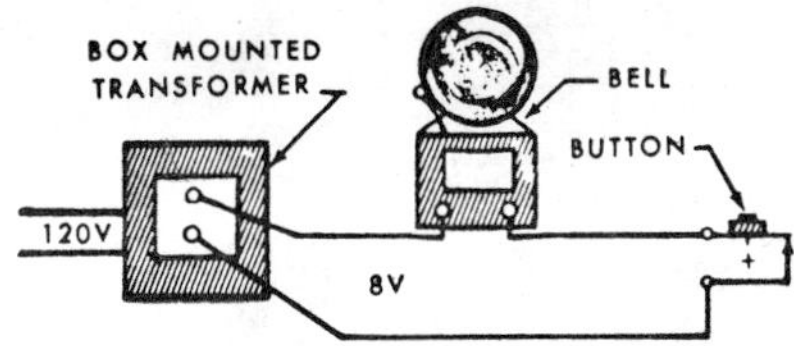

Figure 17-5.—Bell and buzzer wiring.

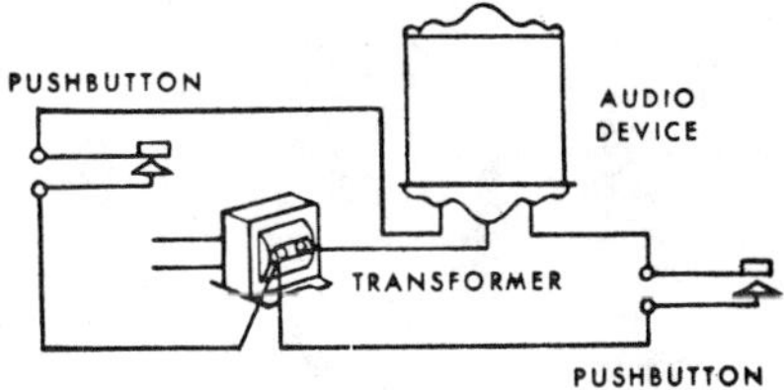

Figure 17-6.—Two pushbutton system.

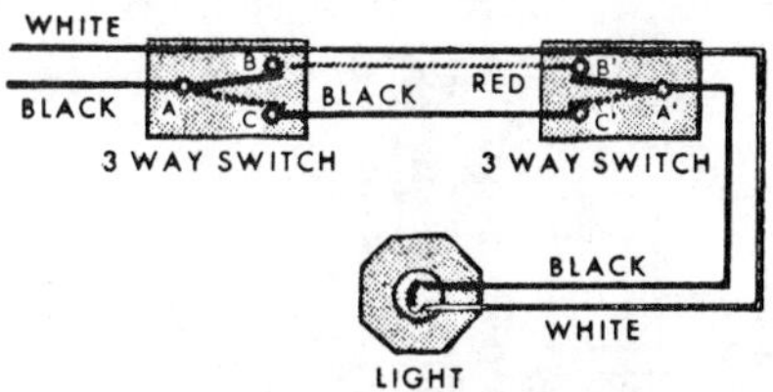

Figure 17-7.—Three-way switch wiring.

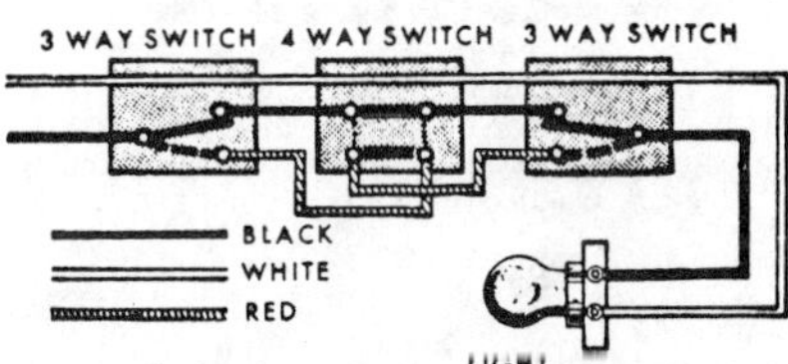

Figure 17-8.—Four-way switch wiring.

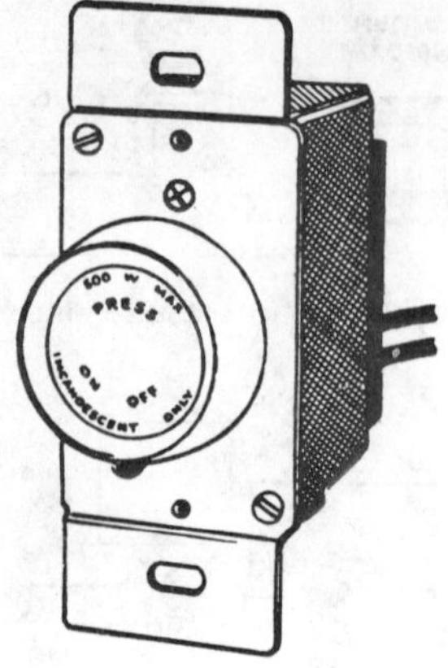

Figure 17-9.—Push-pull rotary switch.

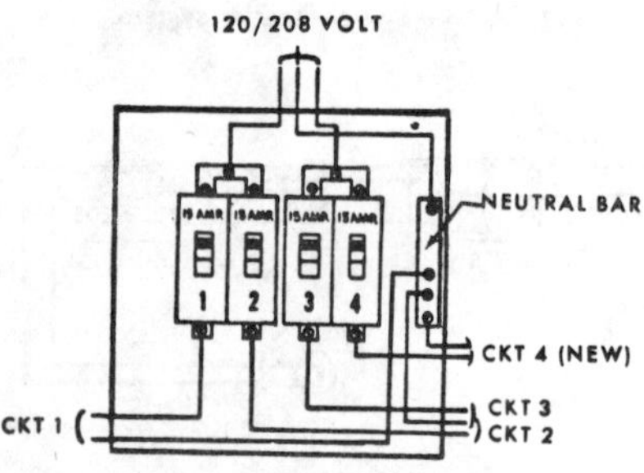

Figure 17-10.—Addition of a new circuit.

wire commonly used is insulated with several layers of paraffin-impregnated cotton or with a thermoplastic covering. Upon installation, these wires are attached to building members with small insulating staples and are threaded through building construction members without insulators.

BATTERY OPERATION

Early installations of low-voltage signal systems were powered by 6-volt dry cells. For example, two of these batteries were installed in series to service a 12-volt system. If the systems involved a number of signals over a large area, one or more batteries were added in series to offset the voltage drop. Though this type of alarm or announcing system is still being used and installed in some areas, it is a poor method because the batteries used as a power source require periodic replacement.

TRANSFORMER OPERATION

The majority of our present-day buzzer and bell signal systems operate from a transformer power source. The transformers are equipped to be mounted on outlet boxes and are constructed so that the 120-volt primary-winding leads normally extend from the side of the transformer adjacent to box mounting. These leads are permanently attached to the 120-volt power circuits, and the low-voltage secondary-winding leads of the transformer are connected to the bell circuit in a manner similar to a switch-and-light combination (figure 17-50). If more than one buzzer and pushbutton is to be installed, they are paralleled with the first signal installation. A typical wiring schematic diagram for this type of installation is shown in figure 17-6.

SPECIAL SWITCHES

Three-Way Switching. A single-throw switch controls a light or a receptacle from only one location. When lights have to be controlled from more than one location, a double-throw commonly called a three-way switch is used. Three-way switches can be identified by a common terminal, normally color coded darker than the other terminals and located alone at the end of the switch housing. A schematic wiring diagram of a three-way switch with a three-wire cable is shown in figure 17-7. In the diagram terminals A and A' are the common terminals, and switch operation connects them either to B or C and B' and C' respectively. Either switch will operate to close or open the circuit, turning the lights on or off.

Four-Way Switching. Occasionally it is necessary to control an outlet or light from more than two locations. Two three-way switches plus a four-way switch will provide control at three locations as shown in figure 17-8. The switches must be installed with the four-way units connected between the two three-way units and the three-wire cable installed between the switches.

Variable Control Devices. Electrical dimmers provide a full range of light from bright to dim for incandescent lighting. Special electronic dimmers exist for use with fluorescent lighting. A turn of the dial adjusts the brightness level, and a push-knob

NOTES:

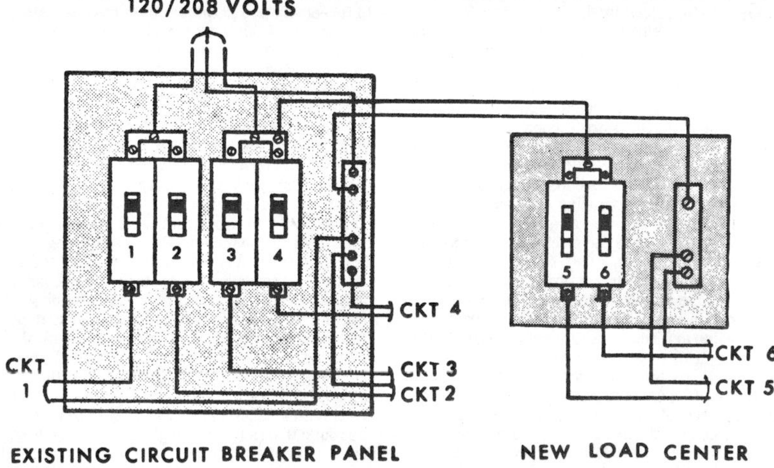

Figure 17-11.—Addition of new load center.

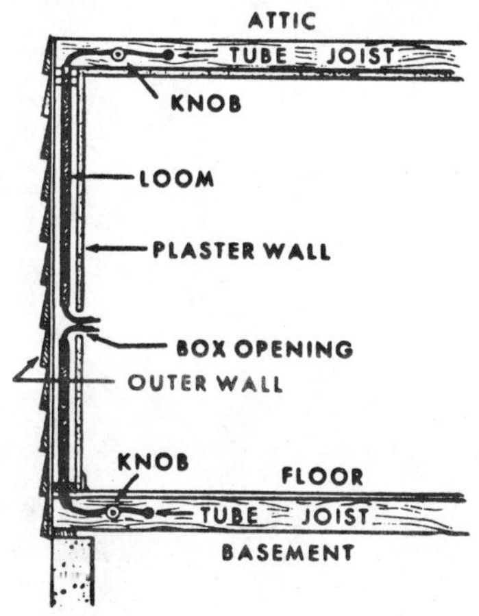

Figure 17-12.—Concealed wire addition.

switch turns the light on or off without changing the brightness setting (figure 17-9). Another type of dimmer switch has a high-low control that provides two levels of illumination: full brilliance at the top position and approximately half brilliance at the bottom position. Variable control devices are also used as speed control devices that control the speed of tools and equipment using standard universal ac-dc type motors. Some variable control devices are electronic in nature since their construction utilizes solid state circuitry and solid state switches.

ADDITIONS TO EXISTING WIRING

Circuit Capacity. In the installation of additions to existing wiring in a building, the electrician determines first the available extra capcity of the present circuits. This can readily be obtained by ascertaining the fused capacity of the building and subtracting the present connected load. If all the outlets do not have connected loads, their average load should be used to obtain the connected load figure. When the existing circuits have available capacity for new outlets and are located near the additional outlet required, they should be extended and connected to the new outlets. Consideration must be given to the additional voltage drop

NOTES:

created by extending the circuit. The proper wire size may then be determined.

New Circuits. When the existing outlets cannot handle an additional load, and a spare circuit has been provided in the local fuse or circuit breaker panel, a new circuit is installed. This is also done if the new outlet or outlets are so located that a new circuit can be installed more economically than an existing circuit extension. Moreover, the installation of a new circuit will generally decrease the voltage drop on all circuits, resulting in an increase in appliance operating efficiency. Figure 17-10 illustrates the addition of a new circuit from the spare circuit No. 4 in the circuit breaker panel.

New Load Center. In many wiring installations, no provisions are made for spare circuits in the fuse panel. Moreoever, the location of the new circuit required is often remote from the existing fusing or circuit breaker panel. In this case the most favorable method of providing service to the circuit is to install a new load center at a location close to the circuit outlets. This installation must not overload the incoming service and service-entrance switch. Should such an overload be indicated, the service equipment should also be changed to suit the new requirements. This sometimes can be accomplished in two-wire systems by pulling in an additional wire from the powerline. This changes the service from two-wire to three-wire at 120 to 240 volts. In these cases the fuse or circuit breaker box should also be changed and enlarged to accommodate the increased circuit capacity. Figure 17-11 schematically illustrates the installation of an additional load center for a new circuit.

Concealed Installations. The addition of outlets in a building with finished interior walls having enclosed air spaces entails the use of a stiff wire called a fish wire and a drop chain. Figure 17-12 shows the addition of a wire run for an outlet accomplished by a drop from the attic space or as a riser from the basement. First an opening is made in the interior finished wall at the desired outlet position. If the attic circuit is to be tapped, holes are drilled in the top plates of the wood studding and a drop chain is lowered inside the wall and pulled through the box opening. The wire to be installed is then attached to the chain and pulled through, completing the rough-in operation for the outlet. Similarly, when a wire is to be pulled in from the basement, a fish wire is used. After drilling through the rough floor and bottom plate of the studding, the fish wire is pushed up from the basement until it is grasped at the box opening. The wire to be pulled is then attached and pulled through the inner wall section.

NOTES:

APPENDIX I
GREEK ALPHABET

Name	Capital	Lower Case	Designates
Alpha.	A	α	Angles.
Beta.	B	β	Angles, flux density.
Gamma.	Γ	γ	Conductivity.
Delta	Δ	δ	Variation of a quantity, increment.
Epsilon.	E	ϵ	Base of natural logarithms (2.71828).
Zeta.	Z	ζ	Impedance, coefficients, coordinates.
Eta	H	η	Hysteresis coefficient, efficiency, magnetizing force.
Theta.	Θ	θ	Phase angle.
Iota	I	ι	
Kappa	K	κ	Dielectric constant, coupling coefficient, susceptibility.
Lambda	Λ	λ	Wavelength.
Mu.	M	μ	Permeability, micro, amplification factor.
Nu	N	ν	Reluctivity.
Xi	Ξ	ξ	
Omicron	O	o	
Pi	Π	π	3.1416
Rho	P	ρ	Resistivity.
Sigma.	Σ	σ	
Tau	T	τ	Time constant, time-phase displacement.
Upsilon.	Υ	υ	
Phi	Φ	ϕ	Angles, magnetic flux.
Chi	X	χ	
Psi	Ψ	ψ	Dielectric flux, phase difference.
Omega	Ω	ω	Ohms (capital), angular velocity ($2 \pi f$).

NOTES:

APPENDIX II

COMMON ABBREVIATIONS AND LETTER SYMBOLS

alternating current (noun) .... a.c.
alternating-current (ajd.) .... a-c
ampere .... a
audiofrequency (noun) .... AF
capacitance C
capacitive reactance .... X c
centimeter .... cm
conductance .... G
coulomb .... Q
counterelectromotive force .... cemf
current (d-c or rms value) I
current (instantaneous value) i
dielectric constant .... K,k
difference in potential (d-c or rms value) .... E
difference in potential (instantaneous value) .... e
direct current (noun) .... d.c.
direct-current (adj.) .... d-c
electromotive force .... emf
frequency .... f
henry .... h
hertz .... Hz
horsepower .... hp
impedance .... Z
inductance L
inductive reactance .... X_L
kilovolt .... kv
kilovolt-ampere .... kva
kilowatt .... kw
kilowatt-hour .... kwhr
magnetic field intensity .... H
magnetomotive force .... m.m.f.
microampere .... ua
microfarad .... uf
microhenry .... uh
microvolt .... uv
microwatt .... uw
milliampere .... ma
millihenry .... mh
millivolt .... mv
milliwatt .... mw
mutual inductance .... M
picofarad .... pf
power .... P
resistance .... R
revolutions per minute .... rpm
root mean square .... rms
time .... t
torque .... T
volt v
volt-ampere .... va
watt .... w

NOTES:

For computing the current or the horsepower of a d-c motor, use the formula given for single phase, but omit the power factor.

To find the value of an alternative current when voltage, power (in watts) and power factor are known, use the formulas:

$$\text{amp} = \frac{\text{watts}}{\text{volts times P.F.}}, \text{ or}$$

$$\text{amp} = \frac{\text{watts}}{\text{volts times 1.73 times P.F.}},$$

according to whether the equipment is single phase or three phase.

NOTES:

APPENDIX III

ELECTRICAL TERMS

AGONIC.—An imaginary line of the earth's surface passing through points where the magnetic declination is 0°; that is, points where the compass points to true north.

AMMETER.—An instrument for measuring the amount of electron flow in amperes.

AMPERE.—The basic unit of electrical current.

AMPERE-TURN.—The magnetizing force produced by a current of one ampere flowing through a coil of one turn.

AMPLIDYNE.—A rotary magnetic or dynamo-electric amplifier used in servomechanism and control applications.

AMPLIFICATION.—The process of increasing the strength (current, power, or voltage) of a signal.

AMPLIFIER.—A device used to increase the signal voltage, current, or power, generally composed of a vacuum tube and associated circuit called a stage. It may contain several stages in order to obtain a desired gain.

AMPLITUDE.—The maximum instantaneous value of an alternating voltage or current, measured in either the positive or negative direction.

ARC.—A flash caused by an electric current ionizing a gas or vapor.

ARMATURE.—The rotating part of an electric motor or generator. The moving part of a relay or vibrator.

ATTENUATOR.—A network of resistors used to reduce voltage, current, or power delivered to a load.

AUTOTRANSFORMER.—A transformer in which the primary and secondary are connected together in one winding.

BATTERY.—Two or more primary or secondary cells connected together electrically. The term does not apply to a single cell.

BREAKER POINTS.—Metal contacts that open and close a circuit at timed intervals.

BRIDGE CIRCUIT.—The electrical bridge circuit is a term referring to any one of a variety of electric circuit networks, one branch of which, the "bridge" proper, connects two points of equal potential and hence carries no current when the circuit is properly adjusted or balanced.

BRUSH.—The conducting material, usually a block of carbon, bearing against the commutator or sliprings through which the current flows in or out.

BUS BAR.—A primary power distribution point connected to the main power source.

CAPACITOR.—Two electrodes or sets of electrodes in the form of plates, separated from each other by an insulating material called the dielectric.

CHOKE COIL.—A coil of low ohmic resistance and high impedance to alternating current.

CIRCUIT.—The complete path of an electric current.

CIRCUIT BREAKER.—An electromagnetic or thermal device that opens a circuit when the current in the circuit exceeds a predetermined amount. Circuit breakers can be reset.

CIRCULAR MIL.—An area equal to that of a circle with a diameter of 0.001 inch. It is used for measuring the cross section of wires.

COAXIAL CABLE.—A transmission line consisting of two conductors concentric with and insulated from each other.

COMMUTATOR.—The copper segments on the armature of a motor or generator. It is cylindrical in shape and is used to pass power into or from the brushes. It is a switching device.

CONDUCTANCE.—The ability of a material to conduct or carry an electric current. It is

NOTES:

the reciprocal of the resistance of the material, and is expressed in mhos.

CONDUCTIVITY.—The ease with which a substance transmits electricity.

CONDUCTOR.—Any material suitable for carrying electric current.

CORE.—A magnetic material that affords an easy path for magnetic flux lines in a coil.

COUNTER EMF.—Counter electromotive force; an emf induced in a coil or armature that opposes the applied voltage.

CURRENT LIMITER.—A protective device similar to a fuse, usually used in high amperage circuits.

CYCLE.—One complete positive and one complete negative alternation of a current or voltage.

DIELECTRIC.—An insulator; a term that refers to the insulating material between the plates of a capacitor.

DIODE.—Vacuum tube—a two element tube that contains a cathode and plate; semiconductor—a material of either germanium or silicon that is manufactured to allow current to flow in only one direction. Diodes are used as rectifiers and detectors.

DIRECT CURRENT.—An electric current that flows in one direction only.

EDDY CURRENT.—Induced circulating currents in a conducting material that are caused by a varying magnetic field.

EFFICIENCY.—The ratio of output power to input power, generally expressed as a percentage.

ELECTROLYTE.—A solution of a substance which is capable of conducting electricity. An electrolyte may be in the form of either a liquid or a paste.

ELECTROMAGNET.—A magnet made by passing current through a coil of wire wound on a soft iron core.

ELECTROMOTIVE FORCE (emf).—The force that produces an electric current in a circuit.

ELECTRON.—A negatively charged particle of matter.

ENERGY.—The ability or capacity to do work.

FARAD.—The unit of capacitance.

FEEDBACK.—A transfer of energy from the output circuit of a device back to its input.

FIELD.—The space containing electric or magnetic lines of force.

FIELD WINDING.—The coil used to provide the magnetizing force in motors and generators.

FLUX FIELD.—All electric or magnetic lines of force in a given region.

FREE ELECTRONS.—Electrons which are loosely held and consequently tend to move at random among the atoms of the material.

FREQUENCY.—The number of complete cycles per second existing in any form of wave motion; such as the number of cycles per second of an alternating current.

FULL-WAVE RECTIFIER CIRCUIT.—A circuit which utilizes both the positive and the negative alternations of an alternating current to produce a direct current.

FUSE.—A protective device inserted in series with a circuit. It contains a metal that will melt or break when current is increased beyond a specific value for a definite period of time.

GAIN.—The ratio of the output power, voltage, or current to the input power, voltage, or current, respectively.

GALVANOMETER.—An instrument used to measure small d-c currents.

GENERATOR.—A machine that converts mechanical energy into electrical energy.

GROUND.—A metallic connection with the earth to establish ground potential. Also, a common return to a point of zero potential.

HERTZ.—A unit of frequency equal to one cycle per second.

HENRY.—The basic unit of inductance.

HORSEPOWER.—The English unit of power, equal to work done at the rate of 550 foot-pounds per second. Equal to 746 watts of electrical power.

HYSTERESIS.—A lagging of the magnetic flux in a magnetic material behind the magnetizing force which is producing it.

IMPEDANCE.—The total opposition offered to the flow of an alternating current. It may consist of any combination of resistance, inductive reactance, and capacitive reactance.

INDUCTANCE.—The property of a circuit which tends to oppose a change in the existing current.

INDUCTION.—The act or process of producing voltage by the relative motion of a magnetic field across a conductor.

NOTES:

INDUCTIVE REACTANCE.—The opposition to the flow of alternating or pulsating current caused by the inductance of a circuit. It is measured in ohms.

INPHASE.—Applied to the condition that exits when two waves of the same frequency pass through their maximum and minimum values of like polarity at the same instant.

INVERSELY.—Inverted or reversed in position or relationship.

ISOGONIC LINE.—An imaginary line drawn through points on the earth's surface where the magnetic variation is equal.

JOULE.—A unit of energy or work. A joule of energy is liberated by one ampere flowing for one second through a resistance of one ohm.

KILO.—A prefix meaning 1,000.

LAG.—The amount one wave is behind another in time; expressed in electrical degrees.

LAMINATED CORE.—A core built up from thin sheets of metal and used in transformers and relays.

LEAD.—The opposite of LAG. Also, a wire or connection.

LINE OF FORCE.—A line in an electric or magnetic field that shows the direction of the force.

LOAD.—The power that is being delivered by any power producing device. The equipment that uses the power from the power producing device.

MAGNETIC AMPLIFIER.—A saturable reactor type device that is used in a circuit to amplify or control.

MAGNETIC CIRCUIT.—The complete path of magnetic lines of force.

MAGNETIC FIELD.—The space in which a magnetic force exists.

MAGNETIC FLUX.—The total number of lines of force issuing from a pole of a magnet.

MAGNETIZE.—To convert a material into a magnet by causing the molecules to rearrange.

MAGNETO.—A generator which produces alternating current and has a permanent magnet as its field.

MEGGER.—A test instrument used to measure insulation resistance and other high resistances. It is a portable hand operated d-c generator used as an ohmmeter.

MEGOHM.—A million ohms.

MICRO.—A prefix meaning one-millionth.

MILLI.—A prefix meaning one-thousandth.

MILLIAMMETER.—An ammeter that measures current in thousandths of an ampere.

MOTOR-GENERATOR.—A motor and a generator with a common shaft used to convert line voltages to other voltages or frequencies.

MUTUAL INDUCTANCE.—A circuit property existing when the relative position of two inductors causes the magnetic lines of force from one to link with the turns of the other.

NEGATIVE CHARGE.—The electrical charge carried by a body which has an excess of electrons.

NEUTRON.—A particle having the weight of a proton but carrying no electric charge. It is located in the nucleus of an atom.

NUCLEUS.—The central part of an atom that is mainly comprised of protons and neutrons. It is the part of the atom that has the most mass.

NULL.—Zero.

OHM.—The unit of electrical resistance.

OHMMETER.—An instrument for directly measuring resistance in ohms.

OVERLOAD.—A load greater than the rated load of an electrical device.

PERMALLOY.—An alloy of nickel and iron having an abnormally high magnetic permeability.

PERMEABILITY.—A measure of the ease with which magnetic lines of force can flow through a material as compared to air.

PHASE DIFFERENCE.—The time in electrical degree by which one wave leads or lags another.

POLARITY.—The character of having magnetic poles, or electric charges.

POLE.—The section of a magnet where the flux lines are concentrated; also where they enter and leave the magnet. An electrode of a battery.

POLYPHASE.—A circuit that utilizes more than one phase of alternating current.

POSITIVE CHARGE.—The electrical charge carried by a body which has become deficient in electrons.

POTENTIAL.—The amount of charge held by a body as compared to another point or body. Usually measured in volts.

NOTES:

POTENTIOMETER.—A variable voltage divider; a resistor which has a variable contact arm so that any portion of the potential applied between its ends may be selected.

POWER.—The rate of doing work or the rate of expending energy. The unit of electrical power is the watt.

POWER FACTOR.—The ratio of the actual power of an alternating or pulsating current, as measured by a wattmeter, to the apparent power, as indicated by ammeter and voltmeter readings. The power factor of an inductor, capacitor, or insulator is an expression of their losses.

PRIME MOVER.—The source of mechanical power used to drive the rotor of a generator.

PROTON.—A positively charged particle in the nucleus of an atom.

RATIO.—The value obtained by dividing one number by another, indicating their relative proportions.

REACTANCE.—The opposition offered to the flow of an alternating current by the inductance, capacitance, or both, in any circuit.

RECTIFIERS.—Devices used to change alternating current to unidirectional current. These may be vacuum tubes, semiconductors such as germanium and silicon, and dry-disk rectifiers such as selenium and copper-oxide.

RELAY.—An electromechanical switching device that can be used as a remote control.

RELUCTANCE.—A measure of the opposition that a material offers to magnetic lines of force.

RESISTANCE.—The opposition to the flow of current caused by the nature and physical dimensions of a conductor.

RESISTOR.—A circuit element whose chief characteristic is resistance; used to oppose the flow of current.

RETENTIVITY.—The measure of the ability of a material to hold its magnetism.

RHEOSTAT.—A variable resistor.

SATURABLE REACTOR.—A control device that uses a small d-c current to control a large a-c current by controlling core flux density.

SATURATION.—The condition existing in any circuit when an increase in the driving signal produces no further change in the resultant effect.

SELF-INDUCTION.—The process by which a circuit induces an e.m.f. into itself by its own magnetic field.

SERIES-WOUND.—A motor or generator in which the armature is wired in series with the field winding.

SERVO.—A device used to convert a small movement into one of greater movement or force.

SERVOMECHANISM.—A closed-loop system that produces a force to position an object in accordance with the information that originates at the input.

SOLENOID.—An electromagnetic coil that contains a movable plunger.

SPACE CHARGE.—The cloud of electrons existing in the space between the cathode and plate in a vacuum tube, formed by the electrons emitted from the cathode in excess of those immediately attracted to the plate.

SPECIFIC GRAVITY.—The ratio between the density of a substance and that of pure water at a given temperature.

SYNCHROSCOPE.—An instrument used to indicate a difference in frequency between two a-c sources.

SYNCHRO SYSTEM.—An electrical system that gives remote indications or control by means of self-synchronizing motors.

TACHOMETER.—An instrument for indicating revolutions per minute.

TERTIARY WINDING.—A third winding on a transformer or magnetic amplifier that is used as a second control winding.

THERMISTOR.—A resistor that is used to compensate for temperature variations in a circuit.

THERMOCOUPLE.—A junction of two dissimilar metals that produces a voltage when heated.

TORQUE.—The turning effort or twist which a shaft sustains when transmitting power.

TRANSFORMER.—A device composed of two or more coils, linked by magnetic lines of force, used to transfer energy from one circuit to another.

TRANSMISSION LINES.—Any conductor or system of conductors used to carry electrical energy from its source to a load.

VARS.—Abbreviation for volt-ampere, reactive.

VECTOR.—A line used to represent both direction and magnitude.

NOTES:

VOLT.—The unit of electrical potential.

VOLTMETER.—An instrument designed to measure a difference in electrical potential, in volts.

WATT.—The unit of electrical power.

WATTMETER.—An instrument for measuring electrical power in watts.

NOTES:

APPENDIX IV

COMPONENT COLOR CODE

COLOR	1ST DIGIT	2ND DIGIT	MULTIPLIER	TOLERANCE (percent)
Black	0	0	1	
Brown	1	1	10	
Red	2	2	100	
Orange	3	3	1,000	
Yellow	4	4	10,000	
Green	5	5	100,000	
Blue	6	6	1,000,000	
Violet	7	7	10,000,000	
White	9	9	1,000,000,000	
Gold			.1	5
Silver			.01	10
No color				20

(A)

④ TOLERANCE (PERCENT)
① 1ST DIGIT
③ MULTIPLIER
② 2ND DIGIT

① 1ST DIGIT
② 2ND DIGIT
③ MULTIPLIER
④ TOLERANCE (PERCENT)

(B)

Resistor color code.

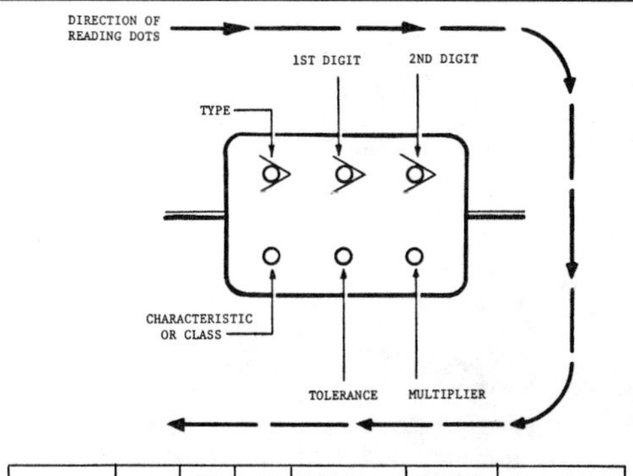

TYPE	COLOR	1ST DIGIT	2ND DIGIT	MULTIPLIER	TOLERANCE (PERCENT)	CHARACTERISTIC OR CLASS
JAN, MICA	BLACK	0	0	1.0	± 1	APPLIES TO
	BROWN	1	1	10	± 2	TEMPERATURE
	RED	2	2	100	± 3	COEFFICIENT
	ORANGE	3	3	1,000	± 4	OR METHODS
	YELLOW	4	4	10,000	± 5	OF TESTING
	GREEN	5	5	100,000	± 6	
	BLUE	6	6	1,000,000	± 7	
	VIOLET	7	7	10,000,000	± 8	
	GRAY	8	8	100,000,000	± 9	
EIA, MICA	WHITE	9	9	1,000,000,000		
	GOLD			.1	±10	
MOLDED PAPER	SILVER			.01	±20	
	BODY					

6-Dot color code for mica and molded paper capacitors.

NOTES:

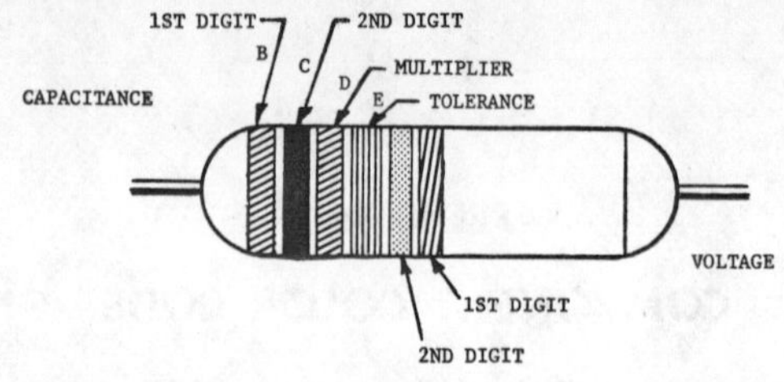

COLOR	CAPACITANCE			TOLERANCE (PERCENT)	VOLTAGE RATING	
	1ST DIGIT	2ND DIGIT	MULTIPLIER		1ST DIGIT	2ND DIGIT
BLACK	0	0	1	±20	0	0
BROWN	1	1	10		1	1
RED	2	2	100		2	2
ORANGE	3	3	1,000	±30	3	3
YELLOW	4	4	10,000	±40	4	4
GREEN	5	5	100,000	± 5	5	5
BLUE	6	6	1,000,000		6	6
VIOLET	7	7			7	7
GRAY	8	8			8	8
WHITE	9	9		±10	9	9

6-Band color code for tubular paper dielectric capacitors.

NOTES:

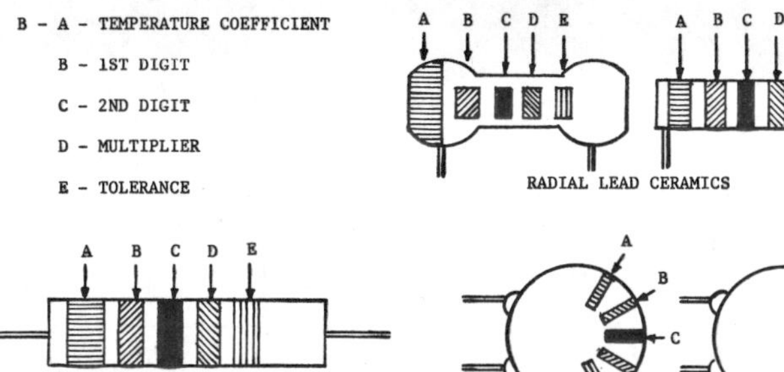

A B C D E

AXIAL LEAD CERAMIC

A B C D E
D B C
5 DOT
3 DOT

CERAMIC DISC CAPACITOR MARKING

COLOR	1ST DIGIT	2ND DIGIT	MULTIPLIER	TOLERANCE: MORE THAN 10 pf (IN PERCENT)	TOLERANCE: LESS THAN 10 pf (IN pf)	TEMPERATURE COEFFICIENT*
BLACK	0	0	1.0	±20	±2.0	0
BROWN	1	1	10	± 1		-30
RED	2	2	100	± 2		-80
ORANGE	3	3	1,000			-150
YELLOW	4	4	10,000			-220
GREEN	5	5		± 5	±0.5	-330
BLUE	6	6				-470
VIOLET	7	7				-750
GRAY	8	8	.01		±0.25	+30
WHITE	9	9	.1	±10	±1.0	+120 TO -750 (EIA)
						+500 TO -330 (JAN)
SILVER						+100 (JAN)
GOLD						BYPASS OR COUPLING (EIA)

*PARTS PER MILLION PER DEGREE CENTIGRADE.

NOTES:

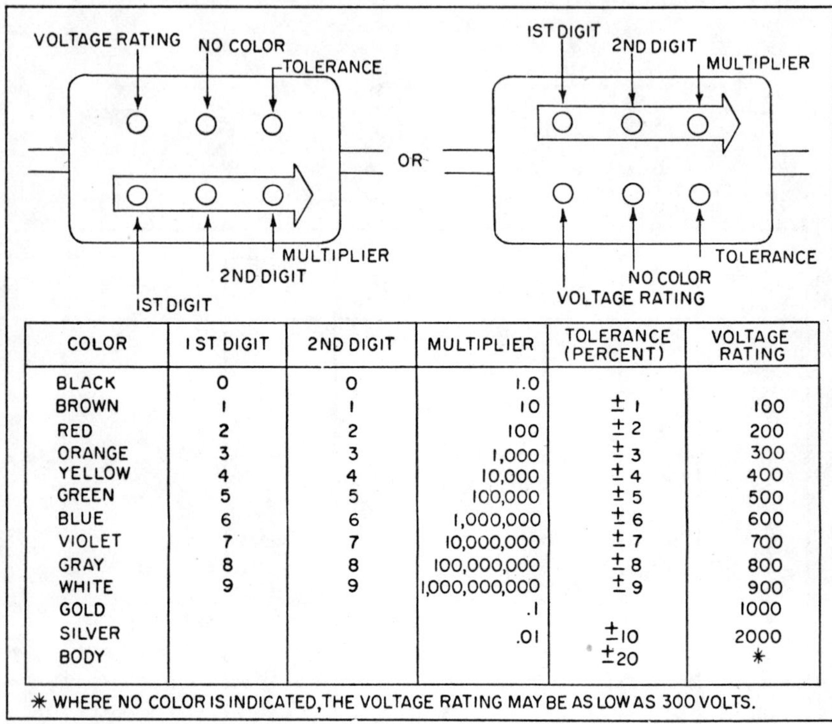

COLOR	1ST DIGIT	2ND DIGIT	MULTIPLIER	TOLERANCE (PERCENT)	VOLTAGE RATING
BLACK	0	0	1.0		
BROWN	1	1	10	±1	100
RED	2	2	100	±2	200
ORANGE	3	3	1,000	±3	300
YELLOW	4	4	10,000	±4	400
GREEN	5	5	100,000	±5	500
BLUE	6	6	1,000,000	±6	600
VIOLET	7	7	10,000,000	±7	700
GRAY	8	8	100,000,000	±8	800
WHITE	9	9	1,000,000,000	±9	900
GOLD			.1		1000
SILVER			.01	±10	2000
BODY				±20	*

* WHERE NO COLOR IS INDICATED, THE VOLTAGE RATING MAY BE AS LOW AS 300 VOLTS.

NOTES:

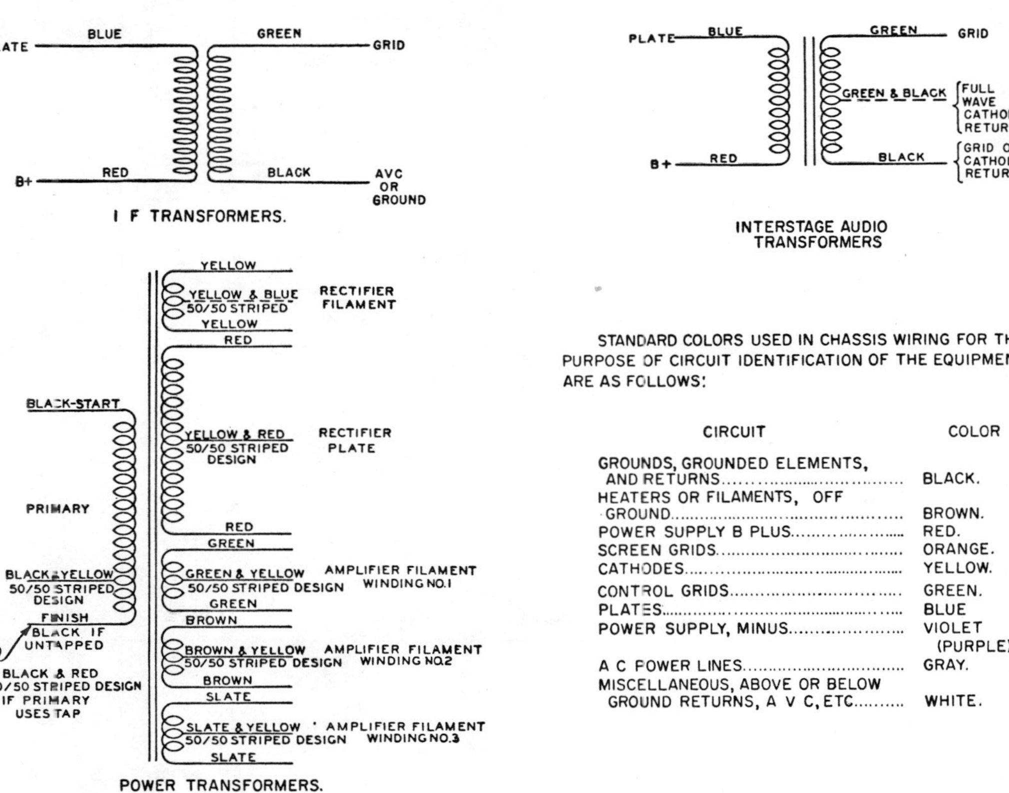

STANDARD COLORS USED IN CHASSIS WIRING FOR THE PURPOSE OF CIRCUIT IDENTIFICATION OF THE EQUIPMENT ARE AS FOLLOWS:

CIRCUIT	COLOR
GROUNDS, GROUNDED ELEMENTS, AND RETURNS	BLACK.
HEATERS OR FILAMENTS, OFF GROUND	BROWN.
POWER SUPPLY B PLUS	RED.
SCREEN GRIDS	ORANGE.
CATHODES	YELLOW.
CONTROL GRIDS	GREEN.
PLATES	BLUE
POWER SUPPLY, MINUS	VIOLET (PURPLE).
A C POWER LINES	GRAY.
MISCELLANEOUS, ABOVE OR BELOW GROUND RETURNS, A V C, ETC	WHITE.

NOTES:

APPENDIX V

CURRENT-CARRYING CAPACITIES OF CONDUCTORS

Based on the National Electrical Code, 1951

Size AWG or Thousand Cir. Mils	Current Capacity in Amperes		
	Rubber or Thermo-plastic Insulation	Paper or Varnished Cambric Insulation	Impregnated Asbestos Insulation
14	15	25	30
12	20	30	40
10	30	40	50
8	40	50	65
6	55	70	85
4	70	90	115
3	80	105	130
2	95	120	145
1	110	140	170
0	125	155	200
00	145	185	230
000	165	210	265
0000	195	235	310
250	215	270	335
300	240	300	380
350	260	325	420
400	280	360	450
500	320	405	500
600	355	455	545
700	385	490	600
750	400	500	620
800	410	515	640
900	435	555	...
1,000	455	585	730
1,250	495	645	...
1,500	520	700	...
1,750	545	735	...
2,000	560	775	...

Notes:
1. The ratings are for not more than three conductors in a cable or raceway, with a room temperature of 30° C or 86° F.
2. The above tables are for copper wires. The ratings for aluminum wires are 84 percent of these values.
3. Consult the National Electrical Code for more information. For example, higher ratings are allowed for single conductors in free air.

NOTES:

NOTES:

APPENDIX VI

FULL LOAD CURRENTS OF MOTORS

[The following data are approximate full-load currents for motors of various types, frequencies, and speeds. They have been compiled from average values for representative motors of their respective classes. Variations of 10 percent above or below the values given may be expected.]

Hp. of motor	Amperes—Full-load current																								
	Direct-current motors			Alternating-current motors																					
				Single-phase motors		Squirrel-cage induction motors										Slip-ring induction motors									
						Two-phase					Three-phase					Two-phase					Three-phase				
	115-volt	230-volt	550-volt	110-volt	220-volt	110-volt	220-volt	440-volt	550-volt	2,200-volt	110-volt	220-volt	440-volt	550-volt	2,200-volt	110-volt	220-volt	440-volt	550-volt	2,200-volt	110-volt	220-volt	440-volt	550-volt	2,200-volt
¼	...	...	...	4.8	2.4	...	...	...	...	...	...	...	...	...	...	...	...	...	...	...	...	...	...	...	...
½	4.5	2.3	...	7	3.5	4.3	2.2	1.1	0.9	...	5.0	2.5	1.3	1.0	...	...	...	...	...	...	...	...	...	...	...
¾	6.5	3.3	1.4	9.4	4.7	4.7	2.4	1.2	1.0	...	5.4	2.8	1.4	1.1	...	6.2	3.1	1.6	1.3	...	7.2	3.6	1.8	1.5	...
1	8.4	4.2	1.7	11	5.5	5.7	2.9	1.4	1.2	...	6.6	3.3	1.7	1.3	...	6.7	3.4	1.7	1.4	...	7.8	3.9	2.0	1.6	...
1½	12.5	6.3	2.6	15.2	7.6	7.7	4.0	2	1.6	...	9.4	4.7	2.4	2.0	...	11.7	5.9	3.0	2.3	...	...	...	...	...	...
2	16.1	8.3	3.4	20	10	10.4	5	3	2.0	...	12.0	6	3.0	2.4	...	12.5	6.3	3.1	2.5	...	14.4	7.2	3.6	2.9	...
3	23	12.3	5.0	28	14	...	8	4	3.0	...	...	9	4.5	4.0	...	...	8.7	4.3	3.5	...	20.2	10	5.0	4	...
5	40	19.8	8.2	46	23	...	13	7	6	...	...	15	7.5	6.0	...	...	13.0	6.5	5.2	...	...	15	7.5	6	...
7½	58	28.7	12	68	34	...	19	9	7	...	...	22	11	9.0	...	...	20.0	10.0	7.6	...	...	25	13	10	...
10	75	38	16	86	43	...	24	12	10	...	...	27	14	11	...	...	24.3	12.1	10.0	...	...	28	14	11	...
15	112	56	23	...	...	...	33	16	13	...	...	38	19	15	...	...	39	19.5	15.6	...	...	45	23	18	...
20	140	74	30	...	...	...	45	23	19	...	...	52	26	21	5.7	...	49	24.7	19.8	...	...	56	28	22	...
25	185	92	38	...	...	...	55	28	22	6	...	64	32	26	7	...	60	30.0	24.0	6.4	...	67	34	27	7.5
30	220	110	45	...	...	...	67	34	27	7	...	77	39	31	8	...	72	36.0	28.8	7.8	...	82	41	33	9
40	294	146	61	...	...	...	88	44	35	9	...	101	51	40	10	...	93	46.5	37.3	9.5	...	106	53	42	11
50	364	180	75	...	...	...	108	54	43	11	...	125	63	50	13	...	113	57	45	12.1	...	128	64	51	14
60	436	215	90	...	...	...	129	65	52	13	...	149	75	60	15	...	135	68	54	14.0	...	150	75	60	16
75	540	268	111	...	...	...	156	78	62	16	...	180	90	72	19	...	164	82	65	17.3	...	188	94	75	19
100	...	357	146	...	...	...	212	106	85	22	...	246	123	98	25	...	214	108	87	21.7	...	246	123	99	25
125	...	443	184	...	...	...	268	134	108	27	...	310	155	124	32	...	267	134	108	27	...	310	155	124	31
150	...	...	220	...	...	...	311	155	124	31	...	360	180	144	36	...	315	158	127	32	...	364	182	145	37
175	...	...	...	...	...	...	...	...	...	...	...	...	...	...	...	...	...	...	...	...	...	...	...	...	...
200	...	...	295	...	...	...	415	208	166	43	...	480	240	195	49	...	430	216	173	44	...	490	245	196	52

APPENDIX VII
FORMULAS

Ohm's Law for D-C Circuits

$$I = \frac{E}{R} = \frac{P}{E} = \sqrt{\frac{P}{R}}$$

$$R = \frac{E}{I} = \frac{P}{I^2} = \frac{E^2}{P}$$

$$E = IR = \frac{P}{I} = \sqrt{PR}$$

$$P = EI = \frac{E^2}{R} = I^2R$$

Resistors in Series

$$R_T = R_1 + R_2 + \ldots$$

Resistors in Parallel

Two resistors

$$R_T = \frac{R_1R_2}{R_1 + R_2}$$

More than two

$$\frac{1}{R_T} = \frac{1}{R_1} + \frac{1}{R_2} + \frac{1}{R_3} + \ldots$$

RL Circuit Time Constant

$$\frac{L \text{ (in henrys)}}{R \text{ (in ohms)}} = t \text{ (in seconds), or}$$

$$\frac{L \text{ (in microhenrys)}}{R \text{ (in ohms)}} = t \text{ (in microseconds)}$$

RC Circuit Time Constant

R (ohms) x C (farads) = t (seconds)

R (megohms) x C (microfarads) = t (seconds)

R (ohms) x C (microfarads) = t (microseconds)

R (megohms) x C (micromicrofarads) = t (microseconds)

Capacitors in Series

Two capacitors

$$C_T = \frac{C_1C_2}{C_1 + C_2}$$

More than two

$$\frac{1}{C_T} = \frac{1}{C_1} + \frac{1}{C_2} + \frac{1}{C_3} + \ldots$$

Capacitors in Parallel: $C_T = C_1 + C_2 + \ldots$

Capacitive Reactance: $X_C = \frac{1}{2\pi fC}$

Impedance in an RC Circuit (Series)

$$Z = \sqrt{R^2 + (X_C)^2}$$

Inductors in Series

$L_T = L_1 + L_2 + \ldots$ (No coupling between coils)

Inductors in Parallel

Two inductors

$L_T = \frac{L_1L_2}{L_1 + L_2}$ (No coupling between coils)

More than two

$\frac{1}{L_T} = \frac{1}{L_1} + \frac{1}{L_2} + \frac{1}{L_3} + \ldots$ (No coupling between coils)

Inductive Reactance

$$X_L = 2\pi fL$$

Q of a Coil

$$Q = \frac{X_L}{R}$$

NOTES:

Impedance of an RL Circuit (Series)

$$Z = \sqrt{R^2 + (X_L)^2}$$

Impedance with R, C, and L in Series

$$Z = \sqrt{R^2 + (X_L - X_C)^2}$$

Parallel Circuit Impedance

$$Z = \frac{Z_1 Z_2}{Z_1 + Z_2}$$

Sine-Wave Voltage Relationships

Average value

$$E_{ave} = \frac{2}{\pi} \times E_{max} = 0.637E_{max}$$

Effective or r. m. s. value

$$E_{eff} = \frac{E_{max}}{\sqrt{2}} = \frac{E_{max}}{1.414} = 0.707E_{max}$$
$$= 1.11E_{ave}$$

Maximum value

$$E_{max} = \sqrt{2}(E_{eff}) = 1.414E_{eff}$$
$$= 1.57E_{ave}$$

Voltage in an a-c circuit

$$E = IZ = \frac{P}{I \times P.F.}$$

Current in an a-c circuit

$$I = \frac{E}{Z} = \frac{P}{E \times P.F.}$$

Power in A-C Circuit

Apparent power: $P = EI$

True power: $P = EI \cos\theta = EI \times P.F.$

Power Factor

$$P.F. = \frac{P}{EI} = \cos\theta$$

$$\cos\theta = \frac{\text{true power}}{\text{apparent power}}$$

Transformers

Voltage relationship

$$\frac{E_p}{E_s} = \frac{N_p}{N_s} \text{ or } E_s = E_p \times \frac{N_s}{N_p}$$

Current relationship

$$\frac{I_p}{I_s} = \frac{N_s}{N_p}$$

Induced voltage

$$E_{eff} = 4.44 \times BAfN \times 10^{-8}$$

Turns ratio

$$\frac{N_p}{N_s} = \sqrt{\frac{Z_p}{Z_s}}$$

Secondary current

$$I_s = I_p \times \frac{N_p}{N_s}$$

Secondary voltage

$$E_s = E_p \times \frac{N_s}{N_p}$$

Three-Phase Voltage and Current Relationships

With wye connected windings

$$E_{line} = \sqrt{3}(E_{coil}) = 1.732E_{coil}$$

$$I_{line} = I_{coil}$$

With delta connected windings

$$E_{line} = E_{coil}$$

$$I_{line} = 1.732I_{coil}$$

NOTES:

With wye or delta connected winding

$$P_{coil} = E_{coil} I_{coil}$$

$$P_t = 3P_{coil}$$

$$P_t = 1.732 E_{line} I_{line}$$

(To convert to true power multiply by $\cos \theta$)

Synchronous Speed of Motor

$$\text{r.p.m.} = \frac{120 \times \text{frequency}}{\text{number of poles}}$$

Comparison of Units in Electric and Magnetic Circuits

	Electric circuit	Magnetic circuit
Force.	Volt, E, or e.m.f.	Gilberts, F, or m.m.f.
Flow	Ampere, I	Flux, Φ, in maxwells
Opposition.	Ohms, R	Reluctance, $\mathcal{R}$
Law.	Ohm's law, $I = \frac{E}{R}$	Rowland's law, $\Phi = \frac{F}{\mathcal{R}}$
Intensity of force	Volts per cm. of length.	$H = \frac{1.257IN}{L}$, gilberts per centimeter of length.
Density.	Current density—for example, amperes per cm^2.	Flux density—for example, lines per cm.^{2}, or gausses.

NOTES:

APPENDIX VIII

TRIGONOMETRIC FUNCTIONS

In a right triangle, there are several relationships which always hold true. These relationships pertain to the length of the sides of a right triangle, and the way the lengths are affected by the angles between them. An understanding of these relationships, called trigonometricfunctions, is essential for solving many problems in a-c circuits such as power factor, impedance, voltage drops, and so forth.

To be a RIGHT triangle, a triangle must have a "square" corner; one in which there is exactly 90° between two of the sides. Trigonometric functions do not apply to any other type of triangle. This type of triangle is shown in figure IX-1.

By use of the trigonometric functions, it is possible to determine the UNKNOWN length of one or more sides of a triangle, or the number of degrees in UNKNOWN angles, depending on what is presently known about the triangle. For instance, if the lengths of any two sides are known, the third side and both angles θ (theta) and Φ (phi) may be determined. The triangle may also be solved if the length of any one side and one of the angles (θ or Φ in fig. IX-1) are known.

The first basic fact to accept, regarding triangles, is that IN ANY TRIANGLE, THE SUM OF THE THREE ANGLES FORMED INSIDE THE TRIANGLE MUST ALWAYS EQUAL 180°. If one angle is always 90° (a right angle) then the sum of the other two angles must always be 90°.

$$\theta + \Phi = 90^\circ$$

and $$90^\circ + \theta + \Phi = 180^\circ$$

or $$90^\circ + 90^\circ = 180^\circ$$

thus, if angle θ is known, Φ may be quickly determined.

For instance, if θ is 30°, what is Φ ?

$$90^\circ + 30^\circ + \Phi = 180^\circ$$

Transposing $$\Phi = 180^\circ - 90^\circ - 30^\circ$$

$$\Phi = 60^\circ$$

Also, if Φ is known, θ may be determined in the same manner.

The second basic fact you must understand is that FOR EVERY DIFFERENT COMBINATION OF ANGLES IN A TRIANGLE, THERE IS A DEFINITE RATIO BETWEEN THE LENGTHS OF THE THREE SIDES. Consider the triangle in figure IX-2, consisting of the base, side B; the altitude, side A; and the hypotenuse, side C. (The hypotenuse is always the longest side, and is always opposite the 90° angle.) If angle θ is 30°, Φ must be 60°. With θ equal to 30°, the ratio of the length of side B to side C is 0.866 to 1. That is, if the hypotenuse is 1 inch long, the side adjacent to θ, side B, is 0.866 inch long. Also, with θ equal to 30°, the ratio of side A to side C is 0.5 to 1. That is, with the hypotenuse 1 inch long, the side opposite to θ (side A) is 0.5 inch long. With θ still at 30°, side A is 0.5774 of the length of B. With the combination of angles given (30° - 60° - 90°) these are the ONLY ratios of lengths that will "fit" to form a right triangle.

Note that three ratios are shown to exist for the given value of θ : the ratio $\frac{B}{C}$, which is always referred to as the COSINE ratio of θ, the ratio $\frac{A}{C}$, which is always the SINE ratio of θ, and the ratio $\frac{A}{B}$, which is always the TANGENT ratio of θ. If θ changes, all three ratios

NOTES:

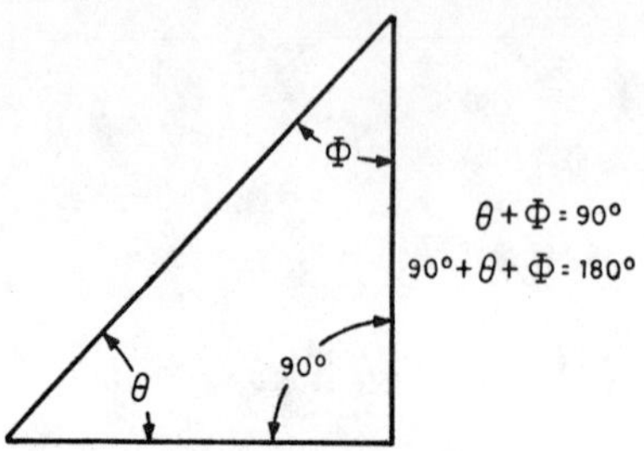

Figure IX-1.—A right triangle.

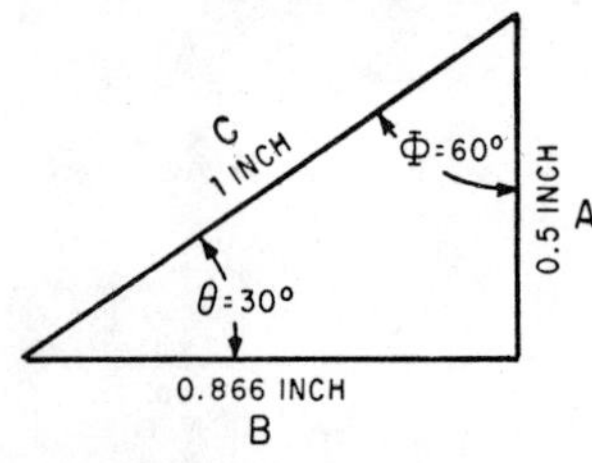

Figure IX-2.—A 30° -60° -90° triangle.

change, because the lengths of the sides (base and altitude) change. There is a set of ratios for every increment between 0° and 90°. These angular ratios, or sine, cosine, and tangent functions, are listed for each degree and tenth of degree in a table at the end of this appendix. In this table, the length of the hypotenuse of a triangle is considered fixed. Thus, the ratios of length given refer to the manner in which sides A and B vary with relation to each other and in relation to side C, as angle θ is varied from 0° to 90°.

The solution of problems in trigonometry (solution of triangles) is much simpler when the table of trigonometric functions is used properly. The most common ways in which it is used will be shown by solving a series of exemplary problems.

Problem 1: If the hypotenuse of the triangle (side C) in figure IX-3 (A) is 10 inches long, and angle θ is 33°, how long are sides B and A?

Solution: The ratio $\frac{B}{C}$ is the cosine function. By checking the table of functions, you will find that

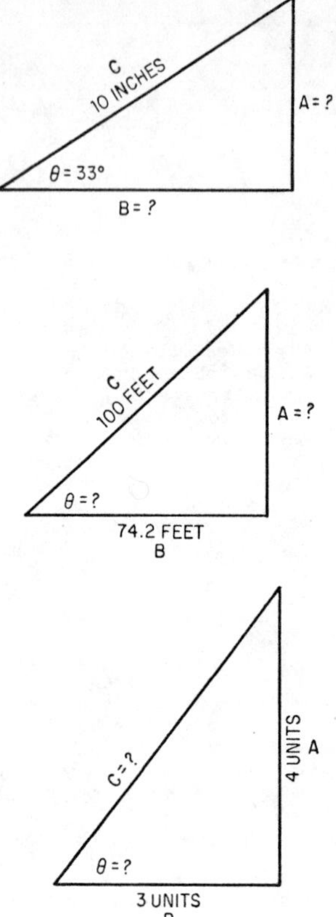

Figure IX-3.—Trigonometric problems.

the cosine of 33° is 0.8387. This means that the length of B is 0.8387 the length of side C. If side C is 10 inches long, then side B must be 10 x 0.8387, or 8.387 inches in length. To determine the length of side A, use the sine function, the ratio $\frac{A}{C}$. Again consulting the table of functions, you will find that the sine of 33° is 0.5446. Thus, side A must be 10 x 0.5446, or 5.446 inches in length.

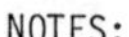

Problem 2: The triangle in figure IX-3 (B) has a base 74.2 feet long, and hypotenuse 100 feet long. What is θ, and how long is side A? Solution: When no angles are given, you must always solve for a known angle first. The ratio $\frac{B}{C}$ is the cosine of the unknown angle θ; therefore $\frac{74.2}{100}$, or 0.742, is the cosine of the unknown angle. Locating 0.742 as a cosine value in the table, you find that it is the cosine of 42.1°. That is, $\theta = 42.1°$. With θ known, side A is solved for by use of the sine ratio $\frac{A}{C}$. The sine of 42.1°, according to the table, is 0.6704. Therefore, side A is 100 x 0.6704, or 67.04 feet long.

Problem 3: In the triangle in figure IX-3 (C), the base is 3 units long, and the altitude is 4 units. What is θ, and how long is the hypotenuse? Solution: With the information given, the tangent of θ may be determined. $\text{Tan}\,\theta = \frac{A}{B} = \frac{4}{3} = 1.33$ Locating the value 1.33 as a tangent value in the table of functions, you find it to be the tangent of 53.1°. Therefore, $\theta = 53.1°$. Once θ is known, either the sine or cosine ratio may be used to determine the length of the hypotenuse. The cosine of 53.1° is 0.6004. This indicates that the base of 3 units is 0.6004, the length of the hypotenuse. Therefore, the hypotenuse is $\frac{3}{0.6004}$, or 5 units in length. Using the sine ratio, the hypotenuse is $\frac{4}{0.7997}$, or 5 units in length.

In the foregoing explanations and problems, the sides of triangles were given in inches, feet, and units. In applying trigonometry to a-c circuit problems, these units of measure will be replaced by such measurements as ohms, ampres, volts, and watts. Angle θ will often be referred to as the phase angle. However, the solution of these a-c problems is accomplished in exactly the same manner as the foregoing problems. Only the units and some terminology are changed.

NOTES:

Natural Sines, Cosines, and Tangents

0°–14.9°

Degs.	Function	0.0°	0.1°	0.2°	0.3°	0.4°	0.5°	0.6°	0.7°	0.8°	0.9°
0	sin	0.0000	0.0017	0.0035	0.0052	0.0070	0.0087	0.0105	0.0122	0.0140	0.0157
	cos	1.0000	1.0000	1.0000	1.0000	1.0000	1.0000	0.9999	0.9999	0.9999	0.9999
	tan	0.0000	0.0017	0.0035	0.0052	0.0070	0.0087	0.0105	0.0122	0.0140	0.0157
1	sin	0.0175	0.0192	0.0209	0.0227	0.0244	0.0262	0.0279	0.0297	0.0314	0.0332
	cos	0.9998	0.9998	0.9998	0.9997	0.9997	0.9997	0.9996	0.9996	0.9995	0.9995
	tan	0.0175	0.0192	0.0209	0.0227	0.0244	0.0262	0.0279	0.0297	0.0314	0.0332
2	sin	0.0349	0.0366	0.0384	0.0401	0.0419	0.0436	0.0454	0.0471	0.0488	0.0506
	cos	0.9994	0.9993	0.9993	0.9992	0.9991	0.9990	0.9990	0.9989	0.9988	0.9987
	tan	0.0349	0.0367	0.0384	0.0402	0.0419	0.0437	0.0454	0.0472	0.0489	0.0507
3	sin	0.0523	0.0541	0.0558	0.0576	0.0593	0.0610	0.0628	0.0645	0.0663	0.0680
	cos	0.9986	0.9985	0.9984	0.9983	0.9982	0.9981	0.9980	0.9979	0.9978	0.9977
	tan	0.0524	0.0542	0.0559	0.0577	0.0594	0.0612	0.0629	0.0647	0.0664	0.0682
4	sin	0.0698	0.0715	0.0732	0.0750	0.0767	0.0785	0.0802	0.0819	0.0837	0.0854
	cos	0.9976	0.9974	0.9973	0.9972	0.9971	0.9969	0.9968	0.9966	0.9965	0.9963
	tan	0.0699	0.0717	0.0734	0.0752	0.0769	0.0787	0.0805	0.0822	0.0840	0.0857
5	sin	0.0872	0.0889	0.0906	0.0924	0.0941	0.0958	0.0976	0.0993	0.1011	0.1028
	cos	0.9962	0.9960	0.9959	0.9957	0.9956	0.9954	0.9952	0.9951	0.9949	0.9947
	tan	0.0875	0.0892	0.0910	0.0928	0.0945	0.0963	0.0981	0.0998	0.1016	0.1033
6	sin	0.1045	0.1063	0.1080	0.1097	0.1115	0.1132	0.1149	0.1167	0.1184	0.1201
	cos	0.9945	0.9943	0.9942	0.9940	0.9938	0.9936	0.9934	0.9932	0.9930	0.9928
	tan	0.1051	0.1069	0.1086	0.1104	0.1122	0.1139	0.1157	0.1175	0.1192	0.1210
7	sin	0.1219	0.1236	0.1253	0.1271	0.1288	0.1305	0.1323	0.1340	0.1357	0.1374
	cos	0.9925	0.9923	0.9921	0.9919	0.9917	0.9914	0.9912	0.9910	0.9907	0.9905
	tan	0.1228	0.1246	0.1263	0.1281	0.1299	0.1317	0.1334	0.1352	0.1370	0.1388
8	sin	0.1392	0.1409	0.1426	0.1444	0.1461	0.1478	0.1495	0.1513	0.1530	0.1547
	cos	0.9903	0.9900	0.9898	0.9895	0.9893	0.9890	0.9888	0.9885	0.9882	0.9880
	tan	0.1405	0.1423	0.1441	0.1459	0.1477	0.1495	0.1512	0.1530	0.1548	0.1566
9	sin	0.1564	0.1582	0.1599	0.1616	0.1633	0.1650	0.1668	0.1685	0.1702	0.1719
	cos	0.9877	0.9874	0.9871	0.9869	0.9866	0.9863	0.9860	0.9857	0.9854	0.9851
	tan	0.1584	0.1602	0.1620	0.1638	0.1655	0.1673	0.1691	0.1709	0.1727	0.1745
10	sin	0.1736	0.1754	0.1771	0.1788	0.1805	0.1822	0.1840	0.1857	0.1874	0.1891
	cos	0.9848	0.9845	0.9842	0.9839	0.9836	0.9833	0.9829	0.9826	0.9823	0.9820
	tan	0.1763	0.1781	0.1799	0.1817	0.1835	0.1853	0.1871	0.1890	0.1908	0.1926
11	sin	0.1908	0.1925	0.1942	0.1959	0.1977	0.1994	0.2011	0.2028	0.2045	0.2062
	cos	0.9816	0.9813	0.9810	0.9806	0.9803	0.9799	0.9796	0.9792	0.9789	0.9785
	tan	0.1944	0.1962	0.1980	0.1998	0.2016	0.2035	0.2053	0.2071	0.2089	0.2107
12	sin	0.2079	0.2096	0.2113	0.2130	0.2147	0.2164	0.2181	0.2198	0.2215	0.2232
	cos	0.9781	0.9778	0.9774	0.9770	0.9767	0.9763	0.9759	0.9755	0.9751	0.9748
	tan	0.2126	0.2144	0.2162	0.2180	0.2199	0.2217	0.2235	0.2254	0.2272	0.2290
13	sin	0.2250	0.2267	0.2284	0.2300	0.2318	0.2334	0.2351	0.2368	0.2385	0.2402
	cos	0.9744	0.9740	0.9736	0.9732	0.9728	0.9724	0.9720	0.9715	0.9711	0.9707
	tan	0.2309	0.2327	0.2345	0.2364	0.2382	0.2401	0.2419	0.2438	0.2456	0.2475
14	sin	0.2419	0.2436	0.2453	0.2470	0.2487	0.2504	0.2521	0.2538	0.2554	0.2571
	cos	0.9703	0.9699	0.9694	0.9690	0.9686	0.9681	0.9677	0.9673	0.9668	0.9664
	tan	0.2493	0.2512	0.2530	0.2549	0.2568	0.2586	0.2605	0.2623	0.2642	0.2661
Degs.	Function	0′	6′	12′	18′	24′	30′	36′	42′	48′	54′

Natural Sines, Cosines, and Tangents

15°–29.9°

Degs.	Function	0.0°	0.1°	0.2°	0.3°	0.4°	0.5°	0.6°	0.7°	0.8°	0.9°
15	sin	0.2588	0.2605	0.2622	0.2639	0.2656	0.2672	0.2689	0.2706	0.2723	0.2740
	cos	0.9659	0.9655	0.9650	0.9646	0.9641	0.9636	0.9632	0.9627	0.9622	0.9617
	tan	0.2679	0.2698	0.2717	0.2736	0.2754	0.2773	0.2792	0.2811	0.2830	0.2849
16	sin	0.2756	0.2773	0.2790	0.2807	0.2823	0.2840	0.2857	0.2874	0.2890	0.2907
	cos	0.9613	0.9608	0.9603	0.9598	0.9593	0.9588	0.9583	0.9578	0.9573	0.9568
	tan	0.2867	0.2886	0.2905	0.2924	0.2943	0.2962	0.2981	0.3000	0.3019	0.3038
17	sin	0.2924	0.2940	0.2957	0.2974	0.2990	0.3007	0.3024	0.3040	0.3057	0.3074
	cos	0.9563	0.9558	0.9553	0.9548	0.9542	0.9537	0.9532	0.9527	0.9521	0.9516
	tan	0.3057	0.3076	0.3096	0.3115	0.3134	0.3153	0.3172	0.3191	0.3211	0.3230
18	sin	0.3090	0.3107	0.3123	0.3140	0.3156	0.3173	0.3190	0.3206	0.3223	0.3239
	cos	0.9511	0.9505	0.9500	0.9494	0.9489	0.9483	0.9478	0.9472	0.9466	0.9461
	tan	0.3249	0.3269	0.3288	0.3307	0.3327	0.3346	0.3365	0.3385	0.3404	0.3424
19	sin	0.3256	0.3272	0.3289	0.3305	0.3322	0.3338	0.3355	0.3371	0.3387	0.3404
	cos	0.9455	0.9449	0.9444	0.9438	0.9432	0.9426	0.9421	0.9415	0.9409	0.9403
	tan	0.3443	0.3463	0.3482	0.3502	0.3522	0.3541	0.3561	0.3581	0.3600	0.3620
20	sin	0.3420	0.3437	0.3453	0.3469	0.3486	0.3502	0.3518	0.3535	0.3551	0.3567
	cos	0.9397	0.9391	0.9385	0.9379	0.9373	0.9367	0.9361	0.9354	0.9348	0.9342
	tan	0.3640	0.3659	0.3679	0.3699	0.3719	0.3739	0.3759	0.3779	0.3799	0.3819
21	sin	0.3584	0.3600	0.3616	0.3633	0.3649	0.3665	0.3681	0.3697	0.3714	0.3730
	cos	0.9336	0.9330	0.9323	0.9317	0.9311	0.9304	0.9298	0.9291	0.9285	0.9278
	tan	0.3839	0.3859	0.3879	0.3899	0.3919	0.3939	0.3959	0.3979	0.4000	0.4020
22	sin	0.3746	0.3762	0.3778	0.3795	0.3811	0.3827	0.3843	0.3859	0.3875	0.3891
	cos	0.9272	0.9265	0.9259	0.9252	0.9245	0.9239	0.9232	0.9225	0.9219	0.9212
	tan	0.4040	0.4061	0.4081	0.4101	0.4122	0.4142	0.4163	0.4183	0.4204	0.4224
23	sin	0.3907	0.3923	0.3939	0.3955	0.3971	0.3987	0.4003	0.4019	0.4035	0.4051
	cos	0.9205	0.9198	0.9191	0.9184	0.9178	0.9171	0.9164	0.9157	0.9150	0.9143
	tan	0.4245	0.4265	0.4286	0.4307	0.4327	0.4348	0.4369	0.4390	0.4411	0.4431
24	sin	0.4067	0.4083	0.4099	0.4115	0.4131	0.4147	0.4163	0.4179	0.4195	0.4210
	cos	0.9135	0.9128	0.9121	0.9114	0.9107	0.9100	0.9092	0.9085	0.9078	0.9070
	tan	0.4452	0.4473	0.4494	0.4515	0.4536	0.4557	0.4578	0.4599	0.4621	0.4642
25	sin	0.4226	0.4242	0.4258	0.4274	0.4289	0.4305	0.4321	0.4337	0.4352	0.4368
	cos	0.9063	0.9056	0.9048	0.9041	0.9033	0.9026	0.9018	0.9011	0.9003	0.8996
	tan	0.4663	0.4684	0.4706	0.4727	0.4748	0.4770	0.4791	0.4813	0.4834	0.4856
26	sin	0.4384	0.4399	0.4415	0.4431	0.4446	0.4462	0.4478	0.4493	0.4509	0.4524
	cos	0.8988	0.8980	0.8973	0.8965	0.8957	0.8949	0.8942	0.8934	0.8926	0.8918
	tan	0.4877	0.4899	0.4921	0.4942	0.4964	0.4986	0.5008	0.5029	0.5051	0.5073
27	sin	0.4540	0.4555	0.4571	0.4586	0.4602	0.4617	0.4633	0.4648	0.4664	0.4679
	cos	0.8910	0.8902	0.8894	0.8886	0.8878	0.8870	0.8862	0.8854	0.8846	0.8838
	tan	0.5095	0.5117	0.5139	0.5161	0.5184	0.5206	0.5228	0.5250	0.5272	0.5295
28	sin	0.4695	0.4710	0.4726	0.4741	0.4756	0.4772	0.4787	0.4802	0.4818	0.4833
	cos	0.8829	0.8821	0.8813	0.8805	0.8796	0.8788	0.8780	0.8771	0.8763	0.8755
	tan	0.5317	0.5340	0.5362	0.5384	0.5407	0.5430	0.5452	0.5475	0.5498	0.5520
29	sin	0.4848	0.4863	0.4879	0.4894	0.4909	0.4924	0.4939	0.4955	0.4970	0.4985
	cos	0.8746	0.8738	0.8729	0.8721	0.8712	0.8704	0.8695	0.8686	0.8678	0.8669
	tan	0.5543	0.5566	0.5589	0.5612	0.5635	0.5658	0.5681	0.5704	0.5727	0.5750
Degs.	Function	0′	6′	12′	18′	24′	30′	36′	42′	48′	54′

Natural Sines, Cosines, and Tangents

30°–44.9°

Degs.	Function	0.0°	0.1°	0.2°	0.3°	0.4°	0.5°	0.6°	0.7°	0.8°	0.9°
30	sin	0.5000	0.5015	0.5030	0.5045	0.5060	0.5075	0.5090	0.5105	0.5120	0.5135
	cos	0.8660	0.8652	0.8643	0.8634	0.8625	0.8616	0.8607	0.8599	0.8590	0.8581
	tan	0.5774	0.5797	0.5820	0.5844	0.5867	0.5890	0.5914	0.5938	0.5961	0.5985
31	sin	0.5150	0.5165	0.5180	0.5195	0.5210	0.5225	0.5240	0.5255	0.5270	0.5284
	cos	0.8572	0.8563	0.8554	0.8545	0.8536	0.8526	0.8517	0.8508	0.8499	0.8490
	tan	0.6009	0.6032	0.6056	0.6080	0.6104	0.6128	0.6152	0.6176	0.6200	0.6224
32	sin	0.5299	0.5314	0.5329	0.5344	0.5358	0.5373	0.5388	0.5402	0.5417	0.5432
	cos	0.8480	0.8471	0.8462	0.8453	0.8443	0.8434	0.8425	0.8415	0.8406	0.8396
	tan	0.6249	0.6273	0.6297	0.6322	0.6346	0.6371	0.6395	0.6420	0.6445	0.6469
33	sin	0.5446	0.5461	0.5476	0.5490	0.5505	0.5519	0.5534	0.5548	0.5563	0.5577
	cos	0.8387	0.8377	0.8368	0.8358	0.8348	0.8339	0.8329	0.8320	0.8310	0.8300
	tan	0.6494	0.6519	0.6544	0.6569	0.6594	0.6619	0.6644	0.6669	0.6694	0.6720
34	sin	0.5592	0.5606	0.5621	0.5635	0.5650	0.5664	0.5678	0.5693	0.5707	0.5721
	cos	0.8290	0.8281	0.8271	0.8261	0.8251	0.8241	0.8231	0.8221	0.8211	0.8202
	tan	0.6745	0.6771	0.6796	0.6822	0.6847	0.6873	0.6899	0.6924	0.6956	0.6976
35	sin	0.5736	0.5750	0.5764	0.5779	0.5793	0.5807	0.5821	0.5835	0.5850	0.5864
	cos	0.8192	0.8181	0.8171	0.8161	0.8151	0.8141	0.8131	0.8121	0.8111	0.8100
	tan	0.7002	0.7028	0.7054	0.7080	0.7107	0.7133	0.7159	0.7186	0.7212	0.7239
36	sin	0.5878	0.5892	0.5906	0.5920	0.5934	0.5948	0.5962	0.5976	0.5990	0.6004
	cos	0.8090	0.8080	0.8070	0.8059	0.8049	0.8039	0.8028	0.8018	0.8007	0.7997
	tan	0.7265	0.7292	0.7319	0.7346	0.7373	0.7400	0.7427	0.7454	0.7481	0.7508
37	sin	0.6018	0.6032	0.6046	0.6060	0.6074	0.6088	0.6101	0.6115	0.6129	0.6143
	cos	0.7986	0.7976	0.7965	0.7955	0.7944	0.7934	0.7923	0.7912	0.7902	0.7891
	tan	0.7536	0.7563	0.7590	0.7618	0.7646	0.7673	0.7701	0.7729	0.7757	0.7785
38	sin	0.6157	0.6170	0.6184	0.6198	0.6211	0.6225	0.6239	0.6252	0.6266	0.6280
	cos	0.7880	0.7869	0.7859	0.7848	0.7837	0.7826	0.7815	0.7804	0.7793	0.7782
	tan	0.7813	0.7841	0.7869	0.7898	0.7926	0.7954	0.7983	0.8012	0.8040	0.8069
39	sin	0.6293	0.6307	0.6320	0.6334	0.6347	0.6361	0.6374	0.6388	0.6401	0.6414
	cos	0.7771	0.7760	0.7749	0.7738	0.7727	0.7716	0.7705	0.7694	0.7683	0.7672
	tan	0.8098	0.8127	0.8156	0.8185	0.8214	0.8243	0.8273	0.8302	0.8332	0.8361
40	sin	0.6428	0.6441	0.6455	0.6468	0.6481	0.6494	0.6508	0.6521	0.6534	0.6547
	cos	0.7660	0.7649	0.7638	0.7627	0.7615	0.7604	0.7593	0.7581	0.7570	0.7559
	tan	0.8391	0.8421	0.8451	0.8481	0.8511	0.8541	0.8571	0.8601	0.8632	0.8662
41	sin	0.6561	0.6574	0.6587	0.6600	0.6613	0.6626	0.6639	0.6652	0.6665	0.6678
	cos	0.7547	0.7536	0.7524	0.7513	0.7501	0.7490	0.7478	0.7466	0.7455	0.7443
	tan	0.8693	0.8724	0.8754	0.8785	0.8816	0.8847	0.8878	0.8910	0.8941	0.8972
42	sin	0.6691	0.6704	0.6717	0.6730	0.6743	0.6756	0.6769	0.6782	0.6794	0.6807
	cos	0.7431	0.7420	0.7408	0.7396	0.7385	0.7373	0.7361	0.7349	0.7337	0.7325
	tan	0.9004	0.9036	0.9067	0.9099	0.9131	0.9163	0.9195	0.9228	0.9260	0.9293
43	sin	0.6820	0.6833	0.6845	0.6858	0.6871	0.6884	0.6896	0.6909	0.6921	0.6934
	cos	0.7314	0.7302	0.7290	0.7278	0.7266	0.7254	0.7242	0.7230	0.7218	0.7206
	tan	0.9325	0.9358	0.9391	0.9424	0.9457	0.9490	0.9523	0.9556	0.9590	0.9623
44	sin	0.6947	0.6959	0.6972	0.6984	0.6997	0.7009	0.7022	0.7034	0.7046	0.7059
	cos	0.7193	0.7181	0.7169	0.7157	0.7145	0.7133	0.7120	0.7108	0.7096	0.7083
	tan	0.9657	0.9691	0.9725	0.9759	0.9793	0.9827	0.9861	0.9896	0.9930	0.9965
Degs.	**Function**	**0′**	**6′**	**12′**	**18′**	**24′**	**30′**	**36′**	**42′**	**48′**	**54′**

Natural Sines, Cosines, and Tangents

45°–59.9°

Degs.	Function	0.0°	0.1°	0.2°	0.3°	0.4°	0.5°	0.6°	0.7°	0.8°	0.9°
45	sin	0.7071	0.7083	0.7096	0.7108	0.7120	0.7133	0.7145	0.7157	0.7169	0.7181
	cos	0.7071	0.7059	0.7046	0.7034	0.7022	0.7009	0.6997	0.6984	0.6972	0.6959
	tan	1.0000	1.0035	1.0070	1.0105	1.0141	1.0176	1.0212	1.0247	1.0283	1.0319
46	sin	0.7193	0.7206	0.7218	0.7230	0.7242	0.7254	0.7266	0.7278	0.7290	0.7302
	cos	0.6947	0.6934	0.6921	0.6909	0.6896	0.6884	0.6871	0.6858	0.6845	0.6833
	tan	1.0355	1.0392	1.0428	1.0464	1.0501	1.0538	1.0575	1.0612	1.0649	1.0686
47	sin	0.7314	0.7325	0.7337	0.7349	0.7361	0.7373	0.7385	0.7396	0.7408	0.7420
	cos	0.6820	0.6807	0.6794	0.6782	0.6769	0.6756	0.6743	0.6730	0.6717	0.6704
	tan	1.0724	1.0761	1.0799	1.0837	1.0875	1.0913	1.0951	1.0990	1.1028	1.1067
48	sin	0.7431	0.7443	0.7455	0.7466	0.7478	0.7490	0.7501	0.7513	0.7524	0.7536
	cos	0.6691	0.6678	0.6665	0.6652	0.6639	0.6626	0.6613	0.6600	0.6587	0.6574
	tan	1.1106	1.1145	1.1184	1.1224	1.1263	1.1303	1.1343	1.1383	1.1423	1.1463
49	sin	0.7547	0.7559	0.7570	0.7581	0.7593	0.7604	0.7615	0.7627	0.7638	0.7649
	cos	0.6561	0.6547	0.6534	0.6521	0.6508	0.6494	0.6481	0.6468	0.6455	0.6441
	tan	1.1504	1.1544	1.1585	1.1626	1.1667	1.1708	1.1750	1.1792	1.1833	1.1875
50	sin	0.7660	0.7672	0.7683	0.7694	0.7705	0.7716	0.7727	0.7738	0.7749	0.7760
	cos	0.6428	0.6414	0.6401	0.6388	0.6374	0.6361	0.6347	0.6334	0.6320	0.6307
	tan	1.1918	1.1960	1.2002	1.2045	1.2088	1.2131	1.2174	1.2218	1.2261	1.2305
51	sin	0.7771	0.7782	0.7793	0.7804	0.7815	0.7826	0.7837	0.7848	0.7859	0.7869
	cos	0.6293	0.6280	0.6266	0.6252	0.6239	0.6225	0.6211	0.6198	0.6184	0.6170
	tan	1.2349	1.2393	1.2437	1.2482	1.2527	1.2572	1.2617	1.2662	1.2708	1.2753
52	sin	0.7880	0.7891	0.7902	0.7912	0.7923	0.7934	0.7944	0.7955	0.7965	0.7976
	cos	0.6157	0.6143	0.6129	0.6115	0.6101	0.6088	0.6074	0.6060	0.6046	0.6032
	tan	1.2799	1.2846	1.2892	1.2938	1.2985	1.3032	1.3079	1.3127	1.3175	1.3222
53	sin	0.7986	0.7997	0.8007	0.8018	0.8028	0.8039	0.8049	0.8059	0.8070	0.8080
	cos	0.6018	0.6004	0.5990	0.5976	0.5962	0.5948	0.5934	0.5920	0.5906	0.5892
	tan	1.3270	1.3319	1.3367	1.3416	1.3465	1.3514	1.3564	1.3613	1.3663	1.3713
54	sin	0.8090	0.8100	0.8111	0.8121	0.8131	0.8141	0.8151	0.8161	0.8171	0.8181
	cos	0.5878	0.5864	0.5850	0.5835	0.5821	0.5807	0.5793	0.5779	0.5764	0.5750
	tan	1.3764	1.3814	1.3865	1.3916	1.3968	1.4019	1.4071	1.4124	1.4176	1.4229
55	sin	0.8192	0.8202	0.8211	0.8221	0.8231	0.8241	0.8251	0.8261	0.8271	0.8281
	cos	0.5736	0.5721	0.5707	0.5693	0.5678	0.5664	0.5650	0.5635	0.5621	0.5606
	tan	1.4281	1.4335	1.4388	1.4442	1.4496	1.4550	1.4605	1.4659	1.4715	1.4770
56	sin	0.8290	0.8300	0.8310	0.8320	0.8329	0.8339	0.8348	0.8358	0.8368	0.8377
	cos	0.5592	0.5577	0.5563	0.5548	0.5534	0.5519	0.5505	0.5490	0.5476	0.5461
	tan	1.4826	1.4882	1.4938	1.4994	1.5051	1.5108	1.5166	1.5224	1.5282	1.5340
57	sin	0.8387	0.8396	0.8406	0.8415	0.8425	0.8434	0.8443	0.8453	0.8462	0.8471
	cos	0.5446	0.5432	0.5417	0.5402	0.5388	0.5373	0.5358	0.5344	0.5329	0.5314
	tan	1.5399	1.5458	1.5517	1.5577	1.5637	1.5697	1.5757	1.5818	1.5880	1.5941
58	sin	0.8480	0.8490	0.8499	0.8508	0.8517	0.8526	0.8536	0.8545	0.8554	0.8563
	cos	0.5299	0.5284	0.5270	0.5255	0.5240	0.5225	0.5210	0.5195	0.5180	0.5165
	tan	1.6003	1.6066	1.6128	1.6191	1.6255	1.6319	1.6383	1.6447	1.6512	1.6577
59	sin	0.8572	0.8581	0.8590	0.8599	0.8607	0.8616	0.8625	0.8634	0.8643	0.8652
	cos	0.5150	0.5135	0.5120	0.5105	0.5090	0.5075	0.5060	0.5045	0.5030	0.5015
	tan	1.6643	1.6709	1.6775	1.6842	1.6909	1.6977	1.7045	1.7113	1.7182	1.7251
Degs.	Function	0′	6′	12′	18′	24′	30′	36′	42′	48′	54′

NATURAL SINES, COSINES, AND TANGENTS

60°–74.9°

Degs.	Function	0.0°	0.1°	0.2°	0.3°	0.4°	0.5°	0.6°	0.7°	0.8°	0.9°
60	sin	0.8660	0.8669	0.8678	0.8686	0.8695	0.8704	0.8712	0.8721	0.8729	0.8738
	cos	0.5000	0.4985	0.4970	0.4955	0.4939	0.4924	0.4909	0.4894	0.4879	0.4863
	tan	1.7321	1.7391	1.7461	1.7532	1.7603	1.7675	1.7747	1.7820	1.7893	1.7966
61	sin	0.8746	0.8755	0.8763	0.8771	0.8780	0.8788	0.8796	0.8805	0.8813	0.8821
	cos	0.4848	0.4833	0.4818	0.4802	0.4787	0.4772	0.4756	0.4741	0.4726	0.4710
	tan	1.8040	1.8115	1.8190	1.8265	1.8341	1.8418	1.8495	1.8572	1.8650	1.8728
62	sin	0.8829	0.8838	0.8846	0.8854	0.8862	0.8870	0.8878	0.8886	0.8894	0.8902
	cos	0.4695	0.4679	0.4664	0.4648	0.4633	0.4617	0.4602	0.4586	0.4571	0.4555
	tan	1.8807	1.8887	1.8967	1.9047	1.9128	1.9210	1.9292	1.9375	1.9458	1.9542
63	sin	0.8910	0.8918	0.8926	0.8934	0.8942	0.8949	0.8957	0.8965	0.8973	0.8980
	cos	0.4540	0.4524	0.4509	0.4493	0.4478	0.4462	0.4446	0.4431	0.4415	0.4399
	tan	1.9626	1.9711	1.9797	1.9883	1.9970	2.0057	2.0145	2.0233	2.0323	2.0413
64	sin	0.8988	0.8996	0.9003	0.9011	0.9018	0.9026	0.9033	0.9041	0.9048	0.9056
	cos	0.4384	0.4368	0.4352	0.4337	0.4321	0.4305	0.4289	0.4274	0.4258	0.4242
	tan	2.0503	2.0594	2.0686	2.0778	2.0872	2.0965	2.1060	2.1155	2.1251	2.1348
65	sin	0.9063	0.9070	0.9078	0.9085	0.9092	0.9100	0.9107	0.9114	0.9121	0.9128
	cos	0.4226	0.4210	0.4195	0.4179	0.4163	0.4147	0.4131	0.4115	0.4099	0.4083
	tan	2.1445	2.1543	2.1642	2.1742	2.1842	2.1943	2.2045	2.2148	2.2251	2.2355
66	sin	0.9135	0.9143	0.9150	0.9157	0.9164	0.9171	0.9178	0.9184	0.9191	0.9198
	cos	0.4067	0.4051	0.4035	0.4019	0.4003	0.3987	0.3971	0.3955	0.3939	0.3923
	tan	2.2460	2.2566	2.2673	2.2781	2.2889	2.2998	2.3109	2.3220	2.3332	2.3445
67	sin	0.9205	0.9212	0.9219	0.9225	0.9232	0.9239	0.9245	0.9252	0.9259	0.9265
	cos	0.3907	0.3891	0.3875	0.3859	0.3843	0.3827	0.3811	0.3795	0.3778	0.3762
	tan	2.3559	2.3673	2.3789	2.3906	2.4023	2.4142	2.4262	2.4383	2.4504	2.4627
68	sin	0.9272	0.9278	0.9285	0.9291	0.9298	0.9304	0.9311	0.9317	0.9323	0.9330
	cos	0.3746	0.3730	0.3714	0.3697	0.3681	0.3665	0.3649	0.3633	0.3616	0.3600
	tan	2.4751	2.4876	2.5002	2.5129	2.5257	2.5386	2.5517	2.5649	2.5782	2.5916
69	sin	0.9336	0.9342	0.9348	0.9354	0.9361	0.9367	0.9373	0.9379	0.9385	0.9391
	cos	0.3584	0.3567	0.3551	0.3535	0.3518	0.3502	0.3486	0.3469	0.3453	0.3437
	tan	2.6051	2.6187	2.6325	2.6464	2.6605	2.6746	2.6889	2.7034	2.7179	2.7326
70	sin	0.9397	0.9403	0.9409	0.9415	0.9421	0.9426	0.9432	0.9438	0.9444	0.9449
	cos	0.3420	0.3404	0.3387	0.3371	0.3355	0.3338	0.3322	0.3305	0.3289	0.3272
	tan	2.7475	2.7625	2.7776	2.7929	2.8083	2.8239	2.8397	2.8556	2.8716	2.8878
71	sin	0.9455	0.9461	0.9466	0.9472	0.9478	0.9483	0.9489	0.9494	0.9500	0.9505
	cos	0.3256	0.3239	0.3223	0.3206	0.3190	0.3173	0.3156	0.3140	0.3123	0.3107
	tan	2.9042	2.9208	2.9375	2.9544	2.9714	2.9887	3.0061	3.0237	3.0415	3.0595
72	sin	0.9511	0.9516	0.9521	0.9527	0.9532	0.9537	0.9542	0.9548	0.9553	0.9558
	cos	0.3090	0.3074	0.3057	0.3040	0.3024	0.3007	0.2990	0.2974	0.2957	0.2940
	tan	3.0777	3.0961	3.1146	3.1334	3.1524	3.1716	3.1910	3.2106	3.2305	3.2506
73	sin	0.9563	0.9568	0.9573	0.9578	0.9583	0.9588	0.9593	0.9598	0.9603	0.9608
	cos	0.2924	0.2907	0.2890	0.2874	0.2857	0.2840	0.2823	0.2807	0.2790	0.2773
	tan	3.2709	3.2914	3.3122	3.3332	3.3544	3.3759	3.3977	3.4197	3.4420	3.4646
74	sin	0.9613	0.9617	0.9622	0.9627	0.9632	0.9636	0.9641	0.9646	0.9650	0.9655
	cos	0.2756	0.2740	0.2723	0.2706	0.2689	0.2672	0.2656	0.2639	0.2622	0.2605
	tan	3.4874	3.5105	3.5339	3.5576	3.5816	3.6059	3.6305	3.6554	3.6806	3.7062
Degs.	Function	0′	6′	12′	18′	24′	30′	36′	42′	48′	54′

NATURAL SINES, COSINES, AND TANGENTS.

75°–89.9°

Degs.	Function	0.0°	0.1°	0.2°	0.3°	0.4°	0.5°	0.6°	0.7°	0.8°	0.9°
75	sin	0.9659	0.9664	0.9668	0.9673	0.9677	0.9681	0.9686	0.9690	0.9694	0.9699
	cos	0.2588	0.2571	0.2554	0.2538	0.2521	0.2504	0.2487	0.2470	0.2453	0.2436
	tan	3.7321	3.7583	3.7848	3.8118	3.8391	3.8667	3.8947	3.9232	3.9520	3.9812
76	sin	0.9703	0.9707	0.9711	0.9715	0.9720	0.9724	0.9728	0.9732	0.9736	0.9740
	cos	0.2419	0.2402	0.2385	0.2368	0.2351	0.2334	0.2317	0.2300	0.2284	0.2267
	tan	4.0108	4.0408	4.0713	4.1022	4.1335	4.1653	4.1976	4.2303	4.2635	4.2972
77	sin	0.9744	0.9748	0.9751	0.9755	0.9759	0.9763	0.9767	0.9770	0.9774	0.9778
	cos	0.2250	0.2232	0.2215	0.2198	0.2181	0.2164	0.2147	0.2130	0.2113	0.2096
	tan	4.3315	4.3662	4.4015	4.4374	4.4737	4.5107	4.5483	4.5864	4.6252	4.6646
78	sin	0.9781	0.9785	0.9789	0.9792	0.9796	0.9799	0.9803	0.9806	0.9810	0.9813
	cos	0.2079	0.2062	0.2045	0.2028	0.2011	0.1994	0.1977	0.1959	0.1942	0.1925
	tan	4.7046	4.7453	4.7867	4.8288	4.8716	4.9152	4.9594	5.0045	5.0504	5.0970
79	sin	0.9816	0.9820	0.9823	0.9826	0.9829	0.9833	0.9836	0.9839	0.9842	0.9845
	cos	0.1908	0.1891	0.1874	0.1857	0.1840	0.1822	0.1805	0.1788	0.1771	0.1754
	tan	5.1446	5.1929	5.2422	5.2924	5.3435	5.3955	5.4486	5.5026	5.5578	5.6140
80	sin	0.9848	0.9851	0.9854	0.9857	0.9860	0.9863	0.9866	0.9869	0.9871	0.9874
	cos	0.1736	0.1719	0.1702	0.1685	0.1668	0.1650	0.1633	0.1616	0.1599	0.1582
	tan	5.6713	5.7297	5.7894	5.8502	5.9124	5.9758	6.0405	6.1066	6.1742	6.2432
81	sin	0.9877	0.9880	0.9882	0.9885	0.9888	0.9890	0.9893	0.9895	0.9898	0.9900
	cos	0.1564	0.1547	0.1530	0.1513	0.1495	0.1478	0.1461	0.1444	0.1426	0.1409
	tan	6.3138	6.3859	6.4596	6.5350	6.6122	6.6912	6.7720	6.8548	6.9395	7.0264
82	sin	0.9903	0.9905	0.9907	0.9910	0.9912	0.9914	0.9917	0.9919	0.9921	0.9923
	cos	0.1392	0.1374	0.1357	0.1340	0.1323	0.1305	0.1288	0.1271	0.1253	0.1236
	tan	7.1154	7.2066	7.3002	7.3962	7.4947	7.5958	7.6996	7.8062	7.9158	8.0285
83	sin	0.9925	0.9928	0.9930	0.9932	0.9934	0.9936	0.9938	0.9940	0.9942	0.9943
	cos	0.1219	0.1201	0.1184	0.1167	0.1149	0.1132	0.1115	0.1097	0.1080	0.1063
	tan	8.1443	8.2636	8.3863	8.5126	8.6427	8.7769	8.9152	9.0579	9.2052	9.3572
84	sin	0.9945	0.9947	0.9949	0.9951	0.9952	0.9954	0.9956	0.9957	0.9959	0.9960
	cos	0.1045	0.1028	0.1011	0.0993	0.0976	0.0958	0.0941	0.0924	0.0906	0.0889
	tan	9.5144	9.6768	9.8448	10.02	10.20	10.39	10.58	10.78	10.99	11.20
85	sin	0.9962	0.9963	0.9965	0.9966	0.9968	0.9969	0.9971	0.9972	0.9973	0.9974
	cos	0.0872	0.0854	0.0837	0.0819	0.0802	0.0785	0.0767	0.0750	0.0732	0.0715
	tan	11.43	11.66	11.91	12.16	12.43	12.71	13.00	13.30	13.62	13.95
86	sin	0.9976	0.9977	0.9978	0.9979	0.9980	0.9981	0.9982	0.9983	0.9984	0.9985
	cos	0.0698	0.0680	0.0663	0.0645	0.0628	0.0610	0.0593	0.0576	0.0558	0.0541
	tan	14.30	14.67	15.06	15.46	15.89	16.35	16.83	17.34	17.89	18.46
87	sin	0.9986	0.9987	0.9988	0.9989	0.9990	0.9990	0.9991	0.9992	0.9993	0.9993
	cos	0.0523	0.0506	0.0488	0.0471	0.0454	0.0436	0.0419	0.0401	0.0384	0.0366
	tan	19.08	19.74	20.45	21.20	22.02	22.90	23.86	24.90	26.03	27.27
88	sin	0.9994	0.9995	0.9995	0.9996	0.9996	0.9997	0.9997	0.9997	0.9998	0.9998
	cos	0.0349	0.0332	0.0314	0.0297	0.0279	0.0262	0.0244	0.0227	0.0209	0.0192
	tan	28.64	30.14	31.82	33.69	35.80	38.19	40.92	44.07	47.74	52.08
89	sin	0.9998	0.9999	0.9999	0.9999	0.9999	1.000	1.000	1.000	1.000	1.000
	cos	0.0175	0.0157	0.0140	0.0122	0.0105	0.0087	0.0070	0.0052	0.0035	0.0017
	tan	57.29	63.66	71.62	81.85	95.49	114.6	143.2	191.0	286.5	573.0
Degs.	**Function**	**0′**	**6′**	**12′**	**18′**	**24′**	**30′**	**36′**	**42′**	**48′**	**54′**

APPENDIX IX

SQUARE AND SQUARE ROOTS

N	N^2	$\sqrt{N}$	N	N^2	$\sqrt{N}$	N	N^2	$\sqrt{N}$
1	1	1.000	41	1681	6.4031	81	6561	9.0000
2	4	1.414	42	1764	6.4807	82	6724	9.0554
3	9	1.732	43	1849	6.5574	83	6889	9.1104
4	16	2.000	44	1936	6.6332	84	7056	9.1652
5	25	2.236	45	2025	6.7082	85	7225	9.2195
6	36	2.449	46	2116	6.7823	86	7396	9.2736
7	49	2.646	47	2209	6.8557	87	7569	9.3274
8	64	2.828	48	2304	6.9282	88	7744	9.3808
9	81	3.000	49	2401	7.0000	89	7921	9.4340
10	100	3.162	50	2500	7.0711	90	8100	9.4868
11	121	3.3166	51	2601	7.1414	91	8281	9.5394
12	144	3.4641	52	2704	7.2111	92	8464	9.5917
13	169	3.6056	53	2809	7.2801	93	8649	9.6437
14	196	3.7417	54	2916	7.3485	94	8836	9.6954
15	225	3.8730	55	3025	7.4162	95	9025	9.7468
16	256	4.0000	56	3136	7.4833	96	9216	9.7980
17	289	4.1231	57	3249	7.5498	97	9409	9.8489
18	324	4.2426	58	3364	7.6158	98	9604	9.8995
19	361	4.3589	59	3481	7.6811	99	9801	9.9499
20	400	4.4721	60	3600	7.7460	100	10000	10.0000
21	441	4.5826	61	3721	7.8102	101	10201	10.0499
22	484	4.6904	62	3844	7.8740	102	10404	10.0995
23	529	4.7958	63	3969	7.9373	103	10609	10.1489
24	576	4.8990	64	4096	8.0000	104	10816	10.1980
25	625	5.0000	65	4225	8.0623	105	11025	10.2470
26	676	5.0990	66	4356	8.1240	106	11236	10.2956
27	729	5.1962	67	4489	8.1854	107	11449	10.3441
28	784	5.2915	68	4624	8.2462	108	11664	10.3923
29	841	5.3852	69	4761	8.3066	109	11881	10.4403
30	900	5.4772	70	4900	8.3666	110	12100	10.4881
31	961	5.5678	71	5041	8.4261	111	12321	10.5357
32	1024	5.6569	72	5184	8.4853	112	12544	10.5830
33	1089	5.7447	73	5329	8.5440	113	12769	10.6301
34	1156	5.8310	74	5476	8.6023	114	12996	10.6771
35	1225	5.9161	75	5625	8.6603	115	13225	10.7238
36	1296	6.0000	76	5776	8.7178	116	13456	10.7703
37	1369	6.0828	77	5929	8.7750	117	13689	10.8167
38	1444	6.1644	78	6084	8.8318	118	13924	10.8628
39	1521	6.2450	79	6241	8.8882	119	14161	10.9087
40	1600	6.3246	80	6400	8.9443	120	14400	10.9545

For numbers up to 120. For larger numbers divide into factors smaller than 120.

NOTES:

APPENDIX X

LAWS OF EXPONENTS

The International Symbols Committee has adopted prefixes for denoting decimal multiples of units. The National Bureau of Standards has followed the recommendations of this committee, and has adopted the following list of prefixes:

Numbers	Powers of ten	Prefixes	Symbols
1, 000, 000, 000, 000	10^{12}	tera	T
1, 000, 000, 000	10^{9}	giga	G
1, 000, 000	10^{6}	mega	M
1, 000	10^{3}	kilo	k
100	10^{2}	hecto	h
10	10	deka	da
.1	10^{-1}	deci	d
.01	10^{-2}	centi	c
.001	10^{-3}	milli	m
.000001	10^{-6}	micro	u
.000000001	10^{-9}	nano	n
.000000000001	10^{-12}	pico	p
.000000000000001	10^{-15}	femto	f
.000000000000000001	10^{-18}	atto	a

To multiply like (with same base) exponential quantities, add the exponents. In the language of alegebra the rule is $a^m \times a^n = a^{m+n}$

$$10^4 \times 10^2 = 10^{4+2} = 10^6$$

$$0.003 \times 825.2 = 3 \times 10^{-3} \times 8.252 \times 10^2$$

$$= 24.756 \times 10^{-1} = 2.4756$$

To divide exponential quantities, subtract the exponents. In the language of algebra the rule is

$$\frac{a^m}{a^n} = a^{m-n} \text{ or}$$

$$10^8 \div 10^2 = 10^6$$

$$3,000 \div 0.015 = (3 \times 10^3) \div (1.5 \times 10^{-2})$$

$$= 2 \times 10^5 = 200,000$$

To raise an exponential quantity to a power, multiply the exponents. In the languague of algebra $(x^m)^n = x^{mn}$.

$$(10^3)^4 = 10^{3x4} = 10^{12}$$

$$2,500^2 = (2.5 \times 10^3)^2 = 6.25 \times 10^6 = 6,250,000$$

Any number (except zero) raised to the zero power is one. In the language of algebra $x^0=1$

$$x^3 \div x^3 = 1$$

$$10^4 \div 10^4 = 1$$

Any base with a negative exponent is equal to 1 divided by the base with an equal positive exponent. In the language of algebra $x^{-a} = \frac{1}{x^a}$

$$10^{-2} = \frac{1}{10^2} = \frac{1}{100}$$

$$5a^{-3} = \frac{5}{a^3}$$

$$(6a)^{-1} = \frac{1}{6a}$$

To raise a product to a power, raise each factor of the product to that power.

$$(2 \times 10)^2 = 2^2 \times 10^2$$

$$3,000^3 = (3 \times 10^3)^3 = 27 \times 10^9$$

To find the nth root of an exponential quantity, divide the exponent by the index of the root. Thus, the nth root of $a^m = a^{m/n}$.

$$\sqrt{x^6} = x^{6/2} = x^3$$

$$\sqrt[3]{64 \times 10^3} = 4 \times 10 = 40$$

NOTES:

INDEX

NOTES:

NOTES:

NOTES:

NOTES:

NOTES:

NOTES:

NOTES:

NOTES:

NOTES: